EVOLUTION
ABOVE THE SPECIES LEVEL

NUMBER XIX OF THE
COLUMBIA BIOLOGICAL SERIES
EDITED AT COLUMBIA UNIVERSITY

EVOLUTION ABOVE
THE SPECIES LEVEL

Bernhard Rensch

Professor of Zoology
in the University of Münster

NEW YORK: MORNINGSIDE HEIGHTS

COLUMBIA UNIVERSITY PRESS

First published in this edition 1959
by Methuen & Co. Ltd., London

Originally published in 1954
by Ferdinand Enke Verlag, Stuttgart
as Neuere Probleme der Abstammungslehre *2nd edition*

English translation © 1959 Columbia University Press
Second printing 1970
ISBN 0-231-02296-4
Library of Congress Catalog Number: 58-13505
Printed in the United States of America

Foreword

It is very fitting to have the English translation of Rensch's *Evolution Above the Species Level* published in 1959, the centenary year of Darwin's classic *On the Origin of Species*. The work of Professor Rensch is one of the great books which have appeared since 1940 and which may fairly be said to have given shape to the modern biological, or synthetic, theory of evolution. The biological theory is the heir to and direct descendant of Darwin's seminal discovery. The intellectual continuity between them is evident and unbroken, but this does not diminish the importance of the radical changes which the century and particularly the last two decades of biological research, thought, and discussion have brought. Apart from the discovery of a great mass of facts, some of them of cardinal importance, there has been a remarkable movement towards synthesis of previously very nearly disconnected lines of investigation and speculation. Instead of the scarcely related 'theories' put forward by systematists, geneticists, paleontologists, and others, we have now a theory of evolution which embraces biology. Together with an attempt by Huxley (1942) which fell rather short of synthesis, we have had the works of Mayr (1942) stemming from zoological and of Stebbins (1950) stemming from botanical systematics and genetics, of Simpson (1944, further developed in 1953) from paleontology, of Schmalhausen (1946 in Russian, 1949 in English) from comparative morphology, of Darlington (1939) and White (1945, 1954) from cytology and genetics, and of Rensch (the two German editions in 1947 and 1954) from systematics, comparative morphology, and paleontology. Notwithstanding their having been based primarily on different bodies of evidence, all these works came to substantially identical conclusions. It is, indeed, a grand synthesis.

This should not be taken to mean that the theory of evolution is now complete except for some emendation. On the contrary, radical changes and major upsets are not only possible but almost certain to occur. The progression of science is, however, uneven; periods of broadening of the evidential base alternate with tides of generalization and forward leaps of understanding. It is our privilege to live in a period of the latter kind, in which the book of Rensch marks one of the forward leaps.

Columbia University   THEODOSIUS DOBZHANSKY
in the City of New York
1 *January*, 1959

Preface

The greater part of this book was written in the last years of the war. My intention was not to present a compilation of known facts brought to light by evolutionary research, but rather to outline the major rules governing the processes of evolution. I wanted to attempt a causal explanation, partly based on new material, with the intention of proving that very probably the major trends of evolution are brought about by the same factors that bring about race and species formation. The first German edition, which came out in 1947, could refer to the works by J. S. Huxley (1942), Simpson (1944), and Mayr (1942), published in England and the United States during the war, only in an addendum after final proof-reading was completed. It was not until the second edition (Stuttgart, 1954) was prepared that the material in these books, as well as in those by Dobzhansky, Haldane, Lack, Edinger, A. H. Miller, and others could be incorporated. The authors were so kind as to place their books at my disposal. I was surprised to find that many scientists, though working independently and using quite different materials as the bases of their studies, had arrived at the same conclusions. For the first time in this century there was a rather general agreement among paleontologists, geneticists, systematists, and comparative anatomists.

The present English translation is based upon the second German edition, and only a few alterations and additions have been made. These consist chiefly of the abridgement of material that is of minor importance and the inclusion of quotations from recent literature. It has not been possible, however, to deal with all the numerous special evolutionary studies available in the most recent literature, as this would have increased too greatly the size of the book.

The rules governing transspecific alterations of the structural type (Chapters 4 and 6) which – in part – have been newly established are treated in more detail, because they represent a sound means of appreciating organic evolution as a whole and because they exemplify the regularities of the major lines of evolution. The chapter on the evolution of phenomena of consciousness has been reworded so as to be intelligible to those readers who are not too well acquainted with philosophical and epistemological considerations. In spite of its somewhat hypothetical character, I have not omitted this chapter because the majority of the readers of the German edition – insofar as their written and spoken comments have reached me – considered it essential and stimulating.

It is my pleasant duty to express my cordial gratitude to Professor Theodosius Dobzhansky for suggesting this translation of my book and for going through the manuscript. Thanks are also due to the Manager of the Publication Department of Columbia University Press, Mr Henry H. Wiggins, for his obliging cooperation in bringing out the American edition. Finally, I wish to thank Dr Altevogt for translating the German text, which was not too easy in places.

B. RENSCH

Münster, Westphalia
March, 1956

Contents

FOREWORD   *page* v

PREFACE   vii

1. INTRODUCTION   1

2. THE CAUSATIVE FACTORS OF INFRASPECIFIC
   EVOLUTION   3
   A. Mutation   3
   B. Fluctuations of population size   9
   C. Selection   10
   D. Isolation   13
   E. Hybridization   14
   F. Interaction of the evolutionary factors   14

3. TYPES OF RACE AND SPECIES FORMATION
   OCCURRING IN NATURE   16
   A. Historical races   16
   B. Geographic races   22
   C. Ecological races   47
   D. Sexually isolated races   50
   E. Genetic isolation   53
   F. Hybrid races   54
   G. Review of all types of races   55

4. UNDIRECTED AND DIRECTED TRANSSPECIFIC
   EVOLUTION   57
   A. The problem of phylogenetic trends and orthogenesis   57
   B. Nondirectedness of evolution   59
   C. Forced development in phylogeny   68
   D. Rules of phylogenetic development   73
   E. Possible evolution of organisms in extraterrestrial bodies   76

5. THE ABSOLUTE SPEED OF EVOLUTION   82
   A. The absolute age of various categories   83
   B. Factors governing the speed of evolution   91

6. THE RULES OF KLADOGENESIS (PHYLO-
GENETIC BRANCHING)   *page* 97

A. Phases of explosive radiation   98
   Historical comments   98
   Examples demonstrating 'explosive phases' and periods of
   explosive radiation   99
   Interpretation of the findings   103

B. The phase of specialization   112
   I. Gradual loss of evolutionary intensity   112
   II. Irreversibility   123
      Transspecific changes of structural systems   126
   III. Rules governing transspecific changes of structural
      types   126
      Constructive genes   129
      Allometric growth and its bearing on evolution   133
      Physiological consequences of allometric growth   160
      Summary of complex differences in species of different
      body size   165
      The gradients of differentiation   169
      Phyletic size limits and consequent alterations of
      the structural type   170
      Further correlations. Compensation of body material   179
      The problem of somatogenic induction   187
      Summary on transspecific alterations of the structural
      type   190
   IV. Parallel evolution   191
      Parallel evolution resulting from similarity of hereditary
      factors   192
      Parallelisms in consequence of parallel selection   198
   V. Orthogenesis   203
      Examples of directed phylogeny and their problems   203
      Cope's rule   206
      Orthogenetic evolution of organs   218
      Reduction of organs   222

C. Overspecialization, degeneration, and extinction   226
   I. Excessive growth and 'overspecialization'   226
   II. Phylogenetic aging and 'degeneration'   232
   III. Phyletic extinction   234

D. The Effects of phylogenetic alterations in various states of the
   individual cycle   239
   I. Causation and classification of phylogenetic alterations
      of ontogeny   239
   II. Archallaxis   241
   III. Permanent deviations and heterochronies   246

IV. Deviations confined to early and intermediate states of ontogeny (coenogeneses) *page* 250

V. Alterations by additional phases of development (anabolies) 253

VI. Palingenesis, proterogenesis, and neoteny 258

VII. Alternation of generations 262

VIII. Frequency of the different types of ontogenetic alterations and their bearing on evolution 263

E. Evolution of new structural types and new organs 266

I. The origin of new structural types 266

II. The origin of new organs 274

III. Summary 279

7. ANAGENESIS (PROGRESSIVE EVOLUTION) 281

A. Historical comments 281

B. Facts of anagenesis 284

C. Analysis of anagenesis 289

I. Increase of complexity 289

II. Increased rationalization 291

III. Special complexity and rationalization of nervous systems 292

IV. Increased plasticity of structures and functions 295

V. Improvement permitting further improvement 297

VI. Increased independence of the environment and increased autonomy 298

VII. Regressions 298

VIII. Summary 299

D. Remarks on human descent 301

8. THE EVOLUTION OF LIFE 309

A. The problem of spontaneous generation 309

B. Characteristics of life 310

C. Series of organisms with successive reduction of vital characters 311

D. The evolution of neucleoproteins 315

9. AUTOGENESIS, ECTOGENESIS, AND BIONOMO-GENESIS 317

10. EVOLUTION OF PHENOMENA OF CONSCIOUS-NESS 322

A. Introduction 322

B. General review of the phenomena of consciousness 323

C. General phylogeny of the phenomena of consciousness 327

D. Phylogeny of the special phenomena   *page* 335
    I. Phylogeny of sensations   335
    II. Phylogeny of conceptions   339
    III. Judgments, conclusions, and actions of insight   344
E. Philosophy and the evolution of psychic phenomena   349
F. Psychic processes and somatic evolution   356

11. CONCLUSION   358

POSTSCRIPT   361

BIBLIOGRAPHY   363

AUTHOR INDEX   404

GENERAL INDEX   412

Figures

1. Spreading of the melanic mutant (*carbonaria*) of the moth *Amphidasis betularia*   page 21
2. New Zealand flycatcher and melanic mutant; New Guinea hawk and white mutant   22
3. Distribution of *Parus major* 'Rassenkreis'   29
4. Range of races in the *Trichoglossus ornatus* 'Rassenkreis'   31
5. Geographic races of the lizard *Ablepharus boutonii*   35
6. Three extreme races of *Lacerta pityusensis* from the Pityusic Islands   37
7. Species of land snail from a small island   38
8. Arboreal snail *Papuina wiegmanni* and variants   39
9. Range of *Papuina wiegmanni* in New Britain   39
10. Average length of wings in populations of serins ($\male$)   41
11. Distribution of size variants in the willow-titmouse *Parus atricapillus*   44
12. Increase of wing length towards the peripheral range in the wren *Troglodytes troglodytes*   45
13. Eggs of central European cuckoo races and their hosts   48
14. Genera of birds of paradise from New Guinea and adjacent islands as examples of undirected transspecific evolution   61
15. Undirected transformation in nonparasitic marine Isopoda (*Is. genuina*)   64
16. Four types of larvae of Nematocera from central European genera   65
17. Convergent evolution of eyes with lenses   72
18. Diagram of the differentiation of systematic categories   93
19. Two proglottids of the tapeworm *Dilpogonoporus grandis* (from man) 105
20. Monster proglottids of a *Taenia* species from a heron   105
21. Drepanididae from the Hawaiian Islands   107
22. Diagram of the number of newly formed and persisting categories of various animal groups during the geological epochs   114
23. Phylogeny of *Mesaxonia*   116
24. Skull of the primeval whale *Zeuglodon* from Alabama (Late Eocene) 124
25. Right foreleg of a domestic horse, showing a polydactylous atavism 125
26. De-differentiation of explants *in vitro* caused by repeated growths and several passages   128
27. Diagram of the mutual formative effect of neighboring tissue explants 128
28. Female of *Drosophila funebris*   129

29. Feather barbules from the upper tail coverts of the peacock (*Pavo cristatus*) *page* 130

30. Effect of prolaction treatment in dove 132

31. Changes of allometrical gradients in the ontogeny of the area dorsalis pallii of *Triturus vulgaris* 134

32. Coincidence of growth gradients in ontogeny and phylogeny 137

33. Different negative allometries in *Evotomys rufocanus smithii* and *Arvicola amph. amphibius* (including *A. a. reta*) 140

34. Negative correlation between relative length of skull and head–body length in juvenile and adult Japanese mice (*Evotomys rufocanus smithii*) ♂ and ♀ 140

35. Allometric shifts in length of facial skull 141

36. Correlation of body weight and relative brain weight in percent of body weight of five races of domestic hens 142

37. Median sections of brains of closely related large and small mammals 143

38. Excessive growth of beak as the result of positive allometry in the black-tailed godwit *Limosa limosa* and the lapwing *Vanellus cristatus* 145

39. Correlation of body weight and relative heart weight in percent of body weight in five races of domestic chickens 147

40. Different negative correlations between head–body length and relative diameter of eyes in the mouse and the rat 148

41. Median sections of the eyes of cow, domestic rabbit, and mouse *Evotomys glareolus* 149

42. Eyes of horse, wild rabbit, and mouse *Apodemus silvaticus* 149

43. Horizontal transverse sections of retina from center of the back of the eye of mouse, albinotic common rat, and domestic rabbit 150

44. Homologous parts from central retina of three newts and two salamanders of different body size 150

45. Relative surface area of Holocortex-2-stratificatus (above) and H.-5-stratificatus (below) in percentage of the whole hemisphere of the white mouse 152

46. Medium-sized pyramidal cells (second layer) from homologous parts of the cortex of two rodents of different body size 155

47. Ganglion cells from the sympathetic nerve of teleosts of different body size 155

48. Epithelial cells from homologous parts of the mid-gut of mosquitoes of different body size 156

49. Homologous sections from protocerebrum (with central body and corpora pedunculata) of three pairs of related insects of different body size 157

50. Increase of sexual dimorphism in the antennae with increasing body size in three species of Scarabaeidae 160

51. Growth curves of body weight in four races of pigeons of different body size 168

52. Cartesian transformation of *Diodon* (left) into *Orthagoriscus* (right) 170

53. Cartesian transformation of various types of crab carapaces 170

54. Oxygen consumption in ccm. per gm. body weight and hour in mice, shrews, and small birds   *page* 173

55. Reduction of symphysis ossis pubis in smallest mammals   174

56. Reconstruction of the largest reptile, *Brachiosaurus*, to show the principle of columnar legs   178

57. Reconstruction of the largest mammal, the hornless Oligocene rhino *Baluchitherium grangeri*, to show the principle of columnar legs   178

58. Correlation coefficients in the antenna joints of the bug *Pyrrhocoris apterus*   179

59. Size increase of a normal hind leg of the stick insect *Carausius morosus* under the influence of a regenerating hind leg or of a regenerating hind leg and contralateral middle leg   182

60. Negative correlation between length of horn and breadth of head in percent of breadth of prothorax in *Copris lunaris*   183

61. Compensatory correlations between number and length of spines in *Prosobranchia*   184

62. Racial differences in the wing tips of two races of the great reed warbler   185

63. Compensatory reduction of first primary in the nightingale and the thrush nightingale   185

64. Skulls of *Adapis parisiensis* and *Megaladapis edwardii*   186

65. Size and position of organs in *Opisthocomus cristatus*   189

66. Comparison of the fourth giant chromosome from *Drosophila pseudo-obscura* and *D. miranda*   192

67. Comparison of gene arrangements in *Drosophila pseudoobscura* (ps) and *D. miranda* (mi)   193

68. Parallel development of flat, crested, and ribbed races and species in various normally round-shelled, dry-land snails from the Mediterranean area   195

69. Parallelisms in the wing patterns of various groups of butterflies   196

70. General scheme of wing patterns in butterflies   197

71. Iterative, directed evolution in the pygidia (from broad to narrow) in the trilobite *Olenus*   199

72. Two parallel proterogenetic evolutionary series of *Clymenia*, a Late Devonian genus of Ammonoidea   199

73. Blunt wing tips in non-migratory birds and pointed wing tips in migratory birds   201

74. Convergent evolution of excessively long canines in the upper jaw and a corresponding bony pouch in the lower jaw   202

75. Hind wing of the large chrysomelid *Melasoma populi* and of the very small chrysomelid *Haltica atra*, showing reduced venation   210

76. Excessive growth in Proboscidians   220

77. Orthogenesis in saber-toothed cats   221

78. Successive reduction of limbs in lizards   223

79. Pelvis of *Anguis fragilis*, showing the remainder of the femur   224

80. Normal hind wing and four reduced hind wings in five species of the ground beetle genus *Poecilus*   224

81. Excessive growth in various families of beetles *page* 227

82. Orthogenetic evolution of excessive horns in a phyletic series of Titanotheria 229

83. Excessive teeth in *Astrapotherium* 230

84. Excessive canines in *Babirussa alifurus* 230

85. Excessive 'Nebenformen' in the Ammonid family Lytoceratidae 232

86. Excessive 'Nebenformen' in the Ammonid family Kosmoceratidae 233

87. Differences in the type of cleavage in two species of Oligochaeta 242

88. Early ontogenetic origin of specific differences in Echinodermata 243

89. Horizontal section of the head of a larval *Triturus cristatus* in which the lens and cornea anlagen of the smaller species *T. taeniatus* were implanted 245

90. Dart sac and mucous glands of the snails *Cepaea nemoralis* and *C. hortensis* 248

91. Three comparable stages of early brain development 249

92. Two different types of eight-cell stage in the turbellarian *Prorhynchus stagnatilis* 250

93. Planktogenic and nereidogenic larvae of the polychaete *Nereis dumerili* 252

94. Early differentiation of mutant of *Drosophila melanogaster* 252

95. Addition of growth phases to the final stage of ontogeny demonstrated in growth of jaws in *Belone acus* 254

96. Skulls of fossil bears 255

97. Proterogenetic increase of coiling in Nautiloidea from the Lower Ordovician 261

98. Diagram of phyletic bifurcation 268

99. Intermediate forms of Crossopterygia and Stegocephalia 270

100. Lower jaw of fetal kangaroo *Macropus* 276

101. Parallel evolution in the anatomy of photoreceptors 278

102. Endocranial casts of the horse line 288

103. Decreasing number of head bones in the course of evolution 292

104. Correlation of limb reduction and increasing number of presacral vertebrae in lizards 299

105. The human-like foot of the mountain gorilla (*Gorilla beringeri*) 302

106. Left hemisphere of human brain, showing Broca's motor area of speech 303

107. *Cystidium gönnertianum*, bronchopneumonia virus of mice 302

108. Tobacco mosaic virus 302

109. Tobacco necrosis virus, showing the regular molecular structure 303

110. Robot 330

111. A sequence of correct choices in sampling experiments with a raven 342

112. Movement of gastral opening to the prey caught by the peripheral tentacles in the jellyfish *Tiaropsis indicans* 343

113. Phylogenetic increase of dendrites per neuron 346

Tables

1. Range of variability of absolute measurements in the ground beetles *Carabus cor. coriaceus* and *C. cor. cerisyi*   *page* 42

2. Range of variability of absolute measurements in the ground beetles *Carabus auronitens auronitens* and *C. aur. festivus*   42

3. Average age of higher categories of various land and water animals of different evolutionary levels   92

4. Number of new genera per geological epoch   100

5. Number of new families per geological epoch   100

6. Number of new orders per geological epoch   101

7. Number of new classes with a good fossil record per geological epoch   101

8. Relative length of skull in individual groups and populations of different body size but of the same geographic race   138

9. Relative length of skull in geographic races of different body size but of the same species   138

10. Relative length of skull in closely related species of west European mammals of different body size   139

11. Relative weight of organs taken from closely related mammals of large and small body size at the same season of the year   146

12. Relative weight of organs taken from closely related birds of large and small body size at the same season of the year   146

13. Relative size of cornea in mammals of different body size   150

14. Relative volume and relative thickness of lens in the eyes of five groups of mammals of different body size   150

15. Total number of neurons and visual cells in the retina of two species of *Triturus* and *Salamandra* differing in body size   152

16. Number of single and double visual cones and number of rods in greatest sagittal sections of the retina in two species of *Triturus* and *Salamandra* differing in body size   152

17. Egg numbers of large and small poikilotherms of the same order or family   158

18. Periods of gestation in mammals of the same orders but of different body size   161

19. Incubation periods in birds of the same orders but of different body size   162

20. Correlation of average life span with body size in mammals of the same orders but of different body size   163

21. Size of red blood corpuscles in related species of birds of different body size   214

1

Introduction

The concepts of mutation and selection developed in modern genetics have supplied an increasingly sound basis for an understanding of race and species formation. Not a few biologists have found this a possible explanation of evolution as a whole. It has been proved, however, that certain rules govern the major trends of phylogeny; these can be traced in the gradual differentiation of genera, families, orders, and categories, and also in the formation of new organs and new anatomical patterns. These rules ('laws') cannot be immediately derived from the genetic, taxonomic, and paleontologic study of race and species formation. Indeed, the progressive evolution of many characteristics suggests the existence of special, not fully understood, phylogenetic agents. Moreover, it was the regularity of the phylogenetic course in evolution in some groups that led some paleontologists to the assumption of such unknown agents. One of the regularities of evolution has been claimed to be an explosive formation of many types at the beginning of a new family or order, then a slowly decreasing speciation that parallels increasing specialization, and finally a degeneration before the extinction of species. These facts were interpreted as due to autonomous factors, creative principles (e.g. Osborn, 1934; Lillie, 1945), or even as an expression of a 'will to individual and free development' (Beurlen, 1937).

With regard to such findings, it has become customary to distinguish between problems of 'macro-evolution' and 'micro-evolution'. As these two terms merely designate 'larger' and 'smaller' events without any clear borderline, and as they are linguistic mixtures of Greek and Latin roots, I prefer to use the terms 'infraspecific' and 'transspecific' evolution. Thus I hope to indicate a little more clearly the difference between those phylogenetic processes that occur within a species or lead to a new species and those that occur beyond the species limit and lead to new genera, families, and lesser divisions, and thus to new constructional types.

At present, transspecific evolution is one of the central problems of evolutionary thinking. A number of biologists and paleontologists say that undirected mutation, selection, and isolation are not sufficient for a workable explanation or the causation of major phylogenetic phenomena and their regularity. Thus far, however, we have not been able to discover or analyze

1

in detail any additional evolutionary agents. Nor can one expect quick agreement on this problem, as those who hold different opinions must finally ask the same important question: have organisms which stand high on the evolutionary scale – including man – arisen through the action of autonomous vital factors, or through undirected mutation and accidents of selection? There can be no doubt that this latter view is somewhat 'unsatisfactory' to those who otherwise look entirely to the cause-and-effect principle in their research work.

The various opinions regarding evolution are, to a certain extent, heterogeneous; the fact is that for valuable evolutionary work a thorough knowledge of different branches of the biological sciences (such as genetics, embryology, comparative anatomy, ecology, and paleontology) is necessary. Because scientific literature is huge in extent and because some important papers are difficult to come by, this desirable universality of knowledge can only rarely be obtained. Thus the discrepancy of opinions arises primarily out of the differing scientific backgrounds of the authors and is not due to the whole bulk of scientifically proved facts.

In contributing the present book to the study of transspecific evolution, I am well aware of the difficulties. I have, however, tried to avoid a one-sided view. My main object has been to examine critically all special factors and rules concerning transspecific evolution to determine their validity and to explore the possibility of interpreting them by evolutionary mechanisms known to science. (I already tried to do this in a brief survey in *Biological Reviews*, vol. 14, 1939, and in a short summary in *Biologia generalis*, vol. 17, 1943.) At the same time, however, I wish to show that these problems can also be studied successfully in recent animals. The methods employed are mainly those of comparative anatomy, because in most cases genetical studies cannot be made; most hybrids of species and genera are sterile, and hybridizing members of families and orders is impossible.

Today one fact can be mentioned which was not obvious when I finished the first edition of this book (1947). A certain optimistic view of the future development of evolutionary theories seems to be justified; there are striking similarities of ideas and results in the works of three authors who worked independently and reached the same conclusions. Wartime conditions prevented sufficient – or indeed any – communication when J. S. Huxley (1942) and I compiled our comparative evolutionary and ontogenetic studies and when Simpson (1944, 1949) prepared his paleontologic treatise. These books and the works of many others seem to indicate by similarity of views and findings that the problems of transspecific evolution rest upon a common ground, with an increasingly firm basis for infraspecific evolutionary studies (Dobzhansky, 1937, 1951; Mayr, 1942; Schmalhausen, 1949; Stebbins, 1950).

To me it seems essential for the understanding of transspecific evolution to realize that there are various ways by which races and species may be formed, and that all possible intermediate links between a race, a species, and a genus can be observed frequently in nature. Hence, two introductory chapters on this subject precede the main treatise.

2

2

The Causative Factors of
Infraspecific Evolution

Numerous studies on race and species formation, especially those based on genetics, have shown that infraspecific evolution can generally be considered as a complex function of the following factors: (1) mutations, (2) recombination of genes and gene flow, (3) fluctuations of population, (4) processes of isolation, and (5) processes of selection. All these factors have been so thoroughly analyzed that fairly generally accepted ideas of race and species formation could be advanced. It has been found, however, that each of these five main factors involves a series of single processes of quite a different character. Due to these special factors infraspecific evolution can proceed along quite different lines. For further studies of these phenomena the excellent surveys by Dobzhansky (1951; 1st ed. 1937), J. S. Huxley (1942), Mayr (1942), and Stebbins (1950) should be referred to, as well as the detailed genetical works by East (1936), N. W. Timoféeff-Ressovsky (1937, 1935), Stubbe (1938), Bauer and Timoféeff-Ressovsky (1943), White (1954), and the evaluation of modern systematics by various authors (J. S. Huxley, ed., 1940). A brief discussion of the main topics with which I shall deal follows.

A. MUTATION

The sudden alteration of inherited characters is the fountainhead of all evolution. It can occur as the mutation of a gene, a chromosome, a genome, a plastid, or a plasmon. Most important in race formation seems to be gene mutation. Normally the effects of genes are only known by mutations of an allele. Genes are considered to be large nucleoprotein molecules or small groups of molecules. They are capable of identical self-reproduction. The main component of the genic material seems to be nucleoproteins capable of many catalytic functions (Butenandt, 1953; Haldane, 1954). They are capable of self-reproduction (see also Chapter 8), and their constancy and stability are extremely great, so that alterations occur only after a relatively long time. This constancy is a necessary condition for the developmental differentiation needing stable interactions of numerous reactions. How spontaneous gene mutations are brought about we do not know. Recent investigations indicate that chemical analysis of these events may not be impossible.

Spontaneous mutations can happen in every gene and can thus produce alterations of every possible morphological and physiological character. Hence, there is no directed mutation at all. But of course the number of mutations of a nucleoprotein is not unlimited, and certain identical or very similar mutations occur from time to time in different populations. Because of this, homologous mutants can be seen in related animal groups, such as the short head in dogs, pigs, and the like.

The mutation rate generally is very low, but different in various genes. (In *Drosophila melanogaster* the rate of lethal mutations is estimated as about 2–$2 \cdot 5 \times 10^{-5}$ per gene per generation. Cf. Dobzhansky, 1951.) The possibility of exact calculations of mutation rates shows that mutation is time proportional. Hence the frequency of spontaneous mutations is higher in older than in just ripened spermatozoa (cf. *Drosophila*, H. J. Muller, 1946). It is of interest that the mutation rates as a whole are similar in bacteria, animals and plants. The instability of a gene can also be caused by a special 'mutator gene' (Mampell, 1946) or even – in *Epilobium* – by certain qualities of the cytoplasm (Michaelis, 1949).

As far as is now known, only one allele is affected in all gene mutations, and the mutants thus arise as heterozygotes only.

It is very probable that genes act primarily as enzymes. Sometimes they affect the earliest stages of development of the whole body or of a special organ. In such cases a mutation may have a quite striking effect, as in homeotic mutants in which a part of the antenna or of the mouth parts of *Drosophila* are changed into leg-like organs (Villee, 1942; Hadorn, 1953; Stern, 1954).

Other mutations affect only a certain step in a chain of later biochemical reactions. This applies to the process of pigmentation of ommatidia in insects, to the formation of anthocyanines in phanerogams, and especially to the synthesis of aminoacids in *Neurospora* (A. Kühn, 1932, 1950; Haldane, 1940; Beadle, 1947; Glass and Plaine, 1950; Hadorn, 1953, and others). On the other hand, it is quite evident that many mutations cause simultaneous alterations of many morphological and physiological traits. Such pleiotropism of gene action apparently is quite frequent (see also Chapter 6, B III) and must be given special consideration in the evaluation of processes of selection.

Manifold and various as a given mutation may be, mutations usually result in pathological alterations, defects, and abnormalities, that is to say, the normal development is more or less severely disturbed by them. In *Drosophila* about 90 percent of all morphologically sharply distinct mutants prove to be lethal (lethal zygotic genes; besides these there are lethal gonic and lethal gametic genes which can hardly be demonstrated because they eliminate themselves immediately after arising). Most mutations cause a decrease in viability (e.g. a general weakness in the struggle for survival) or fertility. One should not think, however, that gene mutations usually result in nothing but the destruction of alleles, as numerous back-mutations to the original type have been reported.

A small number of mutants shows full or even increased viability or superiority in special fields of competition. There is, for instance, a melanistic mutant (At) of the moth *Ptychopoda sericata* in which the mortality during the development is lower than in the normal type. Many other melanistic mutants show an increased viability. Some mutants are less viable than the normal type but have advantages under special conditions (preadaptation). The white mutant of *Drosophila* is more resistant to higher and lower temperatures than the normal type, and hence it is favored during strong shifts of climatic conditions and in possible migrations and expansions into different climatic areas (see also A. Kühn, 1935). Usually only a few favorable mutants arise under laboratory conditions, because in nature nearly all favorable mutations possible within the reaction limits of the genic material concerned have already occurred previously and have been incorporated into the 'normal' set of genes. There is no decrease of viability in certain micro-mutants, which apparently occur quite frequently. They produce no perceptible morphological alterations; yet often they can be detected by physiological methods. In the process of speciation they seem to play quite an important part.

A mutation can be either recessive or more or less dominant. Recessive mutations are considerably more frequent, as mutation usually causes an irregularity in the normal process and very often renders the gene inactive. Fully expressed dominance is definitely rarer than was previously supposed. It should be noted that similar phenotypic traits can be brought about by dominant as well as by recessive genes. There is, for instance, a dominant and a recessive white in fowl and pigs, and a dominant and a recessive black in mice. The intensity of the dominance of an allele can be altered by changes in the genotypic environment or by external conditions. Probably such an alteration in the degree of dominance is usually brought about by the effects of genetic modifiers during various processes of selection (Fisher, 1931). It can also be due, however, to a crossing-over or to any other dislocation by which the gene is brought into a different genotypic environment. Such a position effect can result in changes of viability and fertility. It can also shift the degree of phenotypic manifestation of some genes ('expressivity': N. W. Timoféeff-Ressovsky) and alter the frequency of the manifestation ('penetrance': Timoféeff-Ressovsky). Such alterations in penetrance and expressivity, however, also depend on alterations in the genotypic and external environment. In other cases alleles producing similar phenotypical effects are located at adjacent loci. Such 'pseudo-alleles' suggest a close biochemical interaction of small complexes of genes in the chromosomes (Lewis, 1954). Characters showing a graded series of intensity, as in color or size, are frequently brought about by multiple alleles acting successively or additively.

As with many other physiological processes, the rate of mutation within biological limits is proportional to temperature. It follows Van't Hoff's Rule to some extent as it is increased by higher temperatures; in *Drosophila* , for instance, a rise of temperature of 10°C. produces two and a half to five times as many mutations as would be produced at lower temperatures. It is not yet clear whether or to what extent indirect effects act upon the genes via the simultaneously altering physiological conditions of the body. Chemical agents can also act upon the germ cells as mutagens (mustard gas, formaldehyde, urethane, ethyl ether, etc.; Auerbach, 1949, summary in Dobzhansky, 1951; Herskowitz, 1955).

Finally, in all animal and plant groups the rate of mutation can be increased by ionizing radiations. It is not the kind of rays but the degree of ionization produced which causes the increase (up to 200 times the normal rate). According to our present knowledge, it is improbable that spontaneous mutations are caused by ionizing radiations occurring under natural conditions. The chitinous tergites of *Drosophila* (or a mash of *Drosophila* bodies, $\frac{1}{2}$ mm. high) will absorb most of the ultraviolet radiation, and the natural radio-active and cosmic rays are by no means intense enough to produce the rates of spontaneous mutation observed in nature (Hanson and Keys, 1930; Rajewsky and N. W. Timoféeff-Ressovsky, 1939) though they may, of course, have their share in effecting them. Species of Copepoda and Ostracoda from radioactive waters did not show any special variants (Pax, 1942).

Thus, a gene mutation can be spontaneous or induced, recessive or more or less dominant; it can result in an absolute loss or in an increase or a decrease of viability and fertility; it can produce marked alterations or minute changes; it can affect mainly a single character or a single process in a genetically caused chain of reactions or it can produce a pleiotropic effect; and finally, it can appear as a reverse mutation causing a restitution of the normal allele. But gene mutation is always a nondirected process limited only by molecular potentiality.

The effects of chromosome mutations are no less manifold than those of gene mutations. Like the latter, they are non-directed and can be spontaneous or induced by external factors, especially by ionizing radiations. They are caused by fragmentations and abnormal rearrangements of parts of chromosomes. The simplest case is the translocation of a part of a chromosome to another chromosome. By reciprocal translocation of two acrocentric chromosomes one metacentric chromosome may arise. Thus the number of chromosomes may be altered. By this process some genes are brought into a new neighborhood (position effect) and new combinations arise (thus, causing change of dominance).

If a chromosome breaks twice and the part between the two breaks rotates through 180° and is then fastened again to the rest of the chromosome, an inversion has taken place. Such inverted chromosomal parts usually disturb the pairing with the homologous chromosome at meiosis, and there is no recombination in the inverted parts, which are thus preserved in their original position. By this means chromosomal compounds with favorable combinations of genes can be kept together.

If a chromosome fraction gets lost in the course of a mitotic division a deletion occurs which usually proves to be lethal – when homozygous – even with very small losses of chromosomal material. Furthermore, there can be multiple breaks of chromosomes and correspondingly much more complicated recombinations. If, for instance, two homologous chromosomes break at different points and exchange their end parts duplications can happen.

Genome mutations have been observed only as a doubling of some chromosomes (heteroploidy), or as a multiplication of all chromosomes (polyploidy). The latter phenomenon is quite frequent in the plant kingdom, but seems to be of less importance in the animal world. In the heterogametic sex of animals polyploidy apparently causes too many disturbances in the balance of sex chromosomes and autosomes; often these disturbances lead to disturbances in ontogenetic development. Polyploid variants do appear in animals with parthenogenetic reproduction, as in some cases of *Artemia salina*, a small euphyllopod of inland salt waters, with di-, tetra-, and octo-ploid variants (F. Gross, 1932), in the isopod *Trichoniscus provisorius*, the moth *Solenobia triquetrella* (Seiler, 1938), some curculionids, such as *Otiorhynchus dubius* (Vandel, 1934; Suomalainen, 1947), and others. In the last two groups polyploid generations can be found only in the northern portions of their distribution area, a fact that proves that there is a geographic race formation. This, however, will not lead to the formation of new species, as all races with normal sexual reproduction are diploid.

On the other hand it must not be forgotten that there are closely related animal groups with normal sexual reproduction in which the total numbers of chromosomes indicate speciation by polyploidy. This applies to some butterflies (Lorkovič, 1941), Dermaptera (Bauer, 1947), Orthoptera (E. Goldschmidt, 1952, 1953), Lumbricidae (Omodeo, 1952), Triclada (Aeppli, 1952; Benazzi, 1949), and possibly to some Salmonidae (Svärdson, 1945; Kupka, 1950; compare also White, 1954).

Since Geitler's first reports in 1939 it has become more and more evident that there is also a 'somatic polyploidy' in which the zygote and the embryonic tissue are diploid as usual, but in which during ontogenetic development certain parts of the body become polyploid. This is brought about by endomitosis, a process in which chromosome division occurs in the nucleus without formation of the normal spindle and with a reduced spiraling of the chromosomes during the process. So, in the pond skater *Gerris* (Heteroptera) the muscles are tetraploid, the fat body is 4–32 ploid, the Malpighian tubes are 32–64 ploid and the salivary glands 1024–2048 ploid. Similar cases of endomitotic polyploidy were found not only in Diptera (nurse cells of the ovaries, salivary glands), in Odonata (fat and connective tissue), and in beetles and butterflies, but also in Ciliata (macronucleus), Radiolaria, and especially in Mammalia (liver and spleen, d'Ancona, 1939). By endomitosis the evolutionary advantages of polyploidy (certain reactions being controlled by several identical genes, possible increase of viability and resistance to certain factors of selection, greater adaptability and preadaptation) can become effective without causing any hazards in the mechanism of sexual reproduction. It cannot be told as yet to what extent the rhythmic nuclear growth in any other animals is also due to endomitosis.

Finally the mutations of cytoplasm and plastids should also be mentioned. The study of these has commenced only recently (Michaelis, 1949, 1954; Sonneborn, 1950; McCarty, 1946; and others). It has long been assumed that there is identical self-replication in certain cytoplasmic organelles. In other words, the cytoplasm is race and species specific. The interest of evolutionists has been aroused by the recent findings that self-replicating plasmatic structures can mutate. In contrast to the chromosomal genes that initiate and control certain chains of reactions, mutations of plasmatic genes result in direct hereditary alterations of the reaction systems proper. Cytoplasmic mutations can be important in hybrids of races and species, because the normal interactions of the genes of the chromosomes and the cytoplasm can be altered or disturbed to such a degree that sterility or death may result (Lamprecht, 1944; Michaelis, 1954).

The multitude of gene, chromosome, genome, and plasmatic mutations is further increased by the fact that some mutative alterations cause only minor, but others quite complex, evolutionary effects, which are often important for basic processes of individual development. Such constructive mutations (B. Rensch, 1947; also compare Chapter 6, B III) can be caused either by a single gene process or by a mutative change in a balanced system of gene combinations (e.g. main gene and modifier genes). In the case of the single gene mutation, various characters can be altered that control whole systems of physiological processes: e.g. mutational alterations in the quantity of active tissue of a hormone gland like the pituitary, or a mutational change of body size which causes a long series of correlated alterations in various proportions and functions of the body.

And, finally, the recombination of genes by sexual reproduction, by crossing over, and by translocations and inversions produces various gene combinations in a population which form a sort of genetic 'individual' of a higher order ('gene pool': Dobzhansky, 1951). By recombination, genes or chromosomes can form new gene combinations and may develop new effects. It can even happen that heterozygotes for lethal and normal chromosomes are in some cases more viable than heterozygotes of normal chromosomes (proved in the case of *Drosophila willistoni* by Cordeiro and Dobzhansky, 1954). Heterosis, i.e. hybrid vigor, is one of these effects. Evolution normally acts upon these genetic systems of higher orders, a fact to which taxonomists first paid attention by describing not 'types' of races or species but the whole variability (B. Rensch, 1929, 1934).

The mutants resulting from these various types of changes can lead to the formation of new races or species only in those cases where they can spread over several populations or at least over one large population in the course of generations. If the spreading process is sufficiently rapid, the variants will finally replace the larger part of the population. This end result is rendered possible by factors of selection and is favored by fluctuations in the size of populations and by processes of isolation.

B. FLUCTUATIONS OF POPULATION SIZE

Through the course of generations singly occurring mutations can spread. If it represents a variant without selective importance, only the polymorphism of the population is increased. High mutation rates can alter small populations considerably in a relatively short time. If the mutation rate is 0·01 percent of the population, the mutation pressure can be more important than the selection pressure (Ludwig, 1942). Mutants with minor selective disadvantages can escape extinction if they happen to find a 'niche' in their biotope which has not yet been taken by competitors (this is called 'annidation' of mutants with characters that are neutral or slightly disadvantageous to selection: Ludwig, 1950). In a rapidly increasing population rare mutants and rare combinations of genes and chromosomes will not encounter too much selection pressure and may spread fairly rapidly (Elton, 1930; Ford, 1949). Hence the genic material of the whole population ('gene pool': Dobzhansky, 1951) is the decisive basis of race and species formation. In this connection it is interesting to mention that song in birds not only denotes territory but also serves as a means of keeping the groups of populations together (Kullenberg, 1946).

The length of time elapsing between the first appearance of a selectively favored mutant and its spread and establishment within a population largely depend on the number of individuals belonging to that population and on the speed of the gene flow, i.e. the possible dispersal of the mutant. The number of generations necessary for a new mutant to reach a frequency of 1 permille in the population is extremely great in very large populations and relatively small in small ones; in the latter a new mutation comes to represent a higher percentage more quickly than in large populations, provided that nothing limits cross fertilization. In a small population random pairings will lead rather quickly to many homozygous alleles, because the risk of extinction is large, and loss of rare and new mutants is extremely great. Normally, reproductive animal populations are relatively small or medium-sized, of the order of 1,000 to 10,000 individuals.

Medium-sized populations also show marked numerical fluctuations (less distinct in pelagic animals). Any notable reduction of the number of individuals accelerates evolution and decreases variability.

Often the fluctuations occur periodically (in so-called 'population waves': Tschetverikov, Timoféeff-Ressovsky), bring about severe reductions in the number of individuals during harsh climatic seasons (winter, dry period in summer), and act as an ever-active force of evolution. Besides, there are irregular fluctuations caused by various factors, such as floods, epidemics, extreme cold or heat waves, and so on. If a population increases greatly, its parasites and natural enemies normally also grow in number (thus, the cycles of abundance in insects, especially pest insects).

Changes in the habitat of a species also frequently cause fluctuations in the size of populations. Such populations become especially important if the inhabited area shrinks and certain insular relict populations are cut off from the main population. It is in these relict populations that the principle of reducing the variants becomes obviously effective. But within the main population similar conditions can prevail, especially on the borders of specialized habitats (e.g. where a forest recedes and a steppe gains ground, or where a lake gradually is transformed into a swamp or solid land). If the process of shifting the habitat in the same spatial direction is continued for several generations, certain alleles of a population's gene pool may be lost, because the individuals migrating across the former habitat borders represent only a small fraction of the whole population; thus, they do not preserve the full number of genes. (This is elimination in Reinig's sense. It is not necessary to accept Reinig's assumption of 'single migrations'). Finally, all these numerical fluctuations of populations caused by temporal or spatial factors may combine in various patterns.

C. SELECTION

The speed and extent of race formation and speciation depend on processes of selection rather than on the variability of the material produced by mutation. This could also be demonstrated in the lines of descent of fossil types (Simpson, 1944). If environmental conditions remain constant, selection eliminates mutant genes in which harmful qualities preponderate. If the environment changes or the inhabited area is shifted to another place, variants and individual traits that were neutral in the preceding state can become quite important with regard to selection. Such a preadaptation may also cause major evolutionary steps, e.g. the loss of wings in insects living near the seashore or on islands (proved experimentally by l'Héritier, Neefs, and Teissier, 1937), or the formation of characters typical of cave dwellers, or the origin of terrestrial animals (see also Kosswig, 1948; Cuénot, 1951).

The selective agencies are quite diversified.

1. Many factors of the inanimate world act as forces of selection by wiping out those individuals that cannot stand extreme or harsh conditions caused by environmental factors (minima and maxima of temperature in winter and summer, droughts and floods, rapid or extreme changes of exposure to light, of salinity and hydrogen-ion concentration, and so on). All these factors may act periodically or only occasionally. The same holds for changes due to spatial shifts of the area inhabited, which are especially effective on borders of a biotope. Besides this negative selection there is also a positive one, as some variants may possess preadaptations which become genuine adaptations in the new situation, and as many animals actively move into suitable habitats. Dobzhansky (1948), for instance, stated that the percentage of three different gene arrangements in the third chromosome of *Drosophila pseudoobscura* differs strongly in populations inhabiting different altitudinal levels. The standard type prevails at low elevations, the arrowhead type at higher

10

altitudes. Other evidence on selection by factors of the inanimate environment is to be found in the papers by Alexander (1941); Errington (1943); Dobzhansky and Spassky (1944); Stalker and Carson (1948); Spiess (1950), etc.

2. Among the selective factors of the biotic environment is it the predators that maintain a steady process of selection by preying more frequently on those individuals with a more conspicuous coloration than the rest of the species, an imperfect state of camouflage and mimesis, an insufficient warning color, poor hiding or flight instincts, poor sense organs, and so on. Numerous relevant experiments proved the paramount importance of selection by predators. One should consult, for instance: Poulton and Sanders (1898); Heller (1928); Cesnola (1904); Sumner (1935); Steiniger (1937 a,b); d'Ancona (1939); Cott (1940); Dice (1947); Blair (1948); Cain and Sheppard (1954); and others. It is interesting to note that there is a mutual relation of predator and prey which causes periodic fluctuations in population size (Lotka, Volterra, and d'Ancona, 1939; Gause, 1935). These again influence the intensity of selection among the variants preyed upon, and lead to a similarly periodic fluctuation of selective intensity.

3. A special type of such 'enemies' is represented by parasites and infectious diseases (animals, bacteria, fungi, viruses). If they act on a large scale, in the form of an epidemic, their selective importance can hardly be underrated. At the same time they contribute to the population fluctuations which are so important in evolution.

4. Furthermore, a continuous selection is at work in the competition for food, territory, nesting sites (e.g. among birds nesting in holes), and so on. This daily 'struggle for life' also causes an interspecific selection. This is especially effective, as usually species differ more markedly from each other than do races of a species. Such a selection by rivalry has been studied experimentally in a number of cases. Usually the inferior type was wiped out, if it could not manage to find a niche not competed for (annidation, see below on p. 119). For further details see l'Héritier and Teissier (1934); d'Ancona (1939); Dobzhansky and Spassky (1947); Park (1948); Birch, Park, and Frank (1951); Birch (1954); and others.

5. Finally, sexual selection is effective in many more cases than was supposed by zoologists three decades ago. Here, too, we may find positive selection by ritualized or non-ritualized fighting, or by preference for certain mating partners (including homogamy), or extinction because of aberrant or less vigorous mating instincts, abnormalities of the sexual organs, or insufficient strength of sexual odor. Sexual selection is especially effective in developing and maintaining characters with a releasing function in display or the recognition of the other sex, the young, the parents, or the members of a social group (Tinbergen, 1951). This is evident, for instance, in closely related sympatric species, e.g. species of 'Darwin's finches' on the Galapagos islands. Here a marked sexual pattern is an important means of attracting and recognizing a sexual partner of the same species. If a species differentiation by such means is not necessary – as is the case on lonely islands inhabited by only

one species of a genus – the sexual difference between male and female can become greatly reduced (Lack, 1947).

All these types of selection can act together on the same species, and the daily struggle for life causes a continuous selection. The intensity of the latter can best be estimated from the enormous overproduction of progeny in nearly any type of organism, which was pointed out convincingly by Charles Darwin. Animals with periodic fluctuations of population size are subjected to a series of selective examinations, as Elton (1930) has impressively demonstrated. The increase of a population causes a more intense infraspecific competition; the decrease is characterized by selection through parasites and diseases, and in the cold season by selection through low temperatures. These changing situations of selection always maintain a certain essential level of polymorphism in the population. Such polymorphism can also be favored by the fact that heterozygous individuals have a greater adaptive value in consequence of an increased vigor (heterosis). On the importance of polymorphism, consult Ford (1945) and Dobzhansky (1951).

Selection processes usually affect certain characters. Thus, for instance, the smallest variants of warm-blooded animals are extinguished by the low winter temperatures, and carnivorous animals prey on those individuals in which fleeing and hiding instincts are less well developed than in others, etc. Most such characters are brought about by combinations of genes rather than by single genes. Hence, selection usually eliminates gene combinations or increases or decreases their frequency. Due to the pleiotropic effects of many of these genes, characters will be affected that do not seem to have any connection at all with the particular factors of selection. This is an important fact, as it explains that unfavorable characters may also arise or be intensified by selection in those cases where these negative traits belong to a gene complex in which the positive selectional characteristics prevail. Such complex effects of gene combinations provide a good interpretation of the causation of hypertrophy of some organs (see Chapter 6, B IV).

In this context it is not necessary to deal in detail with the much discussed mathematical foundations of selectional processes and results (compare the papers by Haldane, 1924, 1954: Fisher, 1930, 1936; Wright, 1932; d'Ancona, 1939; Ludwig, 1954). It should be mentioned, however, that with a normal mutation rate (1/100,000) and a selectional advantage of only 1 percent a mutant dominant allele will have totally displaced the original one after about 4,800 generations in an infinitely large population. With recessive alleles, a very long time will pass before the mutation reaches a 1 permille frequency (about 100 million generations). As we have seen, however, these figures are subject to strong alterations with diminishing size of populations, and hence increase and decrease of allele frequency may be influenced by random factors as well as by mutation and selection rates. In smaller populations the time to reach a certain level of frequency (1 permille) is shorter, lasting from about some hundred to a few thousand generations for a mutation with a 1 percent selective advantage, and thus even recessive mutants may gain complete preponderance within a relatively short time. Normally, the mating populations in nature are fairly small in number of individuals, or at least they show marked fluctuations in numbers, during which a temporary numerical minimum is reached. Furthermore, the selective advantage of a mutation is often more than 1 percent. This is especially so in interspecific competition, where the characteristics of the rival species are definitely more marked. Hence – except among the fairly large populations of pelagic animals and in the stable environment of the deep seas – mutations may reach certain levels of frequency within a fairly short time, and selection leads to evolutionary changes within periods that are in accord with those proved for race and species formation by paleontology (see Chapter 5).

Haldane (1954) pointed out that the intensity of natural selection for a special character can be defined as the logarithm of the ratio of the fitness of the optimal phenotype to that of the whole population. Sometimes these values, ranging from near zero to about 12 percent, can be determined from frequency distributions.

D. ISOLATION

The evolutionary factors dealt with in the foregoing chapters may give rise to new races and species in the course of generations. In addition to this historical evolution there is a complicating differentiation into races, due to spatial or physiological segregation. Young phylogenetic differentiations cannot be kept apart from continuous gene flow (panmixia) unless historical isolation is displaced by different isolation types of races *in statu nascendi*. This can happen in various ways.

1. Populations or parts of populations become separated spatially. This is especially clear in those cases where one large area of distribution is torn apart by climatic shifts or other environmental changes, or where an expansion beyond separating barriers takes place (e.g. if terrestrial animals spread or are transported to islands, or if water birds carry a new species to a lake where it was previously unknown simply by having eggs or larvae on their legs. This spreading process leads to the formation of geographic races. However, such isolation need not be complete. Quite a number of animal types show relatively little vagility, which may be due to limited mobility (in land snails, Myriapoda, wingless insects) or to ecological and behavioral (or 'psychic') linkage to the biotopes (e.g. territoriality in birds and mammals). In such cases there is often a continuous area of distribution over a wide range, but the gene flow will be limited to a narrow zone near the border line of the single races. In such relative geographic isolation the reproductive communities are relatively small in number of individuals, a fact which is often neglected in mathematical treatises on selection. Quite often this type of race formation – the arrangement of various races in a sort of geographic mosaic or a chain with more or less marked differences – can be observed in nature, and in many groups of birds, mammals, and insects it is the rule.

Spatial separation can also occur in microhabitats, as various factors of ecological or physiological importance differ within the general habitat. Thus, the small alternating areas of forest, steppe, mountain peaks, isolated lakes, colonies of food plants (in monophagous animals) and of hosts of parasitic species sometimes bring about spatial separation within the borders of fairly small habitats, and may lead to the formation of new races. Hence ecological races can arise within a geographic distribution area.

2. Furthermore, gene flow (panmixia) can be markedly reduced or rendered impossible by sexual isolation. This can be due to differences in mating seasons, different behavior patterns of mating and display (such as special calls to attract the sex partner, or methods of copulation), differences in sex odors, or anatomical deviations of copulatory organs. Incomplete, but quite efficient in the long run, is the sexual isolation provided by the phenomenon of assortative mating (homogamy), which means that similar partners prefer each other for copulation, as has been shown to occur in some cases. Sexual isolation is important, too, as an additional means of forming ecological and geographic races.

3. Genetic isolation can be caused by mutations which prevent fertile interbreeding of the variants, for instance, by severe disturbances in certain stages of development (e.g. after a mutative alloploidy). However, there may be hybrids of normal or even increased viability which are sterile or less fertile than the ancestors. Such types of genetic isolation are often combined with geographic race formation. Hence, the end links of chains of geographic races sometimes show a decreased fertility.

On genes causing sterility in hybrids, consult above all Dobzhansky (1951), Steiner (1942, 1945), Hadorn (1949). For interspecific genes and the inability of hybrids to reproduce possible 'progenes', see the papers by Lamprecht (1948, 1949).

E. HYBRIDIZATION

Sexuality supported by the processes of crossing over has a great revolutionary effect by continuously causing gene recombination. Even without mutation, recombination may produce new races. Furthermore, hybridization of races occurs when formerly separated types come into secondary contact. This can be seen from the examples of some European races of animals that originated during the Ice Age and took refuge in western and southeastern Europe, coming into secondary contact in central Europe in postglacial times. Normally, such events lead to the formation of hybrid races, but sometimes additional types with new characters arise by selection of special hybrid types.

F. INTERACTION OF THE EVOLUTIONARY FACTORS

Up to now there has been no reason to assume that further factors are active in infraspecific evolution. The joint action of mutation, gene recombination, changes of the population size, processes of selection and isolation, and occasional secondary hybridization is a sufficient explanation of all known types of race and species formation. There is quite a number of cases in which this fact has been proved by careful analysis.

As has been pointed out, the evolutionary factors mentioned usually act together in various combinations. Mutation produces new evolutionary materials; by fluctuations of population size it is sometimes given a certain preponderance, but usually only to such a degree that selection can have a more rapid effect on the frequency of the varients. By isolation swamping of any racial differentiation which may have begun is prevented. Depending on the efficiency of these factors, the evolutionary effect may be quite different. If there is a considerable number of new mutants due to a high mutation rate ('mutation pressure') a strong individual variability will result. If there is an extreme reduction of population size, a quite uniform race will emerge. If selection is very harsh and intense, highly specialized types will result. Even in cases where race formation is checked mainly by isolation and where population shifts and selection are less effective, undirected mutation leads to a marked variability not dependent upon environmental conditions, and eventually to the excessive development of certain structures. This is often found to apply to evolutionary processes on small islands.

Finally, we must not forget that animals are not only passively subjected to selection and isolation but that they find their choice of habitat actively guided by inherited releasers for habitat and food. Normally such releasers are only adapted to certain very limited characters of the habitat. Hence, if such characters also occur in a neighbouring area, where climatic conditions may be rather different, the species may enlarge their normal territory.

3

Types of Race and Species Formation occurring in Nature

Judging from the manifold effects of the various evolutionary factors and their combinations, it seems to go without saying that we should not expect to find a universal mode of infraspecific evolution, such as some geneticists have occasionally tried to establish. On the contrary, quite a number of different types of speciation exist (first outlined: B. Rensch, 1939). So we should try to demonstrate the various modes of speciation as they can be traced in free nature, and should evaluate their approximate effect on the sum total of evolutionary processes.

In spite of the theoretical possibility of finding a large number of combinations of evolutionary factors and interactions, it is relatively simple to find a suitable order in which to arrange the types of race formation, because only two of the five evolutionary factors mentioned above are strictly necessary: mutation and isolation. The special types of mutation effective in race formation have been analyzed in only a few cases, and in most animal forms it is hardly possible to analyze them, as many animal species are difficult to breed in captivity or their generations are too long to permit experiments. Since the types of isolation, on the other hand, can more readily be observed, we can distinguish six main types of races: (1) historical races, prevented from panmixia by their temporal succession; (2) geographic races in more or less discrete geographic areas; (3) ecological races living in the same geographic area but separated at least during the mating season; (4) races separated by physiological mechanisms affecting sexual activities; (5) genetic races which cannot produce fully viable or fully fertile hybrids; (6) hybrid races, which result from the secondary contact of formerly isolated (normally geographic) races.

A. HISTORICAL RACES

All well-known fossil animal groups show that the species have a relatively limited period of existence in time. Recent species of mammals and birds can usually be traced back only to the Pleistocene, whereas the respective types of the Pliocene must be considered as species of their own. Some species of Prosobranchia have remained unchanged since the early part of the Tertiary, but in still lower geological horizons they are replaced by different older species. Thus, all animal types show a continuous change of certain features, and from this general rule one can infer that historical race formation and speciation is by far the most frequent and important type of evolution. Considering the effects of time-proportionate mutation and of processes of selection, population shifts, and isolation, one arrives at the same conclusion. All favorable mutants, especially those causing an increased viability or fertility, will increase in frequency as time passes, and hence races and species as a whole will change with time. Of necessity there is also a time factor in all processes of geographic, ecologic, or sexual race formation.

In contrast to possible expectation, the number of paleontologic series proving such a historical race formation and speciation is not very large. The main reason for this fact seems to be that four premises must be successfully met: (1) a totally undisturbed succession of sediments; (2) organisms with structures capable of fossilization; (3) organisms with a sufficiently large number of individuals; and (4) able paleontologists taking the trouble to carefully collect and evaluate hundreds and thousands of fossil specimens.

There is a model example of such work based on the study of more than 4,000 specimens and illustrating the problems of historical race formation: the races of the Ammonite genus *Kosmoceras* from the Callovian (*Jurassic*) (Brinkmann, 1929). Here, in three parallel lines of descent a gradual change of races could be demonstrated, showing differentiations of single ancestral forms into several species as a phylogenetic bifurcation. Usually the changes of one race into the other and of races into species occurred gradually, the variability of characters in one form overlapping that of the successive form to a certain extent. Nevertheless, it seems clear that morphologically stable types lasted longer than unstable forms. Thus it is easier than might be expected to arrive at a correct and adequate grouping of the various forms of historical races. This is also the opinion of the experienced Swiss paleontologist E. Kuhn (1948). In *Kosmoceras*, furthermore, Simpson (1949) could prove, as the coefficients of variation in five characters bear evidence, that the differentiation of the *pollux-ornatum* from the *castor-aculeatum* line was probably due to the divergence of a population which had only some of the genes typical of the ancestral forms. In this divergent group individual variability seems to have been increased later on. It is not known whether this divergence was caused by temporary geographic isolation and possible remigration, or by some ecological or sexual isolation.

There is one more carefully analyzed example: the Trilobite genus *Olenus*. It is fairly probable that in this group repeated geographic isolations and remigrations occurred. R. Kaufmann (1933, 1934) studied six successive species of this genus in an Upper Cambrian profile of alum slate near Andrarum in Southern Sweden (see Figure 71). All species show a change from 'Artfrüh-formen' (early species forms) to 'Artspätformen' (late species forms), and in one case a whole transformation of three species in temporal succession could be demonstrated: *O. transversus–truncatus–wahlenbergi*. Otherwise, the species are separated by evolutionary gaps ('saltations'). Some paleontologists consider these gaps to be evidence of corresponding gaps in the phylogenetic course ('macromutation', see Chapter 6, A and E). However, Kaufmann seems to be more correct in assuming that these 'gaps' were caused by re-immigrations and differentiations of a 'conservative strain' from a habitat located elsewhere. In expounding this, he drew attention to the fact that all geographic 'Rassenkreise' show horizontal race and species formation. The shifts of population habitats in the case of *Olenus* may have been caused by the oceanic transgressions of the Upper Cambrian.

As it is difficult properly to judge geographic variation in fossil animals, paleontologists are often unable to determine whether the successive changes in a series of fossil types from a certain succession of sediments are due to historical race and species formation or to secondary immigration (or re-immigration). Usually the dating of geological horizons is done by using the fossils contained in them and, if the fossils of two horizons are alike, the horizons themselves are considered to be alike. Hence, with geographic variation taken into account, one is never sure whether the horizons in question are older or younger. This uncertainty is increased by the fact that primitive and more advanced geographic races may live at the same time, which can be seen from various examples in the Recent fauna.

It is fairly certain that there were geographic races in the fossil horse *Anchitherium aurelianense*, the evolutionary tempo of which was increased during the Miocene. There was a larger race in France, a smaller race in Switzerland, and a dwarf type in southern Germany (Wehrli, 1938).

It will remain difficult for research to prove that gaps in geological series of fossils from a certain stratigraphy are due to geographic race formation. One may, however, infer from the common fact of geographic race formation in our present times that this process has also happened in former eras. Moreover, one should not forget that some of the evolutionary gaps from undisturbed horizons (Waagen's 'mutations') may be a result of gaps in collecting or of disturbances and gaps of sedimentation. Such irregularities in sedimentation apparently are more frequent than has hitherto been assumed. Hennig (1932) states that the more one studies the details of paleontology and stratigraphy, the more probable it seems that there are irregularities in sedimentation. This is so even in profiles with an apparently undisturbed stratigraphy.

On the other hand, it is self-evident that geographic race formation involves a time component and that this type of speciation is linked with historical race formation. I wish only to emphasize that cases of a strictly historical race and species formation – one species per geological era without splitting into races – seem to be relatively rare. Further examples of historical race formation may be treated in a more cursory way.

The phylogenetic stages in ammonites discovered by Waagen in 1869 from the Brown Jurassic which belong to the group of *Ammonites subradiatus* are not now considered as racial stages, but at best as a successive line of species.

From these series the paleontological term 'mutation' was coined to distinguish historical from spatial race formation (variation). In the modern view the 'racial stages' of Waagen represent different genera ranging from *Stephanoceras* to *Parkinsonia-Perisphinctes-Oppelia* and *Macrocephalites* (see H. Schmidt, 1935). There is a gradual change of characters also in the successive species found in the Liassic Zones 2*c* to 1*d* which cover the species from *Scamnoceras angulosum* to *Sc. angulatum*, on to *Saxoceras praecursor–S. costatum–S. angersbachense*, and at last to *S. schroederi* (Lange, 1941).

In 1856 Hoernes discussed a phylogenetic lineage in the snail *Cancellaria cancellata* L. (Prosobranchia), which starts with small Miocene types (Vienna, and Tortona, Italy) having narrow and flat shell structures and leading to the Recent race (see Abel, 1929, pp. 3–5, Figure 1). In 1866 Hilgendorf demonstrated lines of descent in the snail *Gyraulus multiformis* collected from the Upper Miocene of the Steinheim area, and in 1875 and 1880 Neumayr studied a similar lineage in forms of *Viviparus* (*Paludina*) from the Upper Pliocene of western Slavonia and the Isle of Kos. Judged in the light of today's knowledge these lines are often considered to be mere local variants. The transformation of flat to more cone-shaped shells in *Gyraulus* was probably due to a sudden or progressive effect of warm waters, and in *Viviparus* the gradual development of smooth shell surfaces into more sculptured ones was possibly due to an increase of salinity caused by brackish water (see Abel, 1929). However, I would not exclude the possibility that this gradual transformation might involve a genetic shift. What has been proved (especially in *Gyraulus*) is merely a causal connection between change of environment and shape of shell. Hence, it may well be possible that the formation of hereditary ecological races was caused by a change in selection in addition to the modificatory effect. Races of this type have been shown to occur in recent fresh-water snails (Rumjancev, 1928). Moreover, varieties of sculpture among Melaniidae which Neumayr chose for comparison can be modified to only a small degree. On the isles of the Malay Archipelago there are four variants of *Melania tuberculata* showing strong differences of shell structures, but only one is usually found in the wide range of distribution across Africa (this is so in 98 percent of all relevant specimens in the collections of the Zoological Museum of Berlin). From this fact it is quite clear that in these cases there is no modification by ecological factors, but that the differences are hereditary (see B. Rensch, 1934, pp. 401–4). As a last example of historical race formation in fresh water, mussels, belonging to the species *Anthrocomyia modiolaris* from the Carboniferous, may be mentioned (Sylvester-Bradley, 1951).

Among the corals, temporal successions of races have been described in the genus *Zaphrentis* from the Carboniferous by Carruthers (1910) and Sylvester-Bradley (1951). An especially elaborate treatise on lineages of races and species in the genus *Bolivinoides* (Foraminifera) from the Senon was presented by Hiltermann and Koch (1950) and Hiltermann (1951). This study, too, based on thousands of specimens, shows that the periods of phylogenetic splitting are relatively short.

Concerning historical races in mammals, a few examples must serve. The rather strongly differing mastodons, *Bunolophodon angustidens* (Middle and Upper Miocene), *B. longirostris* (Lower Pliocene), and *B. arvernensis* (Middle Pliocene and Lower Upper Pliocene), are perfectly connected by intermediate types, morphologically and temporally. Hence, one can agree with Abel's opinion (1929) that this is a genuine line of descent. A still clearer situation shows up with the racial stages of the cave bear (pre-, eu- and hyper-spelaeoid stages) as they were discovered in the Drachenhöhle of Mixnitz (Styria, Austria) in layers of fossil bones of about 12 m. depth (Ehrenberg, 1928). A probable racial lineage can also be seen in the ancestry of the red deer from the recent *Cervus elaphus elaphus*, which is linked by *C. e. angulatus* with the early Pleistocene *C. e. priscus*, and possibly with the still earlier *C. acoronatus*.

As regards the ancestry of man, there is a historical race formation at least in the transformation of more or less Cro-Magnon-like types (*Homo sapiens fossilis*) to the recent *H. s. sapiens*. The same is possibly true in the racial change of *Pithecanthropus* into the types of Ngandong (Java)–Cohuna (Australia). It seems probable that the situation will prove to be quite similar in the general lineage of the Neanderthal type from the *Pithecanthropus* forms (including *Sinanthropus* and perhaps *Africanthropus*), though the direct lines of descent have not yet been elucidated in detail.

Numerous as the cases of historical race formation may seem to be, their number is not yet as large as would be expected taking into account the huge number of generic and species lineages. Perhaps this may best be explained by the assumption that in former geological periods the bulk of speciation happened in the same way as in our present time: by geographic variation, i.e. by spatial isolation of types.

In some cases the historical transformation of geographic races and the gradual change of a recent race can be observed even during the relatively short time of a few decades. In the moth *Pachys* (*Amphidasis, Biston*) *betularia* a dominant melanic mutant originated (*carbonaria*) which spread so rapidly that in some parts of the range in northwestern Europe it is the prevalent or even the only type (Ford, 1954, 1955). On the isles of Rügen, Usedom, Wolin, in the neighboring coastal districts, in some parts of central and southern Germany, and in the Brandenburg provinces this mutant appeared as late as between 1890 and 1925, and today this spreading is still under way (see Schneider, 1935; Schneider and Wörz, 1939). The map in Figure 1 showing this was drawn after the findings by Ule (1925) and some other entomologists. There is a similar situation in the gypsy moth *Lymantria monacha L.*, in which an increasing number of melanic variants (caused by three genetic factors) could be observed during the last 30 to 40 years (Gäbler, 1939). About 50 species of moths are now known in northwestern Europe in which the normal color phases have been replaced by melanic mutants (Ford, 1955).

Among birds this phenomenon may have occurred in the New Zealand flycatcher *Rhipidura flabellifera* (Figure 2) and in the sugar bird *Coereba*

normal type, melanic type (after Seitz, 1915).

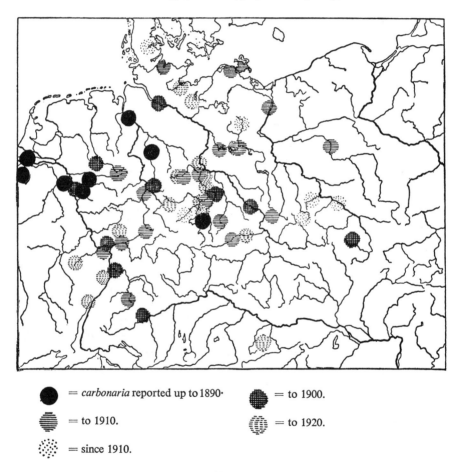

● = *carbonaria* reported up to 1890·   ● = to 1900.

▬ = to 1910.   ⦙ = to 1920.

⦙ = since 1910.

FIGURE 1. Spreading of the melanic mutant (*carbonaria*) of the moth *Amphidasis betularia*, after reports by Ule, Blasche, A. Bergmann, and others.

flaveola, which show a predominant melanic mutant in some parts of their areas. But the supposed spread of these blackish mutants during the last century is not sufficiently proved, as the observational data of earlier reports are incomplete (Mayr, 1942). There are similar situations among birds of

21

various orders, but here the details of spreading have received even less study (melanic mutants in the blackcap *Sylvia atricapilla heineken* on the Canary Islands and Madeira, in the shrike *Lanius schach schach* in southern China, in the weaver bird *Coliuspasser ardens* in western and central Africa, leucistic mutants in the hawk *Accipiter novae-hollandiae* of Australia – the exclusive type of Tasmania [Figure 2] – in the goose *Chen hyperboreus*, and others [Stresemann, 1926]). There has been a spreading of a melanic mutant in the hamster *Cricetus cricetus* since the end of the eighteenth century, with the center at the lower part of the Belaja River in the southern Ural region and the

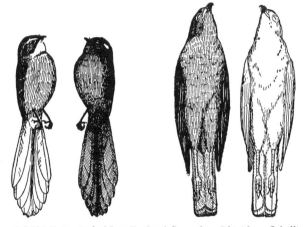

FIGURE 2. *Left:* New Zealand flycatcher *Rhipidura flabellifera* and its melanic mutant, '*fuliginosa*' (after Stresemann, 1926). *Right:* New Guinea hawk *Accipiter novae-hollandiae etorques* and its white mutant, '*leucosomus*' (after Stresemann, 1926).

main spreading direction towards the west, finally crossing the Dnieper to take its course along the northern boundary of the area inhabited (N. W. Timoféeff-Ressovsky, 1939). A remarkably quick spreading of a gene could also be reported in a colony of the butterfly *Panaxia dominula* by Fisher and Ford (1947). It is a striking fact that in all these cases more or less aberrant and dominant (often melanic) types are concerned, which are clearly beyond the normal range of variation ('exotypes' in the sense of Remane, 1928); possibly this type of a relatively quick racial transformation is restricted to certain vital forms of mutations.

B. GEOGRAPHIC RACES

In contrast to the cases of strictly historical race formation there is a quite different situation to be observed when the geographic races are studied. Here we need not search for suitable examples, as geographic variation can be observed in nearly every group, and there are even extremely large numbers of geographic variants, at least in those groups of terrestrial and fresh-water animals that present a great diversity of forms. (It should be noted that usually the races that have been studied have differed in morphological traits only, and that the numerous physiological variations have not yet been dealt with sufficiently.)

In many cases it has proved useful to designate large polymorphic species comprising several geographic races as 'Rassenkreise' (mosaics of races; B. Rensch, 1926, 1929, 1934). This term was introduced because strongly differentiated members of a 'Rassenkreis' do not differ less from each other than do closely related 'good species' (in the old sense) without geographic variation, and because such races when getting into secondary contact either do not hybridize or produce a more or less infertile offspring. Nevertheless, 'Rassenkreise' and species differ only by degrees, and in most cases the terms are synonymous.

The systematic study of geographic race formation has not yet been completed in any animal group, but in several cases the situation is now fairly clear, as the following data will show. When Hartert and Steinbacher finished their well-known work (1938), 1,715 geographic races of the palearctic Oscines had been described (doubtful forms not included), which were grouped into 363 'Rassenkreise' (comprising from 2 to 31 races each, with an average of 4·7 per 'Rassenkreis'). There were 153 more 'species' in the old sense which could not be grouped into 'Rassenkreise' (8·2 percent of all forms mentioned), many of which, however, represent borderline cases between race and species. A similar situation is found on other continents.

Species of larger birds with a wider distribution often covering several continents show a considerably weaker geographic variation. So the five bird families of herons (Ardeidae), storks (Ciconiidae), ibises (Ibididae), bustards (Otididae), and cranes (Gruidae) of the palearctic region show a total of only 32 races in 20 species or 'Rassenkreise' (making an average of 1·6 races per 'Rassenkreis' if the extra-palearctic races are excluded) and 24 species without any geographic race differentiation. The decrease in geographic race formation in such larger birds is probably caused by their greater vagrancy. In the species of smaller birds, migrating and nonmigrating birds also show typical differences as to their number of geographic subspecies. The palearctic families of ravens (Corvidae), tree creepers (Certhiidae), nuthatches (Sittidae), titmice (Paridae), wrens (Troglodytidae), and woodpeckers (Picidae), most species of which are more or less sedentary, comprise a total of 586 races in 81 'Rassenkreise' and only 34 species without geographic variations. The migratory families of shrikes (Laniidae), flycatchers (Muscicapidae s. lat.), hedge sparrows (Accentoridae), swallows (Hirundinidae), wagtails (Motacillidae), and orioles (Oriolidae) present 550 races in 173 'Rassenkreise' besides 115 species not varying geographically. Thus, the average number of species per 'Rassenkreis' is 7·2 in the sedentary and 3·2 in the migratory bird families, and the percentage of geographically nonvarying species is 5·5 in the nonmigratory and 17·3 in the migratory form.

The European mammals have not yet been studied sufficiently as to their 'racial systematics'. Up to 1937, however, the critical list by Oekland enumerated 399 mammalian races in 88 'Rassenkreise' (an average of 4·5 races per 'Rassenkreis'). Unfortunately Oekland did not mention the 'species' that showed no geographic variation. They amount to 90, which is 18 percent of all forms concerned. In studying Eurasiatic and Indo-Australian bats Tate (1941 a,b) applied the 'Rassenkreis' principle successfully (up to 18 races per 'Rassenkreis'). Mertens and Müller (1940), working on the European reptiles, could enumerate 205 geographic races forming 54 'Rassenkreise' (up to 24 races per 'Rassenkreis') and 50 nonvarying species (19·6 percent). In the European amphibians they could establish 20 'Rassenkreise' comprising 62 races. Twenty-one species (25·3 percent of all forms described) did not show geographic variation. (It is interesting to note that in 1928, when the first edition of the check list was published, 26·5 percent of the reptiles and 28·4 percent of the amphibian forms were considered to be species.)

In a zoogeographic survey of the European fresh-water fishes, Berg (1933) classified 207 geographic races (besides numerous subraces and ecologic races) in 64 'Rassenkreise' and 186 species without geographic variation (47 percent of the total number of forms mentioned in the study, and 40 percent of this total if the said subraces are included).

Breuning's monograph (1932–6) on the ground beetles of the genus *Carabus* enumerates 366 geographic races in Europe (Caucasian and East Russian border countries excluded) in 62 'Rassenkreise' (comprising from 1 to 42 races each) and 27 (6·9 percent) geographically nonvarying species. (Besides the geographic races mentioned by Breuning there are many types referred to as 'natio' which can well be considered as genuine geographic races with relatively small distribution areas.) Inside the German borders (northern Alps included) there are 63 races in 22 'Rassenkreise' and only 3 nonvarying species (4·6 percent). Evidently, then, these ground beetles which have lost their ability to fly and show only a very limited vagrancy generally tend to geographic race formation. And if in Breuning's survey the Asiatic forms are nearly all considered as species, this is due only to our incomplete knowledge of the systematic and faunistic characters of these types.

Quite a similar situation prevails in many other insect groups. Numerous 'Rassenkreise' have already been established as the result of studies on beetles, butterflies, Hymenoptera (especially bumblebees and ants), Orthoptera, and others. It is certain, however, that we are far from knowing enough about the geographic variation (except in bumblebees and in some groups of butterflies). The collections often do not include material that represents the series of specimens from all possible localities, and in some cases the specialist's interpretation points in a different direction from the principle of geographic 'Rassenkreise'. In some groups, such as the Drosophilae, genetically clear-cut races cannot be separated morphologically from each other (Dobzhansky, 1951). Therefore the extent to which speciation via the geographic variation occurred in the other insect groups cannot yet be defined precisely. Geographic variation in European butterflies cannot be surveyed as completely

24

as that in Indo-Australian forms, because in describing the latter types the authors, K. Jordan *et al.*, in the great work on butterflies by Seitz (1927) applied the principle of geographic 'Rassenkreise' systematically. Among the Papilionidae, Pieridae, and Danaidae they listed 2,268 geographic races in 412 'Rassenkreise' and 283 nonvarying species (11 percent of all forms mentioned). So we may now say that in nearly all insect groups the geographic race is probably the most frequent stage of formation of new species.

That the taxonomic method applied in studying a certain animal group is quite decisive for the evaluation of the systematic situation can clearly be seen from the revision of the Cypraeidae (Prosobranchia) by F. A. and M. Schilder (1939). Here formerly only a few 'variants' that could be considered as geographic races had been described. The monograph by Schilder and Schilder revealed 279 geographic races forming 84 'Rassenkreise', and only 77 species in the former sense (21 percent of all forms mentioned). Such a surprising change in the classification of an animal group of great geological age should inspire us to caution when evaluating other animal classes in which geographic variation has not yet been studied sufficiently, if at all.

Thus, it may not be superfluous to give a brief mention of those animal groups in which geographic 'Rassenkreise' have recently been established and in which formerly only species and nongeographic 'variants' had been described: Tunicata (Eisentraut, 1926: 3 'Rassenkreise' in *Ciona* and *Botryllus*), Echinodermata (Döderlein, 1902; Engel, 1934; Heding, 1942; Vasseur, 1952; Mayr, 1954), Cephalopoda (Adam, 1941: e.g. 4 races in *Sepia officinalis*), Brachiopoda (Helmcke, 1940), Phyllopoda (Wesenberg Lund, 1904–8, in Cladocera; Colosi, 1923, in *Apus cancriformis*), Copepoda (Ekman, 1917–20, in *Limnocalanus*; Baldi, 1941, even in pelagic Diaptomidae within the same lake; Tonolli, 1949, in diaptomids from high altitudes; Pirocchi, 1951, in cosmopolitan Copepoda and Cladocera; Kiefer, 1952, in many African and often pelagic species), Cumacea (C. Zimmer, 1930, 6 races in *Diastylis glabra*, 3 in *D. rathkei*), Amphipoda (d'Ancona, 1942, in Italian races of *Niphargus*), Hydracarina (Viets, 1926), Scorpionidae (Meise, 1933: 8 'Rassenkreise' in the genus *Rhopalurus*), Diplopoda (Karl, 1940), Trematoda (Erhardt, 1935), Turbellaria (Benazzi, 1945), Anthozoa (Pax, 1936: 4 European and American [Pacific coast] races of *Metridium senile* form one 'Rassenkreis'; Jaworski, 1938, in *Actinia*; Frenzel, 1937, in *Alcyonium*), and salt-water Spongiae (W. Arndt, 1943: 17 'Rassenkreise').

In the plant kingdom geographic race formation is not a rare phenomenon, though systematists do not yet seem to pay sufficient attention to it and sometimes get into trouble, since the recognition of geographic speciation can be difficult because of hereditary ecologic race formation and hybridization. For further study, see Von Wettstein, 1898 (e.g. *Gentiana*, *Euphrasia*); B. Rensch, 1929, 1939; Du Rietz, 1930 (e.g. *Pinus*, *Picea*, *Silene*, *Celmisia*, etc.); Geyr von Schweppenburg, 1935 (*Larix*, *Abies*); W. Zimmermann, 1935 (*Pulsatilla*); O. Schwarz, 1936 (*Pinus*); Stebbins, 1950; Baker, 1953 (*Armeria maritima*).

Summing up, we may state that geographic race formation has been proved to occur in nearly all animal groups and in several plant groups. It is evident that the number of examples studied recently and showing this mode of speciation is rapidly increasing, and that in many groups with an abundant variety of forms the taxonomic situation is ruled by this principle. This means that the large number of existing species may be considered to be mainly the result of geographic speciation.

Such a conclusion, of course, requires that the characteristics of the geographic races be hereditary. Modification and mutation often act in the same direction, giving similar phenotypic results (phenocopy), and they – in spite of their totally different causation – often seem to react to the same environmental conditions (because modification and mutation are the two possible ways by which the organisms adapt themselves to their environment). Therefore, a safe judgment of the possible heritability of the phenotype *in situ* cannot generally be given. Experimental tests on geographic races have been carried out in a relatively small number of cases, but these cases dealt with very different groups of animals. And all these cases showed clear cut genotypic differences in their essential racial characteristics. This has been proved in the North American rodents of the genus *Peromyscus* (Sumner, 1930, 1932; Dice, 1935; Svilha, 1935), in the European mouse *Arvicola scherman* (Müller-Böhme, 1935), and in the European hedgehogs (*Erinaceus europaeus* and *E. roumanicus*), which must be considered as strongly differing races (Herter, 1935). It was shown to be true in the European and American bison (*Bison europaeus* and *B. americanus*; the latter is a borderline case, see below) (Iwanow and Philipschenko, 1916), in red and silver foxes (*Vulpes*) (e.g. Demoll, 1930, p. 42), in the races of the African ostrich, *Struthio camelus*, (Duerden, 1919), in the races of the parrot, *Agapornis* (Duncker, 1929), in doves of the genus *Streptopelia* (Whitman, 1919), and in the races of the pheasant *Phasianus colchicus* (e.g. Cronau, 1902; Poll, 1911; Thomas and J. S. Huxley, 1927; and others: some races are still termed species). Additional examples are found in races of frogs (Porter, 1941; Moore, 1954) and of newts (Wolterstorff and Radovanovic, 1938; Callan and Spurway, 1951; Spurway, 1953); in races of the lady beetle *Epilachna* (K. Zimmermann, 1934, 1936); in bees and bumblebees (Armbruster, Nachtsheim, and Roemer, 1917); in flies, genus *Volucella* (Gabritschewsky, 1924); in the geographic races of butterflies, genus *Lymantria* (R. Goldschmidt, 1924–33), *Callimorpha* (R. Goldschmidt, 1924), *Spilosoma* (Federley, 1920; R. Goldschmidt, 1924), *Phragmatobia* (Seiler, 1925), *Colias* (Lorkovič, 1928), *Dicranura* (Federley, 1937: here the numbers of chromosomes vary in different races), *Leucodonta* (Suomalainen, 1941), *Zygaena* (Bovey, 1941), and *Pieris* (Petersen, 1947); in Diptera, genus *Drosophila* (in which the detection of racial differences is based on the pattern of the giant chromosomes, since in other structural and morphologic details there are only slight differences: Dobzhansky, 1950); and finally, in races of the land snail, *Murella* (B. Rensch, 1937).

This short survey of the relevant literature could be considerably increased by carefully examining all entomological and game-management periodicals, and those published by zoological gardens (for example, see Beninde, 1940, on hybridization in the red deer). Furthermore, one should mention the observations on the constancy of characteristics in different races kept under equal environmental conditions, as in all zoological gardens, where the strikingly different races of giraffes, zebras, lions, ostriches, cassowaries, and pheasants keep their respective differences through many generations. How far, however, modificatory alterations can proceed is shown, for instance, by the coregonid fishes (*Coregonus*), which, if they have been transported from their native lake to another one, develop quite a different pattern of gill structure after a few generations (Surbeck, 1920; Kreitmann, 1927; Thienemann, 1928).

Another proof of the heritability of geographic racial characteristics can be given if two races get into secondary contact under natural conditions (hybrid populations). This can be studied in central Europe in the hooded and the carrion crow (*Corvus corone corone* and *C. corone cornix*), and in the races of longtailed titmice (*Aegithalos c. caudatus* and *Ae. c. europaeus*); in southern Europe in the sparrows *Passer domesticus* and *Passer hispaniolensis*; in North America in two races of *Melospiza melodia* near San Francisco Bay (A. H. Miller, 1949), in two races of *Junco caniceps* in Arizona (A. H. Miller, 1941), and in other forms. Such cases are very numerous (for detailed study see also Meise, 1928, 1936; G. Dementiev, 1936, 1938; B. Rensch, 1936, 1945, pp. 42–3; Mayr, 1942; Johannsen, 1944). A typical feature of such regions with hybrid populations due to secondary contact is the extent of variation, which comprises the variability of both contacting races.

Finally, heritability of geographic characters can be considered to be fairly certain in the numerous cases where marked racial boundaries are not accompanied by any conspicuous environmental alterations, as is so often found where uniform races are distributed over a large area with strongly differing environmental conditions.

All such experiments and studies on populations have emphasized the fact that the characteristics used by systematists as typical features of geographic races generally have a genotypic background. Nevertheless, one should never forget that in all cases which have not been analyzed genetically one can deduce the possible heritability of characters only by analogy, although the probability may be fairly great. Concerning the problems dealt with in the present chapter we may state that geographic race formation based on genotypic variation can be a prelude to speciation.

In most cases geographic races differ from each other in several genes. By the formation of various geographic races, each of which is a more or less harmonious unit, a species becomes a very complicated genetic system capable of adapting in various directions, and thus possessing an enormous evolutionary plasticity (cf. Wright, 1939). It is apparently because of this plasticity that geographic variation became such a general phenomenon in speciation.

A further requirement for the assumption of geographic races as the most frequent forerunner of new species is to prove that there are intermediate stages between geographic race and species. This is of importance because in some recent papers the opinion is favored that geographic variation should be considered as a secondary branching of species which originated due to other causes (see especially R. Goldschmidt, 1935, 1940, 1952; compare also Chapter 3, G of this book). Concerning this opinion, it should be noted that in geographic variation fundamentally the same characters which are typical of a species can be altered, and that not only phylogenetically nonessential features (such as differences in coloration, size, and proportion) are concerned. As I have pointed out in more detail (B. Rensch, 1929, 1934, 1943) many geographic races of mammals differ in skull structure, in dental formula, in the number of toes (races of the rodent *Dipodomys heermani* have four or five toes: Dahl, 1939), in relative size of organs, and in number of young. Bird races show different osteological structures, organ proportions, egg colors, structure of eggshells and pores, number of eggs, songs, and fixed hereditary requirements as to habitat and migration. Races of fishes differ in the number of vertebral and fin bones and their migration habits. Insect races can differ in the number of generations produced per year and (often strongly) in shape and structure of genital appendages and organs. Occasionally the same holds true in Mollusca, where the shape of the radula can also differ between the races, etc. (B. Rensch, 1943; Mayr, 1943).

Extreme members of geographic 'Rassenkreise' (species) may show such striking differences in the characteristics mentioned above that, judging them morphologically, one can consider them as 'good species'. (Hence, for a more precise classification the term 'Rassenkreis' is not superfluous.) It can be assumed that, in some cases, extreme races would behave like good species, i.e. that normally they would not hybridize if they should come into secondary contact in the same habitat, as is the case in the races *Parus m. major* and *P. m. wladiwostokensis* of the great tit in the Amur area (B. Rensch, 1934, 1945). This excellent example is again presented in a detailed map of distribution which was drawn according to the late revision of the 'Rassenkreis' by Delacour and Vaurie (1950). It can be seen from Figure 3 that in Persia the European-Asiatic forms with green backs and yellow bellies gradually shade into the South-Asiatic type with grey backs and white bellies (*intermedius*), which type turns into the intermediate form *commixtus* in China, and finally into the East-Asiatic races with pale green backs and white bellies. In the central Amur region (black on the map) one of the races concerned (*minor*) lives side by side as a 'good species' with the green-backed, yellow-bellied race of the *major* group, which probably came from the west in a secondary shifting of distribution. A similar behavior of two such races has been observed in some cases when the races *intermedius* and *bokharensis* or *turkestanicus* and *major* came into contact (black areas on the map).

The same type of non-mixing contact in some places only, as described above, has also been observed in the races *Pyrrhula p. pyrrhula* and *P. p. cineracea* of the bullfinch in the Sajan, Altai, and Kusnetzk Alatau (only

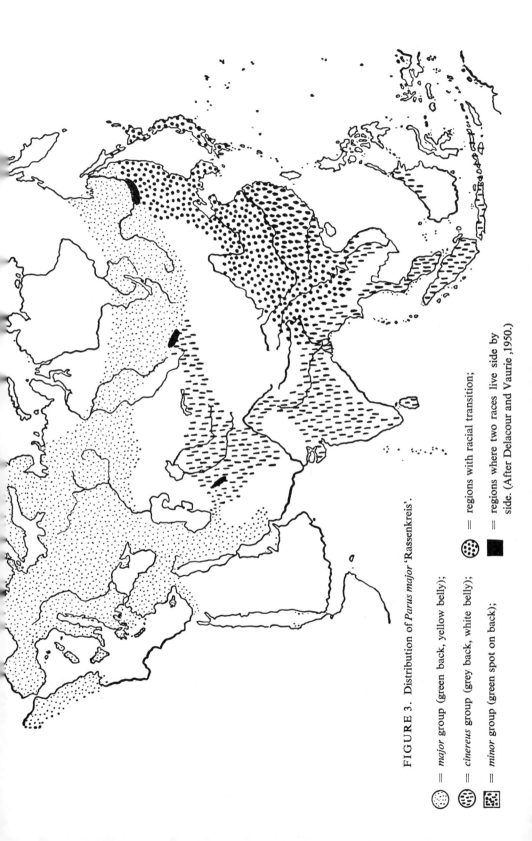

FIGURE 3. Distribution of *Parus major* 'Rassenkreis'.

○ = *major* group (green back, yellow belly);

⊛ = *cinereus* group (grey back, white belly);

▦ = *minor* group (green spot on back);

⊛ = regions with racial transition;

■ = regions where two races live side by side. (After Delacour and Vaurie, 1950.)

occasionally do hybridizations occur here, but a hybrid race lives in the Jenessei area: see Johannsen, 1944), in the warbler races *Phylloscopus trochiloides viridanus* and *Ph. tr. plumbeitarsus* of the upper Jenessei region (Ticehurst, 1938; Mayr, 1942), in the Mexican towhees, *Pipilo erythrophthalmus* and *P. ocai* (Sibley, 1954), in the mice *Peromyscus maniculatus artemisiae* and *P. m. osgoodi* on the eastern border of the Glacier National Park of Montana, in *P. m. austerus* and *P. m. oreas* of the Puget Sound region in Washington, in *P. m. bairdii* and *P. m. gracilis* of Michigan (ecologic isolation, hybrids rare: Dice, 1931). According to Stegmann's investigations, the gulls *Larus argentatus* and *L. fuscus* living side by side as 'species' in a large part of Europe can be looked upon as races of a large circumpolar 'Rassenkreis', in which they represent the clear-cut and well-defined end-links overlapping in a certain part of their ring-shaped area (Europe) in secondary contact. Stresemann and N. W. Timoféeff-Ressovsky (1947) suggested classifying the nineteen dubious forms of this borderline case into three or four species (compare also Voipio, 1954). Finally, the 'Rassenkreis' of the butterfly *Junonia lavinia* in North and South America presents a clear racial organization, while in the West Indies the southern and northern types live side by side without mixing.

Besides such extreme racial differentiations within a 'Rassenkreis', there are many other borderline cases between geographic race and species. They are represented by the frequent examples which the systematist cannot clearly define as either a race or a species. Frequently forms are concerned in which the morphologic divergence is so marked that one might just be able to term them species, but their distribution areas replace each other in various geographic regions. For instance, this is true in some island forms (see Figure 4), in some marine 'twin species' on both sides of the Panama region, or in closely related species and 'Rassenkreise' in the northern parts of the Old and New World, such as red deer and wapiti, European and American bison, stone marten and pine marten, European and North American badger, etc. I have designated such cases as 'Artenkreise' (mosaics of species) or geographic subgenera (B. Rensch, 1928, 1929, 1934), and more recently Mayr named them superspecies. In other cases, young species replacing each other occupy slightly overlapping areas but show hybrid populations only in some spots – similar to the above-mentioned races – whereas in other regions the forms live side by side without mixing. This is the case in the sparrows *Passer domesticus* and *P. hispaniolensis*: hybrids live only in Italy, Algeria, Corsica, and Crete; in the titmice *Parus caeruleus* and *P. cyaneus*; in the Egyptian desert snails *Eremina desertorum* and *E. hasselquisti*; and in the arboreal snails *Amphidromus contrarius* and *A. reflexilabris* of the Lesser Sunda Island of Timor.

Another and more advanced stage of such borderline cases can be seen in forms that are relatively young species and still fairly similar to each other, replacing each other in different geographic areas, but not showing any hybridization in the regions of secondary contact (examples: pairs of species which differentiated during glacial periods in the refuges of western and southeastern Europe or western Asia and now live side by side in central Europe, such as the woodpeckers *Picus viridis* and *P. canus*, the nightingales *Luscinia megarhynchos* and *L. luscinia*, the ground beetles *Carabus purpurascens* and *C. violaceus*, etc. For further examples of such borderline cases, see B. Rensch, 1929, 1934, 1943). In the latter cases the physiologic distance probably is relatively small, and in captivity hybrids can be obtained, as is

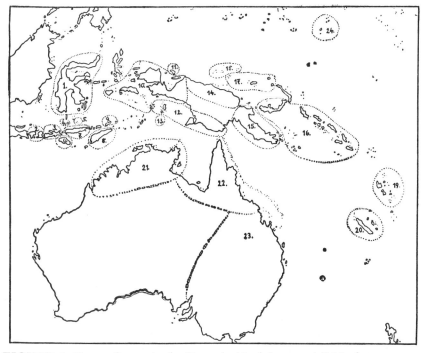

FIGURE 4. Range of races in the 'Rassenkreis' of the parrot *Trichoglossus ornatus*: 1. *T. o. ornatus*; 2. *mitchelli*; 3. *forsteri*; 4. *djampeanus*; 5. *stresemanni*; 6. *fortis*; 7. *weberi*; 8. *capistratus*; 9. *flavotectus*; 10. *haematodus*; 11. *rodenbergi*; 12. *nigrogularis*; 13. *brooki*; 14. *intermedius*; 15. *micropteryx*; 16. *aberrans*; 17. *flavicans*; 18. *nesophilus*; 19. *massena*; 20. *deplanchii*; 21. *rubritorquis*; 22. *septentrionalis*; 23. *moluccanus*; 24. *rubiginosus*.

known from *Luscinia megarhynchos* × *L. luscinia*. Dice (1940) also published such an example: the rodents *Peromyscus leucopus* and *P. gossypianus* are very closely related but clearly separated species living in an area of secondary contact – Dismal Swamp, Virginia – without hybridization. In captivity, however, they can be hybridized and will yield fertile offspring. Hence we see that the important thing in speciation is not whether hybridization is possible, but whether it does or does not take place freely in nature. From this point of view the borderline cases last mentioned above can really be considered species.

Moreover, there is evidence that the fertility of hybrids of extreme geographic races is decreased. From experiments on the hybrids of the strongly different pheasant races *Phasianus colchicus formosanus* (Formosa) $\times$ *Ph. c. versicolor* (Japan) by Thomas and J. S. Huxley (1927), it is known that only 87 percent of the hybrid eggs were fertile, and only half the chickens reached maturity. Hybrids of the butterfly races *Smerinthus populi populi* (Europe) $\times$ *S. p. austauti* (northwest Africa) proved to be partly sterile in both sexes in F_1 and F_2 (Standfuss, 1909). In hybridizing markedly different geographic races of the moths *Lymantria dispar* or *Lasiocampa quercus*, the physiological disturbance becomes manifest in the appearance of intersexes (see R. Goldschmidt, 1920). Analyzing the hybrids of extreme newt races of the *Triturus cristatus* 'Rassenkreis', Callan and Spurway (1951) brought forward the important evidence that chromosomal conjugation during meiosis is severely disturbed, and that inversions sometimes occur. Studying hybrids of the newts *Pleurodeles waltl* and *P. hagenmülleri* (belonging to an 'Artenkreis' in Spain and Tunisia) Steiner (1942, 1945) found lethal segregants in the F_2 generation. Hybrids of geographic races of *Rana pipiens* from more distant regions show a retardation in rate of development and morphological defects. Hybridization between races from geographically extreme regions results in a high degree of hybrid inviability (Moore, 1946). Crosses of the frog races *Hyla aurea aurea* from New South Wales and *H. au. raniformis* from southwest Australia resulted in complete hybrid inviability. Crosses of the similarly distributed races of *Crinia signifera* gave more or less the same results (Moore, 1954). On the other hand, Breider (1936) found that fertile hybrids can be obtained from fish species (genus *Limia*) which replace each other geographically.

It is important that such borderline cases can be found in all animal groups showing geographic variation. This, however, is revealed only if the systematist concerned pays special attention to these intermediate forms and labels them accordingly, because terminologically they are counted as 'still being' geographic races or 'already being' good species. Working on the birds of the Lesser Sunda Islands of Lombok, Sumbawa, and Flores (1931), I could show 47 such borderline cases beside 160 definite 'Rassenkreise'. Stresemann (1931) when reviewing the white-eyes (Zosteropidae) found six borderline cases in addition to 22 'Rassenkreise' and 30 undivided species. Among the Fringillidae of western Siberia, Johannsen (1944) found six such cases ('species-groups') in addition to 38 'Rassenkreise' and species, and among the Corvidae there were three such examples in addition to eight 'Rassenkreise' and species. From these few examples it may be seen that borderline cases of geographic race and species are as frequent as could be expected in the course of a speciation which is due mainly to geographic isolation. (One should also study the 'Artenkreise' of Tenebrionidae of the Sahara analyzed by Knoerzer, 1940.) There are further examples of borderline cases between race and 'good species' in the works by B. Rensch (1928, 1929, 1931, 1934 *a,b*, 1937, 1943); Meise (1928); Herre and Raviel (1939); Fitch (1940); Von Boetticher (1940, 1941, 1944); Ripley and Birckhead (1942); Mayr (1942); Hubbs (1943); Hubbs and R. R. Miller (1943); Hovanitz (1949); Nobis

(1949); De Lattin (1949); Herter (1954); Kauri (1954); E. O. Wilson (1955), etc.

Considering the geographic race formation as the most frequent and, in many animal groups, quite usual precursor of a species, we still have to ask whether Recent animal forms without geographic variation (species in the older sense) have passed stages of a *different* kind. A sure answer can rarely be given. It is possible that recent species are the only remains of former 'Rassenkreise'. Another means of evading the geographic variation – especially typical for geologically old species – may have been the capacity for modificatory adaptation to changing environmental conditions of the habitats. Many small fresh-water organisms capable of forming permanent cysts, statoblasts, spores or eggs, etc., and of modificatory changes in growth and structure, show such adaptive mechanisms. Such species may have been developed by historical alteration alone or by a combination of ecological and historical race formation.

Furthermore, it is important to ask which modes of selection are primarily effective in geographic race formation. This can be answered fairly well by studying the parallelism between certain characters and some environmental factors. Thus, we may establish three evolutionary types of geographic races: (1) those in which the characters do not show any corresponding relations to their environment and may have emerged at random; (2) those in which the characters show a gradual succession of steps according to the geographic distribution, but in which selection due to environmental factors is unlikely; and (3) those in which the characters and certain environmental factors correspond and in which the characters probably resulted from natural selection. Let us have a brief look at these three types.

1. Undirected geographic variation can be observed mainly in those cases where the effect of isolation is stronger than that of selection, i.e. where 'mutation pressure' is more effective than 'selection pressure', or where the range of variation is reduced due to intense fluctuations of the population size or to the establishment of new populations by a few specimens. Hence, random race formation is especially frequent in polytypic species scattered over a range of islands with nearly the same environmental conditions (such as tropical islands). So the conspicuously colored birds, such as fruit doves, parrots, pittas, flower peckers (Dicaeidae), and so forth, vary quite remarkably as to the extent and patterns of colors, and there is no apparent correlation of these differences with any obvious environmental factor on the various isles. Thus, for instance, in the 'Rassenkreis' of the parrot *Trichoglossus ornatus* (*Tr. haematod*, Peters, 1937) ranging from Celebes and Bali via New Guinea and its surroundings to New Caledonia and Australia (Figure 4), we find the race *mitchelli* on the Lesser Sunda Islands with a dark-purple head, a red breast, and a purple-black belly with green lateral parts. On the neighboring Isle of Sumbawa (race *forsteri*) only the center of the belly shows the dark-purple color, and the flanks are yellow. In the race *weberi* inhabiting

33

the Isle of Flores, red colors are totally lacking in the plumage, and the belly is yellow-green. On neighboring Timor we find the race *capistratus* with an orange-yellow breast and a blackish-green belly. On the Isle of Celebes, in the race *ornatus* the dark-purple color of the head is bordered by a red neck stripe with black scales, the red breast feathers show black edges, and the belly is dark green, so that this markedly different form is often considered to be a species. Probably all these different color patterns have the same protective function and camouflage the bird equally well when it feeds between red blossoms (*Erythrina* and others). So it is apparently of no importance at all whether a certain spot of the plumage is red, green, or yellow. Hence, it seems that in these cases either random mutation gave rise to new races without any auxiliary effect of selection, or the color variations are the by-product of pleiotropic genes, meaning that color is linked with a trait subject to certain factors of selection. In this context it is remarkable that in the one extreme race (*weberi*, see above) zooerythrin is totally lacking, while in the other extreme this red pigment is so intense that it bars all green and yellow shades. It is understandable that systematists consider this dark brown and red bird as a species of its own (*Tr. rubiginosus*), especially as it inhabits the peripheral regions of the *Trichoglossus* range, i.e. the Carolines (a borderline case). Quite instructive, too, is the synopsis of the color characters in the races of the shrike *Pachycephala pectoralis* of the Melanesian Archipelago, the races of *Monarcha castaneoventris* of the Solomon Islands, or the south-Asiatic races of *Microscelis leucocephalus*, given by Mayr (1932, 1942). In all these cases, the races originated from different combinations of alternatively varying characters.

Such transformations of racial characteristics, apparently not correlated at all with any environmental factor, may also be found in some other animal groups. Thus, four races of the deer mouse *Peromyscus maniculatus* inhabiting British Columbia may be considered as the possible combinations of three different characters, two of which are linked:

1. Relatively short tail+relatively large skull+dark coloration=race of Bowen Island.

2. Relatively short tail+relatively large skull+light coloration=race of Satura Island.

3. Relatively long tail+relatively small skull+dark coloration=race on the continent.

4. Relatively long tail+relatively small skull+light coloration=race of Vancouver Island (Engels, 1936).

Mertens (1931), in a monograph on the 'Rassenkreis' of the lizard *Ablepharus boutonii*, comprising more than 40 races from Hawaii and Australia to East Africa and the Bonin Islands, concludes that this subspeciation occurred by random mutation and that the combination of 'bipolar' characteristics often resulted in certain extreme forms, so that there are giant types of 47–54 mm. in length and dwarfs of about 35–38 mm., melanic and pale forms, specimens with coarse or fine stripes (Figure 5), with a flat or a high head, with 30–34 lines of scales or with only 20–22 lines. None of these varying characters reveals any correlation whatever with environmental factors.

Some of the bird mutants analyzed by Stresemann (1926) also show random color variations independent of environment. The same applies to the poly-typic species of the ground beetle *Carabus monilis* ranging from France and England to southern Russia and southeastern Europe, and displaying quite a divergent surface structure of elytra in the various races. In *preyssleri*

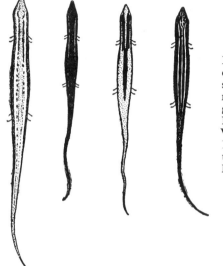

FIGURE 5. Geographic races of the lizard *Ablepharus boutonii* showing extreme variants due to random mutation. Left to right: giant race from the Isle of Juan de Nova (*caudarus*), dwarf race from Viti Levu (*eximius*), race from Lombok (*cursor*), race from an unknown locality, probably an East African island (*quinque-taeniatus*). (After Mertens.)

(Bohemia, Silesia, and northern Hungary) the surface of the elytra is practi-cally smooth; in other races it shows a relief of chains and longitudinal stripes (primary, secondary, and tertiary intervals) of a more (race *monilis*) or less (races *illigeri* from Bosnia, *semetrica* from Serbia, Banat) distinct character. This sculpture may also be variable (Rumanian race *kollari*), and so on.

Judging from the cases mentioned here, one might raise the objection that characters of only minor evolutionary importance are concerned. We should not forget, however, that quite frequently anatomical random differences also arise, capable of serving as essential steps in speciation. This may be seen in the genitalia of Carabidae and Sphingidae. In the latter family, K. Jordan (1905) found differences in the copulative organs in 131 out of 276 geographic races. In some cases, as in *Hyloicus pinastri*, no differences of coloration or wing venation are correlated with these differences of the genital appendages (K. Jordan, 1931). In this context mention should be made of the 'Arten-' and 'Rassenkreise' of the *Hipparchia semele* group that have been analyzed recently by De Lattin (1949). These forms, often closely resembling each other and possibly representing one single 'Artenkreis', differ markedly in the genital structures of both sexes. Two sympatric forms of the northern and central Balkans (*H. semele danae* and *H. aristaeus senthes*) which did not differ in their wing patterns revealed marked differences in their genital structures.

We cannot properly estimate the share of such random variation, independent of factors of environment, in the formation of geographic races in some animal groups. All we can state is that apparently in colorful birds of the tropics this random race formation is more frequent than in birds of the temperate zones, and in polytypic species ranging from the colder regions to the tropics. In these latter forms, the race formation by characters parallel with some environmental factors is definitely more conspicuous than in tropical birds. This is true at least as regards coloration, relative length of certain parts of the plumage, and body size. In the ground beetles of the genus *Carabus* already referred to above, the racial variation seems independent of environment in types from similar climatic environment, e.g. from the Mediterranean countries. But in forms living in different climatic zones, many racial differences prove to be correlated with environment: length and proportions of legs and antennae, relative height of elytra, and coloration and sculpture of the integument to some extent, whereas the structures of the genital appendages are exempt (B. Rensch, 1934). It seems possible that such differences in the type of race formation also exist in mammals, fishes, butterflies, other groups of beetles, and some other animal groups. In other cases, too, the difference between characters dependent on or independent of the environment is rather striking. According to Hellmich (1934 a,b), in the lizards of the genus *Liolaemus* from Chile, geographic variation is correlated with environment as regards the proportions of tail and legs, the relative width of the body, the number of scales, the feeding habits (percentage of vegetable food), and the reproductive and escape instincts. There is no correlation with environment as far as the special arrangement of scales on the head or special sexual dimorphisms are concerned.

If very small areas, e.g. small islands, are invaded, usually the first population is small in number, and hence these few individuals cannot represent the whole gene pool of the species. Thus a separate race will originate even without any further processes of selection, as variability will be reduced and homozygosity increased in a very short time. Such splinter races frequently show certain extreme characters in size or pattern, characters which often did not show up at all in the larger former range, as, due to the continuous panmixia, the formation of extreme traits was invariably prevented there. Examples of such 'new' and extreme characters are provided by the types of the 'Rassenkreis' *Ablepharus boutonii* (see above) and the polytypic species of *Lacerta lilfordi* and *L. pityusensis* (Figure 6) inhabiting the small islands near the Balearic Islands. Some of these forms are extremely small or extremely large, especially quick or remarkably slow, conspicuously slender or definitely plump, and often extremely dark, if not totally black, on the dorsal side (Eisentraut, 1929). In his thorough synopsis of these island variants, Eisentraut (1950) does not discuss the possibility that the extreme characters, with the exception of the black color, were brought about by an increase of random homozygosity due to the small populations (Sewall Wright effect).

In lizards of small islands, melanism is a frequent phenomenon. But it has not been clarified whether this extreme coloration is only the result of a homozygosity in a small population, or whether it is the result of selection (pigmentation serving as a means of protection from irradiation: Kramer, 1949).

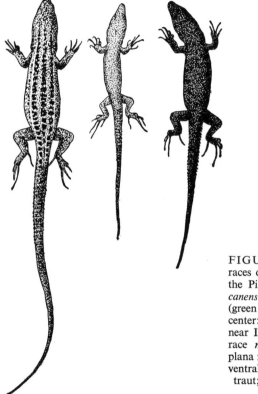

FIGURE 6. Three extreme races of *Lacerta pityusensis* from the Pityusic Islands. Left: Race *canensis* from Cana, east of Ibiza (green with brown lateral sides); center: race *grueni* from Trocados near Ibiza (sand-colored); right: race *maluguerorum* from Bleda plana near Ibiza (black with blue ventral side). (After M. Eisentraut; two-thirds natural size.)

Examples of such cases of island melanisms have been provided by T. Eimer (1874) and Kramer (1949): Faraglioni Rocks and some islets near Capri; by Mertens (1924): rocky islands of Kamik and Melisello near Lissa in the Adriatic Sea, the Isle of Filfalo near Malta and the Glenan Isles off Brittany; by Eisentraut (1925, 1930, 1950): Balearic Islands; by Kramer and Mertens (1938): some islets off the Istrian coast; and by Mertens (1932): Isola Madre in Lago Maggiore. Melanism may be supposed to be a means of protection from insolation, as there are further cases of island melanism in other animal groups. The terrestrial snails *Helix cincta* and *Otala vermiculata* on some islets off the Istrian coast are definitely melanic variants (La Figarola Grande near Rovigno; B. Rensch, 1928). In evaluating such examples of variation possibly correlated with environment, we should, however, account for the fact that, due to the reduced gene pool of the small populations, certain characters, especially melanism, which so often is dominant, may be found quite frequently. In the same manner other parallel characters of related species may emerge in consequence of the reduced gene pool of small populations. In the polytypic species of the land snail *Murella muralis* from western Sicily, for instance, the 'normal' shells show a smooth surface and a more or less spherical shape, whereas in some races inhabiting places that were separate islets in former times (during the late Tertiary) the shells are flat and display solid ridges on their surface.

Such alternative racial characters – spherical and smooth or flat and sculptured shells – have also evolved in the related polytypic species of *Tyrrheniberus villica* from Sardinia, *Rossmaessleria subscabriuscula* from northern Morocco, *Iberus gualterianus* from eastern Spain, *Eremina hasselquisti* from the Libyan desert, and in some forms of *Levantina* (B. Rensch, 1937. See Figure 68). Such limitations of original random variation and parallel subspeciation may easily be supposed to be evidence of definite evolutionary trends which, however, can hardly be true (compare Chapter 6, B III, on the rules of transspecific changes of structural types).

FIGURE 7. Above: left, *Helix cincta cincta* from the Dalmatian mainland; right, *H. c. melanotica* from the island La Figarola Grande. Below: left, *Otala verm. vermiculata* from the Dalmatian mainland; right, *O. v. figarolae* from La Figarola Grande. (After B. Rensch, 1928; two-thirds natural size).

(2) Geographically graded variation may occur without selection by the successive loss of alleles in the course of migrations and expansions of small peripheral populations or even single individuals (compare the statements on population fluctuations on p. 9). The most direct consequence of such migrations is a marked reduction of variability in the forms concerned. I could demonstrate this as highly probable in the terrestrial snail *Papuina wiegmanni* from New Britain of the Bismarck Archipelago (B. Rensch, 1939). Apparently this polytypic species spread from New Guinea, as the race *P. w. kubaryi* and some related races and species inhabit that island. The remotest part of New Britain (about one-third) has not yet been reached by the migrations, although the habitat as such would perfectly suit the snails. On the southern coast of New Britain facing New Guinea, we find a race with a relatively large range of variation (*P. w. wiegmanni*). Here the normal form with two black longitudinal stripes is found, but there are also variants in which these stripes are reduced to two cuneiform spots or are united to form one broad stripe. In other variants of this region the whole shell is covered by this black stripe, while in still others – with the normal pattern of the two stripes – the lip is

white instead of black or dark brown (Figure 8). The easternmost part of the southern coast, including the regions between Wide Bay and Cape Quoy, which obviously have been invaded last, is inhabited by only one variant with two normal bands (*P. w. disjuncta*). On the north coast the situation is quite similar: here the common type is *conjuncta*, in which the two dark bands usually unite with the dark lip. In the center of its range (near Talassea) this type is accompanied by a purely white variant with hyaline bands (frequency of this type: about 15 percent), and towards the eastern end of the range only one type is to be found (Figure 9).

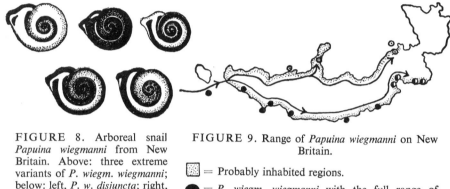

FIGURE 8. Arboreal snail *Papuina wiegmanni* from New Britain. Above: three extreme variants of *P. wiegm. wiegmanni*; below: left, *P. w. disjuncta*; right, *P. w. conjuncta*. (After I. Rensch, 1934.)

FIGURE 9. Range of *Papuina wiegmanni* on New Britain.

▨ = Probably inhabited regions.

● = *P. wiegm. wiegmanni* with the full range of variability.

◉ = *P. w. conjuncta.*

◑ = *P.w. disjuncta.*

Arrows indicate probable direction of spreading. (After B. Rensch, 1939.)

As far as polygenic traits are concerned, successive loss of alleles in the course of continuous spreading may result in a geographically graded variation of a character. Such genotypic variation without any influence of selection was termed 'elimination' by Reinig (1938). Reinig presented various examples of such happenings and tried to prove that – especially in warm-blooded vertebrates – the rules of climatic variation (Bergmann's, Allen's, and Gloger's Rules) were the result of such 'elimination', as in these cases there is a successive gradation of a character, for instance, body size, with geographic latitude. For several reasons, however, I was convinced of the effectiveness of climatic selection in the formation of such gradations, which led to the establishment of the rules concerning the correlation of body size with climate. In a debate with Reinig (Rensch, 1938; Reinig, 1938, Rensch, 1939), I tried to demonstrate that several examples cited by Reinig were not relevant and that others were of doubtful validity. And there are cases in which the largest race will be found in the areas colonized last by the population concerned. Moreover, examples have been found providing ample proof of the effects of climatic selection (see the next paragraph). According to my opinion, concluding the debate, only certain cases remain which are – for the time being – more reasonably comprehensible on the principles of elimination than of selection (compare also Henke, 1938).

To test the hypothesis of elimination, N. W. Timoféeff-Ressovsky (1940) suggested the special study of those cases in which the spreading occurred within historical times and in which its course can be traced fairly precisely. Thus, he studied the yellow-breasted bunting (*Emberiza aureola*) in its westward course in Russia during the last century. This species, however, does not support the hypothesis of elimination, as body size and similar measurements are not decreased in those specimens inhabiting the areas conquered last. On the contrary, there is a distinct increase of these measurements in the types from the 'youngest' areas of the range, and this is well in accord with Bergmann's Rule, and hence it may be regarded as the result of selection by climatic factors. I myself (1941) analyzed the spreading of the serin (*Serinus canaria serinus*) that has occurred since about 1800 and is still under way in central Europe. Here, too, there is no decrease of size and relevant measurements in the types inhabiting the recently occupied parts of the range (central, southern and southeastern Germany), but there is a slight (though not statistically valid) increase in Franconia and the central Rhine areas (Figure 10).

In five cases of Fenno-Scandinavian butterflies, Petersen (1947) could demonstrate that in relatively young postglacial distributions no decrease of size is detectable and that there are geographic gradations of a different kind, thus contradicting Reinig's hypothesis of elimination.

Nor can the successive gradation of melanin pigmentation (Gloger's Rule) have been brought about by successive losses of alleles in the course of postglacial distributions, as it is impossible to comprehend why the coloration towards the northeast should tend to a gradual loss of brown pigments (phaeomelanins) and finally to a complete disappearance of the eumelanins, while towards the northwest of Europe (England) an increase of both types of melanins should occur. As the gradation is parallel and correlated with temperature and humidity, a much more reasonable interpretation is provided by the assumption that climatic selection was the important effect. There are further cases and climatic rules, such as the egg rule, that cannot be interpreted at all by the idea of elimination, as those bird races inhabiting young postglacial areas of their range in the temperate and cooler zones tend to lay more eggs per clutch than their southern relatives.

Quite frequently, races of warm-blooded animals inhabiting high altitudes show a larger body size than their relatives in closely neighboring valleys (for details, compare the findings on birds from western China and eastern Tibet by E. Schäfer and Meyer de Schauensee, 1939). This fact, too, seems to contradict elimination, as according to Reinig's opinion a decrease of size was to be expected, and as on the short way from the valley to the higher elevation there could hardly be such a general effect of elimination. Hence, it is more probable that there was a selection of size variants according to Bergmann's Rule. The variability of shape, color, and size of wings, and of genital appendages (vulva and aedeagus) in races of butterflies is decreased in some types of the high mountains, as stated by Eller (1936, 1937) in *Papilio machaon sikkimensis, P. m. ladakensis,* and others. But here, too, elimination can hardly be held responsible, as in the (spatially) short process of spreading from the valleys to the mountains, climatic selection of preadapted variants or increased

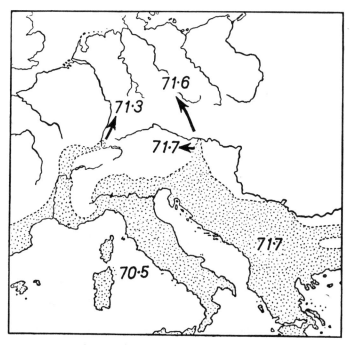

FIGURE 10. Average length of wings in populations of serins (♂). Figures pertain to the populations from the countries of the Western Mediterranean; from Franconia and the central Rhine regions; from upper and lower Bavaria to upper Austria; and from Bohemia, Saxony to western Prussia. Dotted: the original area inhabited before 1800. Arrows indicate direction of spreading. (After B. Rensch, 1941.)

homozygosity due to the small populations of the high altitude habitats is more probable than loss of alleles in the course of elimination.

Elimination might be suspected in European races of the ground beetles, genus *Carabus,* as the general variability decreases towards the northern part of the range, i.e. in places occupied last in the course of postglacial migration. (Tables 1 and 2). In some species, such as *C. ullrichi, C. auronitens,* and *C. nemoralis,* the number of individual color variants also decreases towards the north. But in this latter case we are not sure whether elimination is responsible, as the reduction of color variability is also correlated with climatic conditions (see below). In this context a statement by Voipio (1950) is of interest. He demonstrates an increase of variability ('polymorphization') in warm-blooded animals, especially on the periphery of their range in northern Finland. Quite correctly the author attributes this phenomenon to the smaller size of the different northernmost populations.

TABLE 1. Range of variability of absolute measurements in the ground beetles *Carabus cor. coriaceus* from central Europe and *C. cor. cerisyi* from Greece and Asia Minor in mm. and as percent of average measurements taken from 31 males. Decrease of variability indicated by −, increase by +. (After B. Rensch, 1943.)

	Range of variability in *Carabus cor. coriaceus*		Range of variability in *Carabus cor. cerisyi*	
	in mm.	as percent of average	in mm.	as percent of average
Width of head	0·9	26·5−	1·1	39·3
Length of elytra	4·3	19·4−	4·6	24·9
Width of abdomen	3·6	27·5+	2·6	23·8
Length of scutellum	1·6	25·0−	1·5	26·3
Width of scutellum	2·1	23·9+	1·7	21·3
Length of antennae	2·8	17·0−	4·2	26·9
Length of femur (1st leg)	1·9	26·8−	2·2	35·5
Length of tibia (1st leg)	1·4	26·9−	1·6	34·8
Length of tarsus (1st leg)	1·6	25·8−	1·5	25·9
Length of femur (3rd leg)	2·4	23·8−	3·1	34·9
Length of tibia (3rd leg)	2·7	25·2−	3·7	39·4
Length of tarsus (3rd leg)	2·2	23·7−	3·4	36·9

TABLE 2. Range of variability of absolute measurements in the ground beetles *Carabus auronitens auronitens* from central Europe and *C. aur. festivus* from southern France in mm. and as percent of average measurements taken from 18 males.

	Range of variability in *Carabus auronitens auronitens*		Range of variability in *Carabus aur. festivus*	
	in mm.	as percent of average	in mm.	as percent of average
Width of head	0·6	28·6+	0·4	20·0
Length of elytra	2·2	16·5−	3·1	24·6
Width of abdomen	1·5	18·1−	1·9	24·4
Length of scutellum	0·7	17·5−	0·9	22·5
Width of scutellum	0·8	13·8−	1·0	18·2
Length of antennae	1·9	15·2−	2·2	18·0
Length of femur (1st leg)	1·2	25·5−	1·3	27·1
Length of tibia (1st leg)	0·6	15·8−	0·7	19·5
Length of tarsus (1st leg)	0·6	13·6−	1·3	30·2
Length of femur (3rd leg)	1·2	17·9−	1·6	23·9
Length of tibia (3rd leg)	1·4	18·7−	2·0	28·2
Length of tarsus (3rd leg)	1·7	25·0−	2·2	31·9

Summarizing our findings on the pro and contra of elimination, we may state that occasionally a successive gradation of a character may originate from a loss of alleles without any effects of selection, but that in most cases only the total range of variability will be reduced successively by elimination. This type of subspeciation is most likely in those cases where the differentiating traits do not depend upon climatic conditions or similar factors of environment. Usually, however, 'elimination pressure proper' will be much less effective than selection pressure (see also Ford, 1949, p. 312).

(3) Very much commoner than the types of race formation just discussed is the formation of geographic races by selection, which can be recognized by the fact that an-inherited character varies parallel with factors of the environment. It is quite obvious that in the course of spreading the preadapted variants will be favored by the new environment and that selection will act continuously upon any new mutations that may emerge after the population has settled in the new habitat. Frequently it is difficult to trace the effects of such processes of selection, as some effective environmental factors may be unknown or not fully known (e.g. certain enemies, diseases, difficulties of getting suitable food, and so on). It is much easier properly to evaluate the effect of inorganic factors of the environment, especially if they are varying gradually. In such cases the racial differences of the subspecies will also form gradients. J. S. Huxley's term, 'clines' (1939), referring to such gradients, has now come into general use in the literature (Kiriakoff, 1947; Braestrup, 1945).

The study of whole animal groups from this point of view leads to the establishment of certain rules of climatic character gradients. In each case, of course, the percentage of exceptions from the rule concerned must be taken into consideration (B. Rensch, 1929, 1936, 1956). For instance, if the climate becomes cooler, the following changes (in the sense of a geographical gradation) will be found in warm-blooded vertebrates: increase of body size (Bergmann's Rule), shortening of relative lengths of tail, ears, bill, and limbs (Allen's Rule, partly though not totally due to allometric shifts and hence a consequence of Bergmann's Rule), reduction of phaeomelanins and finally also of eumelanins (Gloger's Rule), increase of relative length of hair and of number of wool-hairs of the underfur (hair rule). Later the wing tips will become more pointed and arm feathers relatively shortened (wing rule, called Rensch's Rule by Allee and Smith, 1951), the relative size of the heart, the pancreas, the liver, the kidneys, the stomach, and the intestine will be increased (B. Rensch, 1956), the number of eggs per clutch or young per litter will be greater, and the migratory instincts will be stronger, in the races of the cooler region (for further details concerning these rules, see B. Rensch, 1936, 1956). All these rules, of course, apply only to geographic variation in those polytypic species whose range covers various climatic zones, and this is especially true in the many examples provided by types inhabiting the temperate and cold zones. It is an important fact that geographic gradations will often be found in characters brought about by several genes or multiple alleles, as is well known in the formation of gradations in coloration or size of body and organs.

The proof that graded differences of body size are due to selection must, of course, rely on more or less indirect evidence. But such evidence is rather convincing, as often a very close parallelism between body size and climate

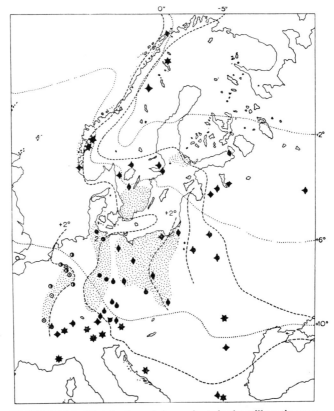

FIGURE 11. Distribution of size variants in the willow-titmouse *Parus atricapillus*. Average length of wings: white circle 57·0–57·9 mm.; circle with central dot 58·0–58·9 mm.; half black circles 59·0–59·9 mm.; black circles 60·0–60·9 mm.; circle with 1 wedge 61·0–61·9 mm.; circle with 2 wedges 62·0–62·9 mm.; circle with 3 wedges 63·0–63·9 mm.; and so forth. Dotted lines: annual isotherms; broken lines: January isotherms. Dotted areas: ranges of 58, 60, and 62 mm. populations. (After B. Rensch, 1939.)

exists. In the willow titmouse (*Parus atricapillus*), for instance, I could demonstrate that the gradation of body size (inferred from measurements of wing length) is closely parallel to the January isotherms, but not to the average isotherms of the whole year (Figure 11). Apparently, the minima of temperatures per year cause the extinction of the smallest variants, as the loss of body heat is greater in the small-sized variants in consequence of their relatively larger body surface, and as vital organs in the absolutely smaller body will more easily be reached by the temperature. Moreover, the physiological functions of limbs and other peripheral organs not reaching the normal blood temperature will usually be more easily impaired in the smaller animals.

There is one more fact supporting the assumption that the minimum temperatures of the winter are a most important factor in size selection: European migratory birds not subjected to the harsh winters follow Bergmann's Rule to a much smaller degree than do resident birds. The comparison of 21 relevant species of migratory birds having subspecies in central Europe

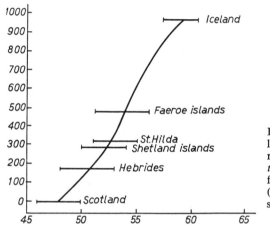

FIGURE 12. Increase of wing length towards the peripheral range in the wren *Troglodytes troglodytes*. Ordinate: distance from Grampian Mountains (northern Scotland) in km. Abscissa: wing length in mm. (After Salomonsen, 1933.)

as well as in northwest Africa (B. Rensch, 1939) revealed that there is an increase of size in only two central European races, whereas in the nonmigratory species 12 out of 22 showed a racial increase of body size in the central European races. The effect of temperature minima is also demonstrated by the following examples: in polytypic species of birds ranging from Morocco and Algeria to Madeira and the Canary Islands, in 12 out of 16 cases the continental races are larger than the island forms, in three races there is no size difference, and in only one case the Madeira Canary race is larger, but this race inhabits a higher altitude than its Algerian relatives. In both regions the average annual temperatures are now about equal, but the average minimum temperatures are definitely lower in northwest Africa (in the coastal regions of Tunisia and Algeria from −4·0° to +4·2° C., on the Canary Islands between 9·1° and 11·0°C.).

Climatic selection is likely to have occurred especially in all those cases where the larger races (contrary to what would be expected on the hypothesis of elimination) inhabit the areas of the range conquered last, and where the temperatures of these areas are lower than those of the central range. The races of the wren (*Troglodytes troglodytes*) show an ever-increasing size as we proceed from Scotland to more northern areas like the Hebrides, the Shetlands, the Faeroes, and Iceland (Figure 12). Birds that have spread from tropical and subtropical zones and invaded Europe in postglacial times, such as the kingfisher (*Alcedo atthis*), the golden oriole (*Oriolus oriolus*), and the turtledove (*Streptopelia turtur*), have developed their largest races in Europe. Moreover in six polytypic species of birds of the Lesser Sunda Islands which may safely be considered as invaders from Java, Sumatra, and Indochina, the largest races are to be found in the peripheral parts of their range (B. Rensch, 1939).

Quite recently Scholander (1954) expressed doubts concerning the explanation of Bergmann's Rule by selection, but he did not discuss the possible extinction by temperature minima and he did not physiologically analyze subspecies or closely related species. The results of Winkel (1951), Salt (1952), and Saxena (1958) contradict Scholander's arguments.

As in the case of the warm-blooded animals, there are rules applying to the poikilothermic forms with regard to body size, relative size of organs, colorations, number of eggs per clutch, special characters in the cycle of development, and so forth. In many of these examples, however, the hereditary character of the traits concerned must still be proved. As far as body size is concerned, only the general rule may be established that size will decrease towards more rigorous regions. There are terrestrial snails, for instance, increasing in size from northern to southern Europe, and others showing a definite decrease in this direction (possibly those especially adapted to a cooler climate, such as *Euconulus trochiformis*, *Vertigo alpestris*, etc.: B. Rensch, 1931). In reptiles and amphibians, Mell (1928) and Schuster (1950) proved a similar correlation between size differences (probably genotypic) and gradations of temperature and humidity. Size gradations of limbs and tails in lizards seem to follow Allen's Rule only as a consequence of negatively allometric growth (Schuster, 1950). Wing length of Fenno-Scandinavian butterflies often decreases towards the northern regions (Petersen, 1947). In races of European ground beetles, size sometimes decreases towards the Mediterranean countries, as in *Carabus coriaceus*, *C. auronitens*, and others, but increases in *C. ullrichi*, for instance. Generally, in various ground beetles inhabiting the warmer countries, antennae and legs are relatively longer (Krumbiegel, 1936) which is not a consequence of allometric growth only (B. Rensch, 1943). There is a more metallic coloration in the Mediterranean races than in types from central and northern Europe. For climatic parallelism of the coloration of butterflies, the papers by Hovanitz (1941, 1950, 1953), Petersen (1947), and others should be studied. For correlation of climate and the number of generations per year, consult the studies by Mell (1943) and Petersen (1947). Climatic selection is also suggested by the close correlation between microclimatic factors and typical colorations in geographic races of bumblebees (Pittioni, 1941). Further rules may be drawn up regarding the number of vertebrae and fin bones in fishes, the number of offspring in several animal groups, physiologically caused optimum temperatures (Moore, 1949, on *Rana pipiens*; Winkel, 1951, on closely related birds from different climatic zones), and so forth (compare also Margalef, 1955).

C. ECOLOGICAL RACES

According to our present knowledge species formation through differentiation of ecological races is considerably less frequent than geographic speciation. Though there are several examples in some animal groups showing the formation of specialized races in narrow biotopes, the importance of such phenomena seems to be very limited. Mayr (1942, 1947) even holds the opinion that in nearly all cases ecological races are nothing but microgeographic ones or even good species differing only in some minor morphological details. I cannot support this view, however, and I should like to point out that all alleged cases of speciation without prior geographic isolation must be considered with great skepticism.

First it should be noted that the term 'ecological race' is used in a narrower or wider sense by the various authors. In 1934 I proposed that this term be applied only to those animal forms which have been selectively adapted to certain biotopes but may live together in the same geographic region, and in which the hereditary character of the respective traits has been proved or is strongly indicated. The spatial separation of the habitats can last either permanently (forest and open country types, mountain and valley forms, fresh and brackish water types, parasites in various animals and plant species, etc.) or only for the mating period (e.g. egg laying by ecological races of insects on different plant species). As such special habitats usually recur in the area inhabited by the species, the territories occupied by an ecological race are normally scattered over a large region.

Minute geographic races, being restricted to a small area and usually also showing ecological adaptations to the requirements of their habitat, may not be detectable in limiting cases (e.g. occurrence of a mountain or beach form in only one place on the edge of the area inhabited). As an example of this kind, the races of Egyptian crested larks can be mentioned, in which the color of the plumage corresponds to the color of the soil prevailing in the biotope of the race concerned: thus, the central estuary of the Nile with its dark alluvial soil is inhabited by the relatively dark *Galerida cristata nigricans*; in the outer regions of the estuary the lighter *G. cr. maculata* can be found; and the sand-colored race *brachyura* (Stresemann, 1943) lives in the drier coastal districts. Further examples of such color adaptations have been studied by Niethammer (1940) in larks, and by Marshall (1948) in *Passerella melodia*. Moreover it is not easy to distinguish ecological races from simple populations showing certain special features which cannot yet be judged properly as to their ecological importance. Furthermore, it is difficult to prove the heritability of a characteristic without experiments, although in a geographic race one can make inferences from the more or less evident constancy of animal characters in the several micro-biotopes of the inhabited area.

Finally, it is important that ecological races very often differ only in minute morphological details or not at all, and that they can be separated with certainty only by physiological differences. So, the autogenous race (especially

47

living in urban regions) of the mosquito *Culex pipiens* differs from the non-autogenous type (which prevails in the open country) in being capable of egg development without previous blood meals, in producing a smaller number of eggs, in producing larvae at a more advanced developmental stage at hatching, and in possessing special mating and hibernation behaviour (see Weyer, 1937). The races *woodi woodi* and *w. externus* of the mite *Acarapis parasitica* on bees, cannot be distinguished by morphological characteristics (the differences of

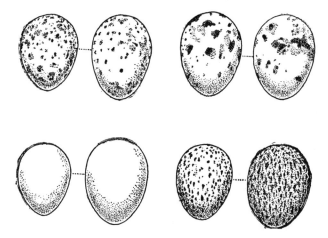

FIGURE 13. Eggs of central European cuckoo races (on right in each pair) and of their hosts. Hosts: above, left: Garden warbler *Sylvia borin*; right: great reed warbler *Acrocephalus arundinaceus*; below, left: redstart *Phoenicurus phoenicurus*; right: white wagtail *Motacilla alba*. (All eggs natural size.)

average measurements not being statistically valid), but will always be found in two quite different places: the first race in the first pair of prothoracic tracheae and the second one on the outside of the bee.

There are clear-cut and morphologically well-defined ecological races in the red ant *Formica rufa*, for instance, in which Gösswald (1941, 1942, 1944) distinguishes a 'small red forest ant', a 'pine ant', a 'dwarf forest ant', a 'sand-forest ant', and so on, besides the large race. All these races differ in their fertility and their inheritable hunting instincts. Quite distinct ecological races can also be found in the European cuckoo (*Cuculus canorus*) insofar as they lay eggs with colors predominantly adapted to those of their hosts. So there are blue eggs in those cuckoo races laying their eggs in the nests of the redstart *Phoenicurus phoenicurus* (e.g. in Finland about 85 percent of all cuckoos do this, according to Rey, 1892), white eggs with coarse spots of the *Acrocephalus* type in those laying in reed warbler nests, and eggs of the shrike type in the cuckoos laying in the nests of the red-backed shrike *Lanius collurio* (Figure 13; see B. Rensch, 1925). In these cases, however, the ecological isolation is not a sharp one as, due to the lack of stark isolation in the

European cuckoos, a frequent genetical mingling takes place (for further examples, see B. Rensch, 1934, 1945).

Another typical example of ecological races is the North American snake *Thamnophis elegans*, which shows an aquatic and a terrestrial group of races both living in the same geographic region (Fox, 1951). For further cases of ecological race formation, consult the instances reviewed by Thorpe (1930, 1940), and also the detailed analyses of the fresh and brackish water forms of the stickleback (*Gasterosteus aculeatus*) by Heuts (1947*a,b*), the discovery of hereditary forms of North Alpine Coregonid fishes by Steiner (1948), the investigation of ecological races in the marine snail *Acmaea* by Test (1946), and others. Above all, however, one should note the genetical studies on ecological *Drosophila* types with special adaptations to various altitudes (Dobzhansky, 1948; Stalker and Carson, 1948), to the different habitats of forest and steppe or urban and rural districts (Dubinin and Tiniakov, 1947), and so forth.

Borderline cases of ecological races representing intermediate links between races and species show that occasionally ecological races play a role in speciation. The highly debated ecological 'races' of *Anopheles maculipennis* seem to be one of the most thoroughly analyzed cases of this kind. These forms often live side by side, some of them showing slight differences in egg structure (Weyer, 1937) and special requirements as to the environmental conditions: the type *atroparvus* prefers a moderate Atlantic climate and its distribution is roughly limited by the isothermal line of $-7°$C. in January; *messeae* is a continental form prevailing in central and eastern Siberia, etc. (Beklemishev and Zhelochovtsev, 1937). Experimentally both these forms proved to be different in their capacity of distinguishing high or low atmospheric humidity. The type *maculipennis typicus* is most resistant to a dry atmosphere, *atroparvus* is the longest living, *messeae* the shortest living race, etc. (Hundertmark, 1938). It is interesting to note that crossing experiments done by M. Bates (1939) showed distinct physiological differences even between those 'races' in which the adults are totally alike in their morphology; hybrids of *atroparvus* × *messeae* did not even develop to the adult stage; those of *atroparvus* × *maculipennis* (*typicus*) were all sterile; among those of *atroparvus* × *labranchiae* several males showed abnormal development and irregular structural traits, etc. Thus it seems to be more correct to regard some of these forms as species (see De Buck, Schoute, and Swellengrebel, 1934; Zumpt, 1941; Mayr, 1942), i.e. as a whole this group of types represents a systematic example showing that ecological races may have been forerunners of these species.

A clearer situation may be seen in the relation of the human head louse and body louse (*Pediculus capitis* and *P. vestimenti*), which are definitely different from each other in their morphology. Some of the occasional hybrids between the two types are intersexes. Hence, most authors refer to the two forms as good species, although others continue to call them ecological races (e.g. Fahrenholtz, 1936). In this case the formation of the two species on an ecological basis is especially plausible, because they can be found on many human races, and they also show a geographic variation, though this is less distinct. Endoparasitic 'species', such as the tapeworms *Hymenolepis nana* (in man) and *H. fraterna* (in Muridae) can be regarded as such borderline cases, and the same holds good for two extremely closely related 'species' of gall mites (*Eriophorus galii*, living on *Galium*, and *E. nudus* on *Geum*; see Nalepa, 1917), for the Psyllid *Psylla mali* and *Ps. peregrina* (Thorpe, 1940), the Flagellata of the *Trypanosoma brucei* group, the *Plasmodium* species in man and apes, *Entamoeba histolytica* and *E. ranarum* (Hoare, 1943), and so forth.

Summing up, we may say that a purely ecological isolation can also serve as a step to speciation, but that these cases seem to be rather limited in number as far as the animal kingdom is concerned. This is because ecological isolation can have only minor evolutionary effects, as usually only small populations are concerned, which are, moreover, imbedded in the same geographic area. Only among parasites does speciation by ecological isolation seem to take place regularly. Frequently, however, ecological isolation is linked with geographic separation.

In the realm of plants, ecological races seem to be far more common than in the animal kingdom, perhaps because such races can rather easily find their scattered special habitats by their well-developed dispersal mechanisms (see Turesson, 1926, 1929; Du Rietz, 1930; Constance, 1937; Gregor, 1946; Stebbins, 1950; Baker, 1953, etc.). Moreover, borderline cases of ecological race and species (e.g. beach or mountain types) occur more frequently in plants. Hence the possibility of polyphyletic speciation arises, as similarly directed selection can act upon the populations of more or less identical habitats.

D. SEXUALLY ISOLATED RACES

Apparently, sexual and geographic isolation are usually connected (e.g. no hybridization takes place between strongly differentiated races in certain borderline cases, as in *Parus major* of the Amur region, in *Passer domesticus-hispaniolensis*, etc.; see pp. 28–30). Probably only in rare cases can sexual isolation alone be sufficiently effective as a step in speciation. The simplest mode of sexual isolation would be the formation of variants with different reproductive seasons within the same habitat. In the sea urchin *Paracentrotus lividus*, for instance, there are two types living in the Bay of Naples. In one type the larvae will develop only at temperatures between 8° and 23°C. (winter spawner), in the other only between 14° and 29°C. (summer spawner) (see Runnström, 1936). There are two morphologically different races in the fish *Ammodytes lancea* near Heligoland, one of which spawns in spring, the other in summer (Jensen, 1941). Such differences in the reproductive seasons can be regarded as a sign of speciation, as such sexual races can spread in different directions according to the degree of adaptation of their reproduction to different temperatures. In this manner sexual and geographical isolation will act together. This has also been shown to be probable in some species of *Rana* and *Bufo*.

Moreover, sexual isolation can be brought about by differences in mating behavior and mechanisms essential for the mutual attraction of the partners, such as olfactory stimuli, locomotory habits, and similar behavior patterns. In this respect especially careful studies have been done on the fruit flies of the genus *Drosophila*. In special containers suited for exact observation, Spieth (1947, 1949) brought together males and females of the *D. willistoni* group from six different geographic localities of tropical and subtropical North and South America. There were 30 possible pairs of combinations among the six groups (in each case 3 ♀ and 3–6 ♂). There were no successful copulations except in one pair, the mating behavior, display, etc., being suddenly interrupted by the ♂ or ♀ (for instance, by ejecting the female genitals), though sometimes a definite drive to accomplish copulation could be clearly observed. In two species (*nebulosa* and *fumipennis*) mating behavior was markedly different in that the males showed a frontal display towards the females, in contrast to the other species, where the males displayed beside or even behind the females. The minute differences in mating behavior causing the failure of copulation attempts could not be analyzed in detail, but may have been due to different sexual odors and their role in mating. This opinion was recently favored by B. Götz's studies (1951) on the influence of olfactory stimuli in male Microlepidoptera. He could attract males to his catching device by exposing females or only by the use of extracted female sexual odors. Stretching out odor-producing organs, making fanning movements to spread the odor, etc., also hint at such possible 'olfactory' isolation of insect species (see Urbahn, 1913, and others).

It is quite important that the closer the spatial neighborhood of closely related *Drosophila* species, the stronger the aversion to interspecific copulation. Strains of *D. pseudoobscura* from regions where they live sympatrically with *D. miranda* show a stronger sexual isolation than those of localities where only one of the two species lives (Dobzhansky and Koller, 1938). The same is true of *D. guarani* and *D. guaru* (King, 1947).

In some cases the beginning of sexual isolation can be detected even before any formation of a sexual physiological race is to be seen. This is due to the so-called 'homogamy' ('assortative mating': K. Pearson, 1903) which was first shown to occur in man and later in some animals, also meaning that similar partners are favored for mating. So Crozier (1918) showed that in *Chromodoris zebra* (a marine snail) individuals of nearly the same body size form mating and copulating pairs more often than do those different in size. The same principle was found to be true in the amphipods *Gammarus locusta* and *Dikerogammarus fasciatus* (1923) and in toads (Willoughby and Pomerat, 1932). Mutual preferences on certain color variants were recorded to occur in the rodent *Peromyscus maniculatus blandus* by W. F. Blair and Howard (1944). It should be stressed, however, that up to now such a homogamy has been reported in only a limited number of cases, and that usually a general panmixia of individuals prevails (Alpatov, 1925; Spett, 1929). In *Drosophila melanogaster* only a very few mutants show signs of such a sexual preference, especially the forms wild-type and white-yellow (Spett, 1932).

Copulation of certain variants can also be reduced or rendered impossible by strictly mechanical obstacles which are, for instance, due to great differences in body size. Just as copulation cannot take place between a heavy Irish horse and a Shetland pony, or a Saint Bernard dog and a dwarf terrier, it is likewise impossible for geographical and ecological races of marked differences in body size which get into secondary contact. In some geographical races, body size is two or more times that of the related race, as in the reed warblers *Acrocephalus arundinaceus arundinaceus* of Europe and *A. a. meyeri* of the Bismarck Archipelago, in the shrews *Sorex obscurus malitiosus* of Alaska and *S. o. parvidens* of California and New Mexico, in the butterfly races *Parnassius mnemosyne nankingi* and *P. m. funkei* of China, etc. In some of these cases the physiological differences may be slight, and artificial insemination may even result in fertile offspring, but this is not the essential point. It is sufficient that in the natural state practically no hybridization occurs, and that the way to further differentiation is opened by this kind of sexual isolation. Besides, size difference of mammals sometimes can be lethal even after a successful copulation of small females and very large males, when the embryos grow too large, causing their mothers' deaths and thus their own. An obstacle to copulation exists in differences in the shape of shells in some families of terrestrial snails (e.g. Partulidae, Achatinellidae of the South Sea Islands), some species or races of which show a certain percentage of shells with sinistral whorls besides dextral ones.

In animals with rigid sexual armatures – especially in insects – copulation can be rendered impossible by differences in the shape of the male or female genitalia. Apparently, however, marked differences in these structures occur only in geographic races of butterflies (K. Jordan, 1905, 1931; Eller, 1936; Warnecke, 1938, and others) and of beetles (H. Franz, 1929, 1949; Meixner, 1939, and others). There can be a marked variability in the shape of these structures in the population of a given locality (see Drosihn, 1933) but, as far as I know, no sympatric insect races have been found showing solely this type of isolation. Moreover, it is rightly doubted that slight structural differences of the genitalia can effectively bar successful copulation (Beheim-Schwarzbach, 1943; Breuning, 1943; Sengün, 1945; Kullenberg, 1948, and others).

Hence, we may state that a purely sexual isolation is by itself a step leading towards speciation only in relatively few cases. Sexual isolation or homogamy can, however, keep apart a certain percentage of extreme variants and thus aid in increasing the differences between geographically and ecologically isolated races, so that in cases of secondary contact mingling can be rendered completely impossible.

Similar conditions seem also to prevail in plants: species which are pollinated only by certain insects or birds show a higher percentage of specific floral adaptations than those which can be pollinated by a large number of creatures (Grant, 1949).

E. GENETIC ISOLATION

Formerly the races A and B of *Drosophila pseudoobscura* were a paradigm of strictly genetic isolation. They could not be distinguished morphologically and lived side by side in the same locality (Lancefield, 1929). Male hybrids of these two 'races' are infertile due to disorders in their spermatogenesis. Later on, however, these 'races' proved to be good species differing in the shape of the penis (*D. pseudoobscura* and *D. persimilis*). Probably they originated from geographic variants, as the larger part of the inhabited area is distinctly separated (Dobzhansky and Epling, 1944; Rizki, 1951). Contrary to the ideas once held by some geneticists, we may now assume that a speciation by genetic isolation only rarely happens. For instance, in the well-analyzed species of *Drosophila* there are some recessive genes which lead to sterility in homozygous males or females, but usually this sterility does not cause an interracial barrier unless there is a simultaneous geographic isolation. Then this hybrid sterility is quite important, as in border regions it prevents (useless) hybridization of related species (see p. 51) and in secondary contact between more strongly differentiated geographic races it bars panmixia and leads to the formation of new species (see p. 14). Probably it is in species of Diptera, which are similar or almost identical as to their morphology, that such a combined effect of geographic and genetic isolation causes hereditary differences. In such species it has been shown that their sets of genes are usually almost homologous, though there are translocations and inversions (for example, in *Drosophila*, see Sturtevant and Dobzhansky, 1936 a,b; East, 1936; Horton, 1939; and, above all, Dobzhansky, 1951; on the very similar *Sciara ocellaris* and *Sc. reynoldsi*, see Metz and Lawrence, 1938; and for further details also compare Chapter 6, B III). In vertebrates a geographic differentiation can also be accompanied by shifts in the chromosomal arrangement, as is indicated by the results of Callan and Spurway (1951) on the meiosis of geographic races in newts.

It is a fact that in phylogenetic differentiation the genetic isolation normally has at least one additional effect, viz., the more or less decreased or even totally lacking fertility of animal species-hybrids. The physiological disorders in such cases can often be seen in the somatic development of the hybrid embryo or in the maturation of the new reproductive cells (see Hertwig, 1936, and Dobzhansky, 1951).

In some cases of mutative polyploidy a more or less complete hybrid sterility can result. This may be inferred from Seiler's studies (1936) on crosses of tetraploid parthenogenetic females of *Solenobia triquetrella* (Lepidoptera) with races of the same species showing normal sexual reproduction. In these cases the triploid offspring was intersexual. Unfortunately, we know far too little about the part of nongenetic factors in isolation in closely related animal species differing in chromosome numbers.

53

F. HYBRID RACES

Finally, isolated geographic races or young species can give rise to new races by hybridization. Examples of such hybrid races in birds are: the Italian house sparrow *Passer (domesticus) italiae*, which arose from hybridization with the closely related 'species' *P. hispaniolensis* (Meise, 1936); various pheasants from the upper Irrawaddy of India, which appear to be derived from *Gennaeus horsfieldi, G. lineatus,* and *G. nycthemerus* (Meise, 1928); the Siberian races *Carduelis caniceps subulata* and *C. can. polyakowi,* which hybridize with the siskin *C. carduelis*; and the bullfinch race *Pyrrhula pyrrhula griseiventris* from Eastern Asia (Johansen, 1944). In North America hybrid populations in the genus *Junco* have been thoroughly studied by A. H. Miller (1939, 1941, 1949) and in the genus *Pipilo* by Sibley (1954).

Ghigi, from extensive experience in the hybridization of birds, holds the opinion that in this way a number of new species may have arisen. Thus he got hybrids of *Lophura rufa* (Malacca) and *L. nobilis* (Borneo), which closely resembled the intermediate *L. sumatrana* from Sumatra (1931). Here, as in his crosses of guinea fowl (genus *Guttera*, Ghigi, 1938), it should be noted, however, that the 'species' used as parents, and the 'species' assumed to be hybrids may well be considered geographic races of the same species.

Recently, hybrids of good species and even of genera have been reported to occur under natural conditions in various groups of fresh-water fishes. This hybridization apparently is fairly frequent and may occasionally serve as a stage in speciation. Thus, in Californian rivers, *Pantosteus santa-annae* hybridized with a species of *Catostomus*, and the resulting hybrids are so frequent that the original species – possibly expelled from its original habitat – is probably extinct as a genuine species (C. L. and L. C. Hubbs and Johnson, 1943). Crosses of *Gila orcuttii* and *Siphateles mohavensis* in the region of the Mohave River proved to be more resistant to ecological changes brought about by a flood in 1938 than genuine *Siphateles* (C. L. Hubbs and R. Miller, 1943).

Besides these examples, it could be shown experimentally that in *Drosophila melanogaster* strains resembling 'incipient species' can be produced by hybridization. By crossing mutants showing translocations in the second and third chromosomes, Kozhevnikov (1936) produced four homozygous strains which could not be crossed with females of the original type. Hence, in some respects, these strains behaved like a new species (*Drosophila 'artificialis'*).

In the plant kingdom, fertile species-hybrids are relatively frequent, and here the evolutionary importance of hybridization is definitely greater. This is especially conspicuous in the genera forming a 'network' of types connected by many hybrids (*Salix, Rosa, Rubus,* etc.). Therefore, in plants those new 'species' which are not isolated genetically could be experimentally produced by crossing (e.g. Heribert-Nilsson, 1936). In some cases new species originated from infertile species-hybrids with asexual reproduction after polyploidization had occurred. By such an allopolyploidy the infertile crosses of *Primula verticillata* and *Pr. floribunda* developed into *Pr. kewensis*, which could be propagated by sexual reproduction but could not be back-crossed with the parental species. Similar conditions prevail in the grass *Spartina townsendi*, in the hybrid of cabbage and radish (*Raphanobrassica*), etc. (See the review by Huxley, 1942, and Stebbins, 1950.)

Summarizing, we may state that hybrid formation is much more important in plants than in animals, as in the latter it usually produces only intermediate geographic races, and rarely new species.

G. REVIEW OF ALL TYPES OF RACES

Summing up our considerations on race formation in nature, we find that this process takes many different paths. It is not possible as yet properly to judge the evolutionary importance of ecological, sexual, and strictly genetic races, as the number of cases studied so far is too small. It is, however, quite certain that geographic races are by far the most common forerunners of speciation, and that there is a great number of animal forms intermediate between geographic race and species. In the differentiation of geographic races towards new species, ecological, reproductive, and – above all – genetic mechanisms of isolation usually play quite an important part, reinforcing each other.

In contrast to this view, R. Goldschmidt (1935, 1940, 1952) stated that no such races are incipient new species, nor could they serve as 'models' of speciation. Goldschmidt's skepticism is partly based on his own findings in the course of genetical studies on related species of *Lymantria dispar*. The two species *L. monacha* and *L. mathura*, he says, differ too much in too many characters, and neither shows any similarity to races of *L. dispar*. My opinion on this is that one cannot expect to find borderline cases still present in every group, but must consider that a good many of them have disappeared in the course of evolution by selection or progressive development. Moreover, I cannot accept his objection (1935) to my statements on the progressive differentiation of geographic races into species, which I published in 1929 and 1934. He says:

> If there is a series of races geographically arranged like a nearly closed ring, and if the most divergent types on both ends of this ring get into secondary contact – which need not necessarily be so, of course – and if there are any features among their differing characteristics which for this or that reason render fertile crossing impossible, and if these particular features have no adaptational value, then the systematist may, indeed, describe the two types concerned as new species. I find it difficult, however, to believe that all these 'ifs' occur simultaneously so often that this phenomenon should have a really essential effect.

It is, however, not at all necessary that all these 'ifs' occur simultaneously in order for speciation to happen via geographic (or ecological) race formation. As we have seen, it is not necessary that a series of races be arranged in a ring-shaped order; differentiation towards speciation can proceed as well in reproductive communities with a more or less linear geographic arrangement as in a two-dimensional situation. This secondary contact of end-links – which is not rare – only proves the fact that forms with strong geographic differentiation can and do behave like good species. Furthermore, there is no need to expect that in cases of secondary contact the characters should have 'no adaptational value': the adaptational value may well be there, but it can act in different ways in the contacting races. Thus, I cannot agree at all with Goldschmidt's skepticism. It is the numerous intermediate types of all kinds between races and species that do quite definitely bear evidence that species originate from races, as does the fact that the features distinguishing one species from another are actually the same as those distinguishing well-characterized geographical races.

In some old lakes, such as Baikal, Ochrida, Tanganyika, and others, there are endemic genera with a relatively great number of species. Woltereck (1931) tried to explain this fact by a 'schizotypic splitting' into species. It is, however, fairly probable that in these cases, too, geographic and ecological isolation played the only important role (compare Brooks, 1950; Greenwood, 1951).

4

Undirected and Directed
Transspecific Evolution

A. THE PROBLEM OF PHYLOGENETIC TRENDS AND ORTHOGENESIS

As shown in the two preceding chapters, it is quite possible to find a satis-factory causal interpretation of the various kinds of race and species forma-tion, though many more analyses are necessary. Now the problem arises, whether these processes and the same factors can be considered a sufficient explanation of transspecific evolution leading to the emergence of new genera, families, orders, classes, etc., and hence to the formation of new organs and of new types of organization, or whether additional or even totally different factors must be assumed. Obviously, Charles Darwin did not consider such additional or new factors necessary, as the title of his main work on evolution as a whole was *On the Origin of Species*. And in the fierce discussion following this revolutionary publication, no autonomous evolutionary factors were mentioned. Nor did Haeckel (1866) think it necessary to postulate unknown evolutionary factors, although he recognized the three main phases of phylogenetic transformation (explosive splitting, specialization, degeneration and extinction; see below) and the evolutionary importance of unspecialized types. Although Von Nägeli (1884) spoke of 'autonomous changes of idio-plasm' (p. 546), nevertheless they seemed to him 'nothing but molecular forces inherent in matter'. On the other hand, Von Baer had assumed a 'directedness' of phylogeny as early as 1876 (p. 49). The increasing number of orthogenetic lines of descent, discovered towards the end of the last century by paleontologists, such as Gaudry, Cope, and others, presented an ever-increasing amount of material apparently rendering impossible any interpretation by principles of selection. Hence, most subsequent paleonto-logists doubted that mutation, selection, and isolation could yield a sufficient basis for transspecific evolution. Dacqué (1935) has stressed again and again the 'autonomous development' of 'types' (i.e. of the various types of organiza-tion); Wedekind (1927) used terms like 'phylogenetic course governed by laws'; and Osborn (1934) coined 'aristogenesis', 'a creative principle causing the development towards a certain end'. Beurlen (1937) likewise assumed the existence of autonomous evolutionary forces and called them 'a will to

57

particular and free assumption of form'. Von Huene (1940, p. 62) was convinced that 'there is a superior and unique principle governing all processes of life and evolution through the ages'. And after discussing some complicated evolutionary adaptations, Hennig (1944) states: 'Nobody can seriously maintain that such complicated and refined adaptations can be the result of random mutation and the extinction of all those competitors in the struggle for existence unable to reach this level of complexity'. Schindewolf (1936, 1939), too, considers phylogeny to be 'a process of differentiation governed by autonomous laws', and in his *Grundfragen der Paläontologie*, 1950, he says: 'Phylogeny does not proceed along the path of race formation, and hence it cannot appear as a product of selection, but rather as a result of orthoevolution' (p. 429). In the same book he states: 'The vital phenomenon of development must be considered a primordial process (*Urphänomen*), a residue which cannot yet be analyzed in more detail and which is not fully accessible to the analytic mind' (p. 430).

Some zoologists also emphasize an autonomy in the processes of evolution leading to the formation of new types of organization and the emergence of new organs. Rosa (1931) advanced a theory of 'hologenesis', according to which dichotomic branchings necessarily proceed in predetermined directions. Berg (1926) coined the term 'nomogenesis', meaning the immanent laws of evolution. Philiptschenko (1927) thought that the increasing complexity in the course of phylogeny is a basic quality of the living substance, and Woltereck (1931), also, believed that there is an 'inner drive to produce diversity', and a 'power' of 'self formation' and of 'self enhancement'. Remane (1939, 1950) thinks that the phylogenetic differentiations into the higher categories, at least, cannot yet be explained by the theory of mutation, as mutations which might cause basic differentiations (in contrast to the 'real mutation' studied experimentally) are not yet known. On the other hand he emphasizes that a 'phyletic drive' cannot be proved and is not to be expected (1950, p. 342). Ludwig (1940), when discussing the theory of selection, also arrived at the conclusion that there are 'numerous characters and patterns in the world of organisms regarding which an interpretation by the principles of selection seems forced'. Such skeptical views, however, have never generally prevailed in zoological phylogeny, and during the last decade they have continued to diminish.

We may well ask, then, if we are bound to assume autonomous factors of evolution in addition to the factors which are a sufficient explanation of race and species formation and which we have found to be mutation, recombination, fluctuations of population numbers, selection, and isolation. It is by no means clear what sort of factors and forces these could possibly be, since they must be localized inevitably in the physiological substance of each individual. Hence, we may well try to avoid the assumption of such hypothetical forces if it is at all possible to find a reasonable explanation of transspecific evolution without assumption of such unknown factors. As has been pointed out in the introduction to this book, the common pattern of many lines of descent shows 'explosive splitting' at the beginning, then differentiation at decreasing rates, and finally degeneration; this has been the main basis for the assumption of autonomous forces in phylogeny. The same is true with the orthogenetic development of many organs, which gives an impression of a directedness towards an 'end', and with the clearly progressive differentiation of organisms, especially that which leads towards man, but also that in many other lineages. Finally it must be said that the assumption of autonomous evolution is attractive to some authors because it helps them to avoid the 'unsatisfactory' conclusion that all phylogenetic differentiation is based on the 'accidents' of mutation and selection. As we see, this problem and its treatment by various authors consciously or unconsciously involve philosophical convictions and feelings (including that regarding the 'ape-ancestry' of man), which are not relevant to discussion in the field of the natural sciences. The new cosmology of Copernicus at first seemed 'unsatisfactory' to most people, though afterwards it became the foundation of a generally accepted view of the universe.

B. NONDIRECTEDNESS OF EVOLUTION

For a critical evaluation of the alleged immanent tendencies of evolution, it is important to know whether there is evidence of absence of direction in transspecific evolution. Such evidence would indicate that even in the differentiation of higher systematic categories the effects of nondirected mutation can be traced. Evidence of such nondirected mutation is most probably to be found in those cases where, in the process of phylogenetic differentiation of certain organs or certain structures, all possibilities of harmonious development are actually encountered. Of course, in many cases it is difficult to judge whether a morphological or physiological character is biologically 'harmonious', i.e. to find out how far the structures are limited by their principles of construction and which disadvantages in selection they might involve. But in spite of this, it is not too difficult to give unquestionable examples of undirected mutation in transspecific evolution; examples are available in various animal groups, and additional cases are known to anyone specializing in any of the respective groups.

First we will choose some examples in which the possibilities of development are so strongly limited that the situation becomes quite clear. There are four possible stages of hibernation in the butterflies of the temperate and cold zones: the insect can spend the winters in the stage of an egg, a caterpillar, a pupa, or an adult. All these possibilities are realized in nature. According to Pagenstecher (1909) about 67 percent of the German Macrolepidoptera spend the winter as a caterpillar (e.g. *Apatura, Aporia, Limenitis,* and *Cosmotriche*), some 28 percent spend it in the pupal stage (most Sphingidae and many Lymantriidae), and 3·5 percent in the egg stage (e.g. the gypsy moth *Lymantria*), while only 1·5 percent of the species hibernate as adults (among others, *Vanessa, Gonepteryx,* and *Plusia*). The situation is similarly clear in regard to the share of work rendered by either sex in the nest building and incubation activity among birds. First, incubation can be a strictly female task, e.g. in the wren (*Troglodytes*), the bullfinch (*Pyrrhula*), the chaffinch (*Fringilla*), the crow (*Corvus*), and the snipe (*Scolopax*). Second, incubation can be assigned to the male alone (in Tinamidae, Turnicidae, *Casurarius*, *Dromaeus*, and others). Third, both sexes can take equal shares in incubation (as in *Sylvia*, *Picus*, *Columba*, and *Struthio*). And fourth, there may be no incubation instinct at all in either sex, as in some cuckoos (*Cuculus*, *Clamator*, *Eudynamys*), some starlings (*Molothrus*, *Cassidix*), and weavers (*Vidua*, *Tetranura*). Hence, in birds all possible types of development can be found, even though some may seem quite aberrant, such as exclusive incubation by males. Nest building activity shows a similar diversity.

An even greater diversity can be seen in strictly morphological characters which also give quite convincing evidence of lack of directedness in their development. The types of horns in antelopes, for instance, cover the whole range of structural possibilities, provided, of course, that one leaves out of consideration those forms that would be quite discordant biologically. Hence, horns are: (1) straight and smooth in *Boselaphus* and *Cephalophus*; (2) straight with transverse ridges on the surface, this type being long in *Oryx*, short in *Oreotragus* and *Madoqua*; (3) slightly curved with a smooth surface in *Connochaetus*; (4) slightly curved with transverse grooves and ridges, the long type being found in *Oryx*, *Hippotragus*, and *Cobus*, and the short one in *Redunca* and *Damaliscus*; (5) twisted like a screw, with a smooth surface, the number of turns being small in *Strepsiceros* and great in *Taurotragus*; (6) short, in *Limnotragus*; (7) with transverse ridges and grooves in the long type of *Antelope* and in the short one of *Antidorcas*; and (8) bent twice, as in *Bubalis*. All these types also show intermediate stages and transitional forms. If we extend our study to the fossil types of horns in antelopes, we find the same fact to be true: nearly all possible forms existed in these series. Of course, these horns serve an important purpose as a weapon and as a means of species recognition, but there can hardly be any selective force necessitating the appearance of the particular characteristics of the surface, of bending and twisting, etc.

In the family of the birds of paradise (Paradisaeidae), the adorning feathers, usually so conspicuously elongated, can arise from various parts of the plumage. They come from the head plumage in *Pteridophora*, *Parotia*, and *Lophorina*, from feathers of the neck in *Diphyllodes*, from the breast and body sides in *Paradisaea*, *Astrapia*, *Cicinnurus*, and *Seleucides*, and from tail feathers on many species (Figure 14 of this book, Figure 7 in B. Rensch, 1943).

There is a careful and thorough study by Kagelmann (1951) on the color patterns in numerous species of wild ducks, showing that special patterns are typical of the genera and subgenera. Apparently, however, in most cases there is no selection force demanding the formation of specific color patterns, as species with quite different colorations and patterns live under similar environmental conditions. Hence, the diversity of the color patterns on ducks is one more example of a full range of possible mutations originating undirectedly in the course of evolution. Certain colors and patterns, then, have

FIGURE 14. Genera of birds of paradise from New Guinea and adjacent islands as examples of undirected transspecific evolution:

1. *Pteridophora;* 2. *Astrapia;* 3. *Lophorina;* 4. *Semioptera;* 5. *Paradisaea;* 6. *Parotia;* 7. *Seleucides;* 8. *Diphyllodes;* 9. *Cicinnurus.* (From Brehm, Wallace, Friedmann after B. Rensch, 1939.)

been utilized as recognition marks of species and sexes, some of them even serving as releasers of display and mating behavior. Here, apparently, sexual selection played its role in maintaining and strengthening the patterns developed undirectedly. Von Mičulicz-Radecki made a similar study of the color patterns in wild pigeons and doves, and found an extraordinary wealth of patterns, only a few of which seem to be of any adaptive importance.

Further examples showing the whole range of biologically tolerable diversity in shape and structure are provided by the shells of Gastropoda. These shells are shaped like a bowl, a cap, a tube, a flat spiral, a more or less tapering cone, a needle, a ball, and so forth. The shell spirals may be regular or irregular, with or without contact with each other, and they may even partly cover each other, as in *Cypraea* and *Conus*. The material of the shells may vary from fragile to solid and from opaque to transparent. Their surface may be covered by longitudinal or transverse ridges and grooves, tubercles, spines, small laminiform structures, hairs, or even more complicated protuberances. Finally, the shell as a whole may be more or less reduced, so that only a small part of the body or none at all is protected by it. This enormous multitude of different forms is further augmented by the diversity of fossil types.

Quite instructive, too, are the stridulating organs in insects to be found on various parts of the body, wherever a 'pars stridens' could be moved against a 'plectrum'. Considering the beetles only, Meixner (1933–6) could distinguish 15 main types. So (1) the vertex is rubbed against the elytra in some Carabidae, Tenebrionidae, Chrysomelidae, and others; (2) the gula (throat) against the prosternum in Ipidae, Anobiidae, and so on; (3) the mesoscutellum against the pronotum in Cerambycidae and Chrysomelidae; (4) the prosternum against the mesosternum in Scarabaeidae; (5) the back against the elytra in Silphidae, Tenebrionidae, Scarabaeidae, and Chrysomelidae; (6) the abdomen against the elytra in Hydrophilidae, Buprestidae, and Carabidae; (7) the elytra against the tergum in Carabidae and Ipidae; (8) the elytra against the abdomen in Carabidae; (9) the elytra against the femur of the second legs in Cerambycidae; (10) the hindwings against the back in Dynastidae and Passalidae; (11) the coxa of the second legs against the belly in Scarabaeidae; (12) the belly against the metacoxa and (13) the femur of the first legs against the pronotum in Carabidae; (14) the femur of the second legs against the belly in Paussidae, and (15) the belly against the metafemur in Heteroceridae. Sometimes the various types of these mechanisms are to be found in one single family, as in Scarabaeidae, where we find five totally different modes: (1) prosternum against mesosternum in *Serica* and *Maladera*; (2) back against elytra in *Trox*, *Copris*, and *Ochodaeus*; (3) abdomen against elytra in *Anoxia* and *Polyphylla*; (4) metacoxa against abdomen in *Geotrupes*, *Ceratophyes*, and *Bolboderas*; and (5) abdomen against metacoxa in *Taurocerastus* and *Trichius*.

If we consider the general shape of the animal body and its various transformations, we shall arrive at the following conclusion: the multitude of different transformations gives the definite impression that 'undirected' differentiation has 'tried' all possible forms in the genera and families. This impression is the same among fishes or beetles, among Cladocera, Copepoda, or Cephalopoda. In any case the 'non-specialist', usually knowing only the more common and 'normal' types, will be surprised at the enormous wealth of different forms and their often grotesque variations. To illustrate this fact, some species of marine Isopoda are shown in Figure 15. I think they show quite convincingly how by simple changes in the relative length and width of the segments, of extremities, and of their parts quite divergent forms are produced. (Still more extreme deviants can, of course, be found in the parasitic and limbless genera *Portunion* and *Danalia*.)

Among plants many similar examples are to be found. One need only think of the differences in type of growth, and in types of leaves and flowers, in Oleaceae (e.g. *Olea, Fraxinus, Syringa, Forsythia, Jasminum*) or in Scrophulariaceae (*Antirrhinum, Digitalis, Pedicularis, Melampyrum, Impatiens, Verbascum, Veronica*, etc.). One should consider the diversity of forms and structures of the flowers in Orchidaceae, the different forms of the thallus in some algae of the order Siphonocladiales, the endless multitude of types in Desmidiaceae and Diatomeae, and so forth.

As the objection might be raised that in all the cases mentioned transformations of only minor importance are concerned, a few examples which have major bearing on more essential physiological systems will be given.

Here, good material for comparative study exists in the different types of gills originating from the most varied organs and parts of the body. So, in Polychaeta gills come from the parapodia, in Isopoda from the legs of the abdomen, and in Decapoda from the maxillary and thoracic legs. There are tracheal gills and genuine gills in insect larvae, and in Mollusca gills may be seen on the pallium or around the anal opening (in Doridae). In asymmetric types of sea urchins we find feathered ambulacral appendages functioning as gills, and in Holothuria the so-called 'water lungs' serve the same purpose. Finally, there is the common type of gill slits in Tunicata, Acrania, and fishes, and in Amphibia we find external gills as appendages of the head.

There are also clear examples of undirected phylogeny in the various mechanisms of oxygen uptake in water insects. Some species take air directly from the water surface, storing it under the elytra (*Dytiscus*), or on the ventral surface of the abdomen (*Notonecta*), and then passing it into the spiracles. Others have developed respiration tubes protruding above the water surface and ending in the tracheal system at various parts of the body (*Nepa*, larvae of *Stratiomys*, larvae and pupae of *Culex*). Some forms, such as Trichoptera, Ephemeroptera, the caterpillar of the butterfly *Paraponyx*, and the larvae of some Dytiscidae, show tracheal gills. There are tracheal gills in the intestine in the larvae of Odonata, and external gills in those of the water beetle *Hygrobia tarda*. In many cases all special respiratory structures are lacking, oxygen uptake being provided by simple diffusion via the skin only (*Ceratopogon, Corethra, Simulium*) or by a 'respiration' through diffuse tracheae of

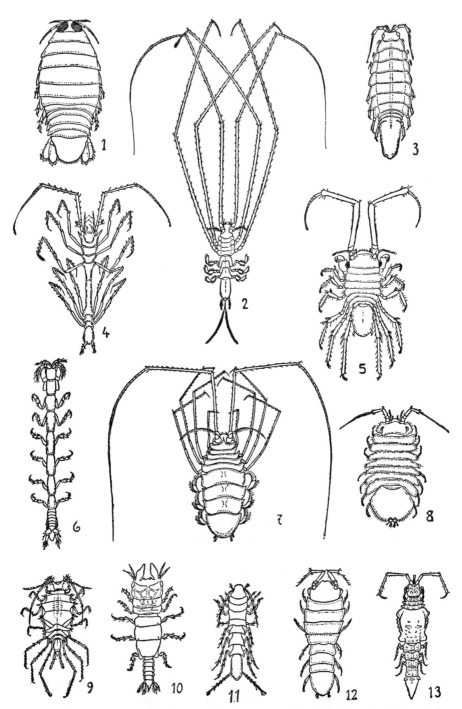

FIGURE 15. Undirected transformation in nonparasitic marine Isopoda (*Is. genuina*).
1. *Rocinela daumoniensis* ♂; 2. *Munopsis typica* ♂; 3. *Idotea baltica* ♂; 4. *Ischnosoma bispinosum* ♂; 5. *Munna palmata* ♂; 6. *Leptanthura tenuis* ♂; 7. *Eurycope cornuta* ♂; 8. *Jaera marina* ♂; 9. *Dendrotion spinosum* ♀; 10. *Gnathia maxillaris* ♂; 11. *Macrostylis spinifera* ♂; 12. *Nannoniscus oblongus* ♂; 13. *Arcturella dilatata* ♀. Various scales of enlargement. (After Sars, 1899.)

the skin (as in the larvae of some Trichoptera). In some cases, several types of these phylogenetic patterns are to be found in a single insect order simultaneously. Thus there are larvae of water beetles with respiration tubes (*Dytiscus*), with tracheal gills (*Hydrophilus*), and with genuine gills (*Hydrobia*), larvae of Diptera with respiratory tubes on the tail end (*Stratiomys*, *Culex*) or on the thorax (pupa of *Culex*), and others with skin respiration only (*Corethra*, *Tendipes*) (Figure 16).

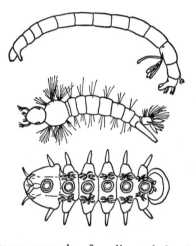

FIGURE 16. Four types of larvae of Nematocera from central European genera, showing marked differences in physical shape and respiratory mechanisms. Top to bottom: 1. *Sayomyia plumicornis* (skin respiration, few tracheae, and four tracheal sacks); 2. *Tendipes sp.* (skin respiration and hemoglobin); 3. *Culex annulatus* (tracheal system and breathing tube); 4. *Liponeura brevirostris*, ventral view (gills and sucking disks). (After Meinhert, Grünberg and Lindner.)

One more example of undirected development of an important physiological system may be found among the various types of hemochromes in the animal world. The red hemoglobin, well known for its optimum efficiency in the process of respiration, has originated independently on various levels of the evolutionary scale and hence these various types show quite different globin components. Hemoglobin has developed in vertebrates, in Nemertina and Annelida, in some species of Mollusca (like *Planorbis*, *Vivipara*, *Aplysia*, *Arca*, *Tellina*), in Echiuroidea (*Urechis*), in Phoronida, in Echinodermata (Holothuria, in Ophiurida: *Ophiactis*), and even in Protozoa (*Paramecium*). There is a hemochrome containing iron, the red hemerythrin, in Sipunculoidea; and in Nematoda (*Chloronema*) and Polychaeta (Sabellidae, Serpulidae, and others) we find the green chlorocruorin, also containing essential iron. On the other hand, copper plays the part of the effective metal in the system of hemocyanin, which turns blue if oxidized and is found in many Mollusca, Crustacea, Scorpionidae, and Xiphosura. In Ascidiaeceae the hemochrome contains the very rare element vanadium, instead of iron or copper. There is a hemochrome in the mussel *Pinna* which is said to contain manganese, but which must be studied further before this statement can be considered to be safe. Finally, in Mollusca some 'achroglobins' without metallic components have been described, capable of binding oxygen (in *Chiton, Patella, Doris*, and others). Reviewing our findings on the hemochromes, we see that there are various ways to satisfy the same organic need. Again, quite a number of different means can be found which may be employed among representatives of a single group, as among Mollusca, where we meet hemocyanin, hemoglobin, pinnaglobin, and achroglobin.

Still more important than the factors mentioned so far is the type of the early development of the egg. Here, too, almost any imaginable method can be found. There are equal, unequal, discoid, and superficial cleavage; and there are various intermediate and modified types (radial, leiotropic, or dexiotropic spiral cleavage); and, finally, there may be a separation of the blastomeres, as in polyembryonic development. Hence, again we find all possible modes of development, and again the same type may have originated independently several times and on various phylogenetic levels. Thus discoid cleavage occurs in Scorpionidea, Cephalopoda, Pisces, some Amphibia (as in *Eleutherodactylus*), Reptilia, and Aves. And the same undirected course of phylogeny can be inferred from the various types of gastrulation: by invagination, polar or multipolar immigration of cells, delamination, or a combined effect of these factors.

Strong evidence of undirected development can also be gathered from a study of the various types of gastrulation within a limited group. So, among Scyphozoa in the Coronatae we find gastrulation by invagination combined with migration of cells from the pole, and in Semaeostomae there is a true invagination, while in Cubomedusae gastrulation is brought about by multipolar migration and delamination. In Stauromedusae no archenteric cavity is formed and the entoderm originates by polar cell migration (compare Thiel, 1936–8; B. Rensch, 1954 *a,b*). Similarly strong differences of the first embryonic stages are to be found, for instance, in Polychaeta (see Chapter 6, D on evolutionary effects).

In Protozoa, almost all vital processes show this undirected 'testing' of nearly all imaginable possibilities. Thus, asexual reproduction may occur by simple fission, schizogony, or sporogony; or there may be sexual reproduction by copulation, conjugation, pedogamy, automixis, or various types of metagenesis and heterogenesis.

Moreover, the course of the ontogeny proper shows an enormous range of phylogenetic alterations in animals and plants. There are permanent alterations right at the beginning of early embryonic development (archallaxis), transitory deviations of development during somewhat later stages (intermediate deviation), permanent deviations originating during intermediate phases of ontogeny or in the final part of it (anaboly). In addition, there are retardations and accelerations of the ontogeny as a whole or of certain parts of it (heterochrony), and a gradual shift of phylogenetic transformation to earlier or later ontogenetic phases, such as palingenesis, proterogenesis, or neoteny (for details, see Chapter 6, D).

We cannot overemphasize the existence of the multitude of types and transformations in animals and plants, as usually we are accustomed to a phylogenetic view stressing homologies and neglecting such undirected diversities. Let us think of the various types of food and feeding to be seen in a single animal group. In insects, for instance, there are carnivorous, herbivorous, and omnivorous types, coprophages, bloodsuckers, keratin and wax eaters, xylophages, nectar suckers and pollen eaters, and finally those eating nothing during their short span of life as an adult. The differences are equally manifold as regards display and mating patterns, structures and appendages of genitalia in bumblebees, butterflies, and so forth. Quite a considerable diversity may also be observed in the study of the mutual relations of animals, such as symbiosis, carposis, commensalism, and parasitism, and the formation of genuine social communities such as states, families, breeding, feeding, and sleeping colonies, and reproductive communities. Excretions may discharge urea, uric acid, ammonia, adenine, guanine, or creatinine, and sometimes all these different physiological mechanisms may be found in the same group, as is the case in Mollusca.

Further light is shed upon the existence of undirected transspecific evolution by the numerous cases of 'grotesque' constructions, regarding which one can hardly avoid the description 'tolerable, but senseless'. Here, too, the number of examples is extraordinarily large, and one need only have a closer look at the structural details of animal types, quite well known to all of us, to see that some of these structures seem, to say the least, unnecessary. Although many of them still serve a certain function, this function could be served equally well by a less grotesque structure or by one showing no excessive traits at all. It is quite clear that the strongly curved tusks of *Babirussa*, bent like a circle and totally useless for rooting and fighting, belong to this category of 'grotesque' organs, though they may still serve a purpose as secondary sex characters. The same holds true as regards the giant beaks of the hornbills and toucans, made up of light and spongy bones and bearing colors and patterns which might serve as a means of species recognition, or as regards the long antennae of longicorn beetles, as they could obviously serve the same purpose (that of bearing sense organs) if they had remained shorter. And the forked antlers of deer, or the twisted horns of antelopes and sheep might better be replaced by simple, straight horns which would obviously be more useful. As will be shown in Chapter 6, C, many of these cases of 'grotesque' growth are tolerable by-products of phylogenetic increases of the body size. Even such an essentially new anatomical trait as the descensus testiculorum seems a disadvantageous but still tolerable acquisition caused by the increasing relative sizes of the intestine and the liver, from which a 'competition' of the organs for space resulted, which the testes 'lost' since they could withstand a lower temperature.

Considerations like these render comprehensible the existence of biologically tolerable 'faults of construction' in nature, which would not have come to exist if evolution were directed. We need only recall the incomplete system of blood circulation in Amphibia, the vagina duplex of Marsupialia, the growth followed by nutritive resorption of embryos at a certain stage of the development in *Salamandra atra*, the formation of atypical, infertile spermatozoa in *Bithynia*, the useless development of the genitalia in bee workers, and the numerous cases of vestigial organs (such as the non-functioning limb rudiments of *Python*, *Seps*, and *Balaena*. Compare also Kramer, 1949).

Other unnecessary though tolerable complications occur in ontogenetic development ('Entwicklungsumwege'). Thus, in the totally terrestrial Gymnophiona, the tadpole stage in the egg is certainly a completely unnecessary phase and can be understood only from Amphibian ancestry. The same applies to the formation in many bird embryos of an '*Archaeopteryx*-tail', possessing separate vertebrae and feather buds arranged in two lateral lines, or to the formation of vestigial teeth preceding the development of the milk teeth (prelactate dentition) in Monotremata, hedgehogs, seals, and others.

All the cases demonstrated in this chapter show that undirected development is a widespread phenomenon in transspecific evolution and in the formation of new anatomical and physiological arrangements. A proper and reasonable understanding of the multitude and diversity of animal forms would be impossible without accounting for this undirected 'testing' of constructive transformations in phylogeny. Hence, we may state that the effects of undirected mutation and selection as the essential factors of race and species formation range into the evolution of transspecific categories, and that they are a fully sufficient explanation of all the cases referred to in this chapter.

C. FORCED DEVELOPMENT IN PHYLOGENY

For a proper understanding of phylogeny, we must try to find out how far basic undirected evolution is limited by environmental factors and by the special organization of the animal or plant body. Such limitations often have such a far-reaching and general effect that they may really be regarded as factors forcing development in certain directions. From this point of view, then, we may investigate the possibility that certain types of phylogeny, especially orthogenetic lines of descent, have originated as a result of directed selection acting upon the material provided by primary undirected variation.

There are examples of necessary development in any animal group, mainly analogous organs or convergencies, of which only a few need be mentioned. If aquatic animals, for instance, turn to planktonic life, they must possess mechanisms for reducing their specific gravity either by enclosures of gas or fats or by the elongation of appendages of the body tending markedly to increase the coefficient of friction resistance. Such elongated structures may originate from various parts of the body (in marine Copepoda, for instance, from the first antennae or the abdominal segments, in Pisces as fins or lateral skin membranes). Hence, in these cases, adaptation or pre-adaptation exist only insofar as they denote the result of selection acting upon the material provided by undirected mutation, which may have supplied various grades of prolongation of different appendages of the body. In a similar manner, nearly every habitat influences the development of its inhabitants, limiting and directing it along certain lines. Nonflying tree dwellers show parallel adaptations in the form of climbing mechanisms, possibly quite different in morphological detail, but all serving the same purpose (claws, adhesive disks, pulvilli, prehensile hands, or tails). There are tubes for sucking nectar, usually with a brush like end and developed from quite different anatomical structures, in animals feeding on nectar, such as Lepidoptera, Hymenoptera, and birds (among these: independent development of such sucking mechanisms in Trochilidae, Nectariniidae, Melliphagidae, Coerebidae, and Drepanididae). The eggs of cave-nesting birds, even of quite different families, are very often simply white, whereas those of birds nesting in the open are provided with diverse patterns and colors.

As can be seen from the last two examples, the factors forcing evolution in certain directions can often have quite definite effects, as only certain organs must be modified. Thus, in cave-dwelling animals, again and again we find long antennae and a reduced development of eyes (Orthoptera, Coleoptera, Amphipoda, etc.). Herbivorous mammals developed molar teeth with relatively flat crowns and strongly folded enamel, as can be seen in kangaroos (*Macropus*), hares (*Lepus*), voles (*Micropus*), perissodactyls such as horses (*Equus*), tapirs (*Tapirus*), rhinoceroses (*Rhinoceros*), ruminants (Ruminantia), and elephants (Proboscidia). In insects, legs adapted for digging burrows developed independently in Carabidae (*Scarites*), Scarabaeidae (*Ateuchus, Geotrupes*), Tenebrionidae (*Gonopus*), Cerambycidae (*Hypocephalus armatus*), and the mole cricket (*Gryllotalpa*), mainly by a broadening of the tibia, which also tends to be armed with spurs and spines. On the other hand, there is a similar widening and arming of the femur in the larvae of Cicadidae (*Cicada* and *Tibicen*), and of the first tarsi in digger wasps (Pompilidae, Sphegidae), while in Ithonidae (a family of subterraneous Neuroptera) the tibia and the tarsus have united to form a sort of shovel.

Most leaves in a tropical rain forest are thick, glossy, and egg-shaped, and have a tapering end so that the water may drip off (in contrast to the less uniform types of leaves in the forests of the temperate zones).

Definite trends also exist in the development of the internal organs of animals. Apparently in all bloodsucking Arthropoda there are special organs known as mycetomes (pseudovitellus), which contain bacteria or Rickettsiae, adding growth-promoting substances and vitamins to the monotonous food of their 'hosts'. The position and ontogeny of these mycetomes vary considerably with the species, which illustrates the fact that these organs, too, originated through undirected development. Mycetomes are situated in the intestine in mites (Gamasinae), in the vasa Malpighii in ticks (Ixodidae), and again in the intestine in Pupiparia (Hippoboscidae). They are not connected with the intestine at all, and lie in the dorsal part of the abdomen in Nycteribiidae, in the wall of the mid-gut in tsetse-flies (Glossinae), and in the fatty tissue in bedbugs (Cimicidae), being attached to the vasa deferentia in the males, or forming an appendage of the stomach in lice (Anophura). Also very different are the types of symbiotic organisms found in these different myce- tomes (rod- and ball-shaped bacteria and rickettsia) and the methods by which these micro-organisms are transmitted from one generation to the other: in *Pupipara* and *Glossina* they reach the larvae in the uterus via the nutritive glands ('milk glands'); in lice they pass to the tubes of the ovary via the blood; and, finally, in the pig louse (*Haematopinus suis*), some symbionts are placed in additional depots – mycetomes of the female – to be used in the 'infection' of the eggs by a complicated process: the bacteria pass the host's skin, enter its body again via the genital opening, and are brought to the tubes of the ovary by special transport cells (Buchner, 1930).

From all these examples it can be seen that the basically undirected trans- formations of organisms are restricted by environmental factors favoring development in certain directions. Thus new characteristics, new organs, or even new structures may arise. So far as we can judge, the selective value of a superior adaptation is decisive. Environmental influence may even have such a far-reaching effect as to alter the whole plan of an animal body, so that a definitely new structural type originates. For instance, stationary ento- parasitism leads to convergent reduction of limbs and locomotory patterns, which may be seen from the lack of epithelial cilia in Trematodes and Cestodes, or from the absence of limbs in parasitic Copepoda (such as Lernaeidae, Penellidae, and Calligidae), Rhizocephala, Isopoda (*Portunion*), Pentastomida, and parasitic insects, such as the females of Strepsiptera and the larvae of Diptera and Hymenoptera. The same is true in parasitic snails (*Enteroconcha*) and mussels (*Enteroxenos*). In all these cases, more conver- gencies may be found: worm-like shape of the body, reduced pigmentation, and reduced or lacking central nervous system and sense organs. This last is especially evident in the case of the eyes. (In larvae of parasitic Hymenoptera there is an almost complete reduction of the nervous system, and in Pentastomida only a subesophageal complex of ganglion-cells remains.) In many cases there is also a great reduction of the intestine, as in Cestodes, Acanthocephala, and Rhizocephala, in the Nematodes *Allantonema* and *Sphaerularia*, and in the sporocysts of Trematodes. As far as gastrointestinal parasites are concerned, the parallel development of a protective system of antienzymes and a solid epithelial cover became necessary to prevent the parasite from digestive destruction by its host. As distributing the offspring and finding a new host are difficult, most parasites show an enormous rate of egg production compared to related nonparasitic types: 350,000 eggs in Pneumonoeces (Trematodes), several millions in *Giganthorhynchus* (Acantho- cephala), about 42 millions in *Taenia solium* (Cestodes), 64 millions in *Ascaris lumbricoides* (Nematodes), and up to 12,000 in *Choniostoma* (Cope- poda), as compared with under 100 in nonparasitic Copepoda. By such or similar transformations in Trematodes, Cestodes, and Acanthocephala, structures have developed which are taxonomically very important, and typical of whole systematic categories.

70

Similar environmental effects more or less 'directing' the transformations can also be studied in ectoparasites of birds and mammals. There is a convergent dorsoventral flattening of the body in Anoplura, Mallophaga, Pupiparia, and Dermaptera (*Arixenia* in bats), and a laterally compressed body in Aphaniptera. Usually, there is a marked reduction of wings and eyes, while the claws are improved to facilitate gripping and adhering to the hairs or feathers of the host. Again, in Anoplura, Mallophaga, and Aphaniptera these adaptive transformations represent characters of taxonomic value.

If there are evolutionary limitations of organs due to the necessities of construction, far-reaching convergences may occur, such as the type of vesicular eye which has a retina, pigment layer, lens, and cornea. This type of eye has been developed in such different groups as Coelenterata (*Charybdaea*), Annelida (*Vanadis*), Echinodermata (*Asterias*), Onychophora (*Peripatus*), Gastropoda, Cephalopoda, and Vertebrata. All these organs function as eyes, but they have a quite diverse ontogenetic descent and development (Figure 17; compare also Novikoff, 1929). In some of these animal groups, however, there are also more primitive types of eyes (flat eye grooves, etc.) as in Coelenterata, Arthropoda, Mollusca, and Echinodermata, and hence we may infer that the undirected evolution of sense organs provided the raw materials from which environmental factors then selected the fittest types, one of these fittest types being a vesicular eye. With regard to the convergent evolution of such eyes, it is remarkable that our instruments of photography are constructed in much the same way and contain the same essential parts (lens, 'retina', pigment layer, 'accommodation mechanism', iris diaphragm). The second 'best' type of eye, the compound eye of the insect, has evolved several times: in Arthropoda, Annelida (*Sabella, Branchiomma*), and Mollusca (*Arca*).

Finally, a limitation of evolutionary lines may be brought about by the fact that, once a favorable 'invention' has occurred, this construction has been maintained in further phylogeny because its selective value is so superior that no less favorable type serving the same purpose could replace or succeed it. The 'invention' of chromosomes, for instance, was favorable, because they proved to be a means of equal distribution of genes in the process of reproduction, and it was only by this genic constancy that more complicated organisms could attain the constancy necessary for adaptation to the environment. Similarly favorable was the 'invention' of the cell as the elementary biological unit of organisms, as in multicellular beings a sufficient constancy of hereditary elements could be guaranteed only if their range of influence was confined to the relatively small range of possible chemical influences. By the 'invention' of a nervous system, the ability to react favorably to stimuli was definitely increased (for instance, by contracting muscles used in fleeing as soon as visual stimuli indicate that there is danger ahead). Accordingly, organs and organ systems such as the brain, blood vessels, nephridia, labyrinth, and others remained principally unchanged once they had been 'invented' in the course of evolution (see also Chapter 7, B).

Summarizing our findings on nondirectedness and on forced trends of transspecific evolution, and carefully taking into account the numerous byways in phylogeny which are so often neglected, we may state: undirected development and 'probing' a great number of diverse types provides the basis of phyletic differentiation. The transformation of these types is usually governed by certain selective restrictions of the development. Not only differentiations of minor importance, but new organs and new structural types as well may arise through this process of undirected development.

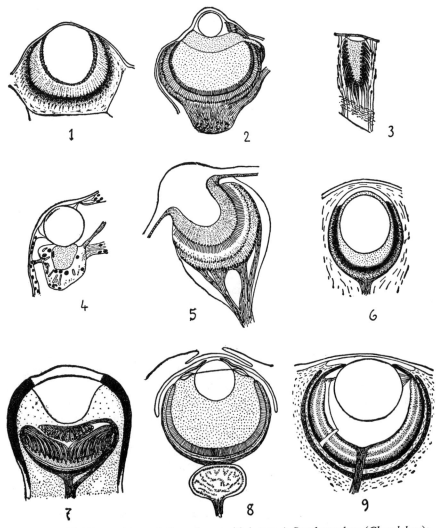

FIGURE 17. Convergent evolution of eyes with lenses. 1. Scyphomedusa (*Charybdaea*); 2. Polychaeta (*Vanadis*); 3. Echinodermata (*Asterias*); 4. Copepoda (dorsal eye of Pontellidae); 5. Spiders (*Avicularia*, main eye); 6. Snails (*Murex*); 7. Mussels (*Pecten*, eye from margin of pallium); 8. Cephalopoda; 9. Teleost fish. (After Plate, Hesse, Grenacher, Bütschli, Schewiakoff.)

Hence, these may reasonably be considered to have their basis in undirected transspecific evolution.

D. RULES OF PHYLOGENETIC DEVELOPMENT

We are now faced with the problem of the extent of determination in the phylogeny of organisms. How far is evolution limited by the specific structure and organization of the living substance and how far by the specific conditions prevailing on the surface of the earth? We shall also have to establish the extent to which the major lines of evolution are affected by these factors (not considering, at this time, the byways of evolution dealt with in the last section).

The numerous ecological studies made during recent decades have proved that the structure of animals is definitely correlated with special modes of life and special habitats. These correlations may be summarized as more or less general rules, often applying to many animal groups. There are further rules pertaining to the functional-anatomical consequences of a certain mode of life. These ecological and functional-anatomical rules may in turn be used to show the directedness of the phylogeny, from which certain lines of future evolution can be predicted. Some examples may illustrate this.

One of these rules states that large terrestrial vertebrates must develop heavy columniform legs with disproportionately large bones, because with an increase in body size the volume of the body (and thus its weight) increases by the third power, while that of the bones as the supporting structures increases only by the second. Hence, such columniform legs can be seen in the large species of mammals in various orders: in elephants, rhinoceroses, large Perissodactyla and Artiodactyla, in the extinct Titanotheria, Amblypoda, and Megatheria, in the largest birds, such as ostriches, and especially in the fossil Dinornithidae, and in the giant reptiles, such as *Brachiosaurus*, *Brontosaurus*, *Iguanodon*, *Triceratops*, and others (Figures 56 and 57) of the Mesozoic. (Giant types of Amphibia, such as the Stegocephalia, had not yet shifted the supporting legs from the original lateral position into a more ventral one, which considerably reduces the frictional resistance.)

Another rule is that the speed of aquatic and aerial locomotion can be increased if the moving body is more or less streamlined. Therefore, the bodies of large and swift swimmers, such as dogfishes, some Amphibia, especially some of the fossil Stegocephalia, crocodiles, penguins, whales, and seals are streamlined, and the same is true in swift and efficient fliers such as bees, Sphingidae, birds, and bats. Ships, torpedoes, aeroplanes, and rockets are designed according to the same rule.

Moreover, one may state that sessile animals could evolve only in the water, because the eddying of food particles and spermatozoa towards the body cannot occur in the air, but only in a liquid medium.

The number of such rules is quite considerable, since almost every 'adaptation type' – and the majority of animals represents such types – provides relevant material. One need only think of the flat body and various mechanisms for adhering to the substratum which are found in animals living on the bottom of rapid waters or in the surf zone. One should remember the similarity and parallels in the structure and shape of legs in jumping animals (grasshoppers, fleas, kangaroos, jumping rodents such as *Jaculus*, jumping Insectivora, and mouse-like jumping marsupials), or in digging forms. One should consider the shape of teeth in animals preying on fish or eating leaves, the convergence of long pointed wings in gliders such as swallows, swifts, gulls, albatrosses, and pteranosaurs of the genus *Pteranodon*, or the necessary shifting of webfeet towards the tail end of the body, and the parallel enlargement and reinforcement of flat locomotive organs (fins, webs). From the large number of such cases, proving that in spite of undirected mutation analogous organs and characters originated from quite various morphological elements by parallel selection, we shall take only a few examples of the more essential trends of phylogeny. They are intended to show how and why the evolution of organisms had to follow certain main lines.

1. Living organisms could originate only on the basis of carbon compounds, as only this element is capable of forming giant molecules with large active surfaces on the principle of carbon-to-carbon bonds. And the auxiliary valences of carbon compounds are one essential fact in many vital processes, especially since, by the phenomena of adsorption, cell respiration and similar processes are rendered possible. (Silicon could also form giant molecules, but these would not be a suitable substratum for the processes of respiration or allied phenomena.)

2. The material constituting the living organisms must be chemically stable (to provide constancy in organisms and their vital processes) and at the same time be accessible to enzymes and catalyzers for relatively quick dissociation and decomposition, so as to release energy required by the processes of life.

3. The evolution of living organisms depends upon the properties of colloidal protein suspensions, as among carbon compounds only the protein systems change easily from the solid to the liquid state, which is necessary for osmosis, metabolism, cell respiration, and reproduction. Hence, life could originate only in water.

4. As proteins coagulate at temperatures of more than 60° C. and most solvents freeze at some degrees below zero (C.), the phenomena of active life could evolve only within this range.

5. As the origin and evolution of living organisms largely depend upon certain states of chemical compounds and their requirements as to water, light, and a fairly limited range of temperature, life definitely depends upon an environment similar to that provided by the surface of the earth during the last one to two billion years.

6. Animals, i.e. heterotrophic organisms, could not originate before the autotrophic organisms (i.e. plants) providing a source of food.

7. Vital processes, such as motion and reaction to stimuli, can be maintained only if there are also processes of metabolism and energy transformation.

8. Organic compounds can be used for food only if decomposing substances (enzymes) are provided (compare Rule 2), as there is no other possible means to bring about metabolism of all vitally essential compounds.

9. Autotrophic organisms must always be provided with a large surface area, i.e. develop a branched structure, when body size increases, because the uptake of nutrition and radiation energy is surface-correlated. Heterotrophic organisms, on the other hand, must be of a more compact shape, so that internal cavities can be provided for the digestion of organic food.

10. As organic food is not omnipresent, animals have had to develop mobile parts or mechanisms of locomotion.

11. Locomotion requires receptors which react to stimuli leading to food.

In this manner quite a number of rules can be formulated, all of them showing that most of the basic qualities of animals and plants – in spite of the numerous specializations – originated of necessity. As only the basic principles are of interest to us at present, only a few more rules will be cited regarding progressive evolution towards higher animals (compare Chapter 7, A and B on anagenesis). We shall refer to this problem in more detail later.

12. In nonparasitic metazoans the efficient utilization of food is facilitated by the development of an anus: this provides a continuous stream of nutritive material, and the various parts of the digestive tract may be specialized for various functions.

13. The evolution of larger bodies in Metazoa is not possible unless there is a transport system for food and oxygen (blood vessels, tracheae). Without such systems the tissues themselves must keep close to those parts where food and oxygen are provided, as the transport of these materials by diffusion can cover only short distances. (Hence, the flat body of platyhelminthes, their rich intestinal branching, and their necessarily lobed gonads, and the long and filiform body of nonparasitic nematodes.)

14. In larger animals appropriate and rapid reaction to environmental changes and stimuli requires a system of stimulus perception and stimulus reaction. Advantageous and more complicated reactions to certain types of environmental stimuli require the development of an extensive system of associations of receptors and effectors and their mutual connections.

15. In terrestrial animals, large body size can emerge only if an internal skeleton is developed, as an external one tends to become prohibitively heavy. Hence, there are not and never have been any really giant types of terrestrial Arthropoda.

16. Terrestrial animals could originate only after the development of the following characters: solid tissues and sturdy organs of locomotion (Archimedes' Principle: in water the body weight is decreased by as much as the weight of the volume of water which is displaced by it), organs for direct aerial respiration, and mechanisms preventing desiccation. As sensory epithelia can function only if kept in a moist condition, these organs had to be 'withdrawn' into the interior of the body.

17. Homoiothermism can develop only in environments providing

temporary or permanent temperatures which are definitely below the average optimum of vital physiological processes. For an efficient homoiothermism the body should have a protective cover of material with low heat conductivity (hairs, feathers, fat layers), and well-developed regulating mechanisms.

18. Reaction to visual patterns (psychic parallel: form perception) will not be possible unless the sense cells are separated optically by pigment layers. If such a receptor is to function efficiently, the light rays must be focused by a lens, and varying environmental situations must be responded to by mechanisms of accommodation and adaptation, and the use of diaphragms (compare Figures 17 and 101).

19. The formation of a 'head' will usually have to begin at the anterior end of the body, as it is this part that first gets into contact with possible food. It proved advantageous to concentrate here the organs of taste, smell, vision, and hearing. Here, too, the association of the different receptors and effectors, i.e. the development of a central nervous system, was favorable.

20. The highly complicated and versatile reactions of a central nervous system can arise only in a 'brain' showing a strong differentiation of sensory, associative, and motor regions.

These rules may be sufficient to show that essential trends in the evolution of structural types and in progressive animal groups are necessitated to a large extent. Hence, certain types of organs highly significant in some progressive animal groups are strictly necessary, because they represent the most efficient or most favorable construction or the sole mode of realizing a certain vital function. It is remarkable that such a determination can clearly be traced in recent and fossil animals in spite of the diversity of patterns of undirected evolution. We may well consider such necessary differentiations as the result of directed selection (orthoselection) acting upon the materials provided by undirected mutation, as the changes favored by selection, lead in more or less the same phylogenetic direction (compare also Chapter 6, B IV). Hence, there is, for the time being, no reason to assume that an autonomous evolutionary force guides these directions of phylogeny.

E. POSSIBLE EVOLUTION OF ORGANISMS
ON EXTRATERRESTRIAL BODIES

The consideration and evaluation of evolutionary determinism may easily lead to the question whether there may be similar prevailing lines in the possible evolution of organisms on cosmic bodies other than the earth. Of course, all considerations concerning extraterrestrial life, its origin, and its development, are extremely hypothetical. Nevertheless, a tentative evaluation of the problem does not seem superfluous to me, as it may have a bearing on our biological and philosophical view of the universe. (Those readers interested only in our 'actual' knowledge of evolution may omit this short digression.)

Organisms (in the terrestrial sense) can exist only on such suns, planets, and large moons as provide surface temperatures between about $-4°$ and

+52°C., a sufficient intensity of light, an atmosphere containing oxygen, and water, the basic requirements of vital processes. As far as our planetary system is concerned, such a situation could possibly be found on Mars, though there seems to be only a little water, and – due to the decreased gravity (0·38 of terrestrial g.) – the atmosphere is thin and not at all rich in oxygen. The atmospheric pressure on Mars is approximately one-seventh of the terrestrial atmosphere. Moreover, temperatures do not seem to be too favorable on Mars, apparently being too cold. Nevertheless, with regard to the latest astronomical investigations, a greenish 'plant' life seems to be possible. For further details on possible life on stars, consult the works by Lundmark (1930) and Beringer (1951). It is not possible to discuss all the publications by other authors, some of them not very reliable. More favorable temperatures are found on Venus, always covered by thick 'clouds' and mist (crystal needles of aldehydes?), so that surface temperatures are about 50°C. during the 'day' and about 0°C. at 'night'. As the solar radiation is reflected by the outer layer of clouds, no oxygen has yet been proved to exist in the atmosphere of Venus. Since on the large moon of Jupiter the temperatures are too low and the atmospheres too thin, it seems quite probable that within the range of our solar system the earth and perhaps Mars are the only celestial bodies inhabited by living beings. There are, however, about 10 billion stars belonging to the system of our galaxy, many of which have their planets (Lavink, 1948; Büdeler, 1951). At present, about 80 million such galactic systems are known, and to this number will be added a great many which have not yet been explored astronomically. Thus, the number of celestial bodies suited for the origin and evolution of living organisms is perhaps not small. Will evolution on extraterrestrial bodies have to be of a similar kind and follow similar rules to that on the earth? May we assume that the principles of evolution are valid in extraterrestrial places, as are the laws of chemistry and physics?

In trying to find approximate answers to these questions, it will be best to outline the probable development of life on cosmic bodies very similar to our earth but differing in size. Let us first consider a body smaller than the earth: there will be a decreased gravitation, and hence the atmosphere will be thinner and will contain less oxygen per unit volume, there will be fewer clouds, stronger ultraviolet radiation, and a more marked daily change of temperatures. Provided that all these differences were not so grave as to prevent the existence of life at all, animal evolution would most probably show the following traits:

1. With decreased gravitation body size could be larger than under terrestrial conditions. The upper size limit in terrestrial animals is due to the fact that the skeleton becomes disproportionately large and heavy, since with increasing body size volume increases by the third power, but skeleton efficiency only by the second, hence leading to disproportionately large bones. Decreased gravitation would mean decreased weight of bones, so that larger

bodies could develop. Similar considerations apply to 'plants' or auto-trophic beings with a branching structure and large surface areas. Due to the thin atmosphere and the stronger wind, however, animals and plants would probably have to remain small.

2. As decreased gravitation facilitates movement and locomotion, absolutely faster animals could evolve. To attain average 'terrestrial' speeds, less strength and fewer muscles and locomotory mechanisms would be needed, so that there would possibly be a surplus of energy which could be used for other functions.

3. Flying would probably be easier, as the decrease of gravitation would be more effective than the decrease of friction resistance due to the thinner atmosphere. The stronger winds, however, due to the rapid and marked daily changes of temperature, would be disadvantageous to flying.

4. The thinner atmosphere and smaller share of oxygen per unit volume would require greater respiratory surfaces or more efficient processes of respiration (for instance, by especially intense activity of special enzymes). Seen as a whole, however, the vital processes in general would be slowed down to a certain extent (compare life at higher altitudes on the mountains of the earth).

5. The more transparent atmosphere would favor the development of photoreceptors (e.g. eyes).

6. The stronger insolation and ultraviolet radiation would require better mechanisms of protection (perhaps by increased pigmentation).

7. The harsh changes of temperature and the low temperatures at night would prevent nocturnal and twilight animal activity, though the strong ultraviolet radiation during the day should favour the evolution of nocturnal types. In addition, morphological and physiological mechanisms of protection against the extremes of temperature would be required (for instance, thermally insulating covers, instincts to spend the hottest and coldest periods in hiding places, and so forth).

Such references could easily be multiplied, but I wanted only to show how changes in the physical and chemical environment favor certain kinds of biological alterations, and how further indirect effects may arise as a result of such changes. It is not easy to summarize the above considerations, but we may say that the world of organisms on a 'minor earth' would definitely be poorer in types and forms than on the earth we know, provided that equal rates of evolution are presumed. This would be due mainly to a poor oxygen supply, extreme temperatures at night, strong insolation at noon, and stronger winds.

If we now consider that there are no terrestrial homoithermal animals living permanently at altitudes above 6,000 m. and that even poikilothermal types are very rare at this altitude, though the diurnal temperatures are quite favorable, we may well doubt that on a celestial body of a definitely smaller size than the earth, organisms similar to terrestrial ones should come into being at all.

Let us now consider the evolutionary consequences of a marked increase of size of a celestial body resembling our earth: with increasing gravitation the atmosphere would be thicker, its pressure higher. There would be more and denser clouds, short-wave radiations would be absorbed to a larger extent, and the change of diurnal and nocturnal temperatures would be less harsh. Mechanisms of locomotion would have to be more solid and sturdy, and possibly there would be no large or flying animal types at all. Tender-skinned animals inhabiting the land would have to develop a certain osmotic pressure (turgor). Respiration would be facilitated because of the richer oxygen supply, and processes of metabolism would possibly be intensified. The less marked extremes of temperature would allow the evolution of many naked or moist-skinned diurnal types, and also of nocturnal animals. On the whole, then, the evolutionary situation would be more favorable on a larger cosmic body than on a smaller one. However, there is an upper limit of size, as at a certain stage the strongly increased gravitation would prevent the locomotion of organisms except in water. One cannot exactly figure out the increase of gravitation necessary to reach that point where extra-aquatic locomotion of any kind is impossible, but when locomotion demands increased efforts a definite increase of nutritive material is required. Hence, there is also an upper size limit caused by the requirements of metabolism.

It is for these reasons that animal types developing on large celestial bodies will have to remain relatively small, and possibly on such bodies no animals or only a few would be able to live on land. Aquatic animals, on the other hand, would be less affected by an increased gravitation, since these do not require such solid structures as terrestrial types. Moreover, animals living in the water can decrease their specific gravity by various means (for instance, by air sacs). They can regulate their osmotic pressure according to varying environmental conditions, and hence they can stand an increase of water pressure, as may be seen in the deep-sea fauna. At any rate, evolution on larger celestial bodies would follow a course different from that on our planet, though some traits and requirements would be similar.

In just the same way, one could evaluate the consequences of marked changes of temperature, of light, of components of the atmosphere, of the total amount of water, of rotation speed, of ecliptic inclination, and so forth. We cannot enter into this matter here.

Only two factors may be briefly dealt with: the importance of atmospheric oxygen and the effects of light. We can imagine terrestrial life without atmospheric oxygen, as oxygen is not an obligatory element in the processes of metabolism, but the evolution of higher animals not depending upon oxygen is not very probable. With regard to a possible settlement of living beings on cooled-down suns, it is interesting to consider whether there may be an evolution of organisms in the absence of solar radiation or with light only from the stars. On our earth can be found certain bacteria not depending on light at all, but using the energy released by oxidation of inorganic compounds, such as the nitrite and nitrate bacteria and the iron bacteria. Higher organisms, however, and especially animals independent of light, such as endoparasites, cave dwellers, and deep-sea types, invariably are heterotrophic and do, in the final analysis, depend upon light, as they consume material originally produced by photosynthesis in autotrophic plants. Why have no organisms independent of light exceeded the evolutionary level of the bacteria? Was it because non-photosynthetic energy can be released only under quite specialized environmental conditions and because this process does not release an adequate amount of energy? Or was the superiority in competition of those types directly or indirectly relying upon photo-synthesized energy too great? These questions can hardly be answered, as we do not know what the environment was like at the times when the above-mentioned bacteria originated. It is probable, however, that the inefficient processes of releasing energy from non-photosynthesized material were decisive. Nevertheless, we cannot deny the possibility that more or less com-plicated organisms independent of photosynthesis may evolve or may already have evolved on extraterrestrial bodies. But it is hardly probable that evolution based on these principles could have progressed so far as to reach the level of arthropods or vertebrates.

Hence, we find that the probability of extraterrestrial evolution is rather slight. Celestial bodies of essentially smaller size than the earth cannot have an atmosphere, and on markedly larger 'earths' the increased gravitation offers serious difficulties to evolution. Lack of carbon, oxygen, hydrogen, and nitrogen prevents the formation of living substance. Lack of light bars the evolution of more complicated organisms. Vital processes in protein com-pounds require temperatures ranging from $+60°$ C. to a few centigrades below zero. All these limitations enormously reduce the number of celestial bodies possibly suited for an origin of life, and cooled-down, dark suns and small bodies (like planetoids and moons) may be disregarded. Hence there will be enormous distances between the relatively few celestial bodies suited for organic evolution. With regard to the origin and evolution of life, our earth is indeed 'the best of all possible worlds' in a fairly large space of the universe, if we may use the term of Leibniz in this quite unphilosophical sense. However, as the number of suns in our galaxy is enormous (about 10 billion), and the number of such galactic systems of equal size is about 80 million (Heckmann, 1943), the 'total' number of 'worlds' inhabited by living beings may be fairly great. Their evolutions will have reached different stages, and will often show different trends from those of our terrestrial one. But as the chemical components of extraterrestrial bodies and of the whole universe are found to be the same as on our earth, and as evolution – in spite of all primary lack of direction – is necessitated by numerous factors, the existence of organisms possessing complicated central nervous systems (such as cephalopods, arthropods, or vertebrates) and, hence, human or supra-human mental capabilities, is not impossible. In considerations like these it is essential that the phylogeny of psychic components (parallel components) should develop in accord with the physical world (in terms of the theory of cognition: reduction components; compare Chapter 10, B). This means that thinking originated and developed in adaptation to the 'realities' of the physical world, and that false abstractions invariably were wiped out, as has been pointed out by Mach. Concerning this problem, the eminent physiologist Verworn (1920) wrote:

> Our process of logical thinking is a mental reproduction of external regularities. . . . If there are organisms capable of thought on Mars or any other celestial body, their type of thinking will be of the same kind as ours.

Statements like these are quite essential with regard to the hylopsychistic theories of different types advanced by natural scientists and philosophers like Thales of Miletus, Heraclitus, Shankara, Spinoza, Paracelsus, Fechner, Wundt, Mach, Haeckel, and Ziehen. Thus human existence becomes a part of the unique pattern of cosmic law, and the importance of this fact may serve as an apology for this digression into the totally hypothetical.

5

The Absolute Speed of Evolution

As we have found in Chapters 2 and 3, the rate of race and species formation may depend upon the mutation rate of the various characters, the effects produced by single mutations, the genetically effective population size and its fluctuations, the more or less efficient mechanisms of isolation and, above all, upon the efficiency of selection. All these factors affect the various animal groups in quite different ways and in diverse combinations. Hence, we may expect that the absolute speed of evolution will vary greatly in various animal groups. It is, therefore, important to examine minimum and average rates of evolution in the different phylogenetic stages, i.e. in animal groups with a more or less rich fossil record. Such strictly paleontological analyses will also provide a means of evaluating the factors causing differences of evolutionary tempo in the various animal groups. Furthermore, such an examination will shed some light upon the problem of possible additional assumptions, as, for instance, 'autonomous drive of development'.

When Darwin discussed the problem of evolutionary tempo in Chapter 11 of his *Origin of Species* (1859), he recognized the existence of striking differences among various animal groups. Therefore he pointed out the brachiopod genus *Lingula* as a typical example of a form that has remained unchanged ever since the Paleozoic. It was also known to Darwin that evolution in marine animals generally is slower than in terrestrial ones and that the tempo of evolution is more rapid in higher forms and slower in more primitive ones.

In calculating the times needed for the formation of new genera, families, orders, and so forth, we can use the geological time scale and date them in terms of years. The methods of dating the past from the evidence of radioactivity have been greatly improved and seem quite reliable within certain limits (the development of isotopes of lead and accumulation of radiogenic helium in radioactive elements such as uranium [U^{238}], thorium [Th^{232}], potassium [K^{90}]: compare Rüger, 1943; Zeuner, 1946, and others). Dating and counting have been done in the following chapters according to Rüger (1943, p. 193) and when more exact data were lacking, the length of the whole period in which the animal group in question appeared or disappeared has been calculated. Hence, subsequent rectifications will be needed if the methods of dating should be improved. For our evaluations of a more general kind, however, the scale of maximum lengths of the periods may be sufficient.

A. THE ABSOLUTE AGE OF VARIOUS CATEGORIES

It is well known that in the Cambrian all animal phyla except the vertebrates already existed and that the latter probably originated during this period or during the Ordovician (here the lancelet-like *Jaymoytius*; compare Colbert, 1955). Hence, we may assume a phylogenetic age of about 500 million years for the phylum Pisces (including Agnatha) and a considerably older one for the other phyla. As most vertebrates produce at least one generation per year, we may state that at least 500 million generations were necessary to shape the type of our present vertebrate fauna.

The majority of our recent animal classes are also extraordinarily old. Most of them can be traced back as far as the Precambrian (Rhizopoda, Porifera, and Onychophora), the Cambrian (Scyphozoa, Chaetopoda, Chaetognatha, Crustacea, Arachnoidea, Xiphosura, Gastropoda, Lamellibranchiata, Scaphopoda, Cephalopoda, Brachiopoda, Crinoidea, Asteroidea, and Holothuroidea), the Silurian (Hydrozoa, Anthozoa, Progoneata, Amphineura, Ophiuroidea, Echinoidea, Bryozoa, and Agnatha), the Devonian (Pantopoda, Insecta, Pisces, and Amphibia), and the Carboniferous (Chilopoda and Reptilia). Hence, these classes are about 540 million (since the Lower Cambrian), 450 million (Silurian), 350 million (Devonian), or 310 million (since the Lower Carboniferous) years old. Only the two homoiothermal classes of birds and mammals (dating from since the Jurassic: about 170 million years) are considerably younger. For other recent animal classes (Flagellata, Ciliata, Turbellaria, Trematodes, Cestodes, Nematodes, Linguatulida, Tardigrada, and others) the fossil records are not sufficiently clear to serve as a reliable source of information about the phylogenetic age. Calculating the average age of 33 recent animal classes providing a sufficiently reliable fossil record, we find an average of 460 million years per class.

The average phylogenetic life span of a class, of course, can be stated only in the case of those forms which originated after the Lower Cambrian and are extinct today. It has not yet been decided whether the animal groups concerned here may be considered as classes or only as subclasses or even orders. Assuming them to be classes, we may enumerate (after Abel, 1920; Hennig, 1932; O. Kuhn, 1938-9): the trilobites (Lower Cambrian-Permian, about 340 million years), the Cystoidea and Carpoidea (Lower Cambrian-Lower Devonian, about 210 million years), the Thecoidea (Middle Cambrian-Lower Carboniferous, about 220 million years), the Auloroidea (Silurian-Devonian, about 170 million years), the Tabulata (Lower Silurian-Cretaceous, about 390 million years), and the Graptolithidae (Upper Cambrian-Lower Carboniferous, about 210 million years). The average age of these eight 'classes' is 236 million years. Quite naturally, this figure is lower than that indicating the age of recent animal classes, the continued existence of which is ample proof of their stability.

This leads us to possible causes affecting the various tempos of evolution. At first, we may state the fact – known to Darwin – that higher animal classes have, on the average, a lower phylogenetic age than more primitive ones, and terrestrial animals a lower age than marine forms. Among Arthropoda, the lower classes of Crustacea and Xiphosura have existed since the Cambrian, the terrestrial classes of Arachnoidea since the same time, the Diplopoda since the Late Silurian, the Pantopoda and Insecta since the Devonian, and the Chilopoda since the late Carboniferous. The lower classes, mainly marine forms, have an average age of about 540 million years, whereas the average is only about 380 million years for the higher terrestrial classes. These figures can also be considered as indicating approximately the numbers of generations needed for the effect of selection. However, in the lower Crustacea the number of generations would probably be still greater. The tempo of evolution cannot be properly judged unless we take into account the number of different types produced within certain periods. A good measure for such an evaluation is provided by the number of orders existing at present. According to Kükenthal-Krumbach's *Handbuch der Zoologie* there are 15 orders of Crustacea, 1 of Xiphosura, 9 of Arachnoidea, 4 of Pantopoda, 7 of Diplopoda, 4 of Chilopoda, and 29 of Insecta. Hence, the four younger (and mainly terrestrial) classes have developed an average of 12 orders in a shorter time than the three older (and mainly marine) classes, which have produced an average of only seven orders within a definitely longer period. A similar situation is met with in the vertebrates. Here, five orders are found in Agnatha (most of them being extinct), and the extinct Placodermi comprise six orders, whereas there are 23 orders in Pisces, 13 in Amphibia, 16 in reptiles, 33 in birds, and 34 in mammals (extinct orders included: compare Colbert, 1955). Hence, the three more primitive and totally aquatic classes of vertebrates have developed an average of only eight orders, while the four higher and mainly terrestrial classes have an average of 24 orders. Thus, there is a generally increased tempo of evolution in the younger terrestrial classes.

Similar results are obtained when one studies the evolutionary rates of the orders. There are nine Recent orders of Crustacea of which the phylogeny is sufficiently known and can be traced back into the Paleozoic. Their age ranges from 240 to 540 million years, which is an average of 410 million years per order. The order Pygocephalomorpha, confined to the Devonian and Carboniferous, existed for about 110 million years (maximum figure). Compared to this, the orders of terrestrial insects are much younger and definitely richer in types: 25 Recent orders with a reliable fossil record have existed from 60 to 350 million years, averaging 180 million years per order (Handlirsch, 1926–38, and H. Weber, 1933), and 13 fossil orders lived from 10 to 100 (average: 45) million years. Among arthropods the younger terrestrial orders show a quicker evolution than the older aquatic ones. In Mollusca there are three Recent orders of Cephalopoda (Nautiloidea, Decapoda, Octopoda) with a sufficient fossil record. These have lived for 540, 200, and 110 million years respectively, averaging 285 million years. Two fossil orders (Ammonoidea and Belemnoidea) lived 350 and 140 (average: 245) million years respectively. Most orders or 'family groups' of terrestrial snails, so rich in forms, have existed since the Cretaceous and the Early Tertiary, i.e. for about 140 or 60 million years.

Among vertebrates, such differences are similarly distinct. But an exact comparison of figures is difficult, as the term 'order' does not have the same meaning in reference to the various groups, and some older types must perhaps be considered as even higher systematic categories. Nevertheless, it is quite clear that in vertebrates, too, the terrestrial orders (and among these the highest forms) evolved more quickly than the fishes. Thus, the average age of recent orders with reliable fossil records in fishes is 270 (175–350) million years, in reptiles 185 (160–200) million years, in birds 65 (20–140) million years, and in mammals 65 (35–150) million years.

This rule applies to the families as well: 40 families of recent Lamellibranchiata have existed for 60 to 500 (average: 225) million years, and 16 extinct families lived for 45 to 390 (average: 195) million years (systematics and age of families according to Wenz, 1938). Ninety-eight families of Recent marine Prosobranchia originated between 15 and 450 million years ago, averaging a phylogenetic age of about 125 million years, and 46 extinct families of this type lasted for 30 to 480 (average: 190) million years. Four families of Recent terrestrial Prosobranchia have existed for 60 to 280 (average: 130) million years, and two fossil families lived for 10 to 40 (average: 25) million years. Thirty-five families of recent terrestrial Pulmonata, however, originated between 20 and 310 (average: 65) million years ago, and one fossil family (Filholiidae) lived about 85 million years. From these figures we may see that in terrestrial Pulmonata the rates of evolution are shorter than in terrestrial Prosobranchia.

To continue our considerations on the age of some families, we find that 24 fossil families of trilobites existed for an average of about 110 million years (40–230 million years). In 18 fossil families of Brachiopoda we find about 160 million years (30–310 million years) as the average period of existence, and in eight Recent families of Brachiopoda the average has been about 340 million years (200–450 million years). In contrast to this, 53 extinct families of hoofed animals averaged 25 million years (from 10 to 60 million years), and 12 Recent families of this type have lived for 15 to 35 (average 30) million years. Five fossil families of Carnivora lasted from 15 to 40 (average 25) million years, and seven Recent families have existed for 20 to 35 (average 30) million years. Similar calculations on the families of mammals have been done by Ekmann (1940), who wished to establish a precise foundation of animal geography by this method. According to him, the average age of a family is about 25 million years.

Studying the differentiation into families and subfamilies, we find a conspicuous phenomenon that can also be seen, though less clearly, in the evolution of genera and species. After a new structural type has originated, there follows a quick evolution of new families and subfamilies. Although most of these do not last very long, those superior in competition may exist for quite a long time (compare the relevant situation in Perissodactyla, Figure 23). We shall refer to this initial 'explosive' evolution in more detail in Chapter 6, A, pp. 98–112.

As regards the rates of evolution in genera, the facts on the whole are the same as those demonstrated above. In genera, however, reliable calculations are difficult because, since the methods based on radioactivity can usually be applied to the estimation of longer periods only, precise estimations of the relatively short periods of life can hardly be made. Other methods of dating the past can be applied only to the Pleistocene and a few other periods. In spite of these difficulties it is quite evident in genera that evolution is slower in the more primitive marine groups than in higher terrestrial animals. Thus, in 30 genera of recent Foraminifera (after Hennig, 1932) the average age is 200 million years (25–540 million years), and nine fossil genera lasted from 25 to 136 (i.e. an average of 90 million years). Recently Schindewolf (1950) calculated the average age of genera in Foraminifera as 71 million years. The 16 known genera of Paleozoic Gigantostraca lived from 30 to 250 (average: about 90) million years, and the 15 known genera of Xiphosura from 15 to 110 (average 40) million years, whereas the Recent genus *Limulus* has existed for about 200 to 240 million years (i.e. probably since the Late Permian, but certainly since the Early Triassic). The phylogenetic age of 26 Recent genera of marine Prosobranchia (belonging to the families Haliotidae, Scissurellidae, Fissurellidae, Patellidae, Acmaeidae, and Lepatidae) ranges from about 15 to 200 million years, with an average life of 75 million years per genus, and in 78 extinct genera of this group of marine snails we find an average of 80 (15 to 300) million years for the genera of the families of Fissurellidae, Patellidae, Acmaeidae, Trochonematidae, Macluritidae, Poleumitidae, Oriostomatidae, Cirridae, Platyacridae, Omphalocirridae, and Euomphalidae, not including genera comprising only one species (systematics according to Wenz, 1938; compare also Table 1 in Schindewolf, 1950).

Most of the Recent genera of terrestrial Pulmonata did not originate until the Tertiary, i.e. less than 60 million years ago. According to the fairly rich fossil records of Helicidae, the genera *Helix*, *Cepaea*, *Fruticicola*, *Helicodonta*, and *Leucochroa* originated during the Oligocene. Some of the related genera, such as *Helicella*, *Helicogona*, *Euparypha*, and *Otala*, appeared during the Miocene, while others did not emerge before the Pliocene (*Isognomostoma*, *Euomphalia*, *Theba*, and *Cochlicella*, for instance). Compare also Simroth and Hoffmann, 1928).

Many of the more primitive genera of Recent birds came into being during the Eocene, i.e. about 50 to 60 million years ago (*Milvus*, *Aquila*, *Haliaëtus*, *Phoenicopterus*, *Charadrius*, *Totanus*, *Numenius*, and others). Some originated during the Oligocene, i.e. about 25 to 35 million years ago (*Sula*, *Pelecanus*, *Puffinus*, *Podiceps*, *Colymbus*, *Leptoptilus*, *Anas*, and the Oscines *Lanius* and *Motacilla*), while others emerged during the Miocene, or about 25 million years ago (*Otus*, *Tringa*, *Uria*, *Ardea*, *Micropus*, and others), or during the

Pliocene, i.e. about one to eight million years ago (*Alauda, Sitta, Anthus, Corvus*, and others). If we consider the age of our Recent genera of mammals, we find that most of them have not emerged before the Pliocene, so that they are no older than about seven to eight million years (*Tragulus, Capreolus, Bos, Bison, Capra, Gazella, Girafa, Elaphus, Camelus, Equus, Dicerorhinus, Rhinoceros, Elephas, Hyaena, Mustela, Meles, Ursus, Felis, Castor, Simia, Pithecus, Semnopithecus, Macacus, Papio*, and others. Compare also Ekman, 1940). Other genera, dating back into the Miocene, have existed 20 to 25 million years (*Sus, Tapirus, Canis, Putorius, Lutra, Geomys*, and *Elyomys*), while *Sorex, Crocidura*, and *Talpa*, for instance, emerged during the Oligocene. Some genera, however, did not appear until about 600,000 to 800,000 years ago, i.e. during the Pleistocene (*Alces, Saiga, Dicrostonyx, Lemmus*, and others, and above all *Pithecanthropus* and related types including *Homo*).

In the main line of descent of horses, eight fossil genera evolved in a period of about 60 million years, so that each genus reached an average phylogenetic age of 7·5 million years. About the same rate of evolution was found in the Chalicotheriidae. The more primitive and geologically older opossums (*Didelphys*), however, were much slower as regards their evolutionary tempo (Simpson, 1944).

Recent species, viz., 'Arten-' and 'Rassenkreise' of terrestrial animals can usually be traced back only into the Pleistocene or Pliocene. For a proper evaluation of their rates of evolution, we should preferably rely on forms which have been found in several geological horizons. So we may be sure, for instance, that the Recent race of the red deer (*Cervus e. elaphus*) has been in existence for only about 130,000 years, as it can be traced back no further than the Third Interglacial Period. In the Second Interglacial, i.e. about 350,000 years ago, a different race existed (*C. e. angulatus*), which was preceded by still another race (*C. e. priscus*) during the First Interglacial, about 500,000 years ago. Immediately before the appearance of the *priscus* race we find the type *C. acoronatus*, which probably belongs to the same ancestry (Beninde, 1937). Hence, the total phylogenetic age of the red deer is about 500,000 years. The recent form of the reindeer (*Rangifer arcticus*) has been existing since the Mindel-Riss Interglacial, i.e. for about 350,000 years (Jacobi, 1931). The cave bear originated during the Riss Glacial as the *deningeri* stage, reached its evolutionary peak during the Riss-Würm Interglacial, and died out during the Würm Glacial, thus reaching an age of about 250,000 years. The brown bear (*Ursus arctos*) came into being about 140,000 to 300,000 years ago, i.e. during the Second or Third Interglacial. The mammoth (*Elephas primigenius*) and the woolly-haired rhinoceros (*Rhinoceros antiquitatis*) can be traced from the Würm to the Riss Glacial, and thus they existed for about 200,000 years. There is a very rapid evolution in the differentiation of the Recent 'Rassenkreis' of *Homo sapiens*. These races – including the extinct Aurignac and Grimaldi types, etc. – originated about 100,000 years ago (250,000 years, if we include the Swanscombe records). The typical *Homo neandertalensis* probably lived for about 40,000 to 50,000 years, as his fossil record ranges from the middle of the Last Interglacial to about the middle of the Würm Glacial. If we include the types found in Palestine (Carmel), the age would be about 200,000 years and date back into the Riss-Würm Interglacial.

The relevant figures pertaining to mastodonts indicate somewhat longer periods of existence. *Bunolophodon angustidens* (Middle and Late Miocene), *B. longirostris* (Early Pliocene), and *B. arvernensis* (Middle Pliocene to early Late Pliocene), which Abel (1929) arranged in phylogenetic succession, lived for about five to eight million years each (Miocene+Pliocene=about 25 million years). We should, however, take into account that the generations of fossil elephants take several years, as in our Recent elephants. We get similar figures for species age if we identify the successive 'stages of specialization' with a line of descent in the horses *Eohippus borealis* (Lower Eocene), *E. venticolus* (Late Lower Eocene), *Orohippus atavus* (Lower Middle Eocene), and *O. progressus* (Upper Middle Eocene). Ekmann (1940) supposed the average species age of Recent mammals to be about 1·4 million years. According to Simpson (1953) North American mammals invading South America some 500,000 to 1,000,000 years ago (latest Pliocene and early Pleistocene) probably have all become specifically distinct.

In some species for which a reliable fossil record is lacking, one can reasonably evaluate the species age by considerations based upon facts of animal geography. So, for instance, in the eastern, southern, and western ranges of the Alps we find numerous endemic species inhabiting quite restricted areas, sometimes a single hill. These areas, apparently, were not covered by ice during the Glacial Epoch and provided refuges in which new species could originate. Examples of such species are the ground beetles *Amara lantoscana* (in the Ligurian Alps and eastern Maritime Alps), *A. frigida* (western Maritime Alps), *A. doderoi* (Pennine Alps), *A. uhligi* (Venetian Alps), *A. alpicola* (Tauern, Gurktaler, Seetaler Alps, Stubalpe), *A. cuniculina* (eastern part of the northern Carnic Alps), and the species belonging to the group *Amara psyllocephala-graja-cardui-alpestris-baldensis-spectabilis-nobilis*. Heberdey (1933) supposed that these species originated after the Würm Glacial, and would thus be only about 100,000 years old. We should, however, be skeptical about this assumption, as there may have been previous racial differentiation, and even today some of the 'species' concerned represent borderline cases between race and species. On the other hand, it is important that many other species, for instance, all cave dwellers surviving the last ice period of the Alps in caves, remained unchanged and do not show any symptoms of the formation of new species (Holdhaus, 1933).

Among the species of fishes that differentiated during the Glacial Epoch and today live in distinct areas, we may enumerate: the anchovies *Engraulis encrasicholus* (North Atlantic) and *E. capensis* (South Africa), or *E. japonicus* (Northwest Pacific) and *E. antipodum* (Australian waters), or *E. mordax* (Californian coast) and *E. ringens* (from the coasts of Peru and Chile: Berg, 1933). According to Berg (1935), pairs of species inhabiting the discontinuous habitats of the North Atlantic and North Pacific ('amphiboreal' species), such as the eels *Anguilla anguilla* (Europe) and *A. japanica* (East Asia) or the flounders *Pleuronectes flesus* (North Atlantic) and *Pl. stellatus* (North Pacific), differentiated in the course of a Pliocene communication, and consequent isolation. Hence, these species would have an age of about 2·7 million years. Much the same phylogenetic age was found in the 'twin species' of fishes living east and west of the Panama region as a result of the Pliocene blocking of the two previously connected oceanic habitats. Such 'twin species' are represented by *Epinephelus adscensionis* of the Atlantic and *E. analogus* of the Pacific, by *Lutianus apodus* (Atlantic) and *L. argentiventris* (Pacific), and so forth. Here one should mention, however, that in some cases only new races developed owing to this disjunction, or that there was no differentiation at all.

In many marine molluscs the species age is still greater. From the multitude of species providing a reliable fossil record, only a few examples of the families Cancellaridae and Fasciolariidae will be mentioned (after Sieber, 1936, 1937). Some species are restricted to a few horizons of the Miocene (usually lasting from Torton to Pont, i.e. from the Middle to Late Miocene) and existed for about 6 to 12 million years, such as *Cancellaria contorta, C. purchi, C. imbricata, C. spinifera, C. lyrata*, and *C. gradata*, or *Lathyrus lynchi, L. crassus, L. rothi, Fasciolaria tarbelliana*, and *Fusus rostratus*. Some species, ranging over several stages of the Miocene and Pliocene, lived for about 15 to 25 million years each, such as *Cancellaria ampullacea, C. scabra, C. mitraeformis, C. parvula, Admete fusiformis*, and *Fasciolaria fimbriata*. Other species, such as *Cancellaria evulsa* and *Euthriofusus burdigalensis*, existed during several epochs of the Oligocene and Miocene, and the recent *Cancellaria cancellata* has been existing since the Miocene, i.e. for about 20 or 25 million years.

Evolution rates seem to be definitely shorter in ammonites. Brinkmann (1929) studied the transformation of *Kosmoceras* during the Callovien (Middle Jurassic) covering four successive species. As the Callovien is one of twelve main periods of the Jurassic, which lasted for about 35 million years, the average species age is something like 250,000 to 350,000 years (according to Brinkmann); at any rate it is less than one million years. In *Manticoceras* Schindewolf (1950) found the average species age to be about one million years. The initial stages of this group, however, again show a very rapid rate of transformation (compare Chapter 6, A, on explosive evolution). Hence, we can state that, generally, the species of marine types and of 'lower' groups persist for a longer time than terrestrial animals, but nevertheless the situation varies a great deal according to the groups concerned and their various phylogenetic epochs.

Finally, the formation of geographic races happened within periods of some 10,000 or 100,000 years. We have already mentioned some cases in Chapter 3, demonstrating a race formation even during historical eras (*Pachys [Amphidasis] betularia*, and others). In these examples, however, only the spreading of certain mutations, which probably differed in only a single gene, was concerned, whereas the 'normal' geographic races usually differ in several genes, as they have been subjected to natural selection for tens of thousands of generations. A reasonable estimation of racial age can well be given in many cases.

As has already been mentioned, the ranges of many European animal forms have been separated into two or more refuges by the glacial epoch. During the Ice Age, this isolation and the different situations of selection caused the formation of numerous new geographic races. So, in the Alpine 'massifs de refuge' of the eastern, southern, and western ranges, a number of new geographic races originated (besides the new species already mentioned). This is especially evident in terrestrial snails of the genus *Orcula* (Zimmermann, 1932) and *Pagodulina* (Klemm, 1939). The more vagile animals were shifted to a larger extent, and glacial refuges developed in larger parts of southeast and west or southwest Europe. With the postglacial reoccupation of central Europe a secondary contact of eastern and western types often occurred, but in many cases there was no hybridization, as the partners already represented new species (as in the nightingales *Luscinia megarhynchos* and *L. luscinia*, the warblers *Hippolais pallida* and *H. icterina*, and so forth). In other cases of this secondary contact intermediate hybrid races originated, or at least a geographic zone of hybridization became evident (in the bullfinches *Pyrrhula pyrrhula*, the long-tailed titmice *Aegithalos caudatus* [Stresemann, 1919; Salomonsen, 1931], the hedgehogs *Erinaceus e. europaeus* and *E. e. roumanicus* [Herter, 1934], the toads *Bombina variegata* and *B. bombina* [Mertens, 1928], and others). Though in all these cases the races may be about 300,000 to 500,000 years old, it must be remembered that racial differences may possibly have existed before the breakup of the species owing to the glacial epoch.

Other geographic races are considerably younger. The numerous races (42) and subraces of the 'Rassenkreis' *Coregonus lavaretus* (Berg, 1933) inhabit lakes of northern Europe and the Alps which cannot be older than about 10,000 to 30,000 years. The race *oernensis* of the fish *Cottus quadricornis* in Lake Oern in Sweden cannot be older than this lake, the age of which is reliably known as about 8,500 years (Lönnberg, 1939). Some races of fishes inhabiting separated areas of the North Atlantic and North Pacific seem to be equally young and must have originated after the glacial epoch (the herrings *Clupea harengus harengus* [Gulf of Biscay – White Sea] and *Cl. h. pallasi* [Bering Strait – Hondo], or the cods *Gadus morhua morhua* [north Atlantic] and *G. m. macrocephalus* [north Pacific], and so forth). The Porto Santo rabbit, supposed to be a species of its own by Darwin, must be considered as a race (the smallest and 'wildest' one) of *Oryctolagus cuniculus*. It originated from domesticated ancestors let loose about 500 years ago. The inheritability of the racial characters has been proved by Nachtsheim (1941).

Some races of birds limited to northern and eastern Europe are probably younger than 10,000 years. The large bullfinch (*Pyrrhula pyrrhula pyrrhula*), for instance, seems to have evolved simultaneously with the immigration of the spruce about 6,000 to 8,000 years ago (G. and J. Steinbacher, 1943). Some less distinct races of *Lacerta muralis* on some of the numerous islets off the coast of Istria cannot be older than these islands themselves, i.e. older than the separation of the islands from the continent, which happened about 9,000 years ago (Kramer and Mertens, 1938). Quite a similar age must be assumed in the case of the numerous races of lizards on the islets of the Balearic Islands. The dark race *nigricans* of the crested lark *Galerida cristata* from the lower Nile cannot be older than 10,000 years (Moreau, 1930). The beetle species *Choleva holsatica* (Silphidae), living only in a post-glacial cave near Segeberg in Holsten, cannot have originated until 16,000 to 20,000 years ago (Benick, 1939). Some endemic races of mammals on young islands of the Great Salt Lake in Utah cannot be older than 20,000 years (Marshall, 1940), and a race of the seal *Phoca vitulina* living in a Canadian lake must be as young as about 3,000 to 8,000 years (Doutt, 1942).

The Recent race of the ibex (*Capra ibex ibex*) separated from the ancestral type *C. i. camburgensis* of the Ice Age about 230,000 years ago, according to the calculations by Zeuner. A similar chronology may be assumed concerning the ancestors of the red deer (*Cervus elaphus angulatus* and *C. e. priscus*) during the glacial epoch. Herre (1948) could prove that a race of the wild boar from the Middle Stone Age was the same as the race *Sus scrofa antiquus* of the post-glacial period. In this case, then, race differentiation took only about 3,000 to 8,000 years. The Recent races of man have probably originated during the last 10,000 to 20,000 years, and equally young must be the geographic races of lice to be found on Europeans, Mongols, and Negroes (Fahrenholtz, 1936).

Finally, we must not overlook the fact that in many cases of special isolation no new races have developed at all. This is proved, for instance, by the beetles *Nebria nivalis*, *Elaphrus lapponicus*, *Amara alpina*, and others, all of which migrated from Scandinavia to the British Isles before the Würm Glacial and have developed no British races (Lindroth, 1935). Nor have any new races appeared in the sparrows *Passer domesticus domesticus*, though they were brought to eastern North America in 1850 and to the west between 1871 and 1873. The sparrows transferred to the Hawaiian Islands during a later period show a minute racial differentiation, which, however, may possibly be a modificatory variation (Lack, 1940).

B. FACTORS GOVERNING THE SPEED OF EVOLUTION

Summarizing our findings on the average age of the various higher systematic categories, we find a gradation similar to that presented in Table 3, which covers some of our Recent animal groups (references, in addition to those mentioned above, after Zittel, Abel, and others). Generally, classes are twice as old as orders, orders are twice the age of families, and families are usually 1·5 times as old as genera. (Of course, there are many exceptions to these generalizations.) Genera are usually about ten times as old as species, and we have found that the average age of the species is somewhere between 100,000 and a few million years. Species, therefore, are about ten times as old as subspecies, as the age of the latter usually is not older than 10,000 to 100,000 years. Hence, in drawing a phylogenetic tree intended to give a true representation of time proportions, the lengths of branches from one ramification to the other should be rendered in geometrical rather than in arithmetical progression. Figure 18 is intended to give an idea of what is meant by this geometrically increasing age of higher categories. (Of course, more lateral branches should normally be filled in to represent the numerous living and extinct orders, families, and so forth.)

TABLE 3. Average age (in millions of years) of higher categories of various land and water animals of different evolutionary levels. (Incomplete calculations and estimated average figures in parentheses.) 'Families' of Crustacea correspond to Beurlen's 'Tribus'.

	Classes	Orders	Families	Genera
Mammalia	190	65	25	(15)
Reptilia	310	185	85	(50)
Pisces	450	270	80	(50)
Insecta	350	180	160	(12)
Crustacea	540	410	(160)	—
Pulmonata ⎱		100	65	(40)
Marine Prosobranchia ⎰	540	(400)	125	75

As has been emphasized several times, all these calculations on the average ages of categories may seriously be restricted by the facts that the appearance of completely new structural types is often succeeded by a period of 'explosive' evolution (tachytely), and that phases of decreased tempo (bradytely) may preserve certain animal types for very long periods (persisting types). Hence, it is the relative differences in phylogenetic age, rather than the absolute times, that should be taken into consideration for a proper understanding of the situation.

As mentioned several times in the preceding chapters, the corresponding taxonomic categories of the more primitive animal groups are older than those of the more progressively developed types (Table 3). This applies to the fishes in comparison to the reptiles, and the reptiles in comparison to the mammals, and holds good as to Crustacea in relation to insects, and marine Prosobranchia to terrestrial Pulmonata. As the different environment of terrestrial and aquatic animals might affect the phylogenetic persistence of a type (different intensity of selection), the ages of categories in strictly terrestrial animal groups should also be compared. Even then, however, the above statement remains valid, as will be seen from the following examples: classes, orders, families, and genera of reptiles have existed for a longer period than those of mammals; the families of terrestrial Prosobranchia average an age of about 130 million years, but those of terrestrial Pulmonata only about 65 million years, and in mammals the more primitive orders date back into the Mesozoic (the Marsupialia originated during the Late Jurassic, i.e. about 150 million years ago), whereas the more advanced orders can be traced only as far as the beginning of the Tertiary. From these facts, Charles Darwin (1859) and Gaudry (1869) concluded that in lower animals the tempo of evolution is generally slow, and Dépérét (1909) thought (p. 153), 'that the tempo of evolution in a certain animal group is inversely proportional to (phylogenetic) longevity'. These opinions, however, can hardly be considered as safe generalizations, since so many groups of 'lower' animals tend to rapid splitting into new forms during certain 'periods of virescence'. (One need only think of the rapid splitting of the *Nummulites* and *Orbitoides* belonging to the phylogenetically old Foraminifera during the Eocene, of the numerous lines of Ammonites, and of the sudden evolution of quite a series of different mammalian types in the Early Tertiary, though the mammals as such had been existing since the Jurassic, and so forth.) The more complicated organization of the higher animal types, however, may well have rendered possible a quicker specialization and adaptation to the respective environmental conditions.

More important seems to be the fact that the lower and phylogenetically older animal groups had more time than the more recent groups to develop various types adapted and adaptable to a changing environment, so that they did not perish when the normal habitat changed. The structural type of tortoises, for instance, unfavorable as it may seem to be by hindering quick locomotion, breathing movements, and so on, has proved a perfect protection from all enemies and therefore has remained unchanged since the

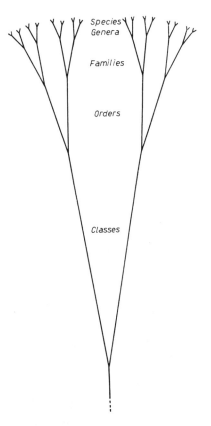

FIGURE 18. Diagram of the differentiation of systematic categories. Distances between ramifications indicate average age of mammalian classes, orders, families, genera, and species.

Permian. It is conspicuous that most such persisting types are not immutable species, but persistent genera, the changing species of which preserve a certain type of adaptation. This can be seen clearly in the genus *Nucula* (Lamellibranchiata), which originated during the Silurian and developed quite a number of species culminating in the Cretaceous adapted to digging by the most diverse mechanisms (Quenstedt, 1930). *Solenomya*, a genus of closely related mussels from the same ancestry, had already reached an ideal level of adaptation in the Carboniferous, and since then no major alterations of the type have occurred. We shall discuss a little later the problem of persisting types and their independence from environmental conditions.

We cannot make the general assumption that higher animal types tend to quicker evolution than lower forms, because the categories of Recent animals have lived longer than the corresponding fossil categories. Twenty-five Recent orders of insects have an average age of about 180 million years, but 13 fossil orders lived for only 45 million years each. Four Recent orders of reptiles have been existing for the last 185 million years, but in 17 extinct orders the average age was about 70 million years. The phylogenetic age of 15 Recent orders of mammals has reached about 65 million years each, but eight fossil orders lived for only 35 million years on the average. Eight Recent families of brachiopods have existed for about 340 million years, but 18 fossil families lived for only about 160 million years. Thus it is evident that the extinct animal groups are branches, which rather early reached stages that could not be adapted to further alterations of the environment and which therefore died out. But the surviving types were those which were less strongly adapted and developed more slowly, i.e. which remained unspecialized (Cope's 'Law of the Unspecialized': compare Chapter 6, C).

The influence of environment on the tempo of evolution is much more important. As may be seen from Table 3, the average age of categories in marine animals is markedly higher than in terrestrial animals. As we have found, the greater phylogenetic age of these marine groups plays a role, but not the only or most important one, as nearly all persistent genera are also marine forms, such as the Foraminifera *Spirillina* (persisting since the Devonian), *Lagena, Ammodiscus, Placopsilina* (since the Silurian), *Dentalina, Agathammina, Saccamina, Trochammina, Lituola* (since the Carboniferous), the Crinoidea *Antedon* (since the Jurassic), the Brachiopoda *Lingula* (at the beginning perhaps rather to be regarded as *Glottidia*) and *Discina* (since the Silurian), *Terebratula* (since the Devonian), the Lamellibranchiata *Avicula, Nucula, Leda, Arca* (since the Silurian), *Pinna, Modiola* (since the Devonian), *Lima, Pecten, Astarte* (since the Carboniferous), the Prosobranchia *Pleurotomaria, Emarginula, Scurria, Acmaea, Delphinula, Natica, Scalaria* (since the Triassic), the Cephalopoda *Nautilus* (since the Carboniferous), the xiphosure *Limulus* (since the Triassic), and many others. Compared to this rich variety of marine forms, the number of persisting terrestrial genera is remarkably small, and usually their age is younger. Examples are the snails of the genus *Glandina* and *Megaspira* (since the Late Cretaceous), the bugs *Belostomum* and *Naucoris* (since the Jurassic), the dragonflies *Anax* and *Aeschna* (since the Jurassic), and the reptiles *Crocodilus* and *Trionyx* (since the Late Cretaceous). Other persistent types, such as *Sphenodon* and *Tarsius*, are relict genera without any fossil record, and the related genera were extinguished in the Jurassic and during the Eocene, respectively.

If so many persisting types are found in marine animals and so few in terrestrial ones, one may reasonably attribute this fact to the influence of environment. Indeed, marine animals are much less subjected to periodical changes of important ecologic factors such as temperature, illumination, oxygen concentration, and so on, than terrestrial forms. Moreover, in most marine animals population boundaries are not very effective, as even in sessile types the reproductive cells and the larvae can readily transgress the limits of the ancestral habitat. Hence, only small fluctuations of the number of individuals in the populations occur, and the effect of natural selection remains relatively constant. It is for this reason that evolution in marine animals is slow, and there is no need to assume a decreased rate of mutation or a special, slow, 'immanent tempo of evolution' (compare also Simpson, 1944). C. Darwin (1859) emphasized the importance of the rather complicated environmental situation of the terrestrial types, and Dépérét (1909, p. 151) assumed that the persistent types so frequent in marine animals were due to their relatively constant environment. In this connection Abel's demonstration (1918, 1929) on the phylogency of Dipnoi, Sirenia and whales is particularly impressive. He clearly showed that slow evolution prevailed if environmental conditions were stable, and that a quick and almost 'saltatory' splitting occurred if there were marked changes in the environment, or in the mode of feeding and so forth. In whales, the steady development was interrupted three times after the Eocene by 'stormy' periods of evolution: first, when the original types preying on large animals (like *Protocetus* from the middle Eocene) changed to living on fishes, and the Squaliodontidae of the Miocene, which had teeth like a shark's, originated; second, when these types shifted to eating the smaller Cephalopoda and changed into the Physeteridae and Ziphiidae with reduced dentition; and, finally, when the Mysticeti with their whalebone mechanism evolved relatively quickly. Hence, Abel stated (1929, p. 298): 'It is only the change of the mode of living that affects the tempo of phylogenetical development.' Ruedemann (1918) and Pompecki (1925) also emphasized that persistence of types depends on constancy of environment.

Finally, that the speed of evolution depends upon the rates of selection may be seen from the fact that in more mobile animals the evolutionary tempo is increased, since such types are able to adapt more easily. As has been shown by Ruedemann, a great number of persistent types are sessile or digging forms, and the persistent genera of mussels and snails also are quite restricted in their mobility. On the other hand, mollusks capable of fairly quick locomotion, such as the ammonites, show a rapid evolution.

Apparently, the rate of evolution also depends on temperatures. In the warmer zones, the number of generations per year is definitely higher in poikilotherms, and to some extent also in homoiotherms, than it is in the temperate and cold zones. Hence, the effectiveness of selection is increased correspondingly. Consequently, in our Recent fauna the number of animal species in warm regions is definitely greater than in temperate and cold zones. Now, all climatic zones were markedly warmer in former geological epochs (at least from the Late Triassic until the Early Tertiary) than today (Schwarzbach, 1946, 1949, 1950). So, during these periods, we may possibly presume a relatively quicker evolution in many animal groups at present confined to temperate and cold zones, and such an increased tempo of evolution may also have prevailed during the Paleozoic before the Permian (B. Rensch, 1952). (Such comparisons of the numbers of generations per year must, of course, be restricted to closely related types, and one cannot compare totally different groups. In elephants, for instance, the number of generations per unit of time is small, but nevertheless the rate of evolution was relatively fast, while in some small mammals, such as rodents, although the number of generations per unit of time was about 40 to 100 times greater than in elephants, evolution in these types is still slow. This fact seems to be due to a greater constancy of the environment and to the larger size of reproductive communities in the rodents.)

Summing up, we may state that the rates of evolution in higher taxonomic categories may reasonably be interpreted on the basis of factors which also bring about infraspecific evolution (i.e. mutation, fluctuations of population size, selection, and isolation). At present, there is no reason to assume autonomous forces causing changes in the rate of evolution. The rates of evolution seem to be influenced primarily by conditions of selection, as will be shown in more detail in the next chapter. Besides this, however, the structural type may have a predisposing influence, as versatile types adapted to very different conditions are less sensitive to the effects of selection.

6

The Rules of Kladogenesis
(Phylogenetic Branching)

A causal analysis of transspecific evolution has to deal with two independent problems: (1) the causes of branchings in the lines of descent and the rules governing this process, and (2) the problem of progressive development toward 'higher' levels in the phylogenetic tree as a whole and in its main lines. Metaphorically: we have to find out not only the basic principles giving rise to the phylogenetic tree, but also the causes for its 'upright' position and the fact that many of its branches tend to an 'upwards' direction. As it seemed desirable to me clearly to separate these two heterogeneous phenomena, I proposed two different terms for them: kladogenesis for the phylogenetic branching and splitting in general, and anagenesis for the development toward higher phylogenetical levels.[1]

As we have seen in Chapter 4, kladogenesis is brought about by the random formation of new branches, the growth of which is limited by environmental circumstances and forced into suitable directions. Besides the special rules already discussed in this chapter, there are also rules of a more general character governing the types of phylogenetic development as a whole. These rules had already been discovered by the end of the last century, but the discussions concerning them have been especially lively during recent years. One of the main items debated is the fact that usually the lines show an intense radiation of types, an 'explosive phase' in the early part of their phylogeny, and that only a limited number of the branches continue to develop, and with decreasing speed. In this second phase the animals acquire numerous adaptations, sometimes maintaining a directed evolution (orthogenesis) for long periods. The formation of parallel structures in related stocks may also occur, or the evolution of certain structures may be repeated (iterations). In a last phase, excessive or 'degenerative' traits tend to emerge, and finally the stock becomes extinct. Normally, the various phases of phylogeny cannot be reversed (rule of irreversibility). Hence, there is a certain regularity in the course of phylogeny and, by analogy with ontogenetic development, many authors have referred to the three main phases of phylogeny as

[1] $\varkappa\lambda\acute{\alpha}\delta o\varsigma$=branch; $\grave{\alpha}\nu\acute{\alpha}$=in an upward direction; $\gamma\acute{\epsilon}\nu\epsilon\sigma\iota\varsigma$=development.

youth, adult stage, and old age of the phylogenetic stock, implying the existence of autonomous, immanent forces of phylogenetic transformation. The assumption of such immanent forces was strongly maintained by Von Baer (1876), Rosa (1931), Dacqué (1935), Beurlen (1937), Schindewolf (1931, 1936, 1950), and Von Huene (1940). Now, we shall have to examine the validity of these rules governing transspecific evolution and find out whether they can reasonably be explained by assuming that transspecific evolution is brought about by the same factors found to be effective in the formation of races and species, i.e. by mutation, fluctuations of population size, and selection, or whether we must assume autonomous tendencies of evolution. The second assumption would, however, represent no explanation at all, as such assumptions would be only words denoting a process or a factor which cannot be analyzed or circumscribed physiologically.

A. PHASES OF EXPLOSIVE RADIATION

Historical Comments. As early as the beginning of the nineteenth century, paleontologists had been struck by the fact that in many animal groups a great number of new types suddenly appeared in certain geological epochs. Cuvier tried to explain this phenomenon by his theory of catastrophes, especially in those cases where in a certain horizon a whole new fauna appeared. Charles Darwin (1859) opposed this opinion and denied such rapid transformations. He pointed out that the fossil records were far from sufficiently complete and that in many cases the 'saltations' in the lines of descent were due to horizontal shifts of single species or whole faunas. Hence he concluded (*On the Origin of Species*, 6th edition, 1876, p. 403) that, as a general rule, the number of species in each group gradually increases toward the maximum, and then, sooner or later, a gradual decrease of the species number occurs. A special 'flowering period' (epacme) in each line of descent was recognized by Haeckel (1866). He wrote (p. 321):

> The epacmatic growth of species and phyla is similar or equal to the anaplastic growth of bionts and consists in an increase of the extent and size of the phylum. In the species the number of individuals and in the phyla the number of subordinate categories (classes, orders, etc.) is increased.

Haeckel made no attempt, however, to characterize these flowering periods in more detail.

So it was not until 1902 that concrete examples of such evolutionary periods were provided. This was done by Jaeckel. Studying the transformation of Crinoidea and Cystoidea, he showed that at certain intervals a 'shaking about' or a 'metakinesis' occurs in the phylogeny, giving rise to the quick formation of diverse and often strongly deviating forms. Jaeckel stated:

> The greatest divergences of types are to be found during the phyletic 'youth' of a branch and may well be considered as typical for this phyletic

stage. . . . On a given basis, however, only few structural plans can be realized, and therefore the final stages usually are much fewer in number than the initial tentative types. . . . These first tentative types of a group usually are quite divergent, and besides the traits typical of the category they often show such strange and atavistic characters that they can easily be distinguished as a division of their own from the bulk of the group.

Walther (1908) introduced the term 'anastrophes' for such phases of rapid transformation, and pointed out that they were to be observed in all animal groups and in genera as well as in families, orders, and classes. Walther emphasized that such anastrophes often do not occur until long periods have passed, during which the animal groups 'live in the form of indifferent, hardly characteristic prototypes showing little transformation. . . .' (p. 551). A similar statement was made by Wedekind (1920), who coined the more appropriate term 'Virenzperiode' for this phase of 'explosive' evolution. Hennig (1929, 1932) pointed out that such phases may be 'late blossoms' of a branch as in teleosts and mammals. Additional examples of explosive evolution were presented by Beurlen (1932, 1937), who considered this 'plastic juvenile phase' a necessary link in the typical 'course of formation', 'governing the whole process of development in the higher and lower categories' (1937, p. 101). I myself tried to interpret these phases of explosive evolution on the basis of increased rates of selection, avoiding the assumption of autonomous factors and macro-mutations (1939, 1943). (One should also compare Heberer's critique of the 'two phases-theory' and the presumed macro-mutations; 1943.) Apparently without knowledge of these papers, Simpson (1944) presented a very similar interpretation of the phenomenon, quoting quite a series of well-founded paleontological records proving this 'quantum evolution', or 'episodes of proliferation', as he later (1953) called it. Without knowledge of the book by Simpson, I gave a detailed account of the problem in the first edition of this book (1947). Finally, Schindewolf (1950) gathered a series of examples of explosive evolution ('typogenesis'), which, however, are considered by him as macro-mutations arising in the course of early ontogeny.

Examples Demonstrating 'Explosive Phases' and Periods of Explosive Radiation. For a proper evaluation of phylogenetical phases showing an increased formation of new types and structures, it is essential to study well-known animal groups quantitatively as to the extent of their splitting into new forms, to find out about the beginning of this period, and to determine which categories are affected by this process. In Tables 4–7 I have tried to gather relevant data for genera, families, orders, and classes of groups with a rich fossil record. The data were taken from the works quoted in the previous chapter regarding the rates of evolution (i.e. Wenz, 1938, on gastropods; Weber, 1928, and Simpson, 1945, on mammals; Lambrecht, 1933, on birds, and so forth). According to common usage, the Silurian was divided into two periods, both very long (70 and 30 million years, respectively) i.e. the Early

Silurian-Ordovician, and the Late Silurian-Gotlandian. The following facts may be inferred from Tables 4–7:

TABLE 4. Number of new genera per geological epoch. (Maximum numbers in bold-face type.)

Families	Cambrian	Ordovician	Gotlandian	Devonian	Carboniferous	Permian	Triassic	Jurassic	Cretaceous	Early Tertiary	Paleocene and Eocene	Late Tertiary	Miocene only
Pleurotomariidae (Gastr.)	—	21	14	17	14	—	16	3	—	—	—	—	—
Rhaphistomatidae (Gastr.)	5	11	6	1	1	—	—	—	—	—	—	—	—
Trochonematidae (Gastr.)	—	7	5	5	4	1	—	—	—	—	—	—	—
Fissurellidae (Gastr.)	—	—	—	—	—	—	1	5	2	6	4	6	2
Euomphalidae (Gastr.)	2	7	4	10	1	1	8	2	2	—	—	—	—
Gigantostraca	—	8	5	1	2	—	—	—	—	—	—	—	—
Xiphosura	3	—	4	3	1	3	1	—	—	—	—	—	—

TABLE 5. Number of new families per geological epoch. (Maximum numbers in bold-face type.)

Orders	Cambrian	Ordovician	Gotlandian	Devonian	Carboniferous	Permian	Triassic	Jurassic	Cretaceous	Early Tertiary	Paleocene and Eocene	Late Tertiary	Miocene
Lamellibranchia	2	12	1	6	2	2	10	9	6	5	5	—	—
Marine Proso-branchia	9	14	9	1	1	3	14	11	36	24	24	5	4
Brachiopoda	6	10	3	2	1	—	3	1	—	—	—	—	—
Trilobita	11	12	1	—	—	—	—	—	—	—	—	—	—
Crinoidea	—	12	6	5	1	—	3	6	2	—	—	—	—
Terrestrial Pulmonata	—	—	—	—	1	—	—	1	6	23	20	3	2
Mammalia	—	—	—	—	—	—	—	7	5	144	102	46	31

TABLE 6. Number of new orders per geological epoch. (Maximum numbers in bold-face type.)

Classes	Cambrian	Ordovician	Gotlandian	Devonian	Carboniferous	Permian	Triassic	Jurassic	Cretaceous	Early Tertiary	Paleocene and Eocene	Late Tertiary	Miocene only
Insecta	—	—	—	1	**12**	7	3	**11**	—	10	10	—	—
Reptilia	—	—	—	—	2	5	**10**	4	—	—	—	—	—
Aves	—	—	—	—	—	—	—	1	6	**16**	**16**	4	1
Mammalia	—	—	—	—	—	—	—	3	2	**25**	19	1	—

TABLE 7. Number of new classes with a good fossil record per geological epoch. (Maximum numbers in bold-face type.)

	Cambrian	Ordovician	Gotlandian	Devonian	Carboniferous	Permian	Triassic	Jurassic	Cretaceous	Early Tertiary	Paleocene and Eocene	Late Tertiary	Miocene only
Animal phyla with a sufficient fossil record	2[1]	**19**	10	2	4	2	—	1	1	—	—	—	—

[1] Possibly more.

1. Periods of rapid radiation can be found in all higher categories, i.e. in genera, families, orders, and classes. As new families cannot be formed until new genera have differentiated, nor new orders until new families have originated, explosive phases of radiation lead to an increase not only of diverse forms but also of new structural types. Hence, these phases represent an extraordinary acceleration of the process of evolution. So, for instance, during the Paleocene and Eocene there originated not only numerous mammalian genera, but also 102 new families and 19 new orders of mammals.

As the animal groups referred to in our study do not represent specially selected examples, it is clearly to be seen that there are hardly any groups without such periods of extensive radiation. Not so clear is the situation in the Pleurotomariidae, a family of Prosobranchia, in which the formation of new genera scarcely decreases from the Ordovician until the Triassic. Nevertheless, we are well justified in considering the regularity of periods of rapid radiation as a rule in the evolution of animals.

2. In about half the number of groups mentioned in the tables, the splitting of types into new genera and new families coincides with the beginning of the groups concerned. In the Ordovician, for instance, the early Pleurotomariidae developed 21 genera, the early Trochonematidae seven, the early Gigantostraca eight, and the early Crinoidea eight. In the remainder of the groups, however, the explosive phase did not occur until a 'starting period' had elapsed, usually covering the preceding geological epoch or a part of it. Though the mammals and birds had existed since the Jurassic, there was no intense splitting into new orders until the Early Tertiary. (This also applies to the families of the teleosts.) In the reptiles, existing since the Carboniferous, the first intense splitting did not occur until the Triassic. These examples of 'retarded flowering periods' (in the sense of Hennig, Wedekind, and others) show that the evolution of higher categories does not always have a cyclic development with a phase of extensive radiation at its beginning.

3. In the differentiation of various categories, especially that of orders into new families, sometimes two widely separated phases of extensive radiation are to be seen: in Lamellibranchiata there is one during the Early Silurian and one during the Triassic and Eocene; in marine Prosobranchia there is one from the Cambrian to the Silurian and another from the Triassic to the Eocene; and in Crinoidea we find one phase of strong radiation in the Silurian and one during the Jurassic. There were two phases of rapid radiation in the Euomphalidae, an extinct family of Prosobranchia, one during the Silurian-Devonian and one in the Triassic. All these cases of double phases of strong radiation are contrary to the assumption that there should be a regular pattern ('Gestalt') in the phylogeny of a category.

4. In only a few cases are the phases of flourishing radiation limited to relatively short geological epochs, viz., to the Paleocene and Eocene, as far as the splitting of mammals into orders and families and of pulmonates into families is concerned. In other groups, however, the periods of stronger radiation lasted considerably longer, through at least two main geological epochs. So, in the marine families of Prosobranchia the first period of radiation lasted from the Cambrian to the Late Silurian (i.e. about 150 to 190 million years), and the second from the Triassic through the Jurassic and Cretaceous to the Early Tertiary (i.e. about 150 to 170 million years). And in Pleurotomariidae the process of rapid splitting into genera lasted from the Early Silurian to the Triassic (with a possible interruption during the Permian). It is evident from these findings that there is no fixed pattern in the course of phylogenetic development, but that instead of short explosive phases relatively long periods of extensive radiation may appear (compare also Stromer, 1944).

As in Tables 4–7 the geological epochs have not been subdivided in a more detailed time scale, it is impossible to see whether this strong radiation is a simultaneous formation or a quick succession of new types. Hence, it may be useful to quote a few examples (after Simpson's classification, 1945) demonstrating the appearance of these flourishing periods. So far as we can judge at present, the suborder Creodonta (primitive Carnivora) begins with 27 genera in the Paleocene. But in the Lower Paleocene there were only six genera, in the Middle Paleocene 13 new genera, in the Upper Paleocene seven new genera, in the Lower Eocene 13 new genera, and so forth. The family of Delphinidae begins in the Lower Miocene with three genera. From the Middle Miocene we know 14 new genera, from the Upper Miocene four new genera. The family of Brontotheriidae began in the Eocene with 28 genera, but in the Lower Eocene there were only two genera, in the Middle Eocene five new genera, and in the Upper Eocene 21 new genera. As these three examples are typical ones, we may state that flourishing radiation is a successive process, not a simultaneous ('explosive') one.

5. In spite of the fact that periods of flourishing radiation may be found in some geological epochs, there are obvious accumulations of such phases in certain geological eras. The Lower Silurian, for instance, is such a period, during which all animal groups studied show an intense transformation, i.e. a definite acceleration of evolutionary processes. In many marine groups a new phase of increased radiation occurred during the Triassic and Jurassic, whereas almost no new types originated during the Carboniferous and Permian. On the other hand, the terrestrial insects show a strong radiation during the Carboniferous and Permian, and the terrestrial pulmonates, mammals, and birds during the Early Tertiary. For the time being, we want merely to state this obvious fact, which suggests the existence of special environmental stimuli. This, however, will later on be discussed in more detail.

Interpretation of the Findings. Is it possible reasonably to interpret the phenomena of explosive radiation by assuming that they are caused by the same factors which bring about subspeciation and species formation? As has been stated on p. 57, this is doubted by some paleontologists, who prefer to assume an autonomous factor of evolution ('formative tendency': Wedekind; aristogenesis as 'a creative principle': Osborn; 'phylogenetic Gestalt as superior principle': Von Huene; 'will to free self-formation': Beurlen; and so forth). By these terms, however, the process in question is given only a name. A causal explanation on the basis of the well-analyzed processes of mutation and selection is, however, possible.

At first it might seem obvious that periods of explosive radiation should be attributed to an increased rate of mutation or to larger mutational steps (macro-mutation). The rate of mutation can be increased by a rise of temperature and the amount of short-wave radiation. And we may well assume that in some former geological epochs either whole ranges or at least certain habitats of organisms were subjected to relatively higher temperatures or exposed to more intense short-wave radiations.

At present the numbers of species inhabiting humid and warm zones are definitely greater than those of the cold and temperate regions, and it is indeed probable that even in recent times the tempo of speciation in warm zones has been rapid. This, however, is probably due to the fact that, because there are no cold seasons, the number of special habitats provided by the great floral diversity is large, and the reproductive communities are relatively small in their numbers of individuals. Hence, the time needed for a favorable mutation fully to express itself in the population is shortened (compare B. Rensch, 1952). Former geological epochs, especially those from the Upper Triassic to the Middle Tertiary and possibly those from the Cambrian to the Carboniferous, were definitely warmer than our recent times (compare Schwarzbach, 1950). This fact, however, cannot provide a suitable explanation of flourishing radiation, as such periods have occurred in some animal or plant group in almost any geological epoch, regardless of climate and temperature. On the other hand, the warm climate might perhaps be considered to be the casual factor preventing the main radiation of mammals from occurring before the Early Tertiary. This could have been due to the initial lack of mammalian superiority over the reptiles, as the latter probably were almost 'homoiothermal' in the warm Mesozoic climates.

As we have seen in Chapter 2, spontaneous mutation at normal rates provides such a wealth of variants that we need assume no further increase of mutation, but only a more intense selection, for a reasonable interpretation of the phases of rapid adaptive radiation. Vice versa, it should be noted that persistent genera need not be phylogenetically immutable, but may reveal quite normal race and species formation rates by normal mutation. The types belonging to the collective genus *Cypraea* (Prosobranchia), for instance, that have existed since the Cretaceous, show a strong geographic variation (F. A. and M. Schilder, 1939: the family Cypraeidae comprises 279 geographic races). In the mussels of the genus *Nucula*, originating in the Middle (or Lower) Devonian, there was a continuous transformation into new subgenera and species throughout the Mesozoic, and even in the Tertiary and in the existing species there are some new phylogenetic acquisitions (Quenstedt, 1930).

A second possible interpretation of explosive phases is provided by the assumption of large steps of mutation (macro-mutation). Schindewolf (1936, 1950 *a,b*), especially favored the opinion that macro-mutations during early stages of ontogeny are decisive factors causing the sudden appearance of many new groups and structural types. In the plant kingdom, mutations which changed characteristics typical of families or orders have indeed been observed (compare the interesting results of Burgeff, 1941, on *Marchantia*; of Stubbe and Von Wettstein, 1941, on *Antirrhinum*; and of Schwanitz, 1955, on *Linaria*). Occasionally such strong alterations of single characters are tolerable, as the plant organism does not vary as much physiologically in the species as the animal. Seen as a whole, however, the major trends of plant phylogeny are probably caused by processes of ordinary mutation and selection, as has been demonstrated by Stebbins in his excellent book on plant evolution (1950).

In the animal kingdom, numerous macro-mutations have also been found and analyzed genetically. In almost every case, however, they proved to be homozygous lethals, or so markedly decreased in their fertility and viability that they were unsuited for successful competition. Examples of this kind, for instance, are supplied by the extreme mutants in *Drosophila*, like the four-winged tetraptera, glass, bithorax, or those with crippled wings like the curled, vestigial, and so forth.

The disappearance due to mutation of the second row of small incisors in the upper jaw of the rabbit (fully or partly dominant: Nachtsheim, 1936) seems to be a macro-mutation, as here a character is concerned which

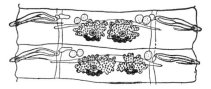

FIGURE 19. Two proglot-
tids of the tapeworm *Diplogonoporus grandis* (from man).
(After Fuhrmann.)

FIGURE 20. Monster proglottids of a
Taenia species from a heron, showing a
duplication of genitalia. (After Fuhrmann.)

distinguishes a whole order (that of the Lagomorpha, the former Duplici-dentata). One should, however, keep in mind that only one character is changed, and that this character is not a very essential one as far as biology and embryology are concerned. Moreover, this character is by no means the essential distinguishing trait in the order, and has been chosen only for convenience. Characters probably more typical of the order are the type of jaw articulation, the foramina optica, the calcaneus articulation, the spiral fold of the cecum, and so on. Speciation by macro-mutation could probably be assumed in some very rare cases, as in the tapeworm *Diplogonoporus*, the anatomical structure of which resembles that of the well-known *Diphyllo-bothrium* (*Bothriocephalus*), but each proglottis is laterally enlarged to about double its size and contains a double set of genitalia (Figure 19). The inter-pretation of this case as an example of macro-mutation is suggested by the fact that occasionally such duplications of genitalia can be observed in terato-logical specimens of tapeworms, viz., in a species of *Taenia* studied by Fuhrmann (1931) (Figure 20). Quite correctly, the author pointed out that due to the self-fertilization of the proglottids the chance of preservation is in-creased for such mutations. Fuhrmann also drew attention to the sudden and possibly mutative disappearance of the vaginal opening in some genera (*Acoleus, Gyrocoelia, Diplophallus*, and others). One more case of macro-mutation might possibly be inferred from the flagellate benus *Lamblia*, in which there are two nuclei and a bilateral symmetry of flagella.

If examples like these are to prove of major importance in evolution, they should occur more frequently and should not be limited to duplications only. Hence, up to now macro-mutation has not been sufficiently proved to represent the initial step of transspecific evolution, though among zoologists this has been suggested, especially by Plate (1933), Remane (1939), and R. Goldschmidt (1940, 1953). In stating this, of course, I do not wish to deny the possibility that some day further evidence of the evolutionary effects of macro-mutation may come to light. Homoeotic mutants can change a whole pattern, sometimes without too much reduction of the viability (compare Stern, 1954, on sex combs in *Drosophila*). In rare cases even 'hopeful monsters' (R. Goldschmidt, 1951) may be the initial stage and cause of speciation. In normal transspecific evolution, however, the normal type of mutation seems to be the causal factor (compare Chapter 6, E). Finally, one should not forget that the effects of temperature and short-wave radiation can result only in an increased rate of mutation and cannot cause larger mutational steps.

To the paleontologist macro-mutations are often suggested by the fact that in a line of successive geological horizons saltatory deviations – Waagen's mutations – are found. As we have seen in Chapter 3, p. 18, even in those cases without gaps in the sedimentation order, these 'saltations' are probably due to horizontal shifts of geographic races or closely related species.

For an understanding of phases of explosive radiation it is not at all necessary to assume an accumulation of macro-mutations or an increase of mutation rates, as normal mutation provides sufficiently diverse variants at any time, a fact which has already been demonstrated in Chapter 2 of this book. It is sufficient to assume an intensified selection during certain times, in most cases lasting for many million years. There are several possible modes by which selection could be accelerated.

Natural selection will become more effective especially in those cases where the geographic range of an animal type is expanded, so that the type is exposed to new conditions of selection. Hence, after the colonization of a new region providing various new habitats not yet occupied by competing animals, the colonizing animal type will tend to a rapid transformation. This process becomes especially clear when oceanic islands are colonized by continental faunas. So, for example, the Hawaiian fauna originated from such accidental continental immigrants. It is probable that the endemic Hawaiian bird family of Drepanididae (honey creepers) originated from species of a North American family of Oscines, which apparently came to the islands millions of years ago. Some of the 22 species constituting nine genera show only slight special differentiation; others have become adapted to quite diverse feeding habits, so that the family looks unusually heterogeneous as far as shape and structure of bill and tongue are concerned (Figure 21). This radiation of Drepanididae by adaptation to very different modes of nutrition offered in the new habitats without competition has been convincingly discussed, first by Mordvilko (1937), and later in a fine monograph by Amadon (1950), and a paper by Baldwin (1953). There are forms with a strong bill, living on seeds and fruits, such as the species of *Psittacirostris* and *Ciridops nana*, while others, such as *Loxops* and *Hemignathus procerus*, live mainly on insects and show a pointed, long, or curved bill. *Hemignathus*, with a long, curved upper and short lower beak, became a specialist in living on insects taken from clefts in the bark of trees. *Drepanis, Vestiaria, Himatione,* and *Palmeria* (and also *Hemignathus*) developed a tubular tongue with a brush-like end in addition to the long beak, and hence they feed mainly on nectar and

FIGURE 21. Drepanididae from the Hawaiian Islands.
1. *Drepanis pacifica;* 2. *Drepanis funerea;* 4. *Himatione sanguinea;* 5. *Loxops virens;* 6. *Loxops coccinea;* 7. *Hemignathus obscurus;* 8. *Hemignathus lucidus;* 9. *Psittirostra psittacea.* (After Porsch and Mordvilko, 1937.)

small insects. As to their habit of visiting flowers, there is only one competitor on each island: a species of Melliphagidae, genus *Moho*, which probably did not reach the islands until the Drepanididae had evolved their various genera. On the island of Hawaii, one more rival of flower birds is to be found: a rare species of *Chaetophila*. In other Drepanididae, competing forms were few, as among small birds there are only two thrushes (*Phaeornis,* one of which is limited to the island of Kauai) and three flycatchers (*Chasiempis*). Hence, strongly differing habitats and feeding patterns led to an 'explosive' splitting of an ancestral type resembling the Coerebidae, while on the American continent no such diversification was likely to occur, as the various habitats had already been occupied by other families.

A very similar splitting into new forms occurred in most Hawaiian insects, 82 percent of which are endemic. (The total number of species is 4,620.) Among the beetles, for instance, there are 204 endemic species of Carabidae in 34 endemic genera, besides six nonendemic species in five nonendemic genera. Buxton (1938), from whom these data are quoted, emphasized that the great number of endemic types is due to the fact that 'many opportunities of evolving' existed and that 'many unoccupied niches' could still be found. A similar statement applies to the snail families Achatinellidae and Amastridae, which show an extraordinary number of diverse types limited to these islands.

Due to the initial lack of competition for habitats on the Galapagos Islands, there was a fairly strong diversification of endemic finches such as *Geospiza*, *Calamarhynchus*, *Certhidea*, and *Pinaroloxias*, though this immigration from the American continent is younger than that of the Hawaiian Drepanididae. In the case of the Galapagos finches, it is again remarkable that all species are separated ecologically, as they require different food or different habitats (compare the thorough study by Lack, 1947).

The bodies of new animal types provide new habitats for a number of parasites. Hence, an explosive phase of evolution of the hosts is often followed by a rapid splitting of parasitic types. When during the Cretaceous and the Early Tertiary an intense transformation of many families and orders of birds and mammals took place, numerous new and isolated habitats with special environments were provided for intestinal parasites. Among the tapeworms (Cestodes), new families such as the Dilepididae (48 genera) and the Hymenolepididae (12 genera), which live in warm-blooded animals only, originated. Many more genera from other families became specialized to inhabit homoiothermal groups, and we now know about 135 genera of cestodes parasitizing birds. The greater part of these genera is confined to single bird orders, and in each genus a corresponding splitting into species took place (data after Fuhrmann, 1932). The effect of peculiarities of the habitats providing special conditions of selection and isolation may be inferred from the fact that 15 of these parasitic species live exclusively in Charadriiformes, and five in Ardeiformes only, five are limited to Accipitriformes, and 15 will be found only in Passeriformes. After species and genera had originated under the influence of isolation, a migration from one bird group to the other became possible, and today there are numerous types of cestodes in one and the same bird species. (In the snipe *Gallinago gallinago* 15 species in five genera have been found: five *Anomotaenia*, one *Echinocotyle*, three *Haploparaxis*, two *Hymenolepis*, and four *Paricterotaenia*.)

In other groups of parasites an equally intense splitting into new types was rendered possible by the rapid evolution in birds. In trematodes, the family Echinostomatidae developed numerous new genera, many of which parasitize birds only (Mendheim, 1943). From the distribution of the species of a genus one can still evaluate the extent to which isolation and selection were effective. So, for instance, the species of the genus *Petasiger* are distributed thus: one species (*exaeratus*) in darters (*Anhinga*), one in cormorants (*Phalacrocorax*), three in grebes (*Podiceps*), and one each in the loon (*Colymbus*), the domestic duck (*coronatus*), and the canary (*nitidus*).

A similar situation prevails among Acanthocephala, of which 14 genera are confined exclusively to birds and 12 to mammals (Meyer, 1932), and we have similar data for Nematoda. So, nearly all groups of worms living in vertebrates, and having perhaps turned into parasites when the fishes originated during the Silurian, reveal a differentiation into numerous new families and genera following the radiation period of homoiotherms during the Cretaceous and the Early Tertiary. As a consequence, the intestine of almost every water bird is inhabited by quite a number of specific parasites.

As new animals represent new habitats to be occupied by parasites, each period of explosive radiation in the hosts tends to be followed by a period of similar radiation in the parasites. And as each host may be inhabited by many parasites, often belonging to the same group, the radiation period in the parasites is often multiplied in intensity. This effect may also be seen from the fact that the percentage of parasites among the total number of animal species is extraordinarily great, amounting to about 25 percent (10,000 to 40,300 in the German fauna (Arndt, 1940): and a similar percentage will apply to the animal kingdom as a whole.

We now have to ask whether some of the periods of explosive radiation mentioned above and given in Tables 4–7 could have been caused by the challenge of unoccupied habitats. In certain cases this is quite possible, and some paleontologists pointed out this probability long ago (e.g. Gaudry, 1896, p. 49). Thus, we must not overlook the fact that in the Upper Devonian no terrestrial fauna originated until the terrestrial cryptogamic flora had emerged. The change from aquatic to terrestrial life, requiring so many new adaptations, led to a simultaneous and rapid transformation in many animal groups. It is for this reason that in the Carboniferous 12 different orders of insects suddenly appeared, besides numerous myriapods, scorpions, and spiders. At the same time, the first terrestrial snails originated – both the terrestrial Prosobranchia (Cyclophoridae) and the basommatophorous Pulmonata (Ellobiidae) – and various families of amphibious or terrestrial Stegocephalia and the first reptiles (Cotylosauria and Pelycosauria) appeared.

A similar period of intense transformation in many branches of the terrestrial fauna began during the Late Mesozoic, as in the preceding epochs, mainly during the Cretaceous, the phanerogams had evolved and gained phylogenetic superiority in a surprisingly short time, due probably to their efficient pollination and dispersal mechanisms. This new flora provided many new habitats with markedly differing food resources (fruits, seeds, softer flowers, nectar, and so forth), as well as a number of living and hiding places (more herbs and shrubs). Hence, the herbivorous terrestrial Pulmonata underwent an intense adaptive radiation (more than 20 families during the Early

Tertiary) during the Late Cretaceous and the earliest Tertiary. Furthermore, the number of insect types increased, and there was a rapid evolution in warm-blooded animals depending upon plant and insect food; six orders of birds and three of mammals originated during the Cretaceous, and 16 more orders of birds and 19 of mammals (with 102 new families and many genera) during the Early Tertiary (compare also Walther, 1908, p. 475; Scott, 1930; and Stromer, 1944).

There are as yet no reliable data concerning the factors that caused the extremely rapid transformation of animal groups during the Early Silurian (compare Tables 4–7). Walther (1908) called attention to the numerous oceanic transgressions, but more important than this expansion of aquatic habitats seems to be the increase of lime supply, as the new animal groups originating during this period show definitely stronger skeletons of calcium components. This may be seen in Silurian corals, bryozoans, mussels, cephalopods, echinoderms, and fishes, as demonstrated by Hennig (1932). On the contrary, many Cambrian groups are distinguished by skeletons poor in calcium compounds and rich in horny substances (Dacqué, 1935). The possibility of using calcareous elements in the structure of the skeletons, of course, opened new possibilities of body organization, which could lead to the appearance of new types. Possibly an important role was also played by an increase in the numbers of predators, as the first vertebrates are represented by armor-plated placoderms and ostracoderms. So there are various possible interpretations of the general period of explosive radiation during the Early Silurian, and it is up to the paleontologist to decide by future analysis which environmental factors of this epoch were most important.

An instructive example – though limited to a much smaller area – of the rapid evolution of a new fauna due to changes of the environment is provided by the history of the Sarmatian Sea, extending from the Vienna plains to the Caspian Sea during the Miocene. During the Late Miocene, this sea had become separated from the Mediterranean and turned into a brackish lake. Consequently, the marine animals, such as corals, echinoderms, cephalopods, and brachiopods, vanished, and in the new environment a rapid differentiation of a few genera into numerous species occurred. Hence, a quite new fauna appears in the Early Pliocene (Pontian stage) comprising snails (*Valenciennesia, Melanopsis*) and mussels (*Monodacna, Adacna, Dreissenomya, Congeria*) as the typical forms.

In a similar manner, the formation of new deserts, new mountain and alpine zones, new lakes from craters and glaciers, and new seas in the course of oceanic invasions provided new habitats with new types of vegetation; and the colonization of such new habitats led to increased animal transformation (Quenstedt, 1929; Ehrenberg, 1939). It is not true, however, that each orogenic phase of earth history was accompanied by a phase of explosive radiation, as was maintained by Schuchert, Kober, Mathew (1915), Szalay (1936), and others (compare the critical reviews of this opinion by Quenstedt, 1929, and Schindewolf, 1937). Changes of topography are not necessarily effective in animal transformation, as the bulk of animal types evades such changes by horizontal shifts of the range inhabited, but the opening of new habitats more or less free from competitors is always important. It is for this reason that the phases of animal radiation mentioned above and depending on the type of vegetation did not occur until a fairly long period (often a whole geological epoch) of plant radiation had elapsed. This fact may be seen from the extensive splitting of the tetrapods during the Late Carboniferous after the bulk of the terrestrial plant types had originated in the Early Carboniferous, from the emergence of mammals in the Jurassic, after the ferns, cycads, and conifers had evolved, and from the explosive phase in many orders of birds and mammals during the Early Tertiary, after the angiosperms had developed in the Cretaceous (compare Von Huene, 1943).

But even if habitats and niches have already been occupied by animals, they can be reconquered by types with stronger competitive superiorities. In such cases the situation is the same as if the habitats were not occupied at all, and a period of strong radiation can follow. So, for example, during the Paleocene and Eocene numerous genera and species of lemurs (Lemuriformes) and tarsiers (Tarsiiformes) appeared as arboreal types in the northern continents, and a period of extensive radiation lasted throughout the whole Eocene. At the end of the Eocene, most types of lemurs and tarsiers disappear in the northern continents, and in the Oligocene they begin to be replaced by monkeys and apes. Hence the assumption is obvious that the lemurs and tarsiers were displaced by types of a higher evolutionary level due especially to their improved brains. Today lemurs are mainly confined to Madagascar, where no competition from monkeys and apes is possible. The remainder of lemurs outside Madagascar are represented by typical relict groups such as *Galago*, *Perodictius*, *Nycticebus*, and *Tarsius*, which have avoided any competition with monkeys and apes by a nocturnal way of living. The period of radiation of prosimians during the Paleocene was probably rendered possible by the superiority of the lemurs and tarsiers over the Mesozoic arboreal mammals belonging to the more primitive types – especially as to the brain – of insectivores and marsupials.

It seems probable that the phase of explosive radiation in the placental mammals (Monodelphia) during the Early Tertiary must also be attributed to the fact that they were able to conquer habitats previously occupied by other animals. The preceding animal types of the Jurassic and Cretaceous, the Multituberculata, Triconodonta, Pantotheria, and Marsupialia, showed a wide range of structural organizations and ways of living. According to their teeth they were carnivorous, insectivorous, and rodent types, and as such they must have inhabited quite diverse habitats and niches. The new Monodelphia were definitely superior in their brain performance and probably also in their care of the young, so that they could easily conquer the habitats previously occupied by inferior types. The modern fauna of the Australian region offers examples of climbing, jumping, running, burrowing, herbivorous, insectivorous, carnivorous, and other types of marsupials. As soon as man introduced higher placental mammals, most of these types were quickly displaced. Correspondingly, the primitive Carboniferous and Permian insect orders were displaced by the rapidly evolving orders of Recent insects, which were superior in their aerotechnical mechanisms and abilities (e.g. solid structure of thorax, more efficient type of wings).

Finally, a period of increased transformation and splitting can be caused by the change of a single vital factor in an otherwise constant habitat, e.g. by new sources of food requiring new types of feeding mechanisms and structures. This may be seen from Abel's findings on the phylogeny of whales, mentioned in Chapter 5, in which the change from feeding on larger animals to eating fishes, or cephalopods, or, finally, planktonic animals was always followed by a relatively short phase of explosive evolution.

Summarizing our findings, we may state that the evolution of new animal groups is often inaugurated by a phase of explosive radiation, but that there are also numerous cases in which a period of rapid splitting occurs after a long and steady phylogeny during several geological epochs. Besides this there are examples of very long periods of extensive radiation. Sometimes periods of strong radiation are repeated several times. Hence, there is no fixed evolutionary pattern ('Gestaltungsplan') in the differentiation of higher categories and, correspondingly, there is no need to assume unknown autonomous factors of evolution. The essential factor in the causation of such periods of explosive radiation is not an increase of the rate of mutation or an accumulation of macro-mutations, but an acceleration of differentiation, brought about by a temporary intensification of selection due to environmental changes, e.g. by new types of vegetation or food resources, or due to the colonization of new ranges with habitats unoccupied or inhabited by types inferior in competition.

B. THE PHASE OF SPECIALIZATION

I. Gradual Loss of Evolutionary Intensity

Periods of rapid adaptive radiation are regularly followed by a relatively long phase of more or less stationary phylogeny during which many of the initial 'random' types die out, while others show a slow, but gradually increasing, adaptation to various aspects of the environment. As early as 1866, Haeckel had recognized this essential period of evolution and termed it 'Blühezeit' (acme, 'flowering season'). He emphasized (II, p. 322) that during this period there is 'not a quantitative, but a qualitative perfection and a versatile adaptation to a wide range of different conditions of existence'. And he continues: 'The genus, family, order, class, and the whole phylum, being at its peak of development, no longer grows in number of types but in perfection' (p. 368). In 1899 Rosa formulated his 'law of decreasing variability', which he believed to be caused by special factors acting within the organisms. On the other side, Plate (1904) thought that there is no such decreasing variability, but pointed out that only the evolutionary effects of variation were gradually decreasing during the periods concerned. Abel (1929, p. 372) argued in favor of Rosa's Law, and tried to interpret it as a result of progressive limitation of possible alterations of the organism. Beurlen (1937) thought that the phase of specialization is a regular part of an autonomous pattern of evolution.

For our own critical review of the phase of specialization, it will be necessary to analyze the phylogeny of some higher categories as to whether, when, and where specializations appeared. From Tables 4–7 it becomes clear that in genera, families, orders, and classes, the periods of explosive evolution are followed by phases during which the transformation is relatively slow and, in most cases, becomes gradually slower. This may also be seen from Figure 22, in which the upper row of each graph represents the number of new sub-categories originated per geological epoch. Looking at these, one might think that during the phases of specialization there is a more or less marked decrease in the number of types per group. This, however, is not true, as may be seen from the lower row of each graph. Here, the total number of subcategories per geological period has been drawn up, the new types per period represented by the white sections and the old types by the dotted sections. From these sum totals of types per epoch it is clearly to be seen that in spite of a decrease in the production of new forms most animal groups grow in their total number of types. Hence, we may well accept Haeckel's term 'acme' in the sense of a phylogenetic 'flowering season'. In groups approaching extinction, such as the Gigantostraca (see Graph 6 in Figure 22), this season is, of course, short and is succeeded by a period of general decrease in the number of sub-categories. The same applies to animal groups of which only a few types escaped extinction, such as the families of brachiopods and crinoids.

If we wish to understand such phases of specialization as an outcome of factors and processes which have been well established by zoology, we will have to take into account the successive decreases of the rates of evolution on one hand and the gradually increasing specialization on the other. Generally there is no doubt about the fact that a progressive adaptation takes place in correlation with environmental conditions. Until maximum adaptation is achieved, a great number of mutational steps and stages of selection (especially of modifying genes) must have passed, requiring shorter or longer periods of time. As many terrestrial habitats were exposed to a variety of changes of climate, surface structure, and vegetation in the course of the history of the earth, in many animal groups the process of specialization will never come to an end. Such 'series of adaptations' have been demonstrated in quite a number of examples by paleontologists. Some of these examples include the successive formation of grinding teeth with complex enamel folds in various orders of hoofed animals when these types turned to feeding entirely on plants and especially on grass. This is particularly clear in the phylogeny of horses and of elephants, in which a parallel reduction of canine teeth and transformation of premolars into true grinding teeth took place. Further examples of successive adaptations to new environments are pro-vided by the gradual reduction of hip bones in Sirenia from *Eotherium* of the middle Eocene to the existing *Halicore*, by the fusion of dorsal and ventral

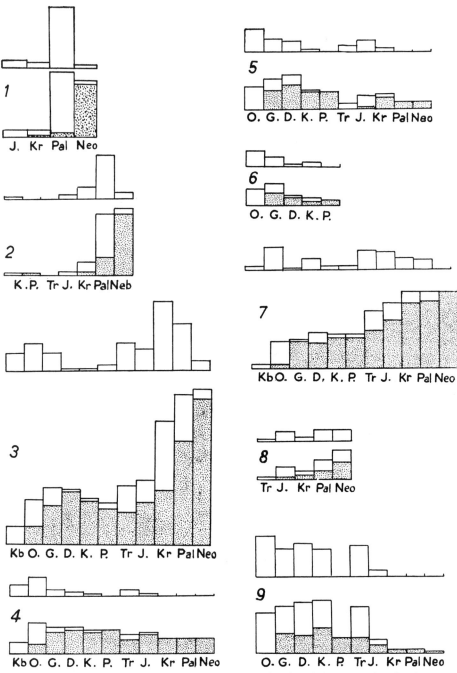

FIGURE 22. Diagram of the number of newly formed and persisting categories of various animal groups during the geological epochs. O=Ordovician, G=Gotlandian, D=Devonian, K=Carboniferous, P=Permian, Tr=Triassic, J=Jurassic, Kr=Cretaceous, Pal=Palaeogene, Neo=Neogene. 1. Mammalian orders. 2. Families of terrestrial Pulmonata. 3. Families of marine Prosobranchia. 4. Families of Brachiopoda. 5. Families of Crinoidea. 6. Families of Gigantostraca. 7. Families of Lamellibranchia. 8. Genera of Fissurellidae. 9. Genera of Pleurotomariidae.

fins with the tailfins in Dipnoi from the Lower Devonian *Dipterus* to *Uronemus* from the Lower Carboniferous, and so forth. Quite instructive, too, are the series of successive adaptations to a burrowing way of life in the mussels of the genera *Solenomya* and *Nucula* (Quenstedt, 1930), already referred to above. In *Solenomya*, a series of alterations resulted in a barrel-shaped body, which apparently was of optimum efficiency in burrowing, as this shape has remained unchanged since the Carboniferous. In *Nucula*, on the contrary, first the shape of the shell was changed, then the vortex was tilted backwards, and finally the shell was rendered more solid by transverse ridges. Hence, in this case there has been a slow but continual change of characters until the present. This last example shows that the time required for optimum adaptation may vary greatly.

In some cases, the trends of adaptations initiated during the first phase of radiation do not lead to optimum anatomical structures and patterns, and may be superseded by types of a superior construction. Such cases have been termed 'fehlgeschlagene Anpassungen' ('adaptive mistakes') by Abel. A well-known example is provided by the different construction of carnassial teeth ('Brechscherengebiss') suited for cracking bones. There are several kinds of adaptation in the primitive Eocene carnivores (Creodonta): in *Sinopa* and *Hyaenodon* the second upper and the third lower molar were enlarged and developed a sharp cutting edge. In *Oxyaena* the first upper and the second lower molar were enlarged. In both cases, however, these specialized teeth could not be further enlarged because their position in the jaws was too far to the rear. Therefore, animals showing this type of dentition disappeared very soon from the phylogenetic scene, and only those remained and continued to evolve in which the fourth upper premolar and the first lower molar specialized as cracking scissors. As these teeth were located farther forward, there was enough space in the mouth of the animals of this type to allow them to develop into larger and more efficient tools, as can be seen in our recent carnivores. One more example of such 'adaptive mistakes', mentioned by Osborn (1906), refers to the molars of Titanotheriidae from the Early Tertiary. As an adaptation to feeding on plants, the molars of these animals developed a zigzag-shaped pattern of enamel folds on the outward side of the crown, while on the inward side the blunt cusps remained. These, however, were definitely lower and could not efficiently be used until the higher zigzag folds were worn down to the heights of the cusps. In other orders of hoofed animals, however, the molars were of equal height on the outward and inward sides, and hence seem to have been superior in efficiency, which was possibly one reason for the extinction of the Titanotheriidae during the Oligocene after a relatively short phylogeny.

Due to similar causes, in almost any animal group many of the initial subcategories will die out, while only a few will continue their course of evolution. This is shown in Figure 23: only two subfamilies of rhinoceroses, the tapirs and the horses, have avoided extinction until today, while all other families and subfamilies have faded away. Such regularities had already been

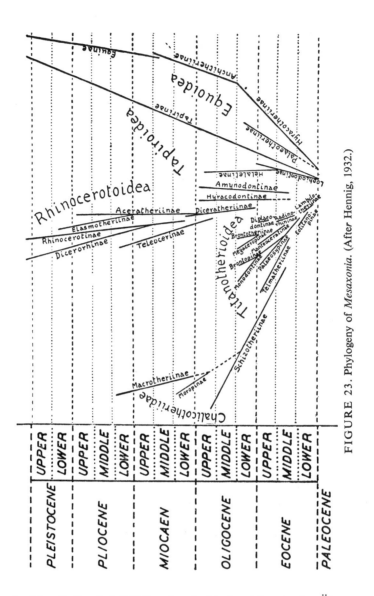

FIGURE 23. Phylogeny of *Mesaxonia*. (After Hennig, 1932.)

recognized by O. Jaeckel (1902), who, in his inspiring study, *Über verschiedene Wege phylogenetischer Entwicklung*, wrote the following sentences (p. 58):

The greatest divergences appear during phylogenetic youth, and they are indeed typical of this stage. From numerous experiments only those types are preserved that achieved harmonious correlation of their parts. On a given basis, however, only a few plans of organization can usually be realized, and hence the large groups usually are few in number compared to the many initial experimental forms.

Schäffer (1952) determined a morphologic evolutionary rate curve for suprageneric characters in dipnoans and coelacanths. These rates increased rapidly during the Late Devonian and Early Carboniferous, since which they have remained low and constant up to the present.

There are still further reasons for the successive decrease of rates of differentiation. From Figure 22 we have seen that in many cases the reduction in the number of new subcategories coincides with an increase of the total diversity of types. This means that with an increasing number of types the competition becomes gradually more severe. More and more habitats will be occupied, and a shift into a new adaptive situation becomes difficult. How far complete occupation of the available habitats by animal types took place is hard to establish. There is strong competition, for instance, between mammals and birds, occupying nesting holes in trees, and also among all birds feeding on nectar, a fact which may be of vital importance in regions like savannahs or steppes, providing only few nectar-yielding blossoms. In most cases, such mutual relations between the members of a biocoenosis need further ecological analysis; however, we can quote certain facts which will allow an evaluation of the influence of complete occupation of a habitat by different species.

Apparently most habitats have been completely occupied in the course of time, so that few opportunities are left to 'new-comers'. This may be inferred from the fact that similar habitats in quite different regions or continents are inhabited by about the same percentages of large, medium-sized, and small animals of a class or an order, though the respective species belong to quite different families or genera.

To illustrate this statement, the following data show the percentage of small, medium-sized, and large birds (not including birds of prey) inhabiting the central European forests (data after C. Zimmer and B. Rensch, 1928), the rain forests of northeastern Brazil (after Snethlage, 1928) and of Kenya in East Africa (after Granvik, 1923), and the rain forests of Flores in the Lesser Sunda Islands (after B. Rensch, 1931). In these tables, birds like finches, tits, and so on have been listed as 'small', starlings, thrushes, and similar types as 'medium', and the remainder as 'large'.

I. CENTRAL EUROPEAN FOREST

Small	Medium	Large
6 Fringillidae	2 Corvidae	1 Picidae
2 Certhiidae	1 Oriolidae	2 Columbidae
1 Sittidae	3 Turdidae	2 Tetraonidae
9 Paridae	4 Picidae	1 Ciconiidae
2 Turdidae	1 Cuclidae	
2 Sylviidae	1 Charadriidae	
5 Muscicapidae	1 Tetraonidae	
1 Troglodytidae		
1 Picidae		
29 species (60%)	13 species (27%)	6 species (13%)

117

II. TROPICAL RAIN FOREST OF NORTHEASTERN BRAZIL

Small	Medium	Large
9 Tanagridae	1 Icteridae	3 Tinamidae
1 Coerebidae	4 Cotingidae	4 Cracidae
1 Vireonidae	13 Dendrocolaptidae	1 Psophiidae
2 Sylviidae	5 Picidae	3 Columbidae
15 Tyrannidae	1 Galbulidae	3 Psittacidae
5 Pipridae	1 Cuculidae	2 Rhamphastidae
16 Formicariidae	2 Trogonidae	
9 Trochilidae	1 Caprimulgidae	
	1 Momotidae	
	1 Alcedinidae	
	2 Psittacidae	
58 species (55%)	32 species (30%)	16 species (15%)

III. TROPICAL RAIN FOREST OF KENYA, EAST AFRICA

Small	Medium	Large
5 Muscicapidae	1 Campephagidae	2 Turacidae
1 Sylviidae	3 Sturnidae	4 Columbidae
6 Timeliidae	1 Oriolidae	1 Psittacidae
2 Laniidae	1 Capitonidae	1 Bucerotidae
1 Zosteropidae	2 Picidae	
1 Campephagidae	1 Trogonidae	
5 Pycnonotidae	2 Upupidae	
1 Ploceidae		
2 Capitonidae		
2 Cuculidae		
26 species (58%)	11 species (24%)	8 species (18%)

IV. TROPICAL RAIN FOREST OF WEST FLORES, LESSER SUNDA ISLANDS

Small	Medium	Large
6 Muscicapidae	1 Sturnidae	1 Phasianidae
2 Timeliidae	1 Oriolidae	3 Columbidae
3 Sylviidae	1 Dicruridae	2 Psittacidae
2 Laniidae	2 Turdidae	
1 Paridae	2 Campephagidae	
1 Ploceidae	1 Pittidae	
2 Melliphagidae	1 Picidae	
1 Nectariniidae	2 Cuculidae	
3 Dicaeidae	1 Alcedinidae	
5 Zosteropidae	1 Loriidae	
26 species (58%)	13 species (29%)	6 species (13%)

In spite of the great differences in bird families inhabiting the various geographic regions mentioned, the percentage of small types does not vary greatly, ranging from 55 to 60 percent, and the same applies to medium-sized birds (24 to 30 percent) and large types (13 to 18 percent). This means that such forests in general provide niches for a rather constant percentage of birds of a certain size. In Europe, the bulk of 'small' birds inhabiting the forests is made up of Fringillidae, Paridae, and Muscicapidae; in Brazil of Tanagridae, Tyrannidae, Formicariidae, and Trochilidae; and in East Africa of Muscicapidae, Timaliidae, and Pycnonotidae, while on the island of Flores the small birds are mainly Muscicapidae and Zosteropidae. From these considerations it becomes clear that with the gradual approach of such final percentages the range of evolutionary possibilities is gradually decreased.

The analysis of other animal groups produces similar results. The layer of decaying leaves in a forest, for instance, provides a suitable habitat for extremely small terrestrial snails, as in this layer the humidity remains more or less constant. In central Europe, these small snails invariably belong to the families of Pupillidae (*Vertigo*) and Zonitidae (*Vitrea, Euconulus, Retinella*), in Brazil to the Endodontidae (*Endodonta*) and Systrophiidae (*Systrophia, Microhappia*), while in central Africa this surface layer is inhabited by Enneidae (*Gulella*) and Stenogyridae (*Opeas, Pseudopeas*). On the island of Flores, finally, these humus layers are the habitats of minute snails belonging mainly to the families of Ariophantidae (*Kaliella, Microcystina, Durgellina*, and others) and Cyclophoridae (*Diplommatina, Palaina, Arinia*). All the forests mentioned also provide habitats for medium-sized and large snails of quite different families, but in these types the percentage of size categories is far less constant than in the small types of the bottom layer. This may possibly be attributed to the fact that the larger snails are mainly herbivorous and inhabit small ranges only. There is hardly any serious competition among the individuals and the species, and other animals do not compete for the habitats of these snails (though a great many higher animals feed on them.) Hence, it is quite possible that in terrestrial snails the number of species, already amounting to some ten thousand, will be increased in future. This is also suggested by the formation of new families during the Later Tertiary (see diagram on p. 114). A suitable habitat for further transformation and splitting of snail genera could be provided by the treetops, still to be occupied in many tropical countries. In snails that have already conquered this habitat, an extremely rich speciation has occurred, as in species of *Cochlostyla* and its allies of the Philippines, in species of *Amphidromus, Xesta*, and *Asperitas* of the Malay Archipelago, in species of *Papuina* from New Guinea and Melanesia, in Achatinellidae and Amastridae of Hawaii, and in Bulimulidae from South America. This conquest of a new habitat, however, will hardly result in the formation of basically new structural types, as terrestrial snails in general have passed their period of flourishing radiation and their present stage is the phase of specialization. Insects, on the contrary, apparently have already occupied all suitable niches in consequence of their greater mobility, their quicker succession of generations, and their higher rate of reproduction.

Such a successive occupation of habitats may also be studied in the remains of the geological past. Ever since the existence of terrestrial faunas began there have been giant animals (the size of bears or larger) besides the small and medium-sized types. At first, these large animals of the fauna were represented by amphibians (*Thinopus*, etc., in the Devonian; *Anthracosaurus*, etc., in the Carboniferous; *Eryops*, etc., in the Permian; *Anaschisma*, etc., in the Triassic), then by reptiles (more than 100 genera of dinosaurs in the Triassic, Jurassic, and Cretaceous). After their extinction the giant dinosaurs were soon replaced by mammals of enormous body size. In the Eocene, the mammals had already evolved more than 20 genera of giant Titanotheriidae and besides these there originated the giant types of Amblypoda (*Uintatherium Loxolophodon*, and others), Notoungulata (*Albertogaudrya, Scabellia*, and others), Pyrotheria, and Proboscidea (Barytheriidae, Moeritheriidae). In the course of the Later Tertiary, these types were superseded by the recent giant types of higher mammals belonging to orders of artiodactyls, horses, rhinoceroses, tapirs, hippopotamuses, elephants, and giant sloths (gravigrads with numerous genera). Hence, it is evident that the suitable habitats were conquered again and again by giant animal types superior to their predecessors, which from then on entered their phase of specialization.

Besides numerous herbivores there have always been carnivores; that is to say, the habitats of herbivorous animals were also suited for a limited number of carnivorous animals. So, among Permian reptiles, there were specialized herbivores, such as *Pareiasaurus* and *Casea*, and typical carnivores, such as *Dimetrodon* and *Edaphosaurus*. Both these types of adaptation continued to exist throughout the whole Mesozoic. In this context it is essential to note that closely parallel to the giant herbivorous dinosaurs of the Cretaceous, such as *Brachiosaurus, Diplodocus, Iguanodon, Plateosaurus, Apatiosaurus*, and *Triceratops* (with horns and armored collar for protection), the largest carnivore of all ages, *Tyrannosaurus*, developed (Osborn, 1930). During the Tertiary, various families of carnivores, such as cats, dogs, bears, and martens, evolved in addition to the more numerous herbivorous animal groups. And here again, the largest types of herbivores, such as antelopes, cows, deer, rhinoceroses, pigs, and elephants, developed horns, antlers, and enlarged teeth as a means of protection against the giant types of carnivores, such as lions, tigers, panthers, ibexes, pumas, bears, and wolves (which certainly is further evidence of the directedness of evolution). Correspondingly, there is a certain number of birds of prey inhabiting the same biotopes as the birds listed in the tables above, this number comprising about two to four species in the forests of each continent. These birds of prey are represented by the hawk (*Accipiter gentilis*), the sparrow hawk (*A. nisus*), the hobby (*Falco subbuteo*), and the small eagle (*Aquila pomarina*) in the central European forests; by *Leucopternis albicollis, Harpagus diodon*, and *Urubitinga urubitinga* in northeastern Brazil; by *Accipiter tachiro* and *Buteo augur* in East Africa; and by *Accipiter fasciatus* and *Hieraëtus fasciatus* on the island of Flores. From this fairly constant percentage of carnivorous animals in the populations of various geographic regions and various geological epochs, one may reasonably infer that the habitats were always 'saturated' with carnivores.

Charles Darwin thought that there was an instable state of balance in the number of organisms in most habitats, and this opinion is shared by many modern authors in their quantitative studies on biocoenoses. Friedrichs (1930), for example, states that:

> many species of a biocoenosis can indeed reach a labile state of balance and maintain it as long as there is no essential change of environmental conditions; in others, however, the numbers of individuals are far from average frequency, and hence there is only a tendency to reach equilibrium.

Similarly Thienemann (1940, p. 31) thinks that:

> such a habitat is inhabited by a number of individuals, not isolated from each other, but living side by side and tied together by biotic relations. Hence, they make up a biological community in which the individuals show a relatively constant numerical relation.

This 'relative constancy', according to Thienemann (1939, p. 14), must be considered as an 'equilibrium that is always attempted, but is never attained'. It is a 'dynamic harmony'. The experimental findings by Gause and co-workers (1934, 1935, 1936) on mixed cultures of various ciliates and mites support this opinion.

Finally, quantitative studies of sample areas show that the numbers of individuals are often large and rather constant, and we may assume that the habitats available and habitable have been occupied fairly completely. Schiermann (1930, 1934), for example, counted the numbers of breeding birds per square kilometer in the swampy forests of the Unterspreewald near Berlin and found that there was an average of 117·5 pairs of breeding birds (maximum figure in some places: 284) per square kilometer, main contributors to this figure being the chaffinch (*Fringilla coelebs*) with 9·2 pairs, the whitethroat (*Sylvia communis*) with 9·0 pairs, the great tit (*Parus major*) with 4·2 pairs, and the robin (*Erithacus rubecula*) with 4·0 pairs. In the drier pine forests south of Berlin, near Kunersdorf, only 106·8 breeding pairs per square kilometer were counted (maximum figure: 236), comprising 36·1 pairs of the chaffinch, 7·2 pairs of the crested tit (*Parus cirstatus*), 7·0 pairs of the tree pipit (*Anthus trivialis*), 6·6 pairs of the great tit (*Parus major*), 5·6 pairs of the willow warbler (*Phylloscopus trochilus*), and 5·5 pairs of the fieldfare (*Turdus pilaris*). In spite of the different species inhabiting these two types of forest, there is a conspicuous similarity of population numbers, which indicates that the population size of a certain habitat is limited by environmental factors such as amount of food, nesting sites, and so forth. It is especially interesting, in this context, to compare the respective numbers of birds in forests of different climatic regions. These figures prove to be generally similar, as is evident from the following data by Soveri (1940): in a pine forest in Finland (*Myrtillus*-type) there was an average of 122 pairs of breeding birds per square kilometer (maximum figure: 150), with the chaffinch amounting to 28 (32) pairs; in a deciduous forest (*Myrtillus*-type) 236 (266) pairs, and in dry pine and heath regions (*Calluna*-type) only 34 (42) pairs. For a proper evaluation of the significance of such figures, observations over a long period are, of course, necessary, but as yet these have been confined to single species and not extended to whole faunas.

However, studies on single species also show that the number of individuals per habitat is limited. Careful studies on the house wren (*Troglodytes aëdon*) by Kendeigh (1934) showed that during a period of 13 years, the number of pairs breeding in a range of 14 acres in Ohio remained nearly constant between 9 and 11. There was a minimum number of six pairs caused by the cold winter in 1926, but the figure had returned to its former value by 1929, and after this the number remained more or less constant, as may be seen from the following data:

1921:	9 breeding pairs		1928:	8 breeding pairs
1922:	9	,, ,,	1929:	11 ,, ,,
1923:	11	,, ,,	1930:	11 ,, ,,
1924:	9	,, ,,	1931:	9 ,, ,,
1925:	9	,, ,,	1932:	10 ,, ,,
1926:	6	,, ,,	1933:	14 ,, ,,
1927:	7	,, ,,		

Counts of the number of animals inhabiting a certain area of soil also show that, apparently, this habitat has been almost fully occupied. Franz (1941) in August counted the number of animals per square meter and down to 3 cm. below the surface of a meadow near the River Enns at 620 m. altitude. He listed 1,560,000 nematodes, 140,000 rotifers, 1,400 enchytraeids, 16 lumbricids, 880 snails, 56 millipedes, 8,920 mites, 1,280 ants, 168 beetles, 72 larvae of beetles and 192 of flies. An equal range in a meadow near Admont at 650 m. altitude was inhabited by 1,800,000 nematodes; a less fertile meadow (Nardetum) near Admont at 710 m. contained 600,000 nematodes; and an equally scanty pasturage (Kaiserau, at 1,160 m.) 500,000 nematodes per square meter. The similar numbers of inhabitants per area in corresponding habitats suggest that the soil, too, and its niches cannot admit more individuals. With regard to evolution this means that under normal natural conditions the habitats are more or less fully occupied by different groups of animals and there will be only limited speciation.

Here we must call attention to the fact that in Recent times some critical authors have denied any equilibrium in the biocoenoses. Elton (1930), for example, states:

'The balance of Nature' does not exist, and perhaps never has existed. The numbers of wild animals are constantly varying to a greater or lesser extent, and the variations are usually irregular in period and always irregular in amplitude.

A similar opinion was favored by Bodenheimer (1930). And, indeed, the introduction and frequently rapid spreading of foreign animal species bears evidence that some habitats are far from being fully occupied, as may be seen from the spreading of muskrat and pheasants in central Europe, or of the European starlings and sparrows in North America. It is true, however, that such spreadings are often accompanied by severe changes of the original biocoenoses. The Chinese woolly-handed crab (*Eriocheir*), for example, has strongly reduced the population of *Sphaerium corneum*, a small mussel inhabiting the harbor waters of Hamburg (Thiel, 1930), and in some parts of New Zealand and Australia, as is well known, the endemic types of biocoenoses have been almost completely destroyed by the introduction of numerous European animals.

Hence, we see that there are habitats which are more or less filled by inhabitants and others which are not yet saturated, and in close correlation with this situation phylogenetic specialization will proceed at slower or faster rates. In terrestrial habitats, the periodical and spatial fluctuations of ecological factors, especially of temperature, humidity, and vegetation, are definitely more marked than in aquatic biotopes. Hence, terrestrial habitats are not usually so saturated as aquatic habitats tend to be, and the processes of evolution are generally more rapid in terrestrial places. This is also evident from the fact that the number of species, genera, and families is greater in terrestrial than in aquatic habitats. After W. Arndt (1940), up to 1939 about 40,300 animal species inhabiting Germany and its coastal waters had become known to science, but only about 10 percent of these lived in water.

Summing up, we may state that the decrease of phylogenetic radiation in the phase of specialization may reasonably be interpreted by the assumption that the habitats available are gradually filled up with different species and that new adaptations become gradually rarer. Mutation and selection seem fully sufficient mechanisms to explain such phases of decreasing phylogenetic development, and we need not assume unknown autonomous factors causing a successive reduction of evolutionary rates.

II. Irreversibility

During the phase of specialization, quite frequently a phenomenon appears that was first fully recognized by Dollo (1893), who named it the 'Law of Phylogenetical Irreversibility', and that has since been often debated. According to Dollo, irreversibility means 'qu'un organisme ne peut retourner même partiellement, à un étàt antérieur, dejà réalisé dans la série de ses ancêtres'. As was later expounded by Abel (1911, compare 1929, p. 310), an organ which has been reduced in the course of phylogeny will never regain its original strength, and an organ which has disappeared will never reappear on the phylogenetic scene though it may be replaced by a similarly functioning but quite different anatomical structure. Abel, Dacqué (1935), Beurlen, and others apparently were convinced that a 'law' was really involved here, and Beurlen referred to it as the fundamental regularity of (phylogenetic) history. Other authors, however, have demonstrated a number of exceptions to this 'law', so that it seems preferable to refer to it as a 'rule' only. The most impressive example of irreversibility, stated by Dollo, is different anatomical origin of armored plates appearing twice in the phylogeny of some turtles. Normally these armored plates originated from bones, but when one group of these reptiles turned to living in the seas, the heavy plates of bone were reduced to a thin layer of bony material during the Jurassic and Cretaceous periods. When a line of this marine group was adapted once again to life in the coastal regions, armored plates again became a necessity, but this time they developed in the dorsal skin and above the vestigial old 'armor'. For further examples regarding this problem the reader is referred to the extensive discussions by Abel (1929), Dacqué (1935), and the references given by these authors.

There are unquestionably, however, exceptions to this rule. So, for example, the toothed whales, characterized by numerous isodont teeth, derived from

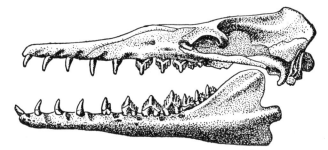

FIGURE 24. Skull of the primeval whale *Zeuglodon* from Alabama (Late Eocene). (After Remington-Kellogg from Abel, 1939.)

the Archaeoceti or primeval whales, which possessed heterodont teeth, as is to be seen in the triangular molars and premolars of *Protocetus* and *Zeuglodon* (Figure 24). Hence, it is evident that in Recent toothed whales the isodont type of dentition has developed a second time, as this type of teeth had originated for the first time in the reptile ancestors of mammals, in Cotylosauria. Another example is provided by the type of vertebral column, which became more and more differentiated in the course of evolution from amphibians to reptiles, developing neck, breast, pelvic, and tail parts, each with special vertebrae. This differentiation was abolished when in snakes and slow-worms (Anguidae, Amphisbaenidae) the limbs became gradually reduced, and the undifferentiated type of vertebral column appeared a second time.

Now, what are the biological reasons underlying the rule of irreversibility, and which factors cause the exceptions? Here the evidence of genetics is important: mutations are reversible. A mutant may mutate back to the ancestral form or to a form indistinguishable from the latter. In some cases, however, such reverse mutation does not bring about the original state of a gene or a gene complex, as after the mutation of a main gene its controlling function in the process concerned may be taken over by modifying genes (compare the summarizing review by H. J. Muller, 1939). Nevertheless it is evident from this genetical point of view that at least minor phylogenetic transformations are fully reversible, and there is good reason to assume that some exceptions to the rule of irreversibility are caused by such reverse mutations. In Recent horses, for example, sometimes a genuine atavism appears, in that small lateral digits with real hoofs on their ends develop (Figure 25). Such cases, not to be confused with embryologic duplications of the toe-bud, could tentatively be interpreted by the assumption that genetically the potentiality of three toes is fixed, but that the growth of the central, or third, digit is markedly accelerated by one or more genes. In the course of such rapid growth the lateral digits do not get enough organic material, as the third digit takes all or nearly all available. Hence, the growth and differentiation of the distal parts of the lateral digits are slowed down or even totally barred, and usually no hoofs will be found on them. If the genes controlling the accelerated growth of the third digit, or some modifying genes, are changed by mutation or by a reverse mutation, and their effect of increasing the growth of the third toe is prevented, the lateral digit will grow and differentiate just as the central digit does. In fact, in polydactyl horses the metacarpal and metatarsal bones of the third digit tend to be weaker and smaller than in normal horses (see also Weidenreich, 1931). A similar interpretation might apply to other cases of such 'spontaneous atavisms' (Plate), like the formation of a fourth toe (which is normally reduced) in guinea pigs (Weidenreich, 1931), of rudimentary hind limbs in whales and dolphins (Andrews, 1921; Sleptsow, 1939), of supernumerary nipples in mammals, and so forth.

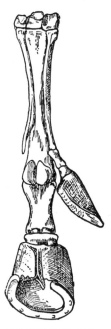

FIGURE 25. Right foreleg of a domestic horse, showing a polydactylous atavism. (After Wood Mason from Boas, 1917.)

From these 'spontaneous atavisms' Plate (1910) distinguished the 'hybrid atavisms', but these may well be interpreted, on the basis of genetics, as reverse processes of phylogeny. In all these cases polygenic characters are concerned. A loss of one or more of these genes causes the partial or total disappearance of the characters in question. However, as soon as the ancestral set of polygenes is restored by hybridizing the respective races, the old characters will of course reappear in the offspring. Plate mentioned as a simple example the atavistic appearance of the wild coloration if silver-gray and white mice were hybridized. A similar case is provided by atavistic colorations in the plumage arising in hybrids of canaries and serins (*Serinus canaria serinus*), canaries and goldfinches (*Carduelis carduelis*), canaries and linnets (*C. cannabina*), and canaries and greenfinches (*Chloris chloris*). These colorations, though typical of a whole group of related species, are not present in the hybridizing partners but appear in their offspring (Klatt, 1901; Mark, 1930).

Thus, spontaneous and hybrid atavisms may well be regarded as exceptions to the rule of phylogenetic irreversibility. Nevertheless, such reversibilities seem to be extremely rare in the major steps of transspecific evolution, and, if we refer once more to the secondary isodont type of dentition in toothed whales and the secondary reappearance of undifferentiated vertebrae in snakes and slow-worms, it is evident that only the general appearance of the character has evolved twice, and not the identical structure. It is obvious, of course, that the isodont dentition of toothed whales is not identical with that of reptiles and that the uniform vertebral column of snakes is not the same as that of fish (compare W. K. Gregory, 1936). Other examples are the secondary lack of shell in snails, the secondary development of a cap-shaped, non-spiralled shell in snails, such as *Ancylus*, the secondary wingless type of insects, and the secondary development of extratracheal gills in insect larvae (e.g. in the rectum of the stone fly *Nemura*), 'blood gills' in the larvae of *Apanteles* and *Pelobius*, and so forth.

Summing up, we may state that the rule of irreversibility proves to be valid for all major steps of phylogeny. Hence, its bearing on the classification of fossil forms into lines of descent must be accounted for. We can understand why the rule is practically valid, though theoretically cases of phylogenetic reversibilities, due to the effects of reverse mutations and the restitution of ancestral gene complements by race or species hybridization, could be imagined. This general validity of the rule is due mainly to the fact that mutation and selection never cease (compare Chapter 2) and that there is a continuous change in all organisms. If in the environment of an animal a situation arises which is similar to or identical with that of some former phylogenetic period, favoring anatomical structures similar to those needed formerly, the new anatomical structures will never be identical with the former ones, as meanwhile the whole organism of the animal has undergone phylogenetic changes. Hence, if the same environmental situation arises a second time, it will never bring about the same construction of an animal, as this has been altered meanwhile, and hence structures only similar to and not identical with those of the first period will be developed. So one can agree with Beurlen's statement that irreversibility is the 'formal condition of historical events' (1937, p. 44), but we cannot agree with his conviction (p. 37) that 'the category of irreversibility proves that phylogeny is governed by laws of its own, not mechanically'. As we have seen, irreversibility can be interpreted on the basis of mutation and selection acting as the essential processes in the course of a continuous phylogenetic development.

III. Rules governing Transspecific Changes of Structural Types

Transspecific Changes of Structural Systems. In discussions on phylogenetic transformation it has often been pointed out that through all levels of evolution the organisms remain well-balanced and harmonious systems, i.e. if there is an alteration of a certain organ, other organs will be altered correspondingly. So, for example, in the parallel evolution of running animals from the various vertebrate orders (Perissodactyla, Artiodactyla, Rodentia,

126

Carnivora, Marsupialia) not only did the legs become longer and the moving muscles stronger, but there was always a corresponding reduction of the number of toes, and a reinforcement of hip and shoulder girdles; the distal parts of the feet were rendered more solid and developed soft pads; the heart and lungs became adapted for better performance; the eyes became relatively larger; and certain patterns of behavior were altered (e.g. strengthening of flight instincts and of social behavior). How are such co-adaptations (or 'synorganizations': Remane, 1952) brought about? Do random mutation and natural selection add one character to the other, or is a certain state of harmonious animal construction being maintained throughout the stages of transformation? Numerous facts suggest that the latter assumption is true. Therefore, many authors emphasize that animals cannot be changed by simply adding new characters, but that each transformation is governed by specific rules affecting the organism as a whole ('ganzheitliche Veränderung').

There can be no doubt, however, that the contrast between 'additive' and 'ganzheitlich' has frequently been overemphasized. No biologist will ever deny that mutual effects of single characters and their alterations will result in new qualities of the organism concerned, and that 'systemic qualities' may arise due to such correlative processes. This is illustrated, for instance, by the development of histological structures. In a tissue culture, some embryonic tissues will grow to clusters of cells with no or only a little specific differentiation, and some differentiated tissues taken from adult animals will de-differentiate *in vitro*. But if other types of tissues are added to the culture, in some cases typical differentiation will be started and maintained by mutual influence of the various types of growing tissue. The influence of chemical (and morphological) agents in this mutual process can be demonstrated, if the specific hormones and trephones are continually rendered ineffective by frequent washing and transferring of the tissue cultures from one jar to another. Under such circumstances de-differentiation of tissue will be especially conspicuous, and even epithelia will grow like diffuse veils (compare Gawrilow, 1941; Figure 26). Hence, numerous morphological characters typical of cells and tissues prove to be such 'systemic qualities' which, in spite of a certain autonomy of development, cannot be understood unless such a complex of growing tissues is considered as a whole. The same applies to the formation of organs. *In vitro*, mammalian nerve tissue tends to grow in all directions with the cell bodies forming a dense peripheral layer and the dendrites growing centripetally. If mesenchymal tissue is added, the nervous tissue will form a 'neural tube' with ganglionic cells in a central position and the dendrites growing centrifugally. Chordal tissue, if placed close to nervous tissue, will cause a cleft of the nervous tissue (Figure 27; compare Holtfreter, 1934). As is well known, such examples of mutual induction have been demonstrated in quite a number of cases, and it is evident that no tissue, and no organ, develops without such mutual influences. Hence, systemic characters are predominant in all organisms.

In the manifestation of genes the mutual influences of chromosomal linkage, position effects, pleiotropism, mutual reactions between the processes brought about by different genes, and modifications during ontogeny produce numerous systemic qualities. The modern way in the biological sciences of looking at organisms as a whole ('ganzheitliche Betrachtung') will prove to

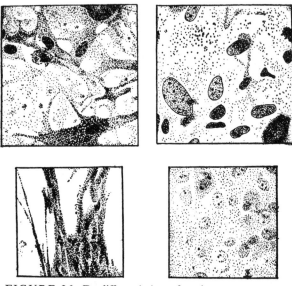

FIGURE 26. De-differentiation of explants *in vitrs* caused by repeated growths and several passages. Above: liver tissue of a hamster, left: six days old; right: fourteen days later. Below: embryonic iris of eight-day-old chicken, left: three-day-old explants; right: same sample after fifteen passages, showing the de-differentiation. (After Gawrilow, 1941.)

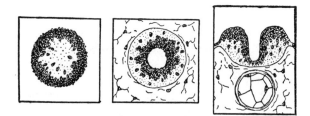

FIGURE 27. Diagram of the mutual formative effect of neighboring tissue explants. Nerve tissue (left), caused by neighboring mesenchyme to form a neural tube (center) and a neural groove under the influence of adjacent chorda tissue (right). (After Holtfreter, 1934.)

be a useful method for a long time to come, as most biological regularities or 'laws' are due to such actions of single elements in mutual relation to the whole ('Gefügegesetzlichkeit'). We should not forget, however, that such regularities of a higher order cannot be studied successfully unless the single elements have been analyzed.

128

Thus there can be no doubt that such a 'study of the whole' is necessary in an analysis of phylogenetic transformation, but this does not mean that special unknown systemic forces, not accessible to analysis, must be assumed, as is maintained by vitalists. There is no reason at all to think that systemic qualities and their causation should involve extracausal factors, as no acausal processes – except psychic components being coordinated lawfully – have been proved in the organic world. (On the three main categories of laws affecting living organisms, see B. Rensch, 1949.) It is, however, probable that the so-called 'systemic qualities' conceal extremely complicated relations of factors diffcult to analyze. This has also been pointed out by Stolte (1937),

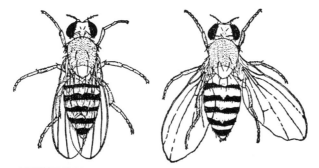

FIGURE 28. Female of *Drosophila funebris*, normal (left) and polyphene (right). (After H. Timoféeff-Ressovsky, 1931.)

H. Weber (1939), and von Bertalanffy (1932, 1942, 1949), though the latter author emphasized an additional 'organismic' lawfulness.

We shall now examine the extent to which such systemic characters and the harmonious patterns resulting from their action can be analyzed in the field of transspecific evolution, and whether some reason is left for the assumption of autonomous factors, phylogenetic self-regulation, or 'active reactions' in the sense of Böker (see below). In the special analyses of systemic transformations we shall have to examine the complex and constructive effects of certain genes, and the effects of selection on the correlation of single characters in the whole system of an organism. As we shall see, genetic pleiotropism, processes of allometric growth and its consequences, and the effects of compensations provide a sufficient explanation of animal transformation even at the level of systemic alterations ('ganzheitliche Änderungen'). Hence, random mutation and selection will be sufficient for an understanding of transformation as a whole, and we need not assume the existence of autonomous factors.

Constructive Genes. Besides a limited main effect, many if not most genes cause pleiotropic effects, as in the course of development the processes caused or controlled by the genes can interact in many ways. In *Drosophila funebris*, for example, the gene polyphene (Figure 28) causes five alterations in the gross morphology of the normal fly: irregular position of ommatidia and their bristles, abnormal pigmentation and thickened hind margins of the abdominal tergites, spreading of wings, incomplete wing veins, reductions and numerical changes in the setae of the head and the thorax (H. Timoféeff-Ressovsky 1931). This gene, however, is a homozygous lethal, and hence has no evolutionary effect.

In another case (which I mentioned in 1925) we can trace the various stages of pleiotropic effects phenotypically. In birds, the iridescent colors of many feathers are caused by a broader distal part of the chain of horny cells of the barbules and a flattening of the granules of melanin. As a result, the small hooks needed for linkage with the adjacent radius are reduced or missing. However, in albino specimens of iridescent birds, differing from the normal type in only one recessive gene (as in the peacock) which controls melanin pigmentation, the structure of feather barbules is 'normal'. In such albinos, then, both pigment and the broad distal parts of the barbules are lacking (Figure 29), and the distal parts are made up of slender, long cells with normal hooks for linkage with the adjoining barbules. Generally, the width of the radius is proportionate to the degree of pigmentation and the intensity of iridescence. This can be seen when the amount of the pigment is experimentally increased by the permanent effect of a relatively great atmospheric humidity or a relatively dark environment. By such stronger melanic pigmentation, iridescence is brought about at the surface of flat granules of melanin by radiation interference (W. J. Schmidt, 1952). Moreover, in the feather germs the cells of a radius are situated one after the other in a row. With more melanin pouring in, these cells can be widened only laterally, i.e. they become broader and flatter, and thus the iridescent surface of the melanin granules is rendered more effective. As the organic material normally used in the formation of linkage hooks is consumed in this process of broadening the radius, it is evident that a single gene mainly controlling the pigmentation produces a pleiotropic effect in the structure of the feather. This change of the feather structure may have a far-reaching effect, as the remiges may be rendered ineffective in flying because the lack of insertion hooks makes it impossible for the feathers to form a solid wing. It is for this reason, apparently, that selection has eliminated all cases of iridescence in remiges, and in strongly iridescent birds, such as humming-birds, honeysuckers,

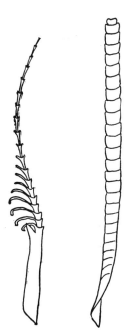

FIGURE 29. Feather barbules from the upper tail coverts of the peacock (*Pavo cristatus*). Left: normal hooked barbule of an albino specimen; right: iridescent barbule of a normal pigmented bird. (After B. Rensch, 1925.)

pheasants, and so forth, the remiges as the essential parts in flying are never affected by this mutation (see B. Rensch, 1927).

Similar structural alterations are caused by genes controlling the differentiation of the various parts of the vertebral column. From a rich material on human twins studied by Fischer (1933) and K. Kühne (1934), it became evident that the caudal limits of the thoracic, lumbar, sacral, and coccygeal vertebrae are determined by genes affecting the tempo of development in whole parts of the vertebral column. As there is a cranio-caudal gradient of differentiation, a general acceleration of ontogenetic development will result in a shift of the limits between the various types of vertebrae in a cranial direction. It is for this reason that in human ontogeny the thirteenth rib fuses with the twentieth vertebra and rapidly disappears. As the development of vertebrae is closely correlated with the formation of corresponding muscles and nerves, it is clear that the genes harmoniously affect the whole structural system of the organism. On this basis, the major changes of constructions recognized by Böker and termed 'Umkonstruktionen' may be interpreted in the way already mentioned by Fischer (1933, p. 216). The reduction of the tail part in the vertebral column of various mammalian orders also seems to be caused by such 'constructive genes', as I have called them. In a summarizing review, Steiniger (1938) emphasized that the tail part of the vertebral column shows little differentiation and that therefore this material is easily accessible to various influences. Moreover, there are no essential parts of the spinal cord in this distal portion of the vertebral column, and hence selection counteracting reduction of the tail region need not be important.

Effects on the organism as a whole will be produced especially by genes affecting the production of hormones by endocrine glands. In pituitary nanism of mice, caused by a recessive gene which bars the formation of eosinophil cells in the anterior pituitary lobe, there arise simultaneous defects of thymus, thyroid, pancreas, and gonads, the animals remain small, and there is no descensus testiculorum (compare G. D. Snell, 1929: Grüneberg, 1943). In domestic chickens, recessive taillessness is a syndrome comprising rudimentation of pygostyle, coccygeal ribs, coccygeal glands, and tail feathers, and deformations of pelvic bones and their decreased calcification. Landauer, who analyzed this mutation in 1945, found that the same syndrome could be produced by insulin injections into the yolk material of normal eggs (1946; 1947). Hence it is quite probable that the recessive mutation affects insulin production in the developing embryo.

The far-reaching effects of hormones and of mutations affecting size, or production of hormone glands, on the formation of organs and on the construction of the whole organism may be demonstrated by some results of studies on prolactin and thyroxin.

1. Prolactin, produced in pigeons by the anterior pituitary, will increase the size of liver, pancreas, intestine, and crop, will cause the secretion of crop milk, raise the level of blood sugar, increase the formation of glycogen in the

liver and the effect of the thyrotropic hormone from the anterior pituitary (Riddle, 1937, 1938), and, finally, will cause the onset of parental behavior (compare Figure 30).

2. Thyroxin, produced by the thyroid gland, accelerates the growth of the heart, liver, and kidney, and increases the general rate of metabolism (though large amounts of thyroxin will depress metabolic rates). It decreases the glycogen contents of the liver and the fat reserves of the body in general, increases the rate of breathing, accelerates the pulse, raises the number of red blood corpuscles, accelerates the growth of hairs and feathers, causes molting and hibernation, and checks the development of the genitalia.

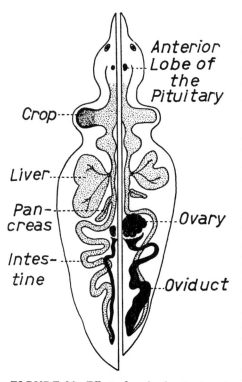

FIGURE 30. Effect of prolaction treatment in dove. Right side not treated. (After Riddle, 1938, combined.)

Harms (1934) made the interesting experiment of administering thyroxin injections to the skipper (*Periophthalmus*), a small fish living outside the water during low tide in the mangroves of the tropics. These injections caused numerous harmonious alterations of organs, which tended to provide better adaptations to a terrestrial mode of life: 1. The pectoral fins were slightly prolonged. 2. The epidermal tissue became thicker and more solid. 3. The animals voluntarily stayed in dry spots for a longer time than the normal ones. 4. The mouth became smaller. 5. The labial and oral skin folds became reduced. 6. The gill covers became smaller. 7. The gills were reduced to about half their size. 8. The frequency of air breathing was increased. 9. The coloration was brighter. 10. The tail fin was rather stout and sturdy. 11. The fin rays were more calcified. 12. The pituitary was enlarged. 13. The mesonephric structure was denser than usual, as the mesonephric tubules were smaller and the blood vessels denser. 14. The intestines were more compact and the mesenteric tissues more solid. 15. The rectal muscles were stronger. 16. The liver was enlarged. 17. The size of the otoliths in the labyrinth was decreased.

A similar multiple effect is caused by all other metabolic hormones (compare, for instance, the effect of thymus hormone on tadpoles: Wurmbach, 1954) and also by the sexual hormones, as these, too, affect the processes of metabolism. Hence, it is evident that all mutations altering the quantity of hormone yield will have a 'constructive' effect. In the case of the skipper a single mutation affecting the size of the thyroid gland may contribute largely to the adaptation of a marine animal to an environment outside the water.

Finally, in the mutation of pleiotropic genes, characters and processes may be affected that are connected only more or less indirectly with the mutating gene. There is an incompletely dominant mutation, 'frizzle-fowl', in the domestic hen, in which the feathers are curled and lack complete cornification, which causes increased loss of body heat. This loss of temperature causes numerous alterations in other organs as indirect results and by-products of the original mutation: the heart rate is increased and the ventricles are hypertrophic; the total amount of blood is increased; the spleen, pancreas, and kidneys are enlarged, and the intestine and cecum are lengthened. Besides these alterations, the white blood corpuscles are altered, the rate of metabolism is increased, the function of the thyroid gland is disturbed, there is a delay in ovulation and the structure of the testes is abnormal, from which decreased fertility and viability result (Landauer, 1946b).

Allometric Growth and Its Bearing on Evolution. From recent studies it seems probable that in the formation of all organs numerous constructive genes are effective, as the shape and size of the single organs, tissues, and structures are interrelated. Especially, complex correlations link those factors which affect or are affected by body size. As in the course of evolution body size is often altered, and as it is successively enlarged in most lines of descent (Cope's Rule; see Chapter 6, B IV), quite complex alterations of proportions will consequently result which may often seem to be 'directed' (orthogenetic). Of course, such alterations of organ and body proportions are accompanied by corresponding physiological changes. Hence, an analysis of all differences caused by a change of body size will offer a good opportunity for us to study the complicated process of transformation as a whole.

Studies of the ontogeny of numerous animals and plants have revealed that the growth ratios of single organs and structures in relation to the whole body remain constant for certain periods. Thus, the genes define growth gradients correlated to the size of the whole body (compare J. S. Huxley, 1924, 1926, 1932; Schmalhausen, 1925, 1949; Teissier, 1936; Von Bertalanffy, 1942). During certain periods, an organ or a structure can grow more quickly than the body as a whole (positive allometry), more slowly (negative allometry), or with the same speed (isometry). Similarly, one may refer to the allometric growth of a certain part of an organ in relation to this organ as a whole. In some cases such allometric tendencies remain constant through long periods of ontogeny. In other cases several different growth gradients follow one another, and then some growth ratios may change from positive to negative allometry or to isometry or vice versa. Such changes often coincide with birth, the end of the larval development, or the onset of sexual maturity, but they may as well occur during any other time. In many mammals and birds, for example, the head grows with positive allometry until birth, and then negative allometry begins. In the newt *Triturus vulgaris*, the area dorsalis pallii grows with negative allometry in relation to the lobus hemisphaericus until the body is about 25 mm. long. Then a period of strongly negative allometry follows, until the animal has reached a body length of 30 mm. and metamorphosis occurs. After this, there is a slight positive allometry, and with the onset of sexual maturity (at a body length of 60 mm.) an approximately isometric period of growth follows (Figure 31).

If the body size of an animal is enlarged by modification or mutation, i.e. if growth is accelerated or certain periods of growth are prolonged, or if both occur, the effects of allometric growth will become more conspicuous,

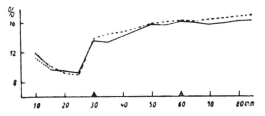

FIGURE 31. Changes of allometrical gradients in the ontogeny of the area dorsalis pallii of *Triturus vulgaris*. Abscissa: body length of larvae. Wedges: metamorphosis and maturity. Ordinate: volume of area dorsalis pallii in percent of volume of lobus hemisphaericus. Full line: right hemisphere; broken line: left hemisphere. (After B. Homeyer.)

and organs with positive allometry will become relatively larger and eventually excessively large, while organs with negative allometry may become smaller and smaller or even vestigial. From this point of view it is clear that large and small specimens of a species or closely related large and small species or genera will hardly ever show the same type of proportions (compare, for example, Mangold and Waechter, 1953). In the course of great phylogenetic increase of body size (Cope's Rule) as, for example, in the line of descent of horses or elephants, a far-reaching change of proportions took place, finally leading to new structural types, although the growth-gradients correlated with body size may have remained unchanged.

It is surprising that this important difference in the proportioning of large and small related organisms has been known for some centuries and yet has not received sufficient attention until the last decades, and, unfortunately, it is commonly overlooked in descriptive systematics. As early as 1718, Galileo (p. 559) drew attention to the fact that for mechanical reasons in large and small animals of a similar type the proportions would not be similar. In 1748, Robinson stated that the hearts of large animals are relatively small, and in 1762 Von Haller formulated a corresponding rule pertaining to the brain and eyes. During the nineteenth century these correlations were studied in more detail and with regard to other organs of mammals. Bergmann (1847, 1855) proved that the proportions of large animals must differ from those of smaller ones, as with increasing body size the efficiency of surface-proportionate organs (such as skin, intestine, respiratory organs) grows by the square, but the body (weight and volume) by the cube. More work was done on the subject by Welcker and Brandt (quoted in 1867, but not published until 1903); Richet (1891); O. Snell (1891); Dubois (1897); Lapique (1898, 1903, 1907, 1908); Maurel (1902, 1903); Magnan (1911, 1912); Klatt (1913, 1919, 1923); D'Arcy Thompson (1917); Hesse (1921); J. S. Huxley (1932), and others, which finally led to the establishment of certain rules governing the proportioning of organisms and their parts. Recently these rules have been extended and analyzed in more detail by my students and myself working on more extensive material.

For a proper evaluation of whether structural alterations in the course of phylogeny must be ascribed to the effect of allometric growth, and of the extent to which the rules of proportionality in large and small animals are due to parallel selection, a reliable method of study is needed. First, the different phases of allometric growth must be carefully studied in the ontogeny of many organs and structures. Then, by comparing the relevant lines of descent of recent species of different size, one tries to establish rules governing the correlations of body size and organs in the course of phylogeny. After this, one attempts to infer from the growth gradients in ontogeny those in phylogeny, as it will be probable that generally only some of the genes will be altered by mutation and hence some main growth gradients will be unchanged. Thus, large species will continue the ontogenetic trend of growth into stages of individual development more advanced than that of their smaller ancestors. Quite often one will find that the general ontogenetic growth tendencies actually remained similar in phylogeny, while the intensity of positive or negative allometry increased or decreased, due to special factors of selection. In many cases only the relative length of single phases of positive or negative allometry will be altered in phylogeny (heterochronism), though the general tendency of the growth ratios remains the same. Finally, there will also be some cases in which the trend of phylogenetic growth ratios is more or less opposed to that of ontogeny, and these cases must be regarded as demonstrating the effects and results of special selection.

Often the fossil records of certain lines of descent are incomplete. In such cases one can establish rules which govern the correlations between size and structure of certain organs and body size by comparing a great number of specimens from related Recent subspecies, species, genera, or families differing in body size. So one will find certain rules concerning the changes of size and structure of organs resulting from changes of body size. Hence, one replaces the phylogenetic series by an anatomical series of Recent related animals. This is a sound method because the rules governing the correlations between body size and organs of recent species, genera, etc., must have been the same in their common phylogenetic ancestors. Otherwise, the establishment of a general rule would have been impossible.

135

Analyses of this kind can be carried out mathematically and quite precisely. As cells, tissues, organs, and the whole body grow three-dimensionally, the process of allometric growth may be mathematically described by a function formula, $y = b \cdot x^{\alpha}$, in which y is the size of the organ under consideration, and x means body size. α is a growth coefficient from which, for example, one can calculate the size of a certain organ at a given body size. This is an important help to paleontologists, as one can find out about the size of organs in animals of a body size exceeding that of the types known so far. Hence, the other way round, one can state the size of a fossil animal of which only some organs have been found (skull, limbs, bones), provided α has been calculated from related forms of the group. In the allometric growth formula, b is a constant of proportion, the value of which depends on the units of measurement employed, on age, sex, state of nutrition, systematic position, and similar typical qualities of the animal under consideration. It will be realized that the above formula may be written $\log y = \log b + \alpha \cdot \log x$, from which it is evident that in a system of logarithmic coordinates this function will be represented by a straight line. The slope of this straight line indicates the degree of positive or negative allometry ($\gtrless 45°$) or isometry ($= 45°$).

If an organ grows mainly in one dimension, this process may be represented by a simple linear regression (see Shepherd, Sholl, and Vizoso, 1949). However, such examples can be considered as borderline cases of allometric growth. What renders the situation more complicated is the fact that a growing organ comprises several parts, each of which may grow at different rates and different periods, and, moreover, there is irregular and rhythmic growth. On the whole, however, the allometric growth formula has proved a very convenient and useful simplification (compare Reeve and J. S. Huxley, 1945; Richards and Kavanagh, 1945; Rensch, 1948; Zuckerman and co-workers, 1950).

The allometric exponent α can be calculated as soon as the absolute size of body or organ of two or more animals of different body size is known. If these animals belong to the same species and sex, and if their physiological state is about equal, b may be omitted in the above formula, hence $\text{organ}_1 : \text{body size}_1 = \text{organ}_2 : \text{body size}_2$, from which equation one can calculate:

$$\alpha = \frac{\log O_1 - \log O_2}{\log BS_1 - \log BS_2} \cdot$$

(See Lapique, 1898; Dubois, 1898; Klatt, 1919, 1923.) If one calculates α at various levels of body size in a certain animal group, one will usually find a considerable range of variation, but the average figures of such 'somatic exponents' (Klatt) or allometric exponents are surprisingly similar for each organ. The same applies to a comparison of related species, but here, of course, the average value will be different from that of the comparison of specimens from one species. In various mammalian species the heart exponent was found to be 0·83 by Klatt (1919), in African hoofed animals 0·85

136

according to Quiring (1939); in small mammals from central Europe (five pairs of different size) I found it to be 0·83 and in central European birds (eight pairs of different size) 0·82 (1948). These figures indicate that the relative heart weight is not strictly correlated with the surface area of the body, i.e. the areas cooling off and losing body heat by radiation (if so, the exponent must be expected to be 0·66), but with some other factors, such as increased intensity of metabolism in smaller animals, quicker circulation of blood with less friction, and so forth. The somatic kidney exponent is 0·87 in large African hoofed animals (Quiring), and 0·76 in small European mammals (B. Rensch); and Schilling (1951) found it to be 0·84 in rabbits and 0·73 in sheep.

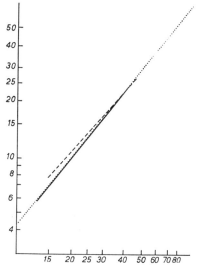

If we wish to know whether, in a certain case, phylogenetic allometry is due to the effect of ontogenetic allometry, we shall have to find out whether the allometric exponent α is the same in both categories.

Such a calculation was made, for example, by Robb (1935) on the increase of the relative length of face bones in horses. Robb found that the ontogenetic growth gradient in the Recent horse (*Equus caballus*) is very similar to that of the whole line of descent leading from *Hyracotherium* (*Eohippus*) to *Mesohippus*, *Miohippus*, *Merychippus*, and *Equus* (Figure 32). So one may state that the phylogenetic transformation of the equine skull bones was predeter-

FIGURE 32. Coincidence of growth gradients in ontogeny and phylogeny. Length of facial skull (ordinate) plotted against total length of skull (abscissa) in the ontogeny of the domestic horse *Equus caballus* (broken line) and the phylogeny of the fossil horses *Hyracotherium* to *Equus* (unbroken line). Dotted line represents the gradient covered by $y = 0·25.x^{1·23}$. (After G. G. Simpson.)

mined by the growth gradient in the face bones of the ancestral type *Hyracotherium*, though this gradient had as yet no conspicuous effects in this relatively small animal. With increasing body size in the course of phylogeny, this growth gradient became more and more effective. As we shall see in Chapter 6, B IV, such findings provide a plausible interpretation of directed evolution (orthogenesis) of certain organs, excessive growth, and rudimentary growth.

In the meantime, however, we shall demonstrate how far cases of complex systemic ('ganzheitliche') transformations, so often maintaining their harmonious state, can be explained by applying the rules of allometric growth, and how far the phylogenetic gradients of various organs and structures differ from the ontogenetic growth gradients. First, morphologic and anatomical characters may be considered.

TABLE 8. Relative length of skull in individual groups and populations of different body size but of the same geographic race. (Absolute data from McTaggart Cowan, Benson, and Hinton; data on *Peromyscus maniculatus* from Dice [1939], who measured one-year-old laboratory material; data on *Sciurus* and *Talpa* from material in the Landesmuseum für Naturkunde in Münster.)

Species	Number of Individuals	Length of Head and Body in mm.	Length of Skull as Percent of Head and Body Length
Peromyscus sitk. sitkensis, Baranof Islands	10 adult	108	27·8
Peromyscus sitk. sitkensis, Chichagof Islands	9 adult	99·4	34·1+
Peromyscus maniculatus Palouse Falls (Laboratory material)	96 adult	96·71	25·5
Peromyscus maniculatus Pullman (Laboratory material)	34 adult	90·00	27·3+
Neotoma lepida devia, Colorado River	12 adult	113	24·2
Neotoma lepida devia, Yuma County	6 adult	80	26·2+
Evotomys ruf. rufocanus (largest 4 out of 12)	4♂	118·3	23·3
Evotomys ruf. rufocanus (smallest 4 out of 12)	4♂	101·5	25·5+
Sciurus vulgaris vulgaris (largest 4 out of 12)	4♂	228	22·8
Sciurus vulgaris vulgaris (smallest 4 out of 12)	4♂	215	23·8+
Talpa eur. europaea (largest 4 out of 12)	4♂	150	23·7
Talpa eur. europaea (smallest 4 out of 12)	4♂	135	25·8+

TABLE 9. Relative length of skull in geographic races of different body size but of the same species ('Rassenkreis'). (Absolute data from G. S. Miller, 1912; J. G. Palmer, 1937; McTaggart Cowan, 1935; Hooper, 1936; Benson, 1935; Hinton, 1926.) The figures for *Neotoma* and *Scapanus* refer to maximum length of skull, those for *Peromyscus sitkensis* to occipito-nasal length, all others to condylo-basal length.

Species	Number of Individuals	Length of Head and Body in mm.	Length of Skull as Percent of Head and Body Length
Scapanus latimanus caurinus	10♂	144·8	25·2
Sc. lat. occultus	7♂	118·2	27·2+
Sorex araneus araneus	20 adult	77·5	24·1
S. ar. granarius	2 adult	64	27·4+
Peromyscus sitkensis isolatus	3 adult	116	24·9
P. sitk. oceanicus	8 adult	108·3	28·1+
Neotoma fuscipes bullatior	4♂	237	18·3
N. fusc. martirensis	8♂	194	19·3+
Neotoma lepida lepida	17 adult	163	23·9
N. lep. flava	2 adult	146	25·8+
Evotomys glareolus reinwaldti	1♂	105	22·4
E. glar. helveticus	1♂	95	24·9+
Evotomys rufocanus regulus	1♂	116	22·9
E. ruf. smithi	11♂	110·5	24·2+

TABLE 10. Relative length of skull in closely related species of west European mammals of different body size (absolute data from G. S. Miller, 1912).

Species	Number of Individuals	Length of Head and Body in in mm.	Condylo-basal Length of Skull in mm.	Condylo-basal Length of Skull as Percent of Head and Body Length
Epimys norvegicus	3♂	243	50·2	20·7
Micromys minutus	7 adult	68·3	17·2	25·2+
Arvicola a. amphibius	2♂	195	42·3	21·7
Pitymys s. subterraneus	5 adult	97	22·3	23·0+
Cricetus cricetus canescens	2♂	215	44·6	20·8
Cricetulus atticus	1♀	87	25·2	29·0+
Glis g. glis	3♂	172	37·2	21·6
Muscardinus avellanarius	4♂	77·8	21·7	27·9+
Marmota marmota	4♂	538	92·9	17·3
Citellus citellus	6 adult	209·1	43·7	20·9+
Lepus eu. europaeus	2♂	670	105·6[1]	15·3
Oryctolagus c. cuniculus	2♂	414	81·0[1]	19·6+
Sorex araneus bergensis	11 adult	78·7	19·4	24·7
Sorex m. minutus	5 adult	55·8	16·4	29·4+
Mustela erminea aestiva	8♂	271	48·8	18·0
Mustela n. nivalis	4 adult	212	39·8	18·8+
Canis lupus signatus	2♂	1230	227·5	18·5
Canis vulpes crucigera	3♂	626	141·7	22·6+

[1] Occipito-nasal length of skull.

In many animal groups, especially in mammals, birds, and reptiles, the head grows with positive allometry during prenatal development, but after birth its growth tends to be negatively allometric (Figure 34). The same applies to numerous groups of heterometabolic (and especially pleurometabolic) insects, in which, consequently, the larvae of the first instar have relatively larger heads than the adult animals (compare H. Weber, 1933, Figure 516; Ågrell, 1948, Figure 1). The effects of this postnatal phase of negative allometry become even more conspicuous if, in the course of phylogeny, body size is increased hereditarily. Hence, in nearly all comparable cases, the larger specimens or subspecies of one species (Tables 8 and 9) and the larger species of one genus or the larger related genera of one family (Table 10) will have a relatively smaller head than the comparable smaller forms. The strongly positive allometry of the head during prenatal development seems to have been favored by selection because in this animal group (Tables 8, 9, and 10) it was especially advantageous to have an efficient central nervous system fully functioning immediately after birth. Hence, the brain must be fully developed at birth, and this requires a certain size. This general allometric trend of ontogeny remained unchanged when, in phylogeny, body size was

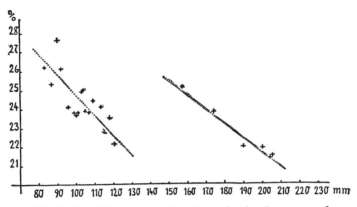

FIGURE 33. Different negative allometries in *Evotomys rufocanus smithii* and *Arvicola amph. amphibius* (including *A. a. reta*). Abscissa: head–body length. Ordinate: condylobasal length of skull in percent of head–body length.

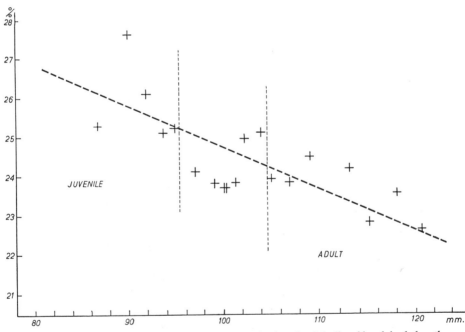

FIGURE 34. Negative correlation between relative length of skull and head–body length in juvenile and adult Japanese mice (*Evotomys rufocanus smithii*) ♂ and ♀. Each cross represents the average data of three specimens. Juvenile specimens: up to 95 mm. head–body length; adults: longer than 105 mm. Abscissa: head–body length. Ordinate: condylobasal length of skull in percent of head–body length.

altered hereditarily, although the allometrical coefficient α and the proportion constant b were usually altered. Hence, if we plot the correlation figures of head and body size of two species of related genera on a system of co-ordinates (Figure 33), the two lines are nearly parallel, but do not form one single line as in the comparison of small and large specimens of one species (Figure 34), where the values of the larger individuals continue the straight line of the small ones.

The shape of the head itself is the result of single-growth gradients acting in different directions during onto-genetic development. So, for example, in mammals and in many birds the facial bones tend to grow relatively faster than the head as a whole, and with increasing age the skull becomes more and more prolonged, and the crests and bone processes of the cranial part of the skull become relatively larger. This ontogenetic trend of posi-tively allometric growth of the facial parts remained more or less constant in the phylogenetic transformations, as we have already seen from the phylogenetic allometry in the skull of horses (p. 137, Figure 32). Usually, however, in contrast to the process in horse ancestry, the growth coefficient α and the constant b do not remain the same in ontogeny and phylogeny, and only a generally equal tendency remains in phylogeny. If we compare large and small related species of the same genus or large and small related genera, we shall find that generally the large types have relatively smaller crania with strong cristae and relatively longer facial parts, often with relatively longer canines and incisors than the small species. Hence, the proportions of the skull in lions and tigers (i.e. in large feline types), more dog-like in appearance than those of smaller cats, are determined by the growth gradients present in the skulls of small feline forms, though in these their effects are not yet apparent. A similar statement applies to the comparison of weasel (*Mustela nivalis*) and marten (*Martes*), of ground squirrel (*Citellus*) and marmot (*Marmota*), of mouse (*Mus*) and rat (*Epimys*), and so forth (Figure 35). (As the distinguishing

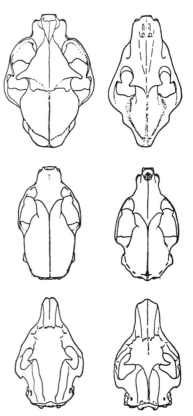

FIGURE 35. Skulls of cat and lion (top), weasel and marten (center), and ground squirrel and marmot (bottom) drawn on same scale to show the allometric shifts in the length of the facial skull.

characters of the genera are often only due to such effects of allometry, one should re-examine the validity of such genera.) There are only rare exceptions to these rules, one being the two-toed sloth *Choloepus*, as can be seen from the illustrations published by Schneider.

The size of the skull is essentially due to the size of the brain, which, correspondingly, grows with negative postnatal allometry. Consequently, the phylogenetic rule of Von Haller (1762), saying that in large species the brain

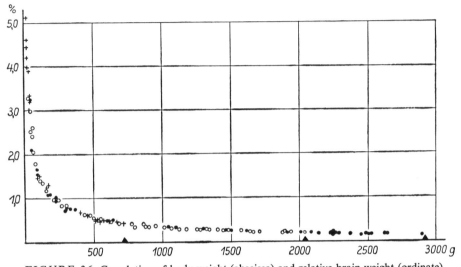

FIGURE 36. Correlation of body weight (abscissa) and relative brain weight (ordinate) in percent of body weight of five races of domestic hens with marked differences of body size. $+$ = dwarf race; $\bigcirc$ = races with medium body size; $\bullet$ = giant races. The wedges indicate the end of ontogenetic growth in the three racial groups; no change of growth gradient. (After E. Schlabritzky.)

is relatively smaller than in related small species, is also valid quite generally. Later authors, like Cuvier (1801), Brandt (1867), and others, confirmed Haller's rule. A corresponding rule has proved valid in insects (compare Goossen, 1949; O. Snell, 1891; Dubois, 1897, 1930; Lapique, 1898; Brummel-kamp, 1946, and others, who calculated the allometric coefficient α as 0·56 [5/9] for related mammalian species and as $\alpha = 0·26–0·22$ for individuals of one species). Sholl (1947) doubted that these figures were correct, but nevertheless it seems clear that in infraspecific comparisons the allometric coefficient α is smaller than in transspecific comparisons. This fact shows again that the ontogenetic trend is altered in phylogenetic transformation, but that the general tendency is maintained. It is an interesting fact that in six races of domestic fowl of very different body size the postnatal growth gradient of the brain remains totally unaltered, i.e. adult dwarf races have the same relative brain weight as juvenile specimens (of the same body weight) of medium or giant races (Schlabritzky, 1953). The plotting of the data from all six races (body weight on the abscissa, relative brain weight on

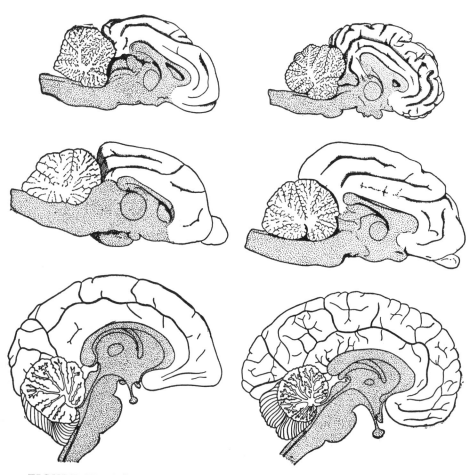

FIGURE 37. Median sections of brains of closely related large and small mammals, drawn on same scale, showing the relatively larger forebrain in the large species. Dotted areas=brainstem. Above, left: sheep; right: cow. Center, left: cat; right: North African lion. Below, left: chimpanzee; right: man.

the ordinate) yields a single uniform growth curve with all data nicely falling in line (Figure 36). This is a rather peculiar fact, as single parts of the brain of fowl show different growth ratios. In the postnatal phase the neostriatum, for instance, grows with positive allometry, and the cortical regions with negative allometry (Krumschmidt). The same holds good for mammals. Here, too, the various parts of the brain and single regions and areas of the pallium grow with different allometry. In large species the brain stem (brain without pallium and cerebellum) is relatively smaller than in related smaller species (Figure 37 and B. Rensch, 1953).

In spite of these complications, there is a general trend in numerous lines to a successive increase of relative brain size, which seems to be due to the fact that a larger brain was a selectional advantage. This phylogenetic change

143

must be accounted for by altering the factor b in our formula of allometric growth. Dubois (1930) thought that this increase of brain size was brought about by 'saltatory' growth caused by doubling the number of nerve cells during the final period of embryonic development. The comparison of a large number of related species by Brummelkamp (1939) revealed that intermediate values of b exist, and Brummelkamp assumed that in the process of increasing brain weight (cephalization) the factor b of the allometric formula was altered in correspondence to a geometrical series with the modulus $\sqrt{2}$. But even so, there are further intermediate values of b which do not coincide with the above geometrical progression. Moreover, it is highly doubtful that these and similar mathematical series are a true representation of the real process, as the various regions and areas of the forebrain grow with quite different growth gradients which may be altered independently in phylogeny (compare Chapter 7, C III). Sholl (1947) has critically reviewed the cephalization hypotheses of Dubois and Brummelkamp.

We cannot discuss in more detail the ontogenetic and phylogenetic allometric growth of all other organs nor their relation to Cope's Rule of successive increase of body size in the lines of descent (for birds, compare Vaugien, 1949). It may be sufficient to point out which characters show the same tendencies in all mammals and birds or in many groups of these two classes and which general rules of proportions can be established. In related species or genera of different body size, the larger specimens will have relatively shorter ears, tails, legs, wings, and bills, provided that in the groups concerned these organs are of 'normal' proportions. This rule, which I called Allen's Rule (see p. 43 of this book) is valid so far as it is not linked to Bergmann's Rule (B. Rensch, 1938; Meunier, 1951). In this case the ontogenetic growth gradients again correspond to those of phylogeny. In species with especially long necks, legs, or bills there is a marked positive allometry during postnatal growth and, correspondingly, in the larger species these organs will be relatively longer or even excessively long. This is demonstrated by birds with especially long bills or legs (Figure 38), by long-necked giraffes (comparing the okapi and fossil types of giraffes to the Recent form), by swans in comparison with geese and ducks, and so forth. In some bird groups, too, the size of certain conspicuous feather tufts depends on body size: in the crested drongo (*Dicrurus paradiseus*) the positive allometry of the feather tuft is quite obvious, as the largest subspecies (*D. p. grandis, D. p. johni*, etc.) have extraordinary large crests, while in the small subspecies of this 'Rassenkreis' (*D. p. brachyphorus, D. p. microlophus, D. p. platurus*) these feathers are extremely short (compare Vaurie, 1949; Mayr and Vaurie, 1948; B. Rensch, in Heberer, 1954). Normal feathers and normal hairs in mammals grow with positive allometry but, nevertheless, these structures are relatively short, even in large types. Hence, in this case, the phylogenetic and the ontogenetic tendencies are opposite to one another. In lizards, the postnatal growth of limbs is negatively allometric, and, correspondingly, in the larger species the limbs tend to be relatively shorter (Schuster, 1950). In Ensifera, a group of grasshoppers, the ovipositor grows with positive allometry, and therefore this organ is relatively longer in the large species of a genus (Ander, 1948).

Regarding the internal organs, one may generally state that the allometric tendency of the main periods of individual growth remains constant in phylogenetic transformations, though the special coefficients of growth may vary. Accordingly, in the larger types of related species or genera of mammals and birds the heart, liver, kidneys (Tables 11 and 12), lungs, thyroid, adrenal glands, pancreas, and pituitary are relatively smaller than in the corresponding smaller types. (Compare the following publications: Manea (1894) on

FIGURE 38. Excessive growth of beak as the result of positive allometry in the black-tailed godwit *Limosa limosa* (above: aged two, eight, and fifty-two days) and the lapwing *Vanellus cristatus* (below: aged three, twenty-one, and fifty-two days). (From skins and from photos by Heinroth.)

kidneys; Jackson (1913) on adrenal glands and thyroid; Hesse (1921) on the heart; Klatt and Vorsteher (1923) on intestine, heart, and muscles; Meissner (1924), Riddle (1927), and Stockard (1934) on the thyroid; Denzer (1935) on the kidneys; Quiring (1939) on heart, kidneys, and liver; B. Rensch (1948) on heart, kidneys, liver, intestine, and other organs; Oboussier (1948) on the pituitary; Padour (1950) on endocrine glands; Schlabritzky (1953) on numerous organs; and so forth. The early papers by Robinson, Von Haller, Bergmann, Welcker and Brandt, Richet, O. Snell, Lapique, Maurel, and Magnan have already been mentioned.) Especially from Schlabritzky's studies on allometric growth in races of domestic fowl with differing body size it becomes clear that phylogenetic transformation is not always a mere continuation of ontogenetic trends. As is shown in Figure 39, the curves of the growth gradients of the heart differ with the races: at first there is a general negative allometry during postnatal development, but then this changes into a positive allometry more or less intense in the various races. Similar differences are found in the growth of liver, kidneys, lungs, and thyroid. With

145

TABLE 11. Relative weight of organs taken from closely related mammals of large and small body size at the same season of the year. (Net weight=body weight minus fat and contents of stomach and bladder.) Plus marks (+) indicate that data correspond with correlation rules. (These data are selected from those given in B. Rensch, 1948.)

Species	Month	Number of Individuals	Net Weight	Liver Weight as Percent of Net Weight	Kidney Weight as Percent of Net Weight	Heart Weight as Percent of Net Weight
Erinaceus europaeus	July, Oct.	2♀	685·6	6·42	10·00	4·30
Sorex araneus	Aug.	2♀	9·3	7·77+	18·13+	9·18+
Epimys norvegicus	Aug.	2♀	263·2	5·48	8·15	4·17
Mus musculus	Aug., Sept.	4♀	19·6	7·42+	17·18+	6·21+
Oryctolagus cuniculus	Oct., Dec.	2♂	1391·0	3·49	6·74	3·54
Mus musculus	Dec.	2♂	16·7	6·68+	17·24+	6·06+
Arvicola amphibius	June	1♂	49·9	6·35	10·01	5·27
Clethrionomys glareolus	Aug.	2♂	18·1	7·97+	13·97+	9·91+
Putorius putorius	Mar., Sept.	2♂	1159·2	5·09	9·88	7·42
Mustela nivalis	Mar., Oct.	2♂	75·0	5·30+	13·39+	11·53+

TABLE 12. Relative weight of organs taken from closely related birds of large and small body size at the same season of the year. (Net weight=body weight minus fat and contents of stomach and bladder.) Plus marks (+) indicate that data correspond with correlation rules. (These data are selected from those given in B. Rensch, 1948.)

Species	Month	Number of Individuals	Net Weight	Liver Weight as Percent of Net Weight	Kidney Weight as Percent of Net Weight	Heart Weight as Percent of Net Weight
Corvus frugilegus	Nov., Jan.	1♂, 1♀	467·5	2·87	9·34	10·13
Coloeus monedula	Sept., Oct.	1♂, 1♀	260·2	3·29+	9·56+	11·03+
Pica pica	Oct.	3♀	205·5	3·70+	12·37+	10·84+
Passer domesticus	Mar., Apr.	3♂, 3♀	28·3	3·67	9·65	14·89
Serinus c. serinus	Mar.	1♂, 1♀	10·8	3·31−	12·42+	15·35+
Sitta europaea caesia	Nov.	1♂, 1♀	22·8	2·69	11·55	12·37
Aegithales caud. europaeus	Sept.	1♀	9·4	4·40+	15·10+	11·70+
Turdus merula	Mar.	1♂, 2♀	98·0	3·06	11·77	12·32
Phylloscopus trochilus	Sept., Oct.	2♀	7·5	5·30+	19·46+	13·59+
Sylvia curruca	June	1♀	11·1	4·88+	22·50+	13·81+
Buteo buteo zimmermannae	Oct.	1♂	642·9	2·33	7·80	7·48
Falco subbuteo	May	1♂	212·9	2·64+	8·14+	13·49+
Cygnus olor	Mar.	1♀	6579	1·85	—	10·3
Anas platyrhyncha	Sept.	1♂	1094	2·54+	—	7·9−
Vanellus vanellus	Apr.	1♂	209·5	2·91	11·31	15·07
Charadrius hiaticula	Sept.	1♀	45·0	4·67+	21·50+	18·89+

regard to these organs, the proportions of an adult dwarf hen are definitely different from those of a juvenile specimen of the medium or giant races of the same body weight. This remarkable fact – in contrast to the proportions and gradients of brain and eyes (Figure 36) – is probably caused by the different kind of metabolism, which is a constructive anabolism in the half-grown giant and a basal metabolism in the adult dwarf of the same body weight. The nervous patterns and requirements of the brain, however, are

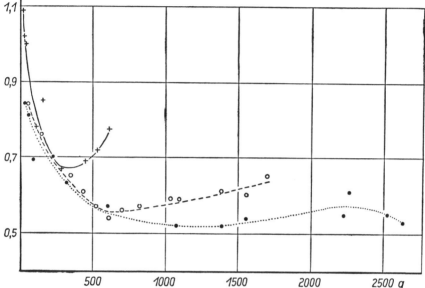

FIGURE 39. Correlation of body weight (abscissa) and relative heart weight (ordinate) in percent of body weight in five races of domestic chickens of different size. +=dwarf race, ○=medium-sized races, ●=giant race. (After E. Schlabritzky.)

more or less the same in all races at the various levels of body size. In small mammals from central Europe I found the allometric exponent of the liver to be α=0·98, while Quiring (1938) stated it as α=0·72 in African hoofed animals. On the other hand, Klatt (1923) found α=0·50 in various races of dogs of different body size, which proves once again that the intensity of relative growth of certain organs in the process of race formation is different from those in speciation.

Contrary to the ontogenetic trends of individual growth, the intestines of large races and species are relatively longer than those of related smaller types. As the physiological efficiency of the intestine grows by the square, and the size of the body to be served increases by the cube, it becomes evident that in larger specimens variants with a relatively longer intestinal tract were favored in selection. In mammals I calculated the allometric exponent of the intestine as α=0·49 (relative length of intestine in percentage of head and body length) and in birds (length of intestine in percentage of $\sqrt[3]{\text{body weight}}$) as α=0·52. (In dogs, Klatt [1923] found 0·45 for the small intestine [ileum]

147

and 0·53 for the large intestine [colon].) Thus one may state that in phylogeny the intestine – as an organ of surface-proportionate efficiency – grows proportionately to the metabolic requirements of the body growing three-dimensionally. This is also evident from the oxygen consumption of large and small related animals, for which I found a similar coefficient of 0·54, using the data on oxygen consumption per minute in rabbit, rat, and mouse as reported by Kestner and Plaut (1924, p. 1109). This positive correlation of the size of intestine and body definitely is not the result of ontogenetic allometrics, as these are negative towards the end of ontogenetic development.

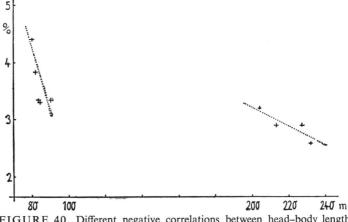

FIGURE 40. Different negative correlations between head–body length (abscissa) and relative diameter of eyes (ordinate) in the mouse (left) and the rat (right).

In rats, for example, the relative weight of the intestine decreases steadily after the sixth postnatal week (Jackson, 1913); in doves this occurs after the eleventh postnatal day (L. Kaufmann, 1927); and with hens the situation is similar (Latimer, 1925). A similar rule applies to the growth of bones, as in these, too, their efficiency with increasing size is proportionate to the square (i.e. to their cross section), but the body to be supported grows by the cube. For further details and data see Dosse (1937) on the weight of skeletons in birds, and Schlabritzky (1953) on the relative growth of bones in races of domestic fowl.

If there is a number of serial organs or parts of organs in an organism and if this number is not precisely fixed for a whole group, in the larger forms these organs will be not larger, but more numerous. According to Schilder (1950), who counted the number of facets in insect eyes, the barbules of bird feathers, the ambulacral feet in echinoderms, the teeth on the radula of snails, the pinnae and pinnules of plant leaves, the sori of ferns, and similar structures, the number of such serial organs increases slowly but steadily with increasing body size. Mathematically, this process can be approximated by the formula $Z^c = a \cdot L$, in which Z means the number of organs, L is body size, and c and a are coefficients, the former accounting for the number of individuals belonging to the category examined and the latter for the absolute size of these individuals. Usually c is greater than 1 (average of 62 cases: 1·72).

Allometric growth and shifts of proportions in consequence of increasing body size also have an important bearing on histological and cytological structures, as we have found in recent years from studies carried out at the Zoological Institute of Münster on numerous objects. A good example of the importance of these results is the allometric correlations of the vertebrate eye. Generally, large mammals, birds, amphibians, and fishes have relatively smaller eyes than related smaller species. This rule was established by Von

FIGURE 41. Median sections of the eyes of cow, domestic rabbit, and mouse *Evotomys glareolus*, drawn on same scale to show the relative size of lens and cornea.

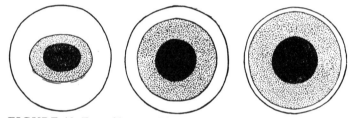

FIGURE 42. Eyes of horse, wild rabbit, and mouse *Apodemus silvaticus*, drawn on same scale to show the increase of relative cornea size in the smaller species (cornea dotted, pupil black).

Haller (1762), and was confirmed and extended later on (B. Rensch, 1948). It also applies to fossil material, as can be learned, for example, from the ancestry of the horse, in which the size of the eyes can be estimated from the size of the orbital cavities. This fact corresponds to the postnatal negative allometry of the eyes which (like the allometry of the brain) remains unchanged in phylogeny, as may be seen from the growth curves in the study by Schlabritzky on domestic fowl. If we compare related species or genera, such as, for instance, the rat (*Epimys norvegicus*) and the mouse (*Mus musculus*) with regard to the relative size of their eyes, a different allometic correlation results (Figure 40). Simultaneously, the proportions of the various parts of the eyes have been altered: in large mammals the cornea tends to be relatively smaller and its longitudinal diameter is often relatively longer than in small related species. Moreover, large types of mammals have a relatively smaller and often flatter lens than the smaller relatives (Tables 13 and 14 and Figures 41 and 42; see also B. Rensch, 1948). This rule does not apply to urodeles

TABLE 13. Relative size of cornea in mammals of different body size.

Species	Number of Eyes Measured	Largest Diameter of Cornea as Percent of Largest Diameter of Eye	Smallest Diameter of Cornea as Percent of Smallest Diameter of Eye
Horse	2	70·8	46·5
Cow	4	69·0	55·9
Sheep	2	68·4	52·8
Pig	4	64·7	56·0
Wild Rabbit	2	75·2	75·0
Squirrel	2	78·5	77·0
Wood Mouse	2	92·7	—
Red Vole	2	86·1	—
Hedgehog	2	93·8	—

TABLE 14. Relative volume and relative thickness of lens in the eyes of five groups of mammals of different body size.

Species	Number of Eyes Measured	Average volume of Lens as Percent of Volume of Eye	Average thickness of Lens as Percent of Lens Diameter
Horse (sex unknown)	2	4·2	62·5
Cow or Bull (sex unknown)	5	6·8	67·3
Sheep (sex unknown)	2	7·0	69·9
Pig (sex unknown)	4	6·3	77·4
Capybara (Hydrochoerus)	1	11·1	—
Wild Rabitt (2♂, 1♀)	5	14·6	69·6
Domestic Rabbit (sex unknown)	3	21·1	74·0
Common Rat (4♂, 1♀)	9	31·9	84·8
Water Vole (1♂)	1	26·7	80·0
House Mouse (1♀)	1	40·0	78·3
Wood Mouse (3♂, 6♀)	16	47·7	84·9
Field Mouse (3♀)	5	30·6	81·2
Red Vole (3♂)	6	38·5	81·8

(Möller, 1950). Such changes in the anatomical substrate, of course, have a bearing on the visual performance, as with eyes having relatively spherical and relatively large lenses only a poor accommodation can be effected, because the large lens occupies most of the interior cavity and no space, or not enough, is left for movements of accommodation.

Visual function is even more impaired by changes of the retina. In mammals, birds, and amphibians the larger forms have a relatively (and in some

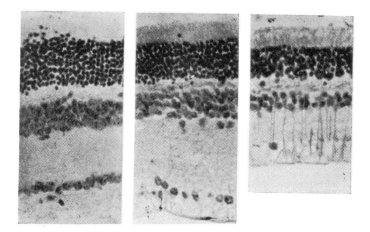

FIGURE 43. Horizontal transverse section of the retina from the center of the back of the eye. Left: mouse; center: albino rat; right: domestic rabbit.

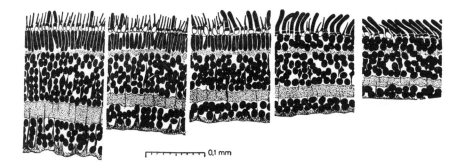

0,1 mm

FIGURE 44. Homologous parts from the central retina of three newts and two salamanders of different body size (enlargement on same scale). (After A. Möller.)

cases even absolutely) thinner retina than related smaller types (Burkhardt, 1931, B. Rensch, 1948; Möller, 1950; see also Figures 43 and 44). In the larger species, the cells in the ganglionic layer and in the inner and outer nuclear layers are less crowded than in smaller relatives. According to Möller, who counted the number of these cells in large and small species of *Triturus* and *Salamandra*, the total sum of ganglion cells in the layers of the multipolar and bipolar cells is about equal in both size categories, but the number of visual cells (i.e. of rods and cones) is considerably greater in the large species (Table 15). Furthermore, in the smaller animals the number of cones and double cones is relatively greater than in the larger kindred forms (Table 16). The relatively thinner retina of large animals corresponds to the normal ontogenetic trend of growth, as the layer of retinal cells is denser in young animals than in adults. The difference in the relative number of rods and cones, however, seems to have been caused by processes of selection (see below on changes of the structural type at the lower size limit). It is interesting to note that in cyprinodonts the situation is quite similar, as H. Müller (1952) proved that in small specimens of *Lebistes reticulatus* the retina is relatively thicker, the layer of cells is denser, and the percentage of cones is relatively greater than in larger species. In contrast to these findings, Bucciante and De Lorenzi (1929, 1930) reported that in the cow and the mouse the number of visual cells per μ^2 is equal. But in the cow, 466 visual cells are connected to one multipolar ganglionic cell on the average, while in the mouse this average is ninety-two visual cells to one ganglionic cell.

The effects of histologic growth gradients become especially important with regard to the brain. Harde (1949) proved that the cerebral cortex of the white mouse does not grow isometrically in postnatal development, but that many regions and areas grow with positive or negative allometry when compared to the forebrain as a whole. These separate allometric tendencies will often change more than once, usually when the eyes are first opened (thirteenth day) and at the onset of maturity. Usually, the phylogenetically older brain parts grow isometrically (e.g. the semicortex) or with negative allometry until the thirteenth postnatal day, after which a period of slightly positive allometry begins, lasting until the onset of sexual maturity, and followed by a final phase of slightly negative allometry (Figure 45). The schizocortex grows with negative allometry until the individual reaches sexual maturity, and then its growth is isometric. The largest and most progressive region, the holocortex 7-stratificatus, grows with mainly positive allometry until the onset of maturity.

To evaluate the influence and persistence of these ontogenetic gradients in the phylogenetic development, C. Schulz (1951) compared adult rodents of different body size. In the series mouse-rat-rabbit (the latter as a lagomorph rodent should be used as a model only, not strictly belonging to the series from the systematist's point of view) the relative size of the holocortex 7-stratificatus (in percentages of the whole pallium) increases from 37·2 to 39·8 and to 42·0 (data statistically significant). The ontogenetic trend of slightly

TABLE 15. Total number of neurons and visual cells in the retina of two species of *Triturus* and *Salamandra* differing in body size (after Möller).

Species	Diameter of eye	Total Number of Ganglion Cells	Total Number of Cells of Inner Granular Layer	Total Number of Distal Ends of Visual Cells
Trit. vulgaris (10♀)	2·7	193,600	511,000	171,300
Trit. cristatus (8♀)	2·2	207,500	566,000	224,300
Sal. atra (4♀)	3·3	316,500	957,000	386,300
Sal. maculosa (4♀)	5·5	308,300	995,000	533,000

TABLE 16. Number of single and double visual cones and number of rods in greatest sagittal sections of the retina in two species of *Triturus* and *Salamandra* differing in body size. Average data calculated from 15(*Tr.*) or 8(*S.*) sections (after Möller).

Species	Single Cones	Double Cones	Rods
Trit. vulgaris (10♀)	259	107	67
Trit. cristatus (8♀)	200	53	202
Sal. atra (4♀)	181	78	308
Sal. maculosa (4♀)	192	54	390

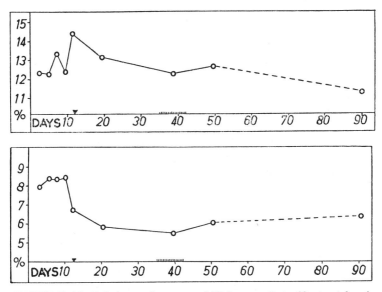

FIGURE 45. Relative surface area of Holocortex-2-stratificatus (above) and Holocortex-5-stratificatus (below) in percentage of the whole hemisphere (ordinate) of the white mouse. Abscissa: age in days. Wedge: day of opening the eyes. Short vertical lines: onset of maturity. (After K. W. Harde.)

negative allometry or even isometry in the semicortex becomes even more negative in phylogeny, which is evident from the decreasing relative size of the semicortex in the above series: mouse 18·7-rat 13·9-rabbit 9·0 percent. In the holocortex-5-stratificatus a complete reversal of ontogenetic tendency takes place in phylogeny. In ontogenetic development this brain part grows with positive allometry until thirteen days after birth (first opening of eyes), and the final growth is negatively allometric. In the phyletic 'model' series mouse-rat-rabbit, however, this region shows a steady increase of relative size: 12·2-13·8-17·0 percent with statistically significant differences. Apparently this region – being a progressive brain part of considerable size – was favored by selection. Similarly to the examples presented above, the other brain areas sometimes continue the ontogenetic growth trends in phylogeny, but in some cases the phyletic tendency is markedly opposed to ontogenetic gradients. The brain of the dwarf mouse (*Micromys*), for example, corresponds only slightly with the ontogenetic trends of the white mouse. An investigation of the brains of Indian squirrels of very different body size (*Ratufa indica, Funambulus tristriatus, E. palmarum*) gave similar results (Harde, 1955): semi-cortex, isocortex and schizocortex show a parallelism of ontogenetic and phylogenetic allometry, whereas the holocortex bistratificus is relatively smaller in giant squirrels (although ontogenetically in mice it grows with positive allometry).

As has been already mentioned, in domestic fowl the neostriatum grows with positive allometry, and correspondingly in larger races adult specimens have a relatively large neostriatum and a relatively small cortical region. Other parts of the brain show a similar parallelism of ontogenetic and phylogenetic allometry (Krumschmidt, 1956). The same holds good for lizards of different body size. Here the neostriatum grows with positive allometry in both ontogeny and phylogeny. The positive allometry of the cerebellum causes a relatively larger size of this part of the brain in the larger species (Rose, 1956).

In amphibians Nolte (1953) found, for instance, a relatively larger area dorsalis in larger species according to their positive allometrical growth (Homeyer, 1951). The striatum growing with negative allometry is relatively smaller in larger newts compared with smaller species, but relatively larger in the larger salamanders (*Salamandra*).

In fishes the smallest species of the Cyprinodontidae (*Heterandria*) has a relatively large tectum opticum corresponding with the positively allometrical growth (Wellensiek, 1953).

Summing up, we may state that the separate parts and regions of the brains of all vertebrates show different growth gradients and that relatively larger races or species have other proportions than smaller ones, although in some cases ontogenetic and phylogenetic allometry are mutually opposed.

In histology, too, differences correlated with body size are to be found. In rodents, the number of cortical cells of the brain is approximately proportionate to the length of the spinal cord in the adult animals (Brummelkamp,

1940). The density of cortical cells decreases with increasing body size, but the number of cells in homologous parts of the cortex remains about the same, which is due to the different cell size (Bok and M. J. van Erp Taalman Kip, 1939). In certain brain regions of amphibians, also the absolute number of cells is greater in larger species (Nolte, 1953). Large species of mammals with larger ganglionic cells have more dendritic ramifications than smaller relatives (Bok, 1936; B. Rensch, 1949, 1953). This becomes especially evident if one compares related species with considerably differing body size. A typical example is provided by *Hydrochoerus capybara* and *Cavia cobaya* (body weight of the former is 107 : 1 of the latter; Figure 46, after Spina Franca Netto, 1951). Such more numerous ramifications evidently are more effective switch mechanisms, and it is probable that with increasing body size and the correlated increase of relative size of progressive brain areas and of absolute brain size, the function of this important central organ became more efficient.

As it is not possible to enumerate all the other histological differences depending on body size which we have found, it will be sufficient to mention a few examples. The number of cells capable of division, such as blood corpuscles, the cells of bones and connective tissues, epithelia, and the like, is greater in large species, and sometimes the size of these cells is a little larger, but not at all in proportion to the body size. Measuring the size of cell nuclei in the coiled parts of the renal glomeruli, I found an average of $5 \cdot 8 \times 5 \cdot 1 \mu$ ($n=52$) in the rat (*Epimys norvegicus*) and of $5 \cdot 6 \times 4 \cdot 8$ μ ($n=64$) in the mouse (*Mus musculus*). Counting the number of cells on homologous cross sections of the jugular-metajugular bones I found an average of 137 nuclei (thickness of cross sections, 8μ) in the great tit (*Parus m. major*) from central Europe, and only 115 in the smaller race *P. m. cinereus* from the Sunda Islands. The areas of these cross sections were 100 : 86 and the relevant cell numbers 100 : 84, i.e. there was no significant difference of cell size between the two races. For further data of this kind the reader is referred to G. Levi (1914, 1925) and Teissier (1939) on vertebrates, and to Conklin (1912), Löwenthal (1923), and Partmann (1948) on invertebrates. The cells of permanent tissues (not capable of further divisions) of vertebrates, such as nerve cells, muscle and lens fibers, and the like, are equal in number, but different in size when related species of different body size are compared. The size of these cells seems to be positively correlated to body size (Figure 47), which was also proved by Linzbach (1950) in the fibers of the myocardium.

Due to such differences in number and size of cells, the histological structure of many organs depends on body size. Small mammals, for example, have relatively larger Langerhans' cells and the relative size of their island tissue (in percentage of the whole pancreas) is larger than in their larger relatives (Padour, 1950). In races of domestic fowl the situation is quite similar (Schlabritzky, 1953). In large animals the number of renal glomeruli is greater than in related smaller species: I counted 279–347 (average: 308) glomeruli on median longitudinal sections in the rat *Epimys norvegicus* and 132–159 (average: 146) on corresponding sections of the mouse (*Mus musculus*). Besides, in large species the glomeruli are composed of more cells than in smaller relatives: in the rat I counted 63–107 (average: 83) and in the mouse 45–69 (average: 54) cells. In races of sheep, an increase of body size

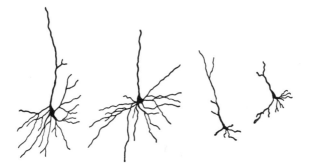

FIGURE 46. Medium-sized pyramidal cells (second layer) from homologous parts of the cortex of two rodents of different body size. Left: *Capybara*; right: guinea pig. Enlargement on same scale. (After A. Spina Franca Netto.)

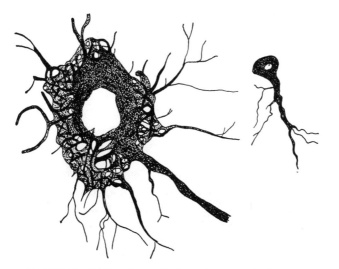

FIGURE 47. Ganglion cells from the sympathetic nerve of teleosts of different body size. Left: *Orthagoriscus mola* (50 kg.); right: *Trigla corax* (340 gm.). (After Levi, 1925.)

is accompanied by an increase of the size of glomeruli, while in larger races of rabbits the number of glomeruli is increased (Schilling, 1951). The type of renal papillae is also correlated with body size (Sperber, 1944).

In large species of insects the structural type of the mid-gut is different from that of related smaller species, the epithelium being of the cylindrical type in the large animals, and of the cubic or flattened type in the small forms

(Figure 48; compare Partmann, 1948; B. Rensch, 1948). This phyletic difference corresponds with the growth gradients of ontogeny. The indirect wing muscles of very small Diptera are a compact mass of fibers, while in large species the groups of fibers are more separate and less compact (Partmann). Above all, there is a strong difference in the histological structure of the brain: in large species the mushroom bodies are far more differentiated than in related smaller species (Figure 49; Goossen, 1949). Obviously, the relative size of surface area of neuropile substance is larger, and (in contrast to brain differences in mammals) the absolute number of globuli cells is greater. If one counts the globuli cells of the corpora pedunculata on maximum longitudinal sections one finds 615 per slide in the water beetle *Dytiscus marginalis* (body length 27 mm.) and only 188 in the related smaller *Ilybius fenestratus* (12 mm. long). The relevant data of the cockchafer *Melolontha vulgaris* (20·6 mm. long) and its smaller relative *Phyllopertha* are 1,176 and 624, and in the hornet *Vespa crabro* (22·2 mm. long) we find about 1,000 globuli cells, while in the related smaller species *Ancristoceros parietinus* (7·9 mm. long) this number is only about 500. Closely parallel to this more advanced differentiation of histological structure, the level of instinctive patterns and performances is different in large and small species (see below). Hence, in contrast to the formerly widespread view, the degree of differentiation in the mushroom bodies is more than a mere index of the phylogenetic level of a certain insect group.

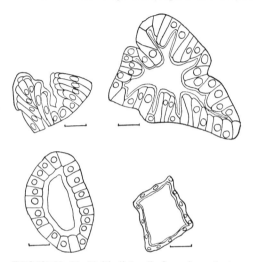

FIGURE 48. Epithelial cells from homologous parts of the mid-gut of mosquitoes of different body size. Above, left: *Ctenophora atra*; right: *Theobaldia annulata*; below, left: *Lycoria spec.*; right: *Contarinia spec.* Scale: 0·02 mm. (After W. Partmann.)

All these rules, of course, cannot claim a general validity. In the Australian *Macropanesthia rhinocerus*, a giant cockroach, Day (1950) could not find any higher level of differentiation in the corpora pedunculata nor in any other organs, compared to medium-sized cockroaches. This may depend mainly upon the different ontogenetic development of the brain structure. In cockroaches the positively allometric phase of the corpora pedunculata is to be found in the first and second stage of the larvae. In later stages this brain structure grows with slight negative allometry (Neder, unpublished). In holometabolic insects, however, such as bees, wasps, and flies, the positively allometric corpora pedunculata appears at the end of larval development.

Especially important as a selective advantage is the absolutely larger number of eggs in the large species of poikilothermic animals in comparison with related smaller species. This rule is important with regard to its bearing on selection. The data in Table 17 were compiled from Brehm's *Tierleben* (4th ed.); Sternfeld (1913); Walter (1913); Vogt and Hofer (1909), and from the work on aquarium fishes by Holly, Meinken, and Rachow, and give a good idea of the differences of egg number in related small and large forms. In this table those forms have been omitted in which the pattern of parental care for the offspring is more marked, as such species generally have a much smaller number of eggs or young than non-caring types. Of course, the material of

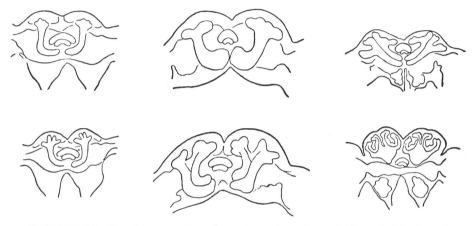

FIGURE 49. Homologous sections from the protocerebrum (with central body and corpora pedunculata) of three pairs of related insects of different body size. Sections of the smaller species (above) much more enlarged than those of the larger types (below). Above, left: *Ectobius sylvestris*, below, left: *Blatta orientalis* (Blattidae); above, center: *Phyllopertha horticola*, below, center: *Melolontha vulgaris* (Scarabaeidae); above, right: *Ancristocerus parietinus*, below, right: *Vespa crabro* (Vespidae). (After H. Goossen.)

the table represents only species with strong differences of body size. A similar rule probably applies to many groups of arthropods. So, for example, Schwerdtfeger (1944) stated that the egg number of butterflies is proportionate to the weight of the pupae: Schütte (1957) confirmed this rule in the moth *Tortrix viridana* by counting the number of ripe eggs in the oviducts. Petersen (1950) proved that in Lycosid spiders the number of eggs depends on the body size of the mother. In *Drosophila* Goetsch (1948) found that large females, fed on a diet of Vitamin T which produced an increase of body size by about 23 percent, laid eight to ten times as many eggs as were laid by normal specimens.

Secondary sex characters of many animals usually represent structures of positively allometric growth because they usually become conspicuous during a more advanced period of ontogenetic development (due in mammals to the increasing influence of sex hormones). Thus, the rule is valid that in numerous animal groups the sexual dimorphism increases with body size (B. Rensch,

TABLE 17. Egg numbers of large and small poikilotherms of the same order or family. (After Holly, Meinken, and Rachow; Vogt and Hofer; Walter, Sternfeld, and Brehm's *Tierleben.*)

Species	Size	Eggs
A. 1. *Small Lacertilia*		
Lacerta agilis	18–23 cm.	4–14
Lacerta viridis	30–40 cm.	4–13
Psammodromus algtrus	27 cm.	8
Eremias guttulata	16 cm.	2–4
Agama stellio	28 cm.	9
Draco volans	11 cm.	3–4
Uromastix acanthinurus	up to 50 cm.	8
Anolis carolinensis	14–22 cm.	2
Tropidurus torquatus	25 cm.	4
A. 2. *Large Lacertilia and Crocodilia*		
Varanus niloticus	170 cm.	24
Varanus griseus	130 cm.	10–20
Varanus flavescens	longer than 100 cm.	24
Iguana tuberculata	140–160 cm.	10–35
Tubinambis teyuixin	1,000 cm.	50–60
Crocodilus niloticus	up to 600 cm.	90–100
Crocodilus americanus	up to 600 cm.	100
Gavialis gangeticus	up to 500 cm.	40
Caiman niger	up to 400 cm.	30–40
Alligator mississipensis	up to 400 cm.	30–60
B. 1. *Small Chelonia*		
Emys orbicularis	30–35 cm.	9–11
Testudo graeca	20 cm.	8–15
Terrupene carolina	13–17 cm.	5–6
B. 2. *Large Chelonia*		
Chelonia mydas	110 cm.	300–400
Dermochelys coriacea	150 cm.	1,000
Testudo elephantopus	150 cm.	10–14
C. 1. *Small Ophidia*		
Thyphlops vermicularis	33 cm.	6
Tropidonotus natrix	120–150 cm.	15–35
Tropidonotus tesselatus	70–80 cm.	5–13
Zamenis gemonensis	150 cm.	8–15
Coluber longissimus	100–150 cm.	5–8
C. 2. *Large Ophidia*		
Python molurus	up to 400 cm.	45
Python reticulatus	up to 1,000 cm.	96
Python sebae	300–500 cm.	*ca.* 100
Eunectes murinus	300 cm.	30–80
D. 1. *Small Cyprinidae*		
Barbus oligolepis	5 cm.	300
Barbus semifasciatus	10 cm.	150–300
Brachydunio nigrofasciatus	4·5 cm.	300
Brachydunio rerio	5–6 cm.	400–500

Species	Size	Eggs
D. 2. *Large Cyprinidae*		
Cyprinus carpio	60–100 cm.	up to 300,000
Barbus fluviatilis	up to 80 cm.	3,000–8,000
Tinea vulgaris	up to 70 cm.	300,000
Abramis brama	50–70 cm.	140,000
Aspius rapax	60–70 cm.	100,000
E. 1. *Small Salmonidae*		
Salmo fontinalis	20–45 cm.	7,000
Salmo trutta	25–40 cm.	1,500–2,000
E. 2. *Large Salmonidae*		
Salmo salar	50–150 cm.	10,000–40,000
Salmo hucho	60–100 cm.	10,000–40,000
F. 1. *Small Osphromenidae*		
Macropodus opercularis	8–10 cm.	300–400
Betta pugnax	10 cm.	100
F. 2. *Large Osphromenidae*		
Osphromenus gourami	up to 100 cm.	800–1,000
G. 1. *Small Cichlidae*		
Haplochromis multicolor	7 cm.	30–100
Apistogramma pleurotaenia	7·5 cm.	40–150 cm.
Etroplus maculatus	8 cm.	100–300
G. 2. *Large Cichlidae*		
Herichtys cyanoguttatus	30 cm.	up to 500
Cichlasoma facetum	30 cm.	300–1,000
Hemichromis fasciatus	30 cm.	1,000
H. 1. *Small Siluridae*		
Corydoras punctatus	7·5 cm.	60–250
H. 2. *Large Siluridae*		
Silurus glanis	up to 220 cm.	60,000

1950). This rule, however, applies only to subspecies of a species, to related species of a genus, or to related genera of a family. In species of birds in which the male is larger than the female, the relative sexual difference increases with body size. If, by way of exception, the females are larger than the males, as among many species of birds of prey, the opposite correlation applies, i.e. the greater sexual difference is found in the smaller species. This fact is probably due to the more negative allometry of sexual characters in the male (B. Rensch, 1953). This rule of sexual differences is especially well demonstrated in some groups of beetles showing an excessive differentiation of the male antennae (as in some Lamellicornia; Figure 50) or of the tarsal sucking discs in the male (as in Dytiscidae). In mammals, the anatomical structures used as a weapon, such as antlers, horns, teeth, and the like, provide more examples of this kind. One need only compare large and small species of deer, pigs, baboons, and so forth.

Physiological Consequences of Allometric Growth. In my statements on the relative sexual differences and the different structure of eye and brain in related species of different body size, I have already pointed out that such changes of the structural type have far-reaching physiological effects. Let us examine some of these consequences in more detail. First, the rate of metabolism depends on body size. As Bergmann (1847, 1855) Rubner (1883), and

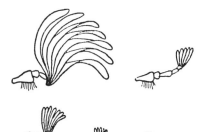

FIGURE 50. Increase of sexual dimorphism in the antennae with increasing body size in three species of Scarabaeidae. Above: *Polyphylla fullo* ♂ and ♀ (24–34 mm.). Below, left: *Melolontha vulgaris* ♂ and ♀ (20–25 mm.); right: *Phyllopertha horticola* ♂ and ♀ (8–12 mm.).

Richet (1885) recognized, in homoiotherms the rate of oxygen consumption depends on the surface area of the animal. With increasing body size, the volume of the body grows by the cube, but the surface radiating, i.e. losing, body heat increases by the square, and hence the loss of body heat is relatively smaller in large animals, and fewer calories are required to maintain the proper body temperature. A different opinion regarding this is held by Kestner (1934) and Blank (1934), who argue that the relatively low oxygen consumption of large animals is due to the relatively small size of oxygen-consuming organs, like liver and kidneys. It is obvious, however, that homoiothermal animals are rather narrowly confined to a certain range of environmental temperatures – Bergmann's Rule bears convincing evidence of this fact, as the body size in geographic races corresponds to climatic zones – and this supports, in my opinion, the view that the intensity of metabolism depends primarily on the surface area. At any rate, there can be no question about the fact that in small homoiotherms metabolism is more intense than in larger types, and the same applies to most poikilothermic animals and also to plants (compare Hemmingsen, 1950; Von Bertalanffy, 1951; and others). In very large homoiotherms, however, the applicability of this rule is limited, as according to Von Buddenbrock (1934) the intensity of metabolism (per unit of weight) rises again in animals exceeding a certain body size. This fact is due to the amount of energy used in muscular performances, locomotion and the like, which is relatively larger in very large animals, so that the loss of energy by heat radiation is less important.

In consequence of the relatively low oxygen consumption, in larger animals the rate of respiration and pulse rate are diminished. In the mouse, for instance, the heart will beat 520–780 times per minute, in the pig 60–80 times, in the horse 34–36 times, and in the elephant 25–28 times. In the sparrow the

TABLE 18. Periods of gestation in mammals of the same orders but of different body size. (After Krumbiegel and Brehm; Heck; Hilzheimer, 1930; and Heinrich, 1930.)

Species	Period of Gestation in Weeks
Erinaceus	6
Talpa	4–6
Sorex	3–4
Ursus	28–30
Procyon	9
Martes	9
Mustela	5
Felis tigris	14–16
Felis pardalis	12–14
Felis ocreata dom.	8
Alces	36
Cervus	33–34
Dama	32
Taurotragus	34–36
Antelope	6
Bos	41
Capra	21–22
Castor	6
Oryctolagus	4
Mus	3
Nemertinus	33
Callithrix	13
Hippopotamus	34
Choeropus	26
Sus	17

heart beats 745–850 times per minute, in the hooded crow about 378 times, and in the turkey about 93 times. For further details see the summaries by Von Buddenbrock (1924), Stresemann (1927–34), and others. In consequence of these facts large animals have a relatively smaller amount of blood sugar and a lower body temperature than kindred smaller species (Erlenbach, 1938). Moreover, in the large forms the blood circulation is slower, but blood pressure is higher, and locomotion is relatively slower than in smaller types.

Due to the less intense metabolic processes in large animals, the periods of ontogenetic development are longer, i.e. their gestation (Table 18) or incubation periods (Table 19) are longer and their postnatal growth is slower than in their smaller relatives. In a mouse, for example, sexual maturity is

TABLE 19. Incubation periods in birds of the same orders but of different body size. (After Heinroth, 1922.)

Species	Body Weight of ♀ in Grams	Incubation Period in Days
Podiceps iristatus	1,000	25
Podiceps ruficollis	150–200	20
Ardea cinerea	1,500	25–26
Ardetta minuta	150	16–17
Cygnus olor	9,000	35
Anas crecca	330	22
Haliaëtus albicilla	5,000	35
Falco tinnunculus	220	28
Meleagris gallopavo	3,500	28
Coturnix coturnix	100	18
Fulica atra	650	22–23
Porzana porzana	80	18
Numenius arquatus	900	$29\frac{1}{2}$
Gallinago gallinago	100	$19\frac{1}{2}$
Larus marinus	1,500	26
Sterna minuta	40	$21\frac{1}{2}$
Goura coronata	2,000	28
Streptopelia turtur	160	$14\frac{1}{2}$
Psittacus erithacus	400	30
Melopsittacus undulatus	30	18
Bubo bubo	2,750	33–36
Athene noctua	175	28
Corvus corax	1,300	20–21
Taeniopygia castanotis	12	11–12

reached in five to six weeks, and in a rat in about two weeks after birth, but in the beaver this maturity is attained one year after birth, and numerous other pairs of related animals provide similar data: red deer one and a half and elk two and a half years, wolf two and a half and brown bear three to four years, wild pig one and a half and rhinoceros five to six years, and so forth. These data indicate that the tempo of development and average age (life expectancy) of an animal type are affected by and depend upon body size, and, generally, that large homoiotherms reach a higher individual average age than related smaller types (Table 20). Backman (1943) could even develop a mathematical formula for the correlation between the tempo of ontogenetic development and average age, by which he could predict the major events of

TABLE 20. Correlation of average life span with body size in mammals of the same orders but of different body size. (Data from Korschelt, 1924; Przibram, 1927; Stein, 1938.)

Species	Average Life Span in Years
Castor	20–25
Arctomys	14
Lepus	7–8
Cavia	4–5
Mus	1½
Bos	20–25
Capra	12–15
Alces	20
Capreolus	15–16
Camelus	25–45
Auchenia	15
Ursus	40–50
Canis	12–15
Felis leo	20–25
Felis pardalis	14
Felis ocreata dom.	9–10

the life cycle (i.e. birth, onset of maturity and senility, and death) using the 'individual time' (*Eigenzeit*) of each organism as the basis of his calculation. The course of 'organic time', then, represents a pattern similar in all animals, but different from the physical time scale. (In this context it may be indicated that in different animals the time sense, i.e. the awareness of individual time spans, may be different. This is a question mentioned by Mach in 1865.)

With regard to possible advantages in selection and to the problem of progressive evolution, it is important to study the functional changes of brain and sense organs brought about by absolute increase of these organs in consequence of changes of body size. Apparently, the absolute number of visual cells in the eyes of large vertebrates is greater than in the eyes of smaller relatives (for precise counts of photosensitive cells in the urodele eye see Möller, 1950). Similarly, large insects have more ommatidia than related smaller species. This may be illustrated by a few data: the blowfly *Calliphora vomitoria* has 4,585 ommatidia, but the small *Drosophila melanogaster* only 688 (Partmann, 1948); in the large ground beetle *Carabus coriaceus* the number of ommatidia is 4,250, in the medium-sized *C. violaceus* 3,200, and in the related smaller *Bembidium rupestre* about 400 (Leinemann, 1904). With a large number of photosensitive cells the resolving power of the visual apparatus is, of course, better than with a few visual cells only, and hence smaller details of visual patterns can be perceived and reacted to.

In an absolutely larger brain the number of ganglion cells is often increased, especially in insects. In vertebrates (compare Figure 46) the size of the ganglion cells increases with body size, and hence such larger neurons tend to have more dendritic ramifications than smaller ones and provide a more efficient associative mechanism capable of more complicated performances. Besides this improvement, a phyletic increase of body size is often accompanied by an increase of the relative size of progressive brain parts, i.e. its most progressive or complicated regions (such as the seven-layered cortex in mammals or the mushroom bodies in insects). Hence it is evident that in the course of phyletic increase of body size a higher level of instinct patterns may be attained.

With regard to instincts we may state that in all small European species of bees and wasps (such as *Andrena*, *Halictus*, *Ancristocerus*, and *Symmorphus*) there are no social instincts such as exist in the large species of *Apis*, *Bombus*, *Vespa*, *Polistes*, and so forth. Among ants the small Ponerinae have simpler social instincts than the larger Camponotinae, and the smallest ant species *Monomorium salomonis* (length 1·8–2·5 mm.) show only poor social instincts. Among the Scarabaeidae the large species like *Scarabaeus*, *Copris*, and *Geotrupes* show complicated instincts in the care of their eggs, and in the construction of breeding pills from dung, subterranean canals, and breeding chambers, whereas the many small species of the related genus *Aphodius* have not developed such instincts (they simply deposit their eggs in dung). The same holds good for the Silphidae, in which the large species of *Necrophorus* show very complex instincts by burrowing in the carcass and feeding the young larvae, whereas the small species of the related genus *Catops* (only 3 mm. long) show no such instincts.

In some cases larger vertebrates also show more complicated instincts than related smaller species in accordance with their greater sexual dimorphism (see above, p. 157). Thus the sexual display of the great bustard (*Otis tarda*) is more complicated than the display of the smaller related *O. tetrax*, and the same holds good in the larger species of grouse (*Lyrurus*, *Tetrao*) compared with the smaller related species (*Tetrastes*, *Lagopus*) or in the crested grebe (*Podiceps cristatus*) compared with the small species (*P. ruficollis*). Young rats show playing instincts; young mice do not play (Eibl-Eibesfeldt, 1950). But with regard to the enormous complication of brain structure in all vertebrates, even in the smallest ones, it seems doubtful whether a rule similar to that for insects can be established. Much clearer have been the results concerning the better capabilities of learning and memorizing in larger species compared with smaller ones.

In view of this possibility I had the brain performance of related animals of different body size examined. Concerning domestic fowl, Altevogt (1951) stated that a large race mastered simultaneously more visual discrimination tasks (up to seven) than dwarf races (up to five). The smaller races were quick in learning simple visual discriminations, but the larger types were better in learning more complicated and difficult discriminations, and, moreover, were superior in memorizing.

The same results are obtained in the study of mice and rats. In the latter species the visual learning process is quicker and the number of tasks finally mastered simultaneously is greater: three out of five rats learned eight tasks, one out of five mice seven tasks (Reetz, 1958). As to retention and transposition ability, the larger rats were again superior to the smaller mice (Von Boxberger, 1953; Reetz, 1958). In a comparison of visual learning ability in cyprinodonts, the smaller *Lebistes* learned more visual tasks (four) than the larger *Xiphophorus* (two), but the data were not sufficiently reliable. In transposing experiments both types were about equal, but invariably the large species was definitely better in memory (B. Rensch, 1953). Continuing this kind of visual discrimination experiment, we were able to show that a mammal with an absolutely large brain (about 6,000 gm.), the Indian elephant, is especially capable of learning and retaining a remarkable number of visual tasks (up to 20 pairs, or 40 different patterns). This animal retained the correct solution for 13 tasks for one year and it was also efficient in transposition and abstraction experiments (B. Rensch and Altevogt, 1953, 1955).

Finally, the experiments on the prelinguistic capability of 'comprehending' unnamed numbers by Koehler and his school showed that large bird species were better in learning and transposing seen or heard numbers than related smaller types. Generally, ravens were more efficient than jackdaws, and large parrots learned more than small parakeets (Koehler, 1939, 1943; Schiemann, 1949; Braun, 1952).

Summarizing our findings, then, we may state that races or species of larger body size show better capabilities of learning, memorizing, and perhaps also transposing (abstracting) than smaller related races or species (compare B. Rensch, 1955, 1956).

Summary of Complex Differences in Species of Different Body Size. We have analyzed the morphological and physiological consequences of allometric growth in some detail because they prove that each phyletic alteration of body size is followed by extraordinarily complex effects and because such an analysis provides a causal understanding of the 'systemic' character (*Ganzheitlichkeit*) and 'harmony' of phyletic transformations. If, according to Cope's Rule, a phyletic increase of body size takes place, quantitative shifts of proportions may lead to complete alterations of the structural type and these may bring to light new qualitative traits and characters. This is evident, for example, from a consideration of the eyes of arthropods by which a perception of visual forms is not possible unless they comprise several hundreds of ommatidia. Similarly, a central nervous system becomes qualitatively different from its previous type as soon as the number of neurons is increased, rendering possible a more distinct division of labor and the acquisition of new and more complicated instinct patterns and discrimination reactions. Teeth of a certain type may serve quite different functions as soon as they grow excessively large (e.g. in Proboscidia), and there are many more examples similar to these.

Quite essential in this context is the allometric growth of endocrine glands, which may have quite complex effects on the development of numerous organs in the course of ontogeny and phylogeny (compare the discussion of constructive genes, p. 129).

As so many consequences of phyletic increase of body size have been analyzed, we can now give a fairly precise evaluation of the complex effects of such alterations. The regularity of such shifts of proportions becomes quite evident from the fact that in many animal groups we can predict the characteristics in a given category of body size with 70 to 90 percent probability. For example, a hypothetical 'rat' of excessive body size, about as large as a beaver, would differ from its smaller relatives in the following characters. This animal compared with related smaller species would have a relatively smaller head, brain case (in relation to the facial bones), brain stem (in comparison to the brain as a whole), ears, and feet; a relatively shorter tail and shorter hairs; and a relatively smaller heart, liver, kidneys, pancreas, thyroid, and pituitary and adrenal glands. The weight of the bones would be relatively heavier, the facial bones relatively longer (in relation to the brain case), and the forebrain relatively larger (in relation to the brain as a whole). The retina of this giant rat would be relatively (and probably absolutely) thinner; the layer of ganglion cells and both granular layers of the eye would be less dense; the number of rods and cones would be relatively smaller. In the forebrain the cortex-7-stratificatus would be relatively larger, and the semicortex relatively smaller. The absolutely larger neurons of the brain would be less dense but would have many more dendritic ramifications. There would be equally large but definitely more numerous blood corpuscles and bone and connective tissue cells, and relatively smaller insulin-producing tissue of the pancreas. Finally, the general metabolism (especially the rate of oxygen consumption, breathing, pulse, and blood circulation) would be decreased; the amount of blood sugar would be less, and the relative speed in locomotion would be slower. In this giant rat, the onset of maturity would be postponed, the gestation period and average length of individual age would be prolonged, and the animal would be superior in learning ability and in memory.

Such predictions regarding morphological, anatomical, histological, physiological, and developmental characters are especially helpful in the evaluation of fossil animals, as many of the above-mentioned correlations contribute reliable clues to be followed in the reconstruction of animals from fossil fragments. In addition, they allow inferences to be made concerning some histological, physiological, and developmental traits of the fossil animal in question.

The more detailed analysis of allometric growth and its physiological consequences has revealed essential connections extending the basis of our understanding of the process of evolution. Now we know that ontogenetic allometry may have far-reaching effects in phyletic alterations of body size, as becomes especially evident in the successive stages of increasing body size according to Cope's Rule (compare Chapter 6, B IV on orthogenetic evolution). However, we have also seen that in numerous lines of descent the special growth coefficients are altered so that only the general trend – irrespective of the more or less intense positive or negative gradients – is maintained in phylogeny. This statement is illustrated especially by the growth coefficients of head, brain, and liver, which differ markedly in the species level and the subspecies level. There are other lines of descent proving that the phyletic trends of 'allometry' may be contrary to the growth tendencies of ontogeny (e.g. relative length of hairs and feathers, relative weight of bones, relative length of intestine, relative size of holocortex-5-stratificatus, and so forth). All these cases show that almost any phase of ontogenetic allometry may be prolonged or shortened, intensified or weakened in phylogeny, as far as such changes involve possible advantages in the process of selection. Apparently, then, each of these phases is controlled or brought about by different genes or – in cases of polygenic characters – modifiers.

We should not forget to stress the important fact that by analyses of the kind mentioned above the study of transspecific alterations has become a study of developmental gradients and that it is no longer a description of characters. Mutation does not affect persistent characters only, but also causes alterations of single gradients, rendering them more or less effective during a more or less prolonged or shortened time. All such changes may be strictly quantitative at first. These quantitative gene differences were rightly emphasized in 1927 by R. Goldschmidt, though in the first publication of his physiological theory of heredity he seems to have somewhat underrated the qualitative differences. It was Goldschmidt who first developed a clear idea of accelerations and retardations of certain gene-controlled developmental processes which are due to quantitative differences and their mutual interaction during phenogenesis. By temporal shifts of the beginning and end of such interactions, morphological (i.e. qualitative) differences will result. Summing up, we may state that mutation and selection provide a sound basis for understanding even the 'systemic' (*ganzheitliche*) changes of whole organisms, the so-called 'coadaptations' (or 'synorganizations': Remane).

Now the question arises how far body size is controlled by single genes and how far by interactions of several genes. Apparently, in most cases body size is controlled by the joint action of several genes, as body size may be considered to be the additive result of the sizes of various organs and structures. Crossbreeding experiments on races of mice of different body size (*Mus musc. musculus*; *Mus musc. bactrinaus*) by Green (1930, 1931, 1934) showed that several genes are jointly active in the control of body size, but there is also a gene limiting body length and weight, and another one determining the length of femur and tibia (1934). These results are well supported by the findings of Clark (1941) on races of *Peromyscus* in which no (or only very faint) correlations exist between the length of tail, foot, femur, skull, and mandibles. From his results when crossing rabbits of different body size, Castle (1931, 1934) opposed the assumption of polymeric factors (1936, 1941), because the genes concerned do not act in the same direction. The rate of ontogenetic growth in the embryo is directly affected by growth genes, while pigmentation genes cause different effects on the growth process: brown pigments increase the growth processes, while gray colors decrease them. Additive effects of several genetic factors acting in the same direction were also proved by crossing experiments with races of fowl of different body size, carried out by Jull and Quinn (1931) and by Lauth (1935).

Other examples proved that body size may be controlled by a single gene. This is the case in pituitary dwarfism in mice, where the production of a growth-promoting hormone by the anterior pituitary lobe is barred (Snell, 1929; Grüneberg, 1943; Smith and McDowell, 1931; and others). According to Schnecke (1941) and Nachtsheim (1943) dwarfism in rabbits may also be

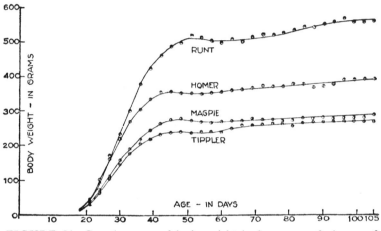

FIGURE 51. Growth curves of body weight in four races of pigeons of different body size. Abscissa: age in days; ordinate: body weight in grams. (After Riddle, Charles and Cauthen, 1932.)

caused by a single gene. Recessive sex-linked genes causing dwarfism have also been found in domestic fowl (see the review by Hutt, 1949).

The main factor in the development of homoiothermic animals is the tempo of postnatal growth, as at birth the differences of body size are relatively small (Figure 51; Castle and Gregory, 1929; Riddle, Charles and Cauthen, 1932). In homoiotherms, the length of the postnatal growth period may be affected by environmental temperature, which has been proved, for example, in mice: at 21°C. their growth period lasted only 60–80 days, but at 9°–13°C. growth continued up to 150–180 days after birth. The mice of the latter type became larger and their onset of sexual maturity was postponed (Kalabuchov, 1938).

In this context a finding by Gregory and Goss (1933) is of interest, that in young rabbits of races with different adult body size the glutathione content immediately after birth varies, but is proportionate to the body size of the adults. Glutathione, a tripeptide, is known to be a regulating factor in tissue metabolism and cell division. Moreover, botanical findings (Rowlands, 1954) suggest that perhaps a different production of desoxyribonucleic acid (correlated with mitosis) may be important. Finally, we must not forget the great influence of water content in all processes of growth (Haardick, 1941).

Hence, we may state that body size is brought about by single genes in some cases and by several genes in others. These genes initiate various growth gradients, often causing a complicated system of coordinations and mutual effects ('genic balance') in certain stages of ontogeny, as was emphasized by Sinnot and Dunn (1935).

Nevertheless, body size must invariably be considered as a *single* character in processes of selection, and if this character is favored, a large number of morphological, anatomical, histological, physiological, and developmental relations will be changed in the process.

The Gradients of Differentiation. The processes of allometric growth and their manifold effects are rendered even more complicated by gradients of differentiation affecting either the body as a whole or only some of its organs. Such gradients of differentiation, especially conspicuous in the shape and organization of extremities, are subject to transspecific alterations in correlation with alterations of body or organ size. J. S. Huxley (1932, Chapter IV, §6) mentioned the 'Law of anterio-posterior development', saying that generally the differentiation of anterior and proximal parts of the animal body is quicker and more advanced than that of the posterior and distal regions and organs. This rule is a special case of allometric growth, demonstrating the temporal sequence of the various periods of allometry by the spatial arrangement of various stages of differentiation, one behind the other. Mutations influencing the general gradients of differentiation may have essential effects in transspecific evolution, as, for example, the acceleration of such a gradient may effect a stronger reduction of posterior or distal organs and regions of the body. It is from this point of view that new light is shed on the shifting of the boundaries of the vertebral regions of thorax, loins, and tail, referred to above and first recognized in its evolutionary importance by Eimer and Fickert (1899) in their studies on the coloration of water birds.

Gradients of differentiation may also result in a more complicated shift of growth gradients, so that certain differences of proportions arise which may be approximated mathematically by Cartesian transformations, as was first demonstrated by D'Arcy Thompson (1917). Such alterations of the systems of coordinates provide a means of understanding some essential changes of the structural type in transspecific evolution, as is shown in Figures 52 and 53, where the shape of the sunfish *Orthagoriscus* arises from that of the *Diodon* type by Cartesian transformation and where several types of brachyuran carapace result from 'normal' ancestors by the same mathematical process. The famous painter Albrecht Dürer (1622; dating from 1527) applied similar methods of transformation, showing that certain differences of proportions in the human body are correlated with certain constitutional types.

Klatt analyzed several types of extreme dog races representing the kind of growth and differentiation mentioned above. Working especially on bulldogs and greyhounds, Klatt could demonstrate that a growth center probably exists in the anterior end of the embryo, which in bulldogs is more active and effective than in greyhounds, so that in bulldogs the skull, brain, eyes, and cranial muscles are developed more vigorously right from the beginning of ontogeny (Klatt, 1948; Klatt and Oboussier, 1951). Hence, this relatively stronger and more intense development of the anterior end of the embryo also affects the growth of such important hormone-producing glands as the

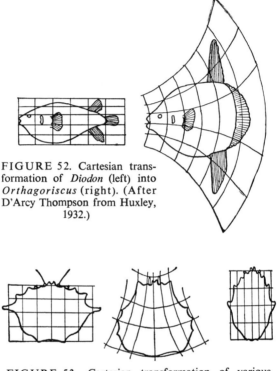

FIGURE 52. Cartesian transformation of *Diodon* (left) into *Orthagoriscus* (right). (After D'Arcy Thompson from Huxley, 1932.)

FIGURE 53. Cartesian transformation of various types of crab carapaces. (After D'Arcy Thompson from Huxley, 1932.)

pituitary and the thyroid, while others in a more posterior position – like the adrenal glands – remain about equal in size in both types of dogs.

As we shall see in our discussion on orthogenetic evolution, the concept of differentiation gradients also provides a reasonable clue to an understanding of vestigial organs and structures which are usually more distinct and obvious in the distal parts of the body.

Phyletic Size Limits and Consequent Alterations of the Structural Type. As we have seen earlier in this chapter, successive alterations of body size as the result of allometric growth and selection often lead to strong alterations in the morphology and physiology of a structural type. Such alterations of body size, however, cannot exceed certain upper and lower limits, reached as soon as the correlated shifts of proportions, reductions, and excessive growth processes in certain organs biologically are no longer tolerable. Phyletic size limits occur especially if shifts of proportion of single parts are progressing at different rates. For instance, the epithelium of intestine or lungs, being effective in proportion to its surface, grows by the square, while liver, brain, etc., being proportionate to volume, grow by the cube. A size limit is also determined by the fact that certain organs or tissues will not function efficiently if the number of cells is increased or decreased beyond a certain critical limit. As such phyletic size limits tend to involve selective disadvantages, they represent the final stage of evolution in numerous lines of descent. The giant forms of animal types especially provide these final evolutionary stages, unless they are altered in their structural type and develop mechanisms favored by selection so that a further increase (or decrease) of body size is rendered possible. We shall discuss some of these size limits in more detail.

In many animal groups, the lower size limit is often determined by the cellular structure, as cells may not be enlarged or reduced in size *ad libitum* because of the fixed functional relation of nucleus and cytoplasm, as regards metabolism and gene action, and because the size of the nucleus is determined by the number and size of chromosomes and genes. Obviously, it is the size of the molecules that finally determines the size of the cell. Hence, the smallest possible cell size also determines the size of those organs requiring a minimum number of cells to function properly or at least sufficiently, and therefore it also fixes the smallest possible body size. In approaching the lower size limit of phylogeny, the organisms usually reduce special structures and special organs and only the indispensable mechanisms are maintained.

A good example of this kind is provided by the minute marine snail *Caecum glabrum* (Opisthobranchia), 1 mm. long and 0·2 mm. wide, the anatomy of which was studied by Götze (1938). In this species, the cells are of about the same size as in larger Prosobranchia like *Littorina littorea*, according to measurements of the epithelial cells of head and intestine. Consequently, the intestinal gland of this minute snail consists of two tubules, and the gonads are represented by a single folded tube, while in other snails these organs are made up of quite numerous tubules forming a solid network of glandular elements. Hence, it is quite evident that a further reduction of these organs is impossible. The ganglion cells, especially those of the cerebral organs, are well developed and relatively very large. No gills are necessary, as due to the very large relative surface a sufficient oxygen intake is provided by skin respiration, and in such a small body the distances to be covered by diffusion of oxygen are short.

It is a common fact that some tarsal joints are reduced or missing in very small insects. The usual five tarsal joints of the larger Orthoptera, Coleoptera, Diptera, and Hymenoptera are reduced to only three in the dwarf beetles, family Ptiliidae, of about 0·25 to 1·4 mm. body length and in the equally small Sphaeriidae, and often one of these tarsal parts is not clearly separated from the other. In the tiny Thysanoptera and Mallophaga, only one or two tarsal elements are to be found, and in Siphunculata there is only one. In Cyclophoridae, a family of terrestrial Prosobranchia, all larger forms (diameter of shell: 1 cm. or more) are provided with a penis, while in the minute *Moulinsia* types from Indo-Australia (length of shell: 4–8 mm.), and in the even smaller species of *Diplommatina* this organ is absent or vestigial (Tielecke, 1939). In the smallest types of Turbellaria (Rhabdocoela) no ramifications of the intestine are to be found, and in very small cestodes the uterus also is not ramified.

The lower limit of body size may also be set when some organs reach a maximum of size in relation to the body as a whole, which cannot be exceeded because no more space is available in the extremely small body. In contrast to larger species of insects, for example, the brain of very small insects occupies all the space of the relatively large head capsule, which has been proved in very small gall midges (Cecidomyidae), flies, and plant lice (Goossen, 1949; B. Rensch, 1948). In these insects, the thorax is completely filled by wing muscles, while there are large gaps between such muscles in the related larger insects (Partmann, 1948; B. Rensch, 1948). The middle intestine of such small species is made up of flat epithelial cells and, apparently, does not function as efficiently as in larger species. Above all, the eggs set a lower limit to body size because their size cannot be reduced *ad libitum*. This fact seems to be the reason why there is only one ripe egg at a time in the smallest Gastrotricha and Rotatoria, in the minute snail *Caecum glabrum*, and apparently also in the smallest frog *Phyllobates limbatus*. There are other such minute types of terrestrial snails, such as *Punctum pygmaeum, Pyramidula rupestris*, and species of *Vallonia* and *Vertigo*, in which only one or two eggs will ripen simultaneously and the smallest cyprinodont, *Heterandria formosa*, usually gives birth to only a very few young at short intervals.

In homoiothermic animals, the lower limit of body size is essentially determined by the relatively rapid loss of body heat, i.e. by metabolic difficulties. Plotting the body weight of small mice, shrews, and birds on the abscissa of a coordinate system, and the corresponding oxygen consumption per gram of body weight on the ordinate, shows that with decreasing body size the oxygen consumption increases more and more, and that – in shrews and birds – the curve finally becomes asymptotic, proving that the lower limit of body weight is something like 3·5 gm. in mammals and about 2·0 gm. in hummingbirds (Figure 54; O. H. Pearson, 1950). Correspondingly, the curves of the relative size of certain organs of very small birds and mammals reveal a similar asymptotic tendency, showing that a certain lower limit of body size cannot be exceeded (for example: the relative size of heart, brain, and eggs of the smallest birds). For corresponding details on the smallest fresh-water fish, *Heterandria formosa*, see Wellensiek (1953).

The lower limit of body size fixed by the factors mentioned above may be lowered further by phyletic alterations of the structural type. Such structural alterations near the lower size limit are fairly common in the animal kingdom. In the smallest mammals, for example, the structural type of the pelvis is

172

altered because the head of the embryo would be relatively too large to pass the pelvic canal without difficulty. As we have seen in one of the above chapters (p. 139), small mammals have relatively larger heads than large types. Moreover, the weight of the young animal at birth is relatively greater in small mammals than in related larger types. A white mouse, for example, weighs 1·47 gm. at birth, i.e. 8·65 percent of the adult weight at the onset of maturity about 10 weeks after birth (16·98 gm.), and 6 percent of the weight reached at 18 weeks after birth (24·51 gm.) (Backman, 1943). A white rat,

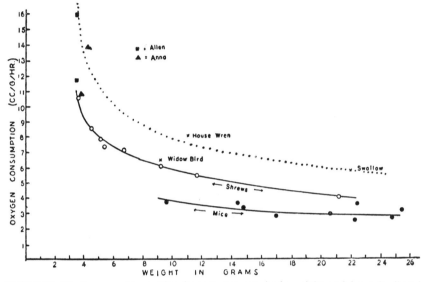

FIGURE 54. Oxygen consumption in ccm. per gm. body weight and hour (ordinate) in mice, shrews, and small birds. Abscissa: body weight in gm. (After O. P. Pearson, 1950.)

however, weighs 5·4 gm. at birth, i.e. only 3·8 percent of the weight reached in the tenth postnatal week (190·7 gm.) (Donaldson, 1924). As the head grows with negative postnatal allometry, the relative size of the head at birth is quite considerable, especially in the smallest types, and the act of parturition becomes difficult unless the pelvic passage is widened by some structural change.

Indeed, there has been structural alteration during phylogeny to which, apparently, attention has not been paid so far. In the smallest mammals the symphysis of the pubic bones shows no complete and solid junction, but is replaced by an elastic ligament or by soft cartilage, or the junction is reduced to a single point of contact so that both pubes may easily be moved aside, widening the passage for birth. This fact becomes readily evident from a comparison of the pelves of various rodents. There is a relatively long solid junction of the pubic bones in the larger forms, such as capybara (*Hydrochoerus*), porcupine (*Hystrix, Erethizon*), hare (*Lepus, Dolichotis*), squirrel (*Sciurus*), and jumping mouse (*Jaculus*), while this junction is reduced to a mere point of contact in the smaller types, such as rats (*Epimys*), and even more reduced in various small species of mice (*Mus, Apodemus, Microtus*; Figure 55), deer mice (*Peromyscus*), and others. In the males of the small pocket gophers (*Geomys*), the pubic bones are connected by a bony bridge which is completely lacking in sexually mature females, where a broad gap separates the bones. According to Hisaw (1925) the formation of this gap is a secondary process, which is due to the effects of follicular hormones and may be evoked in the male by injecting the hormones. A similar structure is found in the guinea pig (*Cavia*), as has been pointed out by R. Owen. In this the pubic bones form a fairly broad symphysis, but nevertheless they can diverge

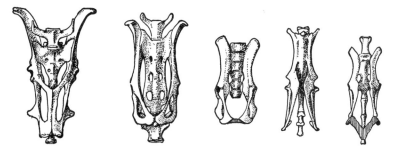

FIGURE 55. Reduction of symphysis ossis pubis in smallest mammals. Pelvic regions drawn to same scale. From left to right: *Dolichotis* with a long symphysis, *Arvicola amphibis*; *Erinaceus, Talpa, Sorex.*

to a remarkable extent so that the large head of the fetus, which is usually much larger in diameter than the bony pelvic passage can pass through (compare X-ray picture 76 and 77 by De Snoo, 1942, pp. 96–7).

There is a similar contrast in Insectivora, as in both sexes of elephant shrews (*Macroscelides*) and tree shrews (*Tupaia*) – animals of rat to squirrel size – the pubic bones are connected by the normal bony symphysis, while in the hedgehog (*Erinaceus*) this function is served by an elastic ligament which is the same structure as in the relatively large tenrecs (*Centetes, Potamogale*) and the Pyrenean shrew *Myogale*. In the common mole (*Talpa*), however, and even more conspicuously in the small shrews (*Sorex*), the pubic bones in both sexes are separated by a broad gap (Figure 55). Finally, it may be mentioned that in most types of bats (Vespertilionidae) a cartilaginous ligament (tending to ossify in older males) serves the function of a symphysis. As is well known, young bats are relatively large at birth, and only one or two young are born at a time. Among lemurs, in the small slender loris (*Loris tardigradus*) the pubic bones are separated by a gap of 1–2 mm. and connected by a soft ligament.

In the smallest insects the thorax does not provide enough space for development of the muscles required for hovering. Hence, a new mechanism was introduced by the development of long wing bristles which serve as a means of hovering and decrease the rate of sinking. Such wing bristles, as a convergent development, appeared in quite different groups: on the fore and hind wings of Thysanoptera, of the minute *Alaptus* and *Prestwichia* (Hymenoptera), which are only 0·2 mm. long, and of the smallest types of Zoraptera and Corrodentia; on the hind wings only in the minute beetles of the families Ptiliidae, Clambidae, and Sphaeriidae, and in tiny types of moths, like *Lyonetta* and others.

In the marine snail, *Caecum glabrum*, of about 1 mm. body length, locomotion is not brought about by muscular contraction as in other snails, but by ciliary movements (Götze, 1938).

In the smallest types of Archiannelida, the circulatory system, normally consisting of branched vessels, as in *Protodrilus*, is reduced to a blood cavity on the dorsal and ventral sides of the stomach. This reduction is to be seen in *Diurodrilus* (0·3 mm. body length) and in *Dinophilus gyrociliatus* (0·7–1·3 mm. long), but in the latter an additional ventral vessel serves the blood supply of the ovaries.

Extremely small animals of the open ocean, such as fish larvae and Crustacea, which cannot overcome water currents and keep themselves from sinking in the water by their means of locomotion, develop hovering mechanisms of various kinds, such as prolonged spines on different parts of the body.

As stated above (p. 150), in small types of newts, like *Triturus vulgaris*, the retina is extraordinarily thick and the cell layers are very dense, but besides this the relative number of visual cones is markedly increased, which seems to have a bearing on the quality of visual orientation in these animals (Möller, 1950). The same kind of retinal alteration is met with when the smallest cyprinodont, *Heterandria formosa*, is compared to its larger relatives such as *Lebistes* and *Xiphophorus* (Wellensiek, 1953).

It is a significant fact that all lungless types of Urodela relying on skin respiration belong to genera of small body size, such as *Manculus*, *Hemidactylium*, *Plethodon*, *Desmognathus*, *Spelerpes*, and *Oedipus* (a species of the latter genus, *Oe. townsendi*, having a body length of 3·8–4·2 cm., representing the smallest urodele form known; compare Mertens, 1933). It is evident that only in small types can skin respiration be sufficient to provide the oxygen supply required, as only in small forms is surface area relatively large in relation to body volume.

Finally, such structural alterations must be assumed in the mostly unknown histological substrata of the genetically determined instincts. In very small animals the number of eggs produced is too small to guarantee the perpetuation of the species; in many such animals an intense care of the eggs and young has been developed and viviparity is also frequent. It is the smallest central European lizard (*Lacerta vivipara*; body length about 15–18 cm.), that gives birth to eight or ten young (or lays eggs that hatch immediately after deposition: ovoviviparity), whereas all larger species lay eggs. The smaller types of snakes are also viviparous: the common viper, *Pelias berus* (60–80 cm. long), and the smooth snake, *Coronella austriaca* (75 cm. long). The larger types of Amphibia lay many eggs, and neither male nor female takes care of the hatching young, but among the smaller types there are numerous species with pronounced brood care: the small European toad *Alytes obstetricans*, the well-known tiny frog *Rhinoderma darwini*, and the small species of *Nototrema, Dendrobates*, and others. There is even a viviparous type: the small *Nectophryne*. In the small central European newt species (*Triturus*), the beginnings of brood care are to be noted, as the eggs are deposited singly on plant leaves, and among the fishes the small Cyprinodontidae are viviparous. Brood care is found in the small sea-horses (*Hippocampus*) and pipefishes (*Sygnathus*), the small sticklebacks (*Gasterosteus*), and the small species of *Rhodeus, Macropodus*, and *Betta*. Among the mouthbreeding fishes, many species are relatively small (*Etroplus* 8 cm., *Haplochromis* 7 cm.). Of course brood care sometimes evolved also in larger animals (as in the female Surinam toad [*Pipa*], about 20 cm. long), but the number of viviparous animals of large body size is relatively small in the respective groups. Hence, we may regard the phylogenetical development of viviparity and more pronounced care of eggs and young as an alteration of the structural type which was caused by the fact that the organisms concerned approached their lower limit of body size.

An adaptive alteration of instincts due to a small body size can also be noted in small birds from the temperate and cold zones, which try to compensate the relatively large loss of body heat by a special type of nest-building. Many types have turned to nesting in tree holes (blue, black, and crested tit, *Parus caeruleus, ater*, and *cristatus*; tree creeper, *Certhia*, etc.); some build spherical nests protected on all sides by insulating material (wren, *Troglodytes troglodytes*; tailed tit, *Aegithalos caudatus*; warblers of the genus *Phylloscopus*; and others); or the nest may provide sufficient insulation by especially thick side walls and a very deep cavity, so that the brooding bird is almost fully protected (e.g. the crested wren, *Regulus*). It seems probable that the continuous movements so typical of small birds and the 'tail clapping' of many small species must also be considered a special alteration of behavior due to the approach to the lower limit of body size, as the general rate of metabolism (though already high in small animals) is thus undoubtedly increased. The same applies to the behavior of small mice and shrews.

In the smallest types of insects, the number of ommatidia is so small that the perception of visual images is difficult or even impossible. Such small types generally do not respond to details of structure and form as releasing mechanisms in their behavior. The smallest workers of the leaf-cutting ant (*Atta sexdens*) of about 0·6 mm. body length have only two to three ommatidia per eye and do not carry building material to the nest as do their larger fellow workers of about 0·9–3·3 mm. The work of these minute specimens is strictly confined to the dark interior of the nest (Goetsch, 1940).

Alterations of proportions and of the structural type in consequence of changing body size have an important bearing on taxonomy, which should not be overlooked. There are whole categories of small animal types which have been and still are considered as special taxonomic units of their own, though their special systematic traits are mere consequences of small body size. This becomes especially clear in the group of Archiannelida, comprising species of 0·25–15 mm. body size (*Polygordius* of 10 cm. body length being the only exception). This group was established as a separate one (see Claus, Grobben, and Kühn, 1932), because in these animals – in contrast to the Polychaeta – the parapodia are strongly reduced or lacking, because their circulatory system is simpler (in some cases consisting of a single intestinal blood sinus and one ventral vessel), because they possess adhesive mechanisms, especially on the distal parts of their body, and because they have no copulatory organs (which many Polychaeta have). All these and other, similar reductions and constructional alterations (ciliary locomotion, more efficient mechanisms of adhesion) are probably caused by the extreme reduction of body size, and the Archiannelida cannot claim to be an order. Indeed, Remane (1932) arrived at the conclusion that three or four parallel lines evolved from the Polychaeta.

Such considerations also shed new light on the class of Gastrotricha, animals of only 0·06–1·5 mm. body length, with even more pronounced symptoms of reduction (e.g. coelom and protonephridia usually missing). In these animals, apparently, the smallest body size possible is reached. This is also evident from the fact that the tiny forms *Ichthydium* and *Lepidoderma* develop no more than one ripe egg at a time, the size of which is more than half the body size (*Lepidoderma concinnum* is about 92 μ and its single egg is 54 μ long). It is significant that in evaluating the possible relationship of Gastrotricha, Remane wrote the following: 'The average type of Annelida is quite different, but those forms approaching the body size of Gastrotricha show an organization remarkably similar to the latter. . . .' In evaluating the taxonomic situation of Mesozoa, one should consider some alterations of proportions and structure as a possible result of strong reductions of body size, and the consequences of a reduced number of cells should be taken into especial account.

The *upper* size limit of a certain structural type is determined in many cases by the disproportionate growth of surface organs compared to three-dimensional structures. The intestine, then, as an organ of surface-proportionate efficiency must become disproportionately long and coiled (or branched) in large animals of the same type if its capacity and efficiency are to be maintained.

Usually, the development of a giant body requires a modified type of feeding. Originally, the small ancestors of hoofed animals were mainly omnivores, or ate insects or other small animals, as can be inferred from their type of dentition. Giant animals of this type, however, could not maintain such a diet of insects and similar small animals, as terrestrial habitats do not provide sufficient amounts of this kind of food. Consequently, all giant mammals are herbivorous (elephants, rhinoceroses, titanotheres, pyrotheres, odd-toed and even-toed ungulates, giant sloths [Amblypoda], and so forth).

More rarely large animals became carnivorous predators, in which the maximum body size is, however, smaller than in the herbivores, because the carnivores must be more mobile and capable of rapid reactions (large felids, bears, and wolves). The same applies to birds and reptiles. All giant types like ostriches, emus, cassowaries, nandus, and swans are herbivores (or – if omnivores – rely mainly on plant food), and the largest birds of prey (like eagles, vultures, and harpy eagles) are considerably smaller than the herbivorous and omnivorous types. Most of the giant Sauria, except for the huge predator *Tyrannosaurus*, were herbivores. It need not be emphasized

FIGURE 56. Reconstruction of the largest reptile, *Brachiosaurus*, to show the principle of columnar legs. (After Abel.)

FIGURE 57. Reconstruction of the largest mammal, the hornless Oligocene rhino *Baluchitherium grangeri* (about 5·30 m. high) to show the principle of columnar legs. (After Granger and Gregory from Abel.)

that the altered method of feeding caused a series of consequent anatomical alterations (grinding teeth, gizzard, prolonged relative length of intestine, new types of sense organs and patterns of behavior, and so forth).

As the aerodynamic efficiency of wings grows by the square, while the body which must be supported grows by the cube, it is clear that a definite upper limit of body size exists in all flying animals, which is fixed mainly by the possibility of strengthening the wings (compare Meunier, 1951). As to vertebrates, it is important to bear in mind that the efficiency of bones as the essential stabilizing parts of the wings grows by the square (i.e. by their cross-section), while the body which must be carried grows by the cube.

The disproportionate relation of two- and three-dimensionally efficient organs is also the cause of a structural alteration concerning the legs of giant vertebrates. Large animals must develop disproportionately large and heavy bones, which involve a certain anatomical clumsiness. The angled joints, joints of knee, elbow, feet, etc., allowing a normal sized leg to be moved elastically, are replaced by heavy column-like extremities. This is evident in all large hoofed animals, especially in elephants, pyrotheres, horses, cows, rhinoceroses, titanotheres, ostriches, in the extinct giant birds *Dinornis*, *Aepyornis*, *Diatryma* (Eocene types of more than 2 m. height), and in the giant Sauria *Brachiosaurus*, *Brontosaurus*, *Diplodocus*, and so forth (Figures 56 and 57). It is obvious that such large types cannot be adapted to a burrowing, arboreal, or aerial way of life, but must remain running types. In large birds the muscles previously required for flying, the sternal crest, and the wings are reduced, because they have surpassed the upper size limit of flying animals, as is shown by ostriches, cassowaries, emus, nandus, moas, *Diatryma*, *Alca impennis*, *Didunculus*, and others.

Aquatic locomotion by ciliate bands or plates does not proportionately increase with the phyletic increase of body size, as the ciliary mechanisms originate from cellular structure, and cells cannot increase beyond a certain size limit. Hence we find enormously complicated and coiled ciliate bands in the larger types of larval Echinodermata, as a comparison of a bipinnaria or a pluteus with the small dipleurula or the excessive auricularia nudibranchiata will easily prove (compare Hesse, 1935, Figure 146, p. 202).

The effects of histological changes in giant animal types are less important than they might seem at first sight, as usually only the number of cells is increased and the increase of cell size is not very conspicuous. There are, however, quite essential structural alterations in the cell-constant tissues, especially in the nerve cells, and in the giant ganglion cells of large teleosts and mammals perforated parts with lobiform ramifications (the so-called 'paraphytes') are found, besides the normal dendrites and neurites (Figure 47). The development of these special structures was analyzed by G. Levi (1908, 1925).

Further Correlations. Compensation of Body Material. Besides the more general correlations due to the pleiotropic effects of genes and growth allometries, there are numerous further correlations of a more special character affecting the organs during their morphogenesis (pleiotropisms in a wider sense). In this context it is important to remember the rule established by K. Pearson (Whiteley and K. Pearson, 1900) which states that the neighboring parts of an organism are more closely correlated than more distant parts. This is especially clear in serial organs, as is demonstrated by Alpatov and Boschko-Stepanenko (1928) on the joints of the antennae in the red bug *Pyrrhocoris apterus* (Figure 58). They found that the correlation coefficients of

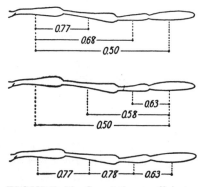

FIGURE 58. Correlation coefficients in the antenna joints of the bug *Pyrrhocoris apterus*. (After Alpatov and Boschko-Stepanenko, 1928.)

neighboring joints were higher (0·63–0·78) than those of more distant parts (0·50–0·68). Furthermore, it is evident that the correlations between the first and the third proximal joint (0·68) and between the first and the second (0·77) were closer than that between the third and the fourth joint (0·63). This means that correlation decreases toward the distal ends, as has also been proved in other cases. Apparently this decrease is caused by the effects of the anterioposterior gradients of differentiation referred to above: the distal parts, developed last, are not so closely correlated and the compensatory correlation, between the nourishing body and the distal parts, especially, is less close toward the tip. Hence, the distal structures are subject to a greater variability.

Such mutual effects of neighboring parts of the body are also met with in studies on the compensation of body material, a phenomenon of possible importance for an understanding of evolution to which sufficient attention has not been paid in recent years. The problem was first mentioned by some ancient Chinese and Indian writers (references compiled by Krumbiegel, 1931) and first recognized more precisely by Geoffroy St. Hilaire (1822), who discussed a 'loi de balancement' according to which the strongly growing parts of an organism consume so much body material that the less strongly growing parts remain small or become more or less reduced. Goethe, too, in the introduction to his *Comparative Anatomy* (1807) ascribed a certain developmental importance to this rule. Charles Darwin (1859) thought that some cases of compensation were merely a result of natural selection wiping out all animal variants with superfluous body material. However, he also compiled some new examples demonstrating the results of compensation of body material; e.g. hens with a large feather crest on their heads tend to have very small wattles. More attention was paid to the problem of compensation by Eimer (1888–1901) in his studies on the origin of species, especially in Part 3, dealing with the skeleton of vertebrates (1901). He pointed out that with the formation of stronger hind legs the lumbar ribs are reduced or disappear and frequently the tail is also reduced (frogs compared with salamanders, baboons with Cercopithecidae and man). Eimer's general view of the problem is summarized in his statement that 'with increasing differentiation of extremities the vertebral column became shorter . . .' (1888, p. 174). Roux, in his hypothesis on 'Intra- and Histo-Selection', and Weismann, in his hypothesis of germinal selection, also postulated compensation of body material. But more recently the whole problem of compensation has been neglected in discussions of evolution, and Krumbiegel (1931) was quite justified in saying that 'from a modern point of view the law of compensation in the sense of Goethe cannot be maintained' (p. 202). In my opinion, however, there are numerous cases which cannot be reasonably explained without the assumption of compensatory correlations between various parts of an organism, and as long ago as 1939 I compiled a series of examples (B. Rensch, 1939, pp. 202–10; 1943*a*, p. 80; 1943*b*, pp. 144–9; 1944, p. 19). Besides, in discussions of individual development, compensation of material was mentioned several times (compare Stockard, 1921; Schmalhausen, 1925; Weiss, 1939; Spiegelman, 1945; Tschumi, 1954).

To evaluate the importance of compensatory processes in transspecific evolution, one can study the phenomena of modification after experimental alterations, because the causal mechanisms governing the process of compensation can be more easily studied in experimental situations. It will be sufficient to cite a few characteristic examples. Pasewaldt (1888) and Hackenbruch (1888) found that in young rabbits and guinea pigs the extirpation of a testis or an ovary causes hypertrophy of the remaining gonad. Ribbert (1894) removed five out of eight mammary buds in rabbits at the age of two months, and the remaining milk glands showed a clear hypertrophic tendency. In adult salamanders, Kochs (1897) reported a compensatory hypertrophy of hind legs and the tail after the forelegs had been amputated. If in juvenile decapod crabs with asymmetrical chelae, such as *Alpheus, Carcinus*, and the like, the more solid 'male-type' chela is amputated, a smaller 'female-type' chela regenerates, and in the course of the following molts the other chela, left intact, becomes larger and larger until finally it represents a 'male-type' chela (E. B. Wilson, 1903; Zeleny, 1905; Przibram, 1907). If one removes one mandible of the larva of the water beetle *Hydrous caraboides*, the remaining mandible will lose one or two teeth in the following molts. In this case, then, regeneration of the removed mandible occurs partly at the expense of the remaining mandible. A similar result was published by Megušar (1907) on the compensatory regeneration in the meal worm (*Tenebrio molitor*): after one hind leg had been amputated, the wing of the respective side was reduced in size; and Przibram (1907) stated a similar phenomenon in the praying mantis, in which the regeneration of an amputated leg caused a decreased growth of the neighboring legs.

However, most of these experiments were made on single animals, or at best on a few, no statistical tests of the results were made, and other possible causes, such as traumatic, hormonal, and nervous stimuli, did not receive sufficient attention in the experiments. Therefore, I had such experiments repeated on a large scale by Wilbert (1953). He worked especially on the stick insect (*Carausius morosus*), to find out whether the increased growth in regeneration affects other organs and whether other organs grow more rapidly or for a longer period if a certain organ is totally lacking. Effects of this kind could reasonably be expected, as Teissier (1934) had found out that the growth ratio of an organ depends on the amount of the material supplied.

Wilbert caused autotomy of one hind leg in 50 young larvae of *Carausius* and measured the relative lengths of femur, tibia, and tarsus in each following instar. He found that, due to the influence of the regenerating leg, the contralateral extremity was shortened by 1–1·5 percent (statistically significant). The strongest effect was to be noted after the third molt. The shortening could be increased by additional autotomy of the collateral middle leg (see Figure 59). If the legs were amputated, including the coxae, i.e. if regeneration was rendered impossible, there was no increased growth whatever in the remaining legs, but there was also no shortening of the contralateral legs as in the first type of experiments, proving that traumatic or nervous stimuli cannot have played a role, and that the first results should be attributed to a compensatory process as the causal factor. If both hind legs were amputated (including their coxae), an increase of the relative weight of the ovaries resulted (in calculating the body weight the weight of the missing legs was added). This proves that neighboring organs, though of quite a different structure and function, may be affected by the process of material compensation. It seems justifiable, then,

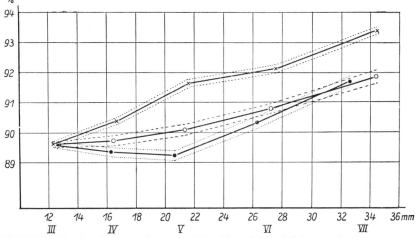

FIGURE 59. Size decrease of a normal hind leg of the stick insect *Carausius morosus* under the influence of a regenerating hind leg (○) or of a regenerating hind leg and contra-lateral middle leg (●). Upper curve (x): relative growth of hind leg in a normal control animal. Ordinate: length of hind leg in percent of length of foreleg (excluding the tarsal joints). Each dot represents the average data of 50 specimens. Mean deviation indicated by dotted and broken lines. Roman numerals=moultings. (After H. Wilbert.)

to assume a similar process of compensation in Ramme's findings (1931) on the grasshopper *Metrioptera roeselii*. Normally, this species has short wings and is unable to fly, but specimens with long wings are sometimes found, in which ovaries and testes show marked symptoms of reduction. A similar compensatory correlation was found by Graber in 1872: in short-winged Locustids the shorter the wings the thicker they are. (I owe this reference to Professor W. Ludwig.)

A fairly clear example of material compensation is also found in Bretscher's and Tschumi's careful studies (1951) on the extremities of Amphibia. These authors proved that after a lack of material was artificially induced (by application of a chemical checking growth) one or more toes remained undeveloped, but that no general reduction of the size of the foot as a whole resulted. Tschumi (1954) explains these results by physiological competition of material. In this context it is interesting to remember that a regeneration blastema implanted into poorly nourished amphibian larvae consumes a great deal of material, due to its stronger growth impetus, irrespective of the fact that the organs of the host have long ceased to grow in consequence of poor nutrition.

The above examples prove that processes of material compensation do occur, and hence we should account for this fact and its consequences in phylogeny, though we must not forget that all evaluation of phyletic transformation and its causes is based on extrapolation, and this should be done with care. If in males of the beetle *Copris lunaris* the cephalic horn is prolonged, the relative breadth of the head capsule will be smaller (Figure 60), which indicates that the horn grows at the cost of the head. Similarly, in individual variants of Prosobranchia the shell spines may be longer than usual, a growth which is invariably accompanied by a reduction of their number (Figure 61). Out of 156 specimens of the Indo-Australian fresh water snail *Melania scabra*, 58 belonged to the long-spined type and had 4–10 spines on the outer whorl of their shells, 54 specimens were short-spined and had 7–16 spines, while in 44 specimens lacking spines there were 10–24 longitudinal ridges. In geographic races of small birds inhabiting cooler regions, the wing tips are more pointed than in races of warmer zones (the wing rule: B. Rensch, 1938), which is due mainly to the fact that the second, third, and fourth primaries are relatively long. This causes, it seems, a marked reduction of the length of the first, fifth, sixth, and seventh primaries (Figure 62). It is for this reason that in the thrush nightingale (*Luscinia luscinia*), which has a more pointed wing tip than its southwestern neighbor, the nightingale *Luscinia megarhynchos*, the second and third wing feathers are relatively long and the reduced first primary is relatively much shorter than in the nightingale (Figure 63).

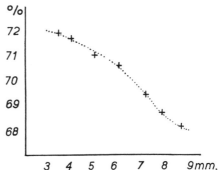

FIGURE 60. Negative correlation between length of horn (abscissa) and breadth of head in percent of breadth of prothorax (ordinate) in *Copris lunaris* ♂. (After B. Rensch, 1943.) Each + is the average value of four or five specimens.

Correlations between wings and gonads, like those of *Metrioptera* referred to above, are also found in butterflies. The wingless females of species belonging to the families Psychidae and Heterogynidae have enlarged ovaries, and the same applies to the wing-reduced species of Lasiocampidae, Lymantriidae, and Arctiidae. In the last three families, slightly enlarged ovaries are found in females having relatively small wings, which consequently do not 'like' to fly. In almost any case of such wing reductions one will find that in the females the proboscis and the tympanal organs are reduced in size or totally lacking. Eggers (1939), working with his school on these and related problems, thought that the enlargement of the ovaries was due to the fact that the wing muscles became reduced and that there was more space in the

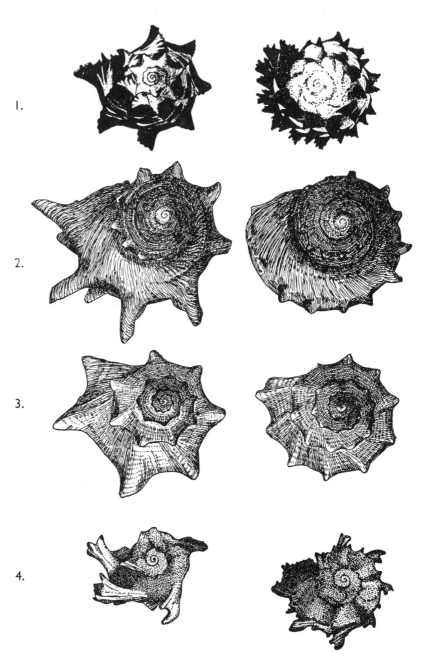

FIGURE 61. Compensatory correlations between number and length of spines in Prosobranchia: 1. *Murex* (*Phyllonotus*) *radix;* 2. *Turbo cornutus;* 3. *Semifusus tuba;* 4. *Angaria laciniata.* (After B. Rensch, 1939.)

interior of the body. Hence, he concluded, the enlargement of the ovaries is the result of a 'phyletic endogeneous correlation', but in my opinion it is more probable that material compensation is a causative agent.

The phenomenon of material compensation is especially well illustrated by some mammalian dentitions. The positive allometry of the teeth – especially that of the second dentition – led to excessive growth of the canines or incisors in some species with phyletically increasing body size. This happened in

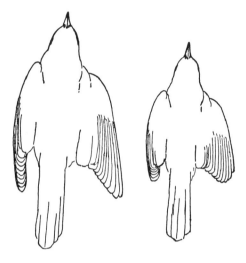

FIGURE 62. Racial differences in the wing tips of two races of the great reed warbler. Left: *Acrocephalus ar. arundinaceus* from central Europe; right. *A. a. meyeri* from the Bismarck Archipelago. Also a demonstration of Bergmann's and Allen's Rule. (After B. Rensch, 1939.

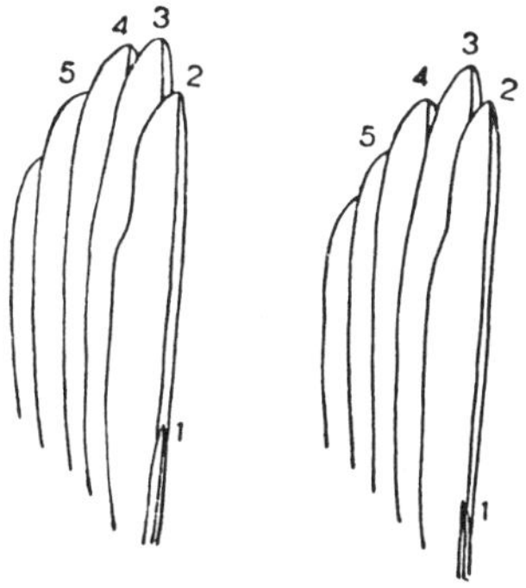

FIGURE 63. Compensatory reduction of first primary in the nightingale (left) and the thrush nightingale (right). (After B. Rensch, 1943.)

genera of quite different orders, and is found, for example, in saber-toothed cats, like *Smilodon* and *Machairodus*, in which the canines are extremely enlarged and, by compensation, the first premolars (which are found in other cats) are lacking, though there is a diastema between C and P_2. Similarly, in walruses (Odobenidae=Trichechidae) the giant canines apparently consumed the material of the first premolars, as the latter are lacking, though they are present in the related seals (e.g. *Phoca*) which have canines of normal size. In the pig *Phacochoerus* the canines are transformed into enormous tusks, but the upper incisors and first premolars are totally reduced, which is not so in the related genera. Among the extinct forms of Amblypoda there is one genus with giant canines (*Uintatherium*), but consequently the incisors are completely reduced, so that the dental formula of the upper jaw is as follows: I0. C1. P3. M3. In other genera of the same order having smaller canines, like *Pantolambda*, three incisors on each side are found and the dental formula is 3.1.4.3. A similar situation occurs in Lemuroidea, as the giant lemurs of the genus *Megaladapis* of the Early Pleistocene have excessively large upper canines but lack the incisors (formula: 0.1.3.3), while the ancestral genus *Adapis* showed canines of normal size and, consequently, normal incisors in the upper jaw (formula: 2.1.4.3; Figure 64). In Proboscidia, two incisors are developed into enormous tusks, but the canines and the remainder of the incisors in the upper jaw are wanting: 1.0.3. In *Moeritherium*, a rather unspecialized Eocene genus, the second incisors are only slightly prolonged and, consequently, there are three incisors and the normal canines: 3.1.3.3. In rodents, too, the incisors are strongly prolonged proximally, i.e. into the jaw-bones, but tend to grow continuously as their surface is worn off. Hence,

FIGURE 64. Skulls of *Adapis parisiensis* (Late Eocene of France) and *Megaladapis edwardii* (Late Quaternary of Madagascar). *Adapis* shown one half natural size, *Megaladapis* one fourth natural size. (Adapted from Stehlin, von Lorenz and Weber.)

there is a steady consumption of nutrient material by these teeth, and it seems that in consequence the canines do not develop, though a diastema is well provided, and that the incisors are reduced to one on both sides (for instance, the formula of the beaver *Castor*: 1.0.1.3). (For Carnivora and Insectivora, compare also Hale, 1940.)

An ever-increasing effect of compensatory processes can also be traced in the various stages of the well-known horse line. The increased growth of the third metacarpalia, metatarsalia, and of the third digit involves a progressive reduction of the second and fourth digits, and in the line *Orohippus* (Middle Eocene), *Mesohippus* (Lower Oligocene), and *Merychippus*, they are more and more reduced in the direction of the well-known splint bones in our Recent horse (*Equus*). In *Equus*, it is interesting to note that the metapodia II and IV are developed before birth, even showing small toes on their ends, but with the onset of the increased growth of the metapodium III, the neighboring digits become reduced. If accidentally these lateral digits do develop (atavism or pseudo-atavism), one finds a weaker development of the central main digit, and it is this correlation that supports the assumption that a process of material compensation is at work (for examples of atavism in horse feet see Abel, 1928, and Krölling, 1934).

Finally, compensatory processes are probable during the phylogeny of those vertebrates in which the tail has become an important organ of loco-motion, as in whales, snakes, slowworms, and the like. In these types the hind limbs are more or less strongly reduced or missing. The process of reduction advances in a direction opposite to that of ontogenetic differentiation, i.e. the most distal parts are reduced first and the pelvic girdle and femora are affected last, if at all. It is this very fact that suggests that compensation is at work because, when the distal parts are differentiated, the whole body has already grown relatively large and the rapidly growing tail region consumes a considerable amount of body material.

All the examples cited above are intended to show the general bearing of compensatory processes in transspecific alterations of the structural type. One should not forget, however, that only a few examples have been sufficiently analyzed as yet, and that it is very difficult to prove that a certain result is due to material compensation, especially in those cases where the material consumed by strongly growing parts is not taken from the adjacent organs. Nevertheless, we shall have to account for processes of material compensation which might also provide a basis for some pleiotropic gene actions. Compensatory phenomena seem to have an especially important bearing on the process of phyletic reductions of organs, as in most cases these cannot be explained by selection. This problem will be referred to below in the discussion of orthogenetic trends.

At any rate, though the manifold correlated processes are sometimes vastly complicated, it seems possible that the study of material compensation might provide a method of causal analysis of many complex phenomena in phylogenetic transformation.

The Problem of Somatogenic Induction. Until quite recently some authors favored the assumption of hereditary transmission of characters acquired during an individual life, i.e. the hypothesis of somatogenic induction. Such a possibility seems plausible especially because hereditary alterations of a structural type and nonhereditary changes caused by environmental factors or by the use or disuse of organs and structures are often phenotypically identical. These authors should have known the grave logical difficulties arising from the Lamarckian hypothesis. As stated by Haecker: a causal chain of reactions a-b-c-d, describing the effect of environmental factors (a) on the soma (b) transmitted by metabolic or similar processes (c) to the reproductive cells in which the alteration (d) is caused cannot have the same effect as another chain of reactions, m-n-o-p, which stands for the ontogenetic development of the adult animal (p) from the germ cells (m) via the embryonic and postnatal stages (n and o). As could be expected, all critical experiments have invariably proved that there is no somatogenic induction.

The often astonishing similarity of modification and mutation explains why Lamarckian ideas are still sometimes held. In some cases of phenocopy not only are the phenotypes identical, but the environmental factors causing the modification are the same as those which are responsible for the special adaptations conditioned by heredity. For instance, races of homoiotherms which inhabited cold countries show increased body size, which is their hereditary feature (Bergmann's Rule); a phenotypically similar modificatory increase of body size, however, can be caused experimentally by lowering the temperature during postnatal growth.

The difficulties of interpreting such cases of phenocopies led me to a Lamarckian point of view some thirty years ago (B. Rensch, 1929). I had not paid sufficient attention, then, to the works by Lloyd Morgan (1900: Chapter XIV) and J. M. Baldwin (1902), who showed that modifications by certain environmental factors may induce evolution of genetic characters in the same or a similar direction. If, for example, the geographic range of a species is enlarged and a neighboring region with different climatic conditions is colonized, special modifications will result (e.g. variants with larger body size in a colder region). However, in such a newly colonized region, mutations which cause an increased body size will also be favorable. Hence, the genetically conditioned norm of reaction includes the potentiality of pre-adapted modifications capable of colonizing a new region, and later on these modifications are replaced by phenotypically similar adapted mutants favored by selection. This 'organic evolution' may also be initiated by use or disuse of an organ and by colonization of a new habitat by these modifications; this is the process rightly observed by Lamarckists but wrongly interpreted as somatogenic induction. It was not until recently that the initial modificatory stages of hereditary transformations were again stressed by Lukin (1936), J. S. Huxley (1942), and Schmalhausen (1949).

Inheritance of modificatory alterations appears to occur also in those cases where certain characters are transmitted not by chromosomes, but by plasmagenes, which are especially important in bacteria, as here the soma and the reproductive cell are the same thing, and each alteration of a cytoplasmic constituent capable of replication may lead to a permanent hereditary alteration. Possibly some of the permanent alterations in strains of Protozoa and bacteria appearing in the course of repeated passages through abnormal host-organisms, may be attributed to such alterations of plasmagenes. Then, too, it is difficult to draw a sharp dividing line between modification and mutation (one should compare the findings by Sonneborn and co-workers, 1950, and Ephrussi, 1953).

Very little is known as yet about the effects of plasmagenes in higher animals, except in cases of maternal effects. There is no reason, however, to assume a 'gradually increasing heritability' of a character due to the use or disuse of an organ in the Lamarckian sense, and all alleged examples of such heritability may more reasonably be interpreted as the effects of mutation and selection.

This may be illustrated by an example taken from the otherwise stimulating publications by Böker (1927, 1936, 1937), concerning the special adaptations of the South American hoatzin, *Opisthocomus* (Figure 65). Böker assumed that this relatively large-winged and long-tailed herbivorous bird evolved from insectivorous ancestors with average wing and tail proportions, and that it turned to eating first soft and then hard leaves. This hypothetical

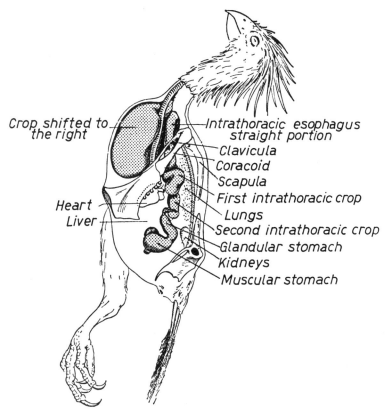

Crop shifted to
the right

Intrathoracic esophagus
straight portion

Clavicula

Coracoid

Scapula

First intrathoracic crop

Heart

Liver

Lungs

Second intrathoracic crop

Glandular stomach

Kidneys

Muscular stomach

FIGURE 65. Size and position of organs in *Opisthocomus cristatus*
(After Böker, 1936.)

transformation seems probable and we will adopt it in our discussion. According to Böker, the phyletic transformation was governed by five causes, each of which effected an alteration of the structural type. Böker's ideas were summarized as follows (1936, p. 274):

First type: normal flight; feeding on insects.

1. Cause: soft plant food. Consequence: true crop.

2. Cause: hard plant food. Consequence: crop S-shaped, very large, . . . very muscular and provided with appendages. Cornified epithelium.

3. Cause: weight of the anterior part of the body too heavy due to the enlarged crop. Consequence: (a) larger wings, (b) longer tail, (c) shifting of the crop towards the center of gravity.

4. Cause: insufficient space for the growing embryos in the egg. Consequence: growth disturbances of carina and wing muscles, progressive reduction of flying ability.

5. Cause: species maintenance risked. Consequence: adaptation to climbing in the young.

Böker assumed that operational stimuli (*Betriebsreize*), i.e. the stimuli arising from the type of habit and activity of the organism, acted as formative stimuli (*Gestaltreize*) effecting immediate anatomic alterations of the structural type as 'purposive, active reactions correlated to the organism as a whole', and that these alterations could not be the result of 'passive random happenings of mutation or selection'.

It is quite justifiable, however, to regard such structural alterations as the result of random mutation and natural selection, and one may reasonably discard the five stages of Böker, as it is not at all probable that first the balance of body weight was disturbed by a heavier anterior body part, and that larger wings and a longer tail were then developed as a sort of compensatory means to regulate the disturbed balance. It seems much more probable that the organism was in a certain state of harmony at all stages of its transformation, as selection by environmental factors affected all structures and characters accessible to it. The phyletic transformations were initiated when it turned to a mixed diet of insects and plants, on which so many birds live. Let us now assume that according to Cope's Rule body size was increased in the course of time. Then we may also assume that the increased amount of insect food required by the larger bird was not available or that plant food became more easily accessible (for, as we have seen above, nearly all large birds are herbivores or predators). Consequently, among the individual variants produced by random mutation those were favored that had a crop-like bag to store for a time more leaves than the stomach could hold. As natural selection proceeded along certain lines (orthoselection), the formation of crops was favored. With regard to the crop, it is quite evident that it is not the result of 'inheritance of acquired characters', as it can already be found in the prenatal embryo. When the development of the crop became more rapid, the weight balance of the body could, of course, never be seriously disturbed, and in each stage of the transformation the size of the wings and the tail remained in correlation with the body as a whole and with the – possibly – heavier anterior body part, because all parts of the body are simultaneously subject to selection. The formation of larger extremities hindered the development of the sternal cristae and wing muscles. A decrease of flying ability was, however, tolerable, as eating plant leaves does not require flight. Finally, it is highly doubtful that the climbing ability of the young should be a consequence of the reduced ability to fly (fifth consequence of Böker), as it is much more reasonable to assume that the prehensile fingers of *Opisthocomus*, enabling the young birds to climb, are a phyletic remainder of its reptilian ancestry. It is quite possible, then, to interpret even such complicated cases of structural alterations by the well-analyzed evolutionary factors of mutation and selection, and there is no need to assume an additional 'active anatomic reaction' effecting a gradually increasing inheritability by a quite unknown and obscure process.

Summary on Transspecific Alterations of the Structural Type. All the causal factors of structural alterations which have been mentioned in the present chapter – i.e. pleiotropic gene actions, allometric shifts of proportion, special selection in consequence of the shifts of proportion, compensation of body material and similar correlations, general alterations of the structural type caused by approaching the upper or lower limit of body size – are 'systemic qualities' of the organisms as a whole – that is to say, factors which can be understood only in the light of their effects on the organism as a whole. The manifold interrelations of such factors (of which many more could be enumerated) are highly complex, and their analysis is still in its beginning. Nevertheless, these factors are quite susceptible of analysis and we need not assume unknown evolutionary factors. There exist, indeed, some 'constructive mutations' which change, more or less harmoniously, whole systems of organs or even whole structural types; however, such mutational alterations are also controlled by natural selection. I do not wish to suggest that only such 'constructive' genes are important factors of transspecific evolution. There are also numerous examples proving that more limited mutations may be gradually compounded together, if they fit harmoniously into a system of characters previously brought about by mutation and selection. Hence, there exist constructive (harmonizing) effects of selection. Clark (1941), for example, proved that there are only slight correlations in the proportions of body and skeleton of various strains of deer mice (*Peromyscus*), and that we have to assume specific growth factors for the size of tail, femur, foot, skull, and mandible.

IV. Parallel Evolution

Paleontological and zoological studies have revealed numerous cases of different animal types showing parallel formation of similar characters, of organs, or of whole structural types. Some of these examples involve similarities and parallelisms in the evolution of closely related forms; however, in others systematically distant types have secondarily become similar (convergence). In some cases parallel forms have evolved from time to time, repeatedly, in the course of the evolution of the same main line (iteration). On the whole, parallelisms are so frequent that Beurlen (1937, p. 87) thought them to be diagnostic of the phase of phylogenetic specialization. Beurlen regarded them as a manifestation of a 'typical systemic regularity' (*Gestalt-gesetzlichkeit*, p. 66) which cannot be interpreted on the basis of mutation and selection because it is an autonomous evolutionary force. Dacqué (1935) pointed out that in some cases parallel developments occurred even in heterogeneous groups during certain geological periods, and he coined the term 'time signatures'. By this statement Dacqué tried to interpret the phylogenetic links, such as the theromorphs, which are intermediate between the reptiles and the mammals, as having only accidental resemblance to the mammals and not being ancestral to them. Consequently, he doubted the phylogenetic relations of the more strongly differing structural types.

Although some parallelisms involve surprisingly large numbers of morphological and anatomical characters, and although some parallelisms, especially among fossil animals, have not yet been found accessible to causal analysis, one can well interpret such phenomena by the principles of mutation and natural selection, provided that one takes into account the complicated correlations referred to in the preceding chapter (i.e. allometric growth, compensatory processes, pleiotropic gene effects, etc.).

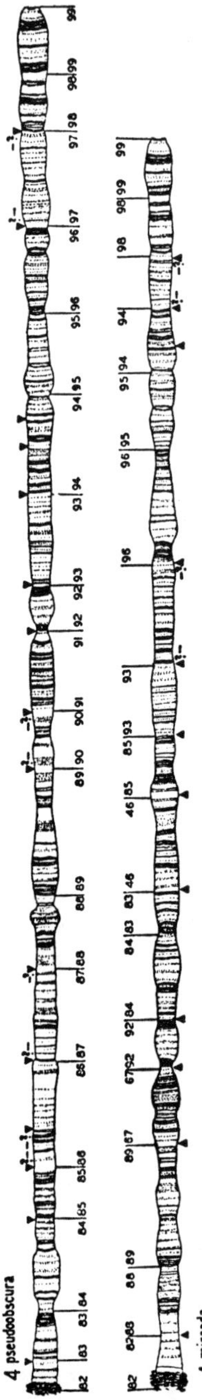

Phylogenetic parallelisms can arise from quite different sources. The three following groups, at least, can be distinguished: (1) parallelisms due to similar hereditary factors, including cases of parallel mutations; (2) parallelisms resulting from parallel selection acting on homologous structures or organs, and (3) parallelisms caused by parallel selection affecting analogous structures and organs. There are also numerous examples of parallel evolution in which two or all three of these types are mixed.

Parallel Evolution Resulting from Similarity of Hereditary Factors. As new races and species arise owing to mutation of a limited number of genes, the bulk of hereditary characters remains unchanged. In species of *Drosophila*, especially, the homology of many genes has been proved by hybridization combined with cytological study of the loci of the genes in question in the giant chromosomes of the salivary glands. In less closely related forms, the homology of certain gene arrangements could only be inferred from the correlation of the gene patterns of the chromosomes. An example is shown in Figure 66, demonstrating the considerable conformity of chromomeres (indicated by equal numbers) in the fourth chromosome of *Drosophila pseudoobscura* and *D. miranda*, in which several inversions are to be observed (inverted parts indicated by black wedges). Homology has not yet been proved in those parts labelled by question marks (? – ?) (Dobzhansky and Tan, 1936). Figure 67 shows the chromosome sets of *D. pseudoobscura* and *D. miranda*, indicating the parts containing homologous genes (partly in inverted or translocated arrangements) and those in which the homology has not yet been proved. As it is difficult – because of

FIGURE 66. Comparison of the fourth giant chromosome from *Drosophila pseudoobscura* and *D. miranda*, showing the corresponding gene loci. Wedges indicate altered serial arrangements of genes, question marks non-homologous portions. (After Dobzhansky and Tan, 1936.)

an identical or a similar phenotypic effect of non-homologous genes – to prove genic homology in species which do not hybridize, Gottschewsky and Tan (1938) transplanted buds of organs in such species. If, after implantation, the gene effect remained unchanged, the genes in question could sometimes be regarded as homologous. By this method the homology of the eye pigmentation genes in genetically rather different species, *Drosophila melanogaster* and *D. pseudoobscura*, could be proved. Spencer (1949) could demonstrate a number of homologous genes in *D. hydei* and *D. melanogaster*, and Spassky,

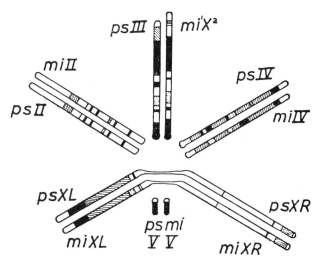

FIGURE 67. Comparison of gene arrangements in *Drosophila pseudoobscura* (ps) and *D. miranda* (mi). Identical parts are white, inverted parts striped, translocations dotted, and non-homologous parts black. (After Dobzhansky, 1937.)

Zimmering, and Dobzhansky (1950) in *D. prosaltans* and *D. melanogaster*. Kosswig (1948) discussed the homology of pigmentation genes of mammalian hairs, and also of various genes in cyprinodonts.

An identical set of genes can, of course, lead to parallel mutations. Hence, the mutations white, yellow, miniature, notch, facet, and the like, which are common in several species of *Drosophila*, can be looked at as being homologous. Homologous variation series, first found in species of cereals by Vavilov (1922), must also be attributed to such identical genes and parallel mutations. The same principle applies to the parallel variants in the European snails *Cepaea hortensis*, *C. nemoralis*, and *C. vindobonensis*, showing no bands or one, two, three, four, or five bands, which may be separated or united, dark-colored, brown, or hyaline.

The occasional presence of identical characters in animal species of an even more distant systematic relationship was termed 'paripotency' by Haecker (1925). He compiled a series of examples demonstrating that phenotypes, such as albinism, melanism, and pituitary dwarfism, occur in vertebrates of quite different species. In many, although not in all, cases these characters may be attributed to homologous genes. A particular mammalian color pattern ('Dutch spotting'), producing strong pigmentation in the anterior and posterior parts of the body, while the central part remains white, apparently is due to homologous genes. This coloration is known as a recessive hereditary character in domestic rabbits as well as in wild gophers of the genus *Thomomys* (Storer and Gregory, 1934).

There are numerous other cases of parallelisms, not yet analyzed genetically, which seem to be due to identical genes. Reference will be made to some examples of special evolutionary interest. When, in race and species formation, several correlated characters are jointly altered because of their correlation, and when such a complex is common to several related species, such parallelisms may seriously impede proper taxonomic judgment. Among Mediterranean dry-land snails of the family Helicidae there are some types belonging to well-differentiated genera, distinguished by their different genitalia, which possess shells of nearly identical shape and structure. It is especially interesting that the formation of geographic races produced remarkably clear but closely parallel alterations. The numerous races of *Murella muralis*, a snail inhabiting the rocky parts of southern Italy and Sicily, usually have a nearly spherical or obtusely conical shell with a relatively smooth surface, similar to that of the central European *Cepaea* (Figure 68). In certain spots of western Sicily, however, very flat shells with sharp keels and with irregular ribs are found, which were formerly regarded as good species but have since been recognized as geographic races (Kobelt, 1881; B. Rensch, 1937). The typical shell structures mentioned above proved to be hereditary (B. Rensch, 1937). In two (if not three) areas definitely separated and surrounded by the geographic range of types with globose-conical shells, the keeled types evolved apparently independently and in a parallel manner: in the northwestern part of the island near Trapani and Calatafimi (and in an isolated spot of Cape San Vito) and on the southwest coast near Caltabelotta (near Marsala). There is a related 'Rassenkreis' in Sardinia providing an astonishing example of an almost identical formation of globose-conical smooth shells and of strongly keeled and ribbed, flat races: *Tyrrheniberus villicus* (being smooth and spherical near Oliena and keeled near Dorgali). The same kind of parallelism is met with in the 'Rassenkreis' of *Rossmaessleria subscabriuscula*, a snail inhabiting the hills near Tetuan in Spanish Morocco (*R. s. boettgeri* being globose-conical and smooth, *R. s. subscabriuscula* flat, keeled, and fine ribbed). If one were to classify the shells, perhaps only on the basis of empty or fossil specimens, without knowing the anatomical details of the snails, one would have no doubt that the globose-conical, smooth shells, on the one hand, and the flat keeled and ribbed types from Sicily, Sardinia, and northern Morocco, on the other hand, should each be grouped as belonging to one species or one 'Rassenkreis', and yet this would be, of course, totally wrong. This case is a warning to paleontologists, who are – unfortunately – bound to base their taxonomic judgments on fossil shells only! The parallel evolution of spherical or conical and of keeled and ribbed shells in geographic races of snails is even more widespread in other Mediterranean genera related to *Murella*: in *Iberus gualterianus-alonensis* on the east coast of Spain, in *Levantina gyrostoma-leachi* of Tripoli, in the Palestinian 'Rassenkreis' of *Levantina hierosolyma*, and in the desert snail

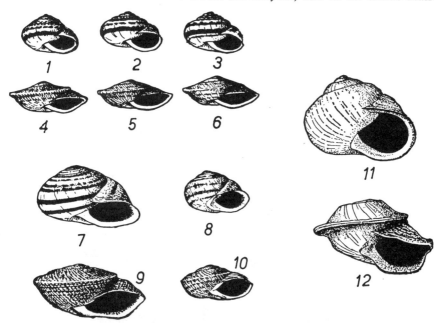

FIGURE 68. Parallel development of flat, crested and ribbed races and species in various normally round-shelled dry-land snails from the Mediterranean area. 1. *Murella muralis muralis* (Girgenti, Sicily); 2. *Tyrrheniberus (v.) villica* (Oliena, Sardinia); 3. *Rossmaessleria (s.) boettgeri* (Tetuan, Morocco); 4. *M. m. segestana* (Alcamo, Sicily); 5. *T. (v.) sardonia* (Dorgali, Sardinia); 6. *R. (s.) subscabriuscula* (Rif, Morocco); 7. *Iberus gualt. alonensis* (Murcia, Spain); 8. *Levantina gyrostoma* (Tripoli); 9. *I. g. gualterianus* (Sierra Elvira, Spain); 10. *L. leachi* (Tripoli); 11. *Eremina hass. hasselquisti* (Libyan Desert); 12. *E. h. zitteli* (Oasis Siwah). (After Rensch, 1937.)

Eremina hasselquisti of the Libyan desert (Figure 68). A blunt keel is also quite frequently found in numerous populations of the Mediterranean dry-land snail *Leucochroa candidissima*, and in the dry-land forms of the related group of *Helicella* the keel and ribs are a typical character of some subgenera (genera *Jacosta*, *Xeroleuca*, *Sphincterochila*). Hence, there is a large group of related Mediterranean Helicidae showing the above-mentioned parallelisms in the alternative evolution of their shell structure.

Among fossil animals the formation of keels and peripheral grooves of Ammonoidea is a significant example of parallelism. The parallel evolution of keels or grooves may be traced in the shells of *Psiloceras*, *Aegoceras*, *Macrocephala*, and *Parahoplites*. *Psiloceras* was transformed into the *Schlotheimia* forms with peripheral groove and, on the other hand, into the keeled forms of *Arietites*. *Aegoceras* evolved into grooved *Phricoderoceras*, and the group of the keeled *Asteroceras obtusum* and the keeled *Ophioceras raricostatum*. That there are parts of identical hereditary elements in numerous Ammonoidea may be concluded from the fact that Salfeld (1921) found,

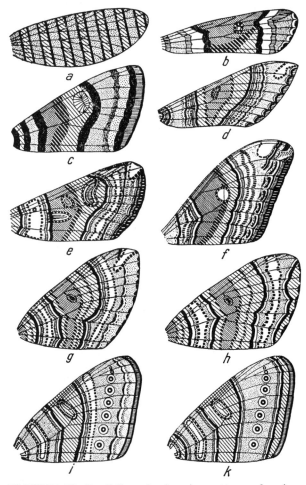

FIGURE 69. Parallelisms in the wing patterns of various groups of butterflies: (a) primitive groups, especially Tineidae and Tortricidae; (b) *Ephestia* and Pyralidae; (c) Arctiidae; (d) Sphingidae; (e) Noctuidae; (f) Saturnidae; (g) and (h) Geometridae; (i) and (k) Nymphalodeae. (After Henke, 1941.)

among thousands of normally keeled *Hecticoceras punctum*, a peripherally grooved specimen of this species in the same locality. It is not surprising that such parallelisms have caused some wrong taxonomic arrangements (family Kosmoceratidae, genus *Hoplites*, and so forth). The parallelisms evolved in Terebratulidae (species of the genera *Pygope*, *Pygites*, and *Antinomia* with narrow holes deriving from types with deep slots) must also be attributed to variable elements common to a fairly large group of related forms (Buckmann, 1906; Dacqué, 1935, pp. 136–7).

Many parallelisms are also found in the patterns of wing coloration in butterflies belonging to quite different families. Judging from the thorough studies by Henke (1935, 1936), Lemche (1937), Henke and Kruse (1941), and others, the wing patterns of this order comprising so many different types can be reduced to a limited number of simple systems which are combined in various manners. The most essential elements of wing patterns are certain spatial arrangements (*Feldgliederungen*) mutually superimposed and over-lapping, and pattern zones (*Musterörter*) in which certain colorations are likely to become manifest. There are some cases from which one may infer that homologous genes underlie both elements mentioned above, as the ontogenetic development of these basic elements of wing patterns proved to be similar ('plastology'). 'Dependent' patterns of wing colorations are also found, i.e. patterns which are more or less 'guided' by other wing structures (veins, for example), but in these at least these guiding structures must be regarded as homologous. The loss of single components of these patterns is governed by certain rules. A type especially frequent in numerous families shows three main stripes with a light central zone bordered by two darker zones (Figure 69). The formation of additional patterns is likely to occur in the lighter central zone. From these

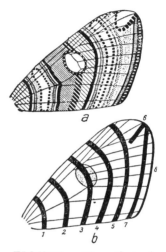

FIGURE 70. General scheme of wing patterns in butterflies: (a) pattern arrangement; (b) fields of origin of patterns in Macrolepidoptera. (After Henke, 1941.)

studies a general schematic representation of wing patterns can be deduced (Figure 70). According to Lemche (1935) there is even an 'archaic' wing pattern applying to all insects and consisting of more or less parallel transverse stripes which are inserted at the distal bifurcations of the wing veins.

Because of these plastologies (probable genic homologies), the patterns of wing coloration are often very similar or even identical in quite different families of butterflies. If, in sympatric species, the wing patterns show such parallelism, and if one of these species is protected from enemies by its unpleasant taste, cases of mimicry may arise and the similarity of both species may be increased by subsequent natural selection. The results of such processes have been known for a long time and are especially frequent in the families of Papilionidae, Danaidae, Heliconidae, Pieridae, and Satyridae (compare, for instance, the coloured plates in Weismann, 1913).

If parallel evolution occurs in characters causing strong correlative effects in other organs and structures, i.e. if there is a parallel increase or decrease in body size, causing correlative changes of numerous proportions of the body, animal types will arise showing marked similarity of many characters. I have referred to such cases in my discussion on alterations of the structural type in Chapter 6, B II of this book. We need only remember that the parallel alterations in four lines of descent of Polychaeta caused by a phyletic decrease of body size led to the establishment of the taxonomic pseudo-category of Archiannelida. I shall refer to some more cases of this kind in the next chapter, which deals with orthogenetic evolution (compare, for instance, the correlative parallelisms arising with an increase of the size of the mammalian skull, Figure 76). It is cases like those mentioned above which prove that complicated parallelisms may well be caused by random mutation and selection, though parallel 'systemic characters' affecting the whole organism are especially concerned.

In some cases, such correlations seem to have been the origin of the so-called 'iterations', i.e. the repeated evolution of parallel structures originating from a generally conservative type. In the Late Cambrian alum slate of Andrarum (southern Sweden), R. Kaufmann (1934) found six species of the Trilobite genus *Olenus* in a temporal succession, which he regards as iterative transformations which evolved several times at intervals, originating from a conservative (persistent) type which he could not discover in the profile studied. In each of the six species one can distinguish an early, an intermediary, and a late form, each of them revealing similar or equal tendencies of development: successive decrease of head and pygidial width (Figure 71), decrease of frontal margin, increase of ocular breadth, and relative prolongation of pygidial spines. As the same alterations of proportions can be observed in the ontogeny, one may assume that these phyletic iterations are based on the same correlations.

Parallelisms emerging in different animal species tend to be similar or identical not only in their final evolutionary stage but also in their intermediate stages, in consequence of their correlative effects on numerous characters or on the whole body (for instance allometries). Such synchronous similarity or identity of parallel evolutionary stages may appear as 'time signatures' (*Zeitsignaturen*), which were studied in more detail by Dacqué (1935). A typical example of this kind is provided by the synchronous and parallel evolution of triangular shells in the Late Devonian genera *Soliclymenia* and *Kamptoclymenia*, deriving from roundish types (*Pachyclymenia* and *Kamptoclymenia endogena*). Later, in the same geological profile, the triangular types synchronously evolved into trilobed forms (*Epiwocklumeria* and *Parawocklumeria paradoxa*, respectively; Figure 72). Schindewolf (1936), who studied these series in detail, rightly stated that the trilobed shape of the shell was potentially contained in the triangular ancestral form. The other cases of so-called 'time signatures' referred to by Dacqué, especially the Theromorpha, represent not parallelisms but phyletic links and intermediate types of transformation which he did not acknowledge.

Parallelisms in Consequence of Parallel Selection. The phylogenetic

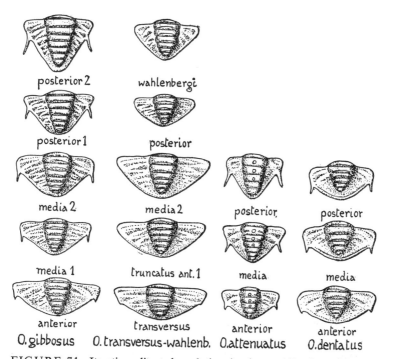

FIGURE 71. Iterative, directed evolution in the pygidia (from broad to narrow) in the trilobite *Olenus*. From bottom to top row in chronological order. (After Kaufmann, 1933, 1934.)

FIGURE 72. Two parallel pro-terogenetic evolutionary series of *Clymenia*, a Late Devonian genus of Ammonoidea, taken from a geological layer of about 2 m. height. From bottom to top: left: *Pachyclymenia abeli, Wocklumeria sphaeroides, Epiwocklumeria applanata;* right: *Kamptoclymenia endogena, K. trivaricata, Parawocklumeria paradoxa.*
(After Schindewolf, 1936.)

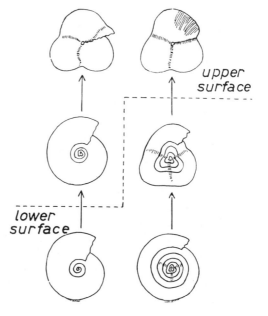

importance of such parallelisms caused by homologous genes is quite different from that of parallelisms which evolved as the result of parallel selection. The most instructive examples proving parallel selection are phyletic convergences arising from analogous rather than from homologous structures or organs. Novikoff (1929, 1930, 1938) referred to such cases as products of homomorphy. When, for example, spiny elongations originate from quite different structures of planktonic Crustacea, i.e. on their carapace, their legs, their abdomen, or their tail fork, or when pinnate gills evolve at various epithelial parts of the body in Mollusca (also anal gills, as in *Doris*), these cases must clearly be regarded as true convergences, which have developed from heterogeneous origins and become more and more similar in the course of phylogeny.

There are some examples of protective coloration and protective shape which seem to be the results of parallelisms that originated not at all or only partly from homologous characters. In the enormous multitude of insect types, for instance, the stick-shaped body evolved independently in quite different groups, as in *Dixippus* (Phasmidae), in *Neides, Limnotrochus* (Hemiptera), and numerous pond skaters (Gerridae), by mere random combination of certain characters. This is rendered probable by the fact that all typical characters of the stick (or leaf-like) body shape are met with as single (i.e. not yet combined) characters in different species of the corresponding families (Just, 1934 *a,b*; 1936 *a,b*). When, by suitable combination of fitting characters, the more or less complete stick-shaped or leaf-like body is approached phyletically, a parallel selection of such 'practical' characters may well have occurred. This applies also to the protecting reflexes and instincts so often linked with leaf-like and stick-shaped types (e.g. cataleptic rigidity of stick insects with legs put together longitudinally). The same principle of random combination of single elements applies to the wing colors and leaf-like patterns of the well-known butterflies *Kallima, Doleschallia, Prepona, Zaretes,* and *Historis,* and less efficient and less complete stages of such leaf-like developments are found in numerous related genera of the nymphalid and satyrid families (*Discophora, Precis, Eriboea, Cymothoë, Charaxes, Zeuxidia,* and *Mycalesis*).

Usually, parallelisms are also referred to as convergences in the case of homologous structures which have passed a more or less distinct stage of dissimilarity, after which they again became more and more similar or equal in the course of time. In such cases the effects of the common genetic background are less important, and parallel selection must be credited as having played the more decisive role. As is well known, in many cases it is not easy to recognize the homologies involved, and various authors have advanced various opinions regarding the criteria of homologous organs. (Remane thought that similarity of the position of organs in the organisms, equality or similarity of ontogenetic development, and the existence of phyletically intermediate forms were reliable critera; for further information on homology connected with problems of developmental physiology, see Baltzer, 1950.)

A characteristic example of parallel selection is provided by the shape of wings in birds. As I could show in 1938, the races of cooler zones have more pointed wings (second, third, and fourth primaries relatively longer, which make them more efficient in flying) than those inhabiting warmer regions, in which the wing tips are less tapering (Figures 62, 63). As this wing rule apparently applies to all corresponding species and as more pointed wings are definitely more favorable in flight than less tapering tips (birds of cooler zones tend to be more migratory than those of warmer zones), it is highly probable that parallel selection brought about this parallelism. The wing rule is also applicable to migratory and nonmigratory species of a genus (Kipp,

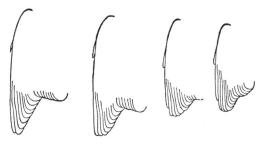

FIGURE 73. Blunt wing tips in non-migratory birds and pointed wing tips in migratory birds. Left to right: two migrants, the blackcap *S. atricapilla* and the garden warbler *S. borin*, and two stationary birds, the warbler *Sylvia undata* from Provence and *S. melanocephala*. (After Kipp, 1942.)

1936) as is illustrated by Figure 73 showing the wings of the central European blackcap (*Sylvia atricapilla*), and garden warbler (*S. borin*) in comparison with those of their Mediterranean nonmigratory relatives *Sylvia undata* and *S. melanocephala*. The same holds good when we compare species of wheatears, redstarts, orioles, and the like. Hence, there are also cases proving the formation of parallelisms resulting from parallel selection in transspecific evolution.

Equally clear are the effects of parallel selection in the transformation processes of Crustacea migrating from salt into fresh waters. The new fresh-water environment required more efficient excretory organs to maintain osmotic balance, and more water had to be removed from the interior of the body, so that relatively longer renal tubules proved advantageous. This difference in the length of renal tubules is met with as a parallelism in most different genera of Ostracoda, Isopoda, Amphipoda, and Decapoda as far as they have developed salt- and fresh-water types. Similarly, the renal glomeruli of fresh-water fishes are larger than those of salt-water types (compare Hesse and Doflein, 1943, II, 159–60).

Parallelisms in the type and structure of molar teeth were proved in four lineages of the rodent families Heteromyidae and Geomyidae by Wood (1937). A parallel formation of similar or equal dentitions evolved in various other groups of rodents (such as Castoridae, and Microtinae) hares and hoofed animals (Bovidae, Equidae), as a consequence of feeding on tough vegetable materials. When in the upper jaw of vertebrates the canines have developed excessive size, we often find that in the distal part of the lower jaw a sort of abutment is formed by a broadened and reinforced part of the jawbone. Thus, the skulls of quite different animal types without any close taxonomic relation-

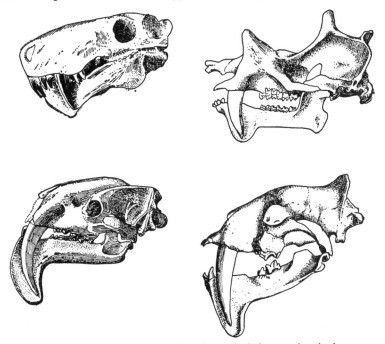

FIGURE 74. Convergent evolution of excessively long canines in the upper jaw and a corresponding bony pouch in the lower jaw. Above, left: *Inostranzewia alexandri* (Sauria), right: *Uintatherium alticeps* (Amblypoda). Below, left: *Thylacosmilus* (Marsupialia), right: *Eusmilus sicarius* (Felidae). (After Hutchinson, Scott, Riggs, and Abel.)

ship become surprisingly similar from this point of view, as in the skulls of the theriodont *Inostranzewia*, the amblypod *Bathyopsis*, the marsupial *Thylacosmilus*, and the large saber-toothed cat *Eusmilus* (Fig. 74). In this context the many parallelisms in the skulls of birds are worthy of mention, such as an ossification of the septum interorbitale in strongly pecking forms (woodpeckers, some types of titmice), a prolongation of the processus orbitalis anterior of the os praefrontale in types whose beaks are adapted to heavy duty (parrots, hawfinches, and similar forms) (B. Rensch, 1924; A. Müller, 1925–6).

All the examples quoted above – and their number could easily be extended – represent more or less parallel transformations of genetically equal or similar constructions. This process was referred to as homophlasy by Ray Lankester (1870) and as 'homodynamic function of homologous organs' by Abel (1929).

Parallel selection may become effective several times at periodic intervals, and repeated formations of similarities will then result, which have been referred to as iterations (Koken, 1896). In the family of pilgrim shells (Pectinidae), for example, repeated parallelisms evolved during the Liassic (*Weyla*), the Cretaceous (*Neithea*), and the Tertiary (*Vola*), invariably representing the same structural type with an arched upper and a flat lower shell ('*Vola*' type), which Abel (1929, p. 307) has rightly considered an adaptation to life on sandy sea bottoms. Similarly, there was a parallel evolution of keeled shells in various lines of Ammonoidea, which seems to have been brought about by parallel selection (in keeled types the hydrodynamic drag is smaller than in non-keeled types). Even with relatively large and rapid phylogenetic alterations, more or less identical iterations of a type may emerge. This is best illustrated by the parallel adaptations of ichthyosaurs and dolphins, both of these groups showing a torpedo-shaped body, a shortening of the neck, the formation of tail fins, dorsal stabilizers, fin-like forelimbs, an increased number of digits, a prolongation of the jawbones, the formation of cusped isodont teeth, and the faculty of being viviparous. (Formerly these parallelisms even gave rise to the opinion that dolphins or whales had immediately evolved from ichthyosaurs.) Similarities and conformities in the mole-like marsupial *Notoryctes*, the genuine mole (*Talpa*), and the golden mole (*Chrysochloris*) provide further examples of parallelisms in consequence of parallel selection.

When we review the cases referred to above (compare also corresponding examples mentioned in Chapter 4), we may state that the interpretation of parallelisms does not require the assumption of special autonomous factors of evolution. Surely, in spite of the undirectedness of mutation, there is often a certain direction in evolution, but this directedness is caused by parallel selection, by common hereditary elements (i.e. homologous or similarly acting genes), and by the manifold interrelations and correlative effects arising from gene interactions. Although many such parallelisms have not yet been adequately analyzed, there are no indications suggesting that unknown forces are at work.

V. Orthogenesis

Examples of Directed Phylogeny and their Problems. Because of the extremely large number of hereditary variations arising as the result of random mutation of all morphological and physiological characters of an animal type, and in view of the multiform conditions of the habitat and their change by climatic alterations, one would expect evolution to be an erratic process following a more or less zigzag course by favoring suitable alterations in changing directions and extinguishing disadvantageous characters of this or that kind. Generally, however, this is not so, and in many lines of descent certain trends of evolution during shorter or longer phylogenetic periods have been recognized. This is one of the most important generalizations of paleontological science.

Von Baer (1876), in his critical studies on Darwin's views, used the term *Zielstrebigkeit* ('purposefulness'), and according to his opinion a 'purposive evolution' would require the existence of consciousness. Von Naegeli (1884) thought that a law of inertia (*Beharrungesgesetz*) existed, applying to organic evolution, corresponding to the physical law of inertia, and being of 'mechanical nature'. It was Cope (1884) and Döderlein (1887) who first tried to demonstrate in special lines of descent that the selection of advantageous types proceeds along straight lines and that such trends sometimes continue in the same direction even when no further progressive adaptations are possible (Döderlein). Then Haacke (1893) introduced the term orthogenesis, denoting this straight-line evolution. By Eimer's studies on the orthogenetic evolution of wing patterns in butterflies (1897) the term became generally known and was then also used by paleontologists. At that time, Gaudry (1896) recognized a general increase of body size in the lines of descent of invertebrates as well as of vertebrates, especially in mammals. Independently, Cope (1896) when establishing his 'Law of the Unspecialized', which will be referred to below, formulated a rule of the general increase of body size in the many lines of descent. Later on Dépérét (1909) established a broader foundation for this rule, which since then has never been seriously doubted.

Orthogenesis, however, remained a serious problem for a long time to come. In 1905, Osborn introduced the term 'rectigradation', denoting directed steps of straight-line evolution, which was later used in a modified way (compare Abel, 1927, pp. 6–10), but no causal analysis of the alleged rectigradations was made. Nor was this done in his later hypothesis on 'aristogenesis' (1934), according to which mutations occur continuously in adaptive direction. Eimer tried to achieve a causal interpretation of the phenomenon by assuming a somatogenic induction ('inheritance of acquired characters'), and Plate (1913) and others followed in his path but, as we have seen, this theory canot be maintained. Plate (1933), however, also pointed out that orthogenesis involved two essentially different types of evolution: cases where in spite of random mutation a steady and constant selection (orthoselection) limits the progress of evolution to a certain point, and those cases – genuine orthogenesis – where 'directed phyletic development occurs without the effects of selection' (p. 508). The first type has already been dealt with in our previous considerations and easily fits into our general ideas on evolution, and we need only try to elucidate the latter type of (genuine) orthogenesis. Our main question will be whether there is no directed selection or whether it might exist, disguised and difficult to analyze.

Some paleontologists and also some zoologists think that the phenomenon of overspecialization, i.e. the continued straight-line evolution which goes beyond maximum adaptation or even reaches disadvantageous stages, disproves the action of selection and proves the existence of autonomous regularities and forces of evolution. Hennig (1932, p. 24), for example, wrote the following:

No new fossil evidence will spoil the firm knowledge that this transformation proceeds in a steady direction which is quite different, for instance, from the transformation of the terrestrial surface, though the latter process certainly is not irregular, but rather unpredictable. Hence, it cannot be the terrestrial environment that directs the course of evolution, but immanent regularities of life itself will be the directive element.

Schwegler (1941), in his consideration of the excessively long belemnite rostrum which evolved several times as the result of convergent evolution, states that 'quite obviously the evolutionary trend leads to the extinction of the respective branches without any environmental effects'. Von Huene (1940) assumed a 'superior principle governing and directing the whole' evolution. Studying the Miocene horses of the genus *Anchitherium*, Wehrli (1938, 1940), stated: 'The process of transformation seems to be directed and not at all caused by random mutation and selection.' Concerning the evolution of the marmots he said:

> The phylogenetic transformation of the Alpine marmot cannot be interpreted as caused by random mutation and selection, either, and one definitely is inclined to assume immanent forces transforming the animal type in the direction which began initially.

Schindewolf's views are similarly skeptical, and in one of his studies on the phylogeny of Cephalopoda (1942, p. 379) he writes:

> Orthogenesis, proceeding regularly along the lines once commenced, is an established fact and by no means an unfounded fiction easily put aside. It cannot be interpreted by the principle of selection.

Regarding overspecialization, Beurlen (1937, p. 72) states:

> It is absolutely clear that adaptation and selection cannot be regarded as the essential principle of phylogeny, as (in these cases) it proceeds from adapted to non-adapted types and it produces disadvantageous organs from useful structures. It is an autonomous regularity of its own governing phyletic transformation according to the laws of Gestalt and regardless of environmental factors.

Correspondingly, Beurlen thinks that the 'directedness of the various types' is the essential symptom of the phase of specialization which also includes the formation of parallelisms. According to this author, specialization is that part of the 'phylogenetic cycle' which succeeds the initial phase of explosive splitting and which tends to fade away in a 'late phase of giant transformations and overspecializations'.

All these considerations, however, did not account for the complicated effects of selection acting on the organism as a whole, though it is quite evident that it is not the characters which are selected (characters are only the tools of selection) but the genes underlying them and causing pleiotropic effects in various other characters, as we have seen in our consideration of the correlation of growth allometries and other complex alterations of the structural type. To illustrate this statement more clearly, we shall once again consider the complex effects arising as a consequence of a successively increasing body size. Hence, we must first discuss Cope's Rule.

Cope's Rule. The rule, established by Cope (1896) and Depéret (1907), applies to the most frequent and general phenomenon of orthogenetic evolution: progressively increasing body size in the lines of descent. Practically all major animal groups provide examples proving the validity of this rule, and there is a large body of material already studied from this point of view (besides Cope and Depéret, see Diener, 1917; Dacqué, 1935; and Newell, 1949). From such studies it became clear that the general phyletic trend of an increase of body size is met with in short lines of minor importance as well as in numerous major categories as a whole, though there are some examples to which this rule is not applicable.

So it will not be sufficient to compile 'examples' but it will also be necessary to find out the approximate percentage of exceptions to the rule in at least one large group of animals. A suitable group for this kind of inquiry is represented by the mammals, as alterations of body size are clearly visible in these. There are a few orders only, such as Monotremata, Dermoptera, Tubulidentata, and Pholidota, the phylogeny of which cannot yet be properly evaluated, as the fossil records are not sufficient. In all other mammalian groups the phyletic trend in body size may well be evaluated on the basis of the more or less complete fossil evidence.

During the Late Cretaceous and the Eocene, the Marsupialia were represented mainly by smaller types about the size of a rat or a hare. Giant types did not evolve until the Pleistocene, when the genera *Diprotodon* and *Nototherium*, which are related to the wombat, reached the size of a rhinoceros. The genera *Koalemus*, *Phascolonus*, and *Sceparnodon* (also related to *Phascolomys*) and the giant kangaroos *Sthenurus* and *Palorchestes* became almost equally large during these epochs. The Recent types of Marsupialia, on the average, have become smaller again. Among Insectivora, rat-sized types are known to have existed during the Late Cretaceous (Deltatheriidae), and the Pantolestidae of the Eocene reached the size of a porcupine, but a distinct increase of body size in single lines of descent cannot be maintained on the basis of present evidence. The same statement applies to *Chiroptera*. The phylogeny of Recent types of *Xenarthra* cannot properly be evaluated as yet, but a clear trend of increasing body size is met with in the fossil ancestries of extinct Gravigrada. The Miocene genera were definitely smaller than the Pleistocene giant sloths, which reached the size of an elephant (*Megatherium*, *Mylodon*) or a cow (*Megalonyx jeffersoni*). In armadillos, rhinoceros-sized types did not evolve until the Pliocene and Pleistocene (*Chlamydotherium*), and the Miocene ancestor, *Hapalops*, was considerably smaller. A definite trend of increasing body size is also obvious in Glyptodontidae, the Miocene genera of which are considerably smaller than their Pliocene and Pleistocene relatives *Doedicurus* (4 m. long) and *Glyptodon* (2 m. long). In Rodentia no exact succession of genera showing increasing body size can be noted as yet, but *Megamys* (Chinchillidae), about the size of a *Hippopotamus*, did not evolve until the Pliocene and the Pleistocene, nor did the giant beavers *Amblyrhiza* and *Elasmodontomys* from the West Indies, and the North American *Castoroides*, which reached the size of a bear. In Carnivora, the number of lines known from rich fossil material is fairly large, and phyletic increase of body size is a common trend, which is especially well demonstrated in bears (Ursidae), hyenas (Hyaenidae), dogs (Canidae), cats (Felidae), and civet cats and their relatives (Viverridae). In dogs a considerable body size was attained in a phyletic side branch during the Late Miocene, when the bear-sized *Dinocyon* evolved. Moreover, it is highly probable that all these families originated from Eocene Miacidae, all types of which were relatively small. Other lines of primitive carnivores (Creodonta), which became extinct during the Eocene, had already reached the body size of a bear (*Pachyaena, Mesonyx, Arctocyon*). A distinct trend toward increased body size is also found in Cetacea, which originated from the primeval type *Protocetus* of the Middle Eocene. The size of this original form may be estimated from the length of its skull, which measured only about 60 cm. *Basilosaurus* of the Upper Eocene from Alabama reached a body length of about 18 m. A general statement applying to all orders of hoofed animals is that they derived from relatively small (about wolf-sized) Eocene Protungulata (especially striking in the ancestry of Perissodactyla and Litopterna), or directly from small Cretaceous Insectivora. It is evident, then, that larger types of hoofed animals did not come into being until a certain time had elapsed. Among the Perissodactyla this is most clearly illustrated by the successive phyletic increase of body size in the well-known horse ancestry, commencing during the Eocene with primitive types of about the size of a fox or a wolf. Quite convincing evidence of phyletic increase of body size has also been proved in the lines of Titanotheriidae and Rhinocerotidae, and in Eocene Lophiodontidae. Dépéret (1907) demonstrated a parallel process of increasing body size in four lines, the increase ranging from the size of a tapir to that of a rhinoceros. In Artiodactyla, Cope's Rule is especially well established, as the Recent types of llamas and camels originated from hare-sized Eocene ancestors, like *Catodontherium*, which evolved into the Oligocene *Anthracotherium* of about the size of a rhinoceros and – via other lines – into the Recent llamas and camels. Among Cervidae, the Miocene ancestors are relatively small (*Merycodus* of the Late Miocene being as large as a roe), and the giant types, like *Megaceros* and *Alces latifrons*, emerged during the Pleistocene. Early forms of Giraffidae are relatively small (e.g. *Palaeotragus* of the Early Pliocene). In Notoungulata a continuous phyletic increase of body size is to be seen from the Oligocene types of *Propyrotherium* to the Miocene *Pyrotherium*. On the other hand, in the Lower Eocene of Patagonia the aberrant type *Carolozittelia* appeared, which was about as large as a tapir. A distinct tendency towards an increase of body size is also evident in the rich fossil record of Proboscidia. The truly ancestral types of this group are hardly known, but the Eocene forms of *Moeritherium* may well be considered as closely related to the original types. Their body size did not exceed that of a tapir. Phyletic increase of body size is demonstrated by Mastodontidae (the Oligocene *Palaeomastodon* being much smaller than the Pliocene *Bunolophodon arvernensis*), by Dinotheriidae (Oligocene species being much smaller than the Lower Pliocene *Dinotherium giganteum* and *D. gigantissimum*) and, finally, by various lines of Elephantidae, the maximum body size of which was attained by the Eurasiatic *Elephas primigenius* and the North American *E. imperator* (Osborn, 1922, 1935). The Recent types of Hyracoidea are relatively small, but *Megalohyrax eocaenus* was as large as a lion. The origin of Sirenia is represented by the Eocene *Eotherium*, which was about 1·5 m. long. Maximum body size was reached by the Pliocene *Felsinotherium*, which was about 3·5 m. long, and body size seems to have remained almost constant ever since, as the Recent types (*Manatus*) are about equally long. The earliest types of Amblypoda of the Lower Eocene were relatively small, but during the Middle and Late Eocene body size was successively increased to that of *Loxolophodon* (about 4 m. long). Finally, in Primates originating from small Cretaceous insectivores, an increase of body size is to be noted in Lemuroidea, which did not happen, however, until the Pleistocene period had begun (skull length of *Palaeopropithecus* about 20 cm., of *Megaladapis* about 31·5 cm.; ancestor *Adapis* about 8 cm.). The Catarrhini began with small Oligocene types and reached the size of modern Anthropoidea during the Lower Pliocene (*Dryopithecus giganteus*). Cope's Rule is well applicable to the human line of descent, as from the tarsioid Eocene ancestors via the Oligocene Catarrhini, the Late Tertiary Anthropoidea, and the *Pithecanthropus* stage, to *Homo*, body size has steadily been increasing. Finally, it may be said that all mammalian orders are derived from the small mammals of the Mesozoic.

Summing up our findings, we may state that in almost every mammalian ancestry a progressive phyletic increase of body size is met with, and that all giant animal types very probably evolved from smaller ancestors. Although some giant types appear in earlier stages of phylogeny, it seems highly probable that they represent the final types of phyletic side branches. At any rate, it is quite impossible to maintain that such giant types represented the evolutionary origins of small relatives emerging in subsequent epochs of evolution. There are only a few known cases which prove that a graded decrease of body size has occurred. Examples showing this phenomenon are provided by that part of the horse ancestry which leads to the small *Nannipus*, by the dwarf elephants from Malta, and by the fact that numerous types of the Ice Age were larger than their Recent relatives (as with *Elaphus*, *Bison*, and the like), but most of these examples do not concern the phyletic ancestors proper, but members of side lines, such as mammoth, giant deer, giant sloths, and similar types.

It is not possible to discuss other animal groups in as much detail as we have given for mammals, but some of the more striking examples provided by Mollusca will be cited. The Cephalopoda originated from very small Lower Cambrian types, like *Volborthella* (shell 5 mm. in diameter), and evolved into the first giant types (1–2 m. in diameter: *Endoceras, Orthoceras*) during the Silurian. Other lines ended with giant body size in the Liassic (*Arietites*, 1 m. in diameter) and the Late Cretaceous (*Pachydiscus seppenradensis*, 2·5 m. in diameter). Similarly, the Gastropoda evolved from relatively small Cambrian ancestors into larger species during the Later Paleozoic (*Pithodea* of the Lower Carboniferous being 15 cm. high and *Tropidostropha* and *Schizostoma* having a diameter of 11 cm. each). The giant species of *Cassis, Strombus, Semifusus*, and others developed only during the Tertiary. The family of Pleurotomariidae began with relatively small types in the Cambrian, while the largest forms exist today. The nominate genus *Pleurotomaria* began with small Triassic ancestors and led to the Recent species *Pl. adamsoniana* of about 17 cm. shell height. Similarly, the earliest types of Cambrian Lamellibranchiata were quite small, and giant forms like *Inoceramus* (shell length about ½ m.) did not evolve until the Cretaceous. The giant *Tridacna gigas* of our present times, the shell of which is more than 1·5 m. long and weighs several hundred pounds, is the largest type of its line (family Tridacnidae), which commenced in the Eocene. With regard to the phyletic increase of body size in Saurian groups, the reader is also referred to Colbert (1948, Figure 3).

A large series of examples taken from the groups of Foraminifera, Corallia, Brachiopoda, Bryozoa, Echinodermata, Mollusca, and Arthropoda was discussed by Newell (1949), who summarized his findings in the statement that in invertebrates, too, a successive phyletic increase of body size is the most common type of directed development.

It is, of course, necessary to discover the frequency of exceptions to this rule. Cope's Rule seems to be disproved by the Recent species of *Cerithium* (Prosobranchia), at least one third as large in body size as their Eocene predecessors of about ½ m. body length (*C. giganteum*); by the giant snail, *Natica leviathan*, which evolved during the Early Cretaceous (Recent relatives essentially smaller); by the giant oysters (*Ostrea gigantea, O. crassissima*, and the like) which existed in the Tertiary; by the giant Orthoceratidae of the Lower Silurian, and so forth. One should not forget, however, that all these giant types probably must be regarded as the final stages of evolutionary offshoots of minor longevity and that in all these cases evolution started from smaller ancestors. At any rate, so far no descent of smaller successors from these giant types has become known. There is an actual decrease of body size, however, in a series of Ammonite species: *Scamnoceras angulosum* (Liassic zone α 2 c) – *Sc. angulatum* (α 2 a,b) – *Saxoceras praecursor* (α 1 f) – *S. costatum* (α 1 f) – *S. angersbachense* (α 1 e) – *S. schroederi* (α 1 d) (see Lange, 1941).

A steady decrease of body size is also certain to have occurred in all those cases where the smallest forms of a structural type are approached, which sometimes causes – as we have seen in the previous chapters – critically excessive proportions and structural alterations. Such forms cannot, of course, have been the initial stages of evolution, which may be illustrated by an example taken from the world of insects. If we compare the venation of the smallest flying insects with that of larger relatives (Figure 75), we find quite distinct symptoms of reduction in the smaller types, brought about by various mechanisms. Similarly, essential reductions are met with in Archiannelida which – in contrast to the literal meaning of their name – probably are not only initial phylogenetic types, but in some cases certainly reduced Polychaeta (see above, p. 177). Secondary reductions of body size have apparently taken place in dwarf shrews, hummingbirds (with a specialized type of hovering

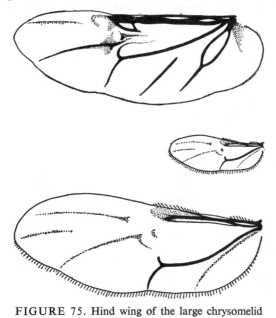

flight; compare Stolpe and Zimmer, 1939), flower-peckers (Dicaediae), small titmice, the smallest frogs and snails, and Crustacea. Seen from a more general point of view, however, such reductions of body size must be regarded as exceptions, as the phyletic increase of body size is far more common. Moreover, reduced body size is met with mainly in those animal groups in which a small body was a favorable character, as in inhabitants of extremely small habitats (Polychaeta, Gastrotricha, insects of mold humus, hair and fur parasites, intracellular parasites, and so forth) or in many flying animals like

FIGURE 75. Hind wing of the large chrysomelid *Melasoma populi* (above) and of the very small chrysomelid *Haltica atra* (center), showing reduced venation. Below: same wing of *Haltica* drawn to same scale as wing of *Melasoma*.

birds and insects whose ancestors seem to have been of a medium body size (*Archaeopteryx*, insects of the Carboniferous). In flying animals larger body size is unfavorable, as the surface area of the wings grows by the square, whereas the body grows by the cube (compare Chapter 6, B III).

Thus, Cope's Rule may be considered as well established and applicable to numerous different animal groups. However, we shall have to see whether a general causal explanation can be found which will not merely attempt to disguise the confession 'Ignoramus' by introducing the term 'autonomous force of development'. A satisfactory explanation should describe the mechanisms underlying Cope's Rule by the known processes of undirected mutation, natural selection, and correlations arising from pleiotropic interactions of characters (which have been referred to in Chapter 6, B II and B III). As in any animal form larger and smaller hereditary variants will arise, our problem may be more concisely formulated as follows: may an increase of body size be regarded as an advantageous character in most animal groups and most situations of selection, so that a successive phyletic increase of body size is effected automatically? Of course, the answer to this question unavoidably has to summarize a great mass of different facts; nevertheless, the question may be answered affirmatively, although with some reservations. I wish to support this affirmation of mine by the following statements proving that increased body size really means an advantage in selection.

1. Larger forms generally are more vigorous and resistant than smaller types, to which they are superior in numerous respects. Quite recently Mangold (1955) could prove that small eggs of *Triturus alpestris* develop much more slowly than larger ones. The same happens in the prenatal development of some snails, the eggs of which are unequal in size. The large eggs will usually develop more vigorously than the smaller ones, and the latter are thought to be eventually consumed by what develops from the former (*Buccinum, Purpura*, and others). A similar process occurs in some dogfishes (*Lamna*) and in the Alpine salamander (*Salamandra atra*). In mammals bearing several young per litter, the weakest embryos have a greater tendency to die prenatally or at birth. After birth an active struggle for existence begins, as – e.g. in birds – the begging movements will be more intense in the larger more vigorous young, which leads to a more frequent feeding of those already more vigorous. Quite often the weaker young will die in consequence. The same applies to weak young mammals in their efforts to reach the source of maternal milk or other food, as they will be pushed aside by the more vigorous individuals of the litter. Smaller individuals are also more likely to die from loss of body heat than their stronger and more resistant litter-mates, as in the former the mechanisms of temperature regulation do not begin to function as early as in the latter. This was proved by Leichtentritt (1919) in kittens, the normally strong individuals of which could well regulate their body temperature right after birth, while the weaker variants were endangered by the lack of sufficient temperature regulation, which was not fully developed until two days after birth. In adult animals there is often keen competition for food, and as regards quick locomotion in fleeing from enemies, trying to find protected places to evade the hardships of nature, and catching prey, the larger and more vigorous specimens will be *absolutely* quicker and, hence, favored by selection (for precise data on the speed of locomotion of small and large animals, see Von Buddenbrock, 1934). Striking examples of the larger variants being favored by selection are also provided by the sexual display and fighting of stags and bulls, heath cocks and blackcocks, salmon and other fighting fishes, stag beetles, and so forth. In species in which such fights are essential elements of the reproductive behavior patterns, only the superior males are given the chance to produce offspring, and the smaller males are often killed. It was Castle (1932) who first tried to interpret Cope's Rule by stressing the effects brought about by pre- and postnatal selection among individual variants of the species.

The importance of such fights and rivalries is not restricted to infraspecific communities but affects transspecific competition between larger and smaller species, as well (for example, large and small bird species competing for feeding and nesting sites). Moreover, climatic selection will wipe out the smaller species in some habitats more readily than the larger ones, so that the latter become the only inhabitants of the biotope. As size differences between the species will usually be greater than between individual variants, the selective advantage of the larger species is relatively important and should probably be rated as more than one percent. (Transspecific selection is often neglected in the discussion of mathematical formulae concerning evolutionary rates.)

2. As we have seen in our consideration of the phylogenetic importance of body size (Chapter 6, B II), the proportions of body and organs of large types differ from those of smaller types, due essentially to the effects of allometric growth. In a number of cases the effects of allometric growth are favorable in selection, as will be illustrated by a few examples.

In many animals, weapons for attack and self-defense represent body parts with a marked positive allometry, i.e. in large variants these organs will be not only absolutely but also relatively larger than in smaller specimens. This applies to the chelae of numerous crabs, the mandibles of many insects (especially those of stag beetles and ants), the antlers and horns of ruminants, and the canines of mammals (especially of carnivores, monkeys, and the like). It is quite evident that this effect of positive allometry can hardly be underrated as to its evolutionary importance (selective value in self-defense, defense of the young, the herd, and the colony, sexual fights, etc.). Excessive growth as in the giant deer *Megaceros*, of course, may at last become disadvantageous (compare Chapter 6, C).

In vertebrates of increased body size, the relatively decreased size of liver, heart, and kidneys may be regarded as an advantageous character because more space is left unoccupied, so that the intestine can grow relatively longer and the intra-uterine development of embryos can be prolonged, which certainly means better care of the young. In large insects, the correlated size decrease of wing muscles and ventral nerve cord leaves additional space in the body cavity to be occupied by tracheal sacs, which provide a means of thermic insulation and a supply of oxygen.

In large mammals, the brain stem is relatively smaller and the pallium is relatively larger than in smaller types. It seems probable that larger animals are capable of more complicated associations and more versatile and plastic reactions to environmental stimuli, and above all, that larger types may have a better memory than smaller relatives. The structure of brain cells can also be more advantageous in larger animals (see below).

3. Histological differences which have selective advantages also arise with increasing body size. As we have seen in the preceding chapter – especially in our consideration of the tiny snail *Caecum glabrum* (Prosobranchia) – a phyletic increase of body size generally is brought about by increasing the

number, rather than the size, of cells. There is also a slight increase of cell size, but this is not at all proportionate to body size. This statement applies to bone cells, epithelial elements, glandular cells, and cells of the connective tissues, but not to the 'permanent' tissues, such as nerves, eye lenses, muscle fibrils, feather barbules, and the like, since the number of such cells usually is more or less fixed and hence the size of these cells is more or less proportionate to body size. Some data by G. Levi (1914, 1925, partly quoted by Teissier, 1939) may serve to illustrate the above statements (for data on invertebrates compare Conklin, 1912; Löwenthal, 1923; and others). The size of liver cells is $19·6 \times 19·6$ μ in the cow (*Bos taurus*) and $15·1 \times 15·1$ μ in the canchil (*Tragulus canchil*) of only 45 cm. body length. The corresponding measurements are $23·7 \times 23·7$ μ in the rabbit (*Oryctolagus cuniculus*), $19·2 \times 19·2$ μ in the rat (*Epimys norvegicus*), $20·4 \times 20·4$ μ in the hedgehog (*Erinaceus europaeus*), and $13·7 \times 13·7$ μ in the shrew (*Sorex araneus*). Size differences of the nuclei are less, and in some cases the smaller species even have slightly larger nuclei: cow $6·1 \times 6·1$ μ, canchil $6·3 \times 6·3$ μ, rabbit $8·03 \times 8·03$ μ, rat $8·7 \times 8·6$ μ, hedgehog $7·7 \times 7·7$ μ, and shrew $5·8 \times 5·8$ μ. Cells taken from the deeper layers of the esophagus epithelium gave the following measurements: rabbit $7·1 \times 12·2$ μ, rat $7·3 \times 6·5$ μ, mouse (*Microtus arvalis*) $5·8 \times 8·5$ μ, hedgehog $8·6 \times 15·1$ μ, shrew (*Sorex vulgaris*) $5·1 \times 10·6$ μ. Nuclei of cells from the coiled part of the renal tubules were, according to my own measurements, $5·8 \times 5·1$ μ in the rat (n=52) and $5·6 \times 4·8$ μ in the female house mouse (n=64). The greater diameter of the mouse nuclei, then, is 95·8 percent of that of the rat, and the smaller diameter of the nuclei of the mouse is 94·3 percent of the rat's nuclei. The length of head and body, however, of the mouse is only 39·5 percent of the rat's length. In 1904 Boveri found that in human giants and dwarfs the size of bone and epithelial cells is equal. Counting the cell numbers on homologous parts of the jugal-metajugal bone (transverse sections cut at 8 μ) I found an average of 137 nuclei in the central European black tit (*Parus m. major*) and of 115 in the smaller race *P. m. cinereus* of the Sunda Islands. The ratio of the transverse section of these bones (of *P. m. major* to that of *cinereus*) is 100 : 86, whereas the ratio of the number of cells is 100 : 84, which means that, practically, the size of the cells is equal in both races.

Generally, then, we may state that large animals normally have more cells than related smaller types, and this fact has several advantageous consequences for the metabolism of the cells. This is well illustrated by the red blood corpuscles, the size of which is not essentially different in large and small related types. I have tried to show this in Table 21, using the data quoted by F. Groebbels (1932, pp. 112–14) after Venzlaff. (As body length is no reliable basis of comparison in birds because of the various type and length of bills and necks, I used as a theoretical 'weight–length' the cube root of the birds' body weight. For corresponding data on large and small races of domestic fowl, see Schlabritzky, 1953.) As will be seen from the table, large types have not only absolutely but also relatively more red blood corpuscles than smaller relatives, which means that the surface to be employed in adsorption is

TABLE 21. Red blood corpuscles in related species of birds. (After Venzlaff by F. Groebbels, 1932.)

Species	Weight–length	Red Blood Corpuscles in μ	Red Blood Corpuscles in Percent of Weight–length
Larus marinus	11·45	15:7·5	0·0013
Larus ridibundus	6·54	15:7·5	0·0022
Cygnus olor	20·41	16:7·5	0·0008
Anas crecca	6·13	14·5:7	0·0023
Ardea cocoi	12·60	16:7·5	0·0013
Ixobrychus minutus	5·25	14·5:7	0·0028
Porphyrio poliocephalus	7·75	16:7·5	0·0021
Porzana porzana	4·40	14·7	0·0032
Pavo cristatus	15·87	16:7·5	0·0010
Coturnix coturnix	4·56	11·3:6·3	0·0025
Aquila chrysaetos	16·63	15·5:7·5	0·0009
Falco tinunculus	6·54	13:7·5	0·0020
Bubo bubo	14·09	16:7·5	0·0011
Otus scops	5·37	14·5:7	0·0027
Corvus corax	11·45	15:6·3	0·0013
Passer montanus	3·11	12·5:6·3	0·0040

relatively larger. It is evident that increasing the relative number of blood corpuscles yields an essentially larger adsorption surface than increasing the relative size, as the relative size of a large body, because surface increases by the square and volume by the cube, is smaller than that of a small body. A more efficient system of red blood corpuscles is, of course, an advantageous character and will be favored by infra- and transspecific selection. The relatively small size of the red blood corpuscles in large species may be regarded as an advantage, as the capillary net of the circulation system can be relatively more delicate and relatively more complicated than in the smaller species. We need not expect, of course, that a basic rationalization of the respiratory mechanism came about with phyletic increase of body size, but it is quite intelligible that extreme physiological efforts placing an overload on respiratory efficiency will more easily be endured by the larger animals of a related group. Hence, the large types can have a lower rate of breathing movements corresponding to the relatively decreased intensity of metabolism in general: a lion will breathe 12 times per minute, a leopard 15, a cat 24 times. The corresponding figures (after Batak, 1921) are 55 in the rabbit, 80–85 in the guinea pig, and 200–300 in the mouse; 10–11 in the camel, 19–22 in the llama, and so forth.

As we have seen in our consideration of the retina of related vertebrates differing in body size (see above, p. 150), the number of visual cells is absolutely larger in the large species. Consequently, a large eye of this type is more efficient in resolving the retinal image into a more detailed nervous impulse pattern than the eye of a smaller relative, and hence, the large animal is capable of reacting to more details of visual stimuli than related smaller types. It need not be stressed that this better visual capacity represents an essential advantage in selection which will be effective in infraspecific competition and much more so in transspecific competition (as in these cases normally the size difference is greater). A similar statement holds true as regards the shape and function of the eye lens to which we have referred in our discussion of the lens rule (see above on p. 149). The relatively smaller and flatter lens of the large forms provides a better mechanism of accommodation than the more spherical lens of the smaller relatives.

As the cells constituting the ommatidia cannot be enlarged *ad libitum*, it is evident that in large Arthropoda the number of ommatidia will be greater, which is found to be true in all insect groups. In the eye of the blowfly *Calliphora vomitoria*, the number of ommatidia is 4,585 (1 ♀), and in that of *Drosophila melanogaster* (2 ♀♀, wild type) 663 and 672 were counted per eye. In the gnat *Culex pipiens* (2 ♂♂) the corresponding numbers were 480 and 498, while in four males of a minute species of Lycoriinae of 3 mm. body length 224–268 ommatidia were counted per eye (Partmann, 1948; for further data see Leinemann, 1904; Schilder, 1950). By an increase in the number of ommatidia per eye the resolving power of the visual system is improved. Small eyes often have so few ommatidia that the formation of true visual images and the perception of visual forms is hardly possible (Von Buddenbrock, 1924, p. 59).

Phyletic increase of body size may involve favorable alterations of the anatomical structure of the kidneys also. On median sections of rat's kidney each glomerulus contained 63–107 (average: 83) cell nuclei (sections cut at 10μ), and in the mouse I found only 45–69 (average: 54) nuclei. The number of glomeruli in the rat is greater than in the mouse. On median sections I counted 279–347 (average: 308) renal glomeruli in the rat and 132–159 (average: 146) in the mouse. This increased number of glomeruli and of glomerular cells may make possible a more efficient functioning of the excretory system, as the excretory surface is larger in the larger animal. This increased efficiency probably would not arise if the number of glomeruli and of glomerular cells remained constant and only the size of the cells increased, because the renal excretory surface would then be relatively smaller. Similar advantages could probably be traced in other tissues and organs, but no precise data pertaining to this question are available as yet.

Finally, some alterations of the histological and cytological structure of the cerebral cortex brought about by increasing body size may be regarded as characters favored by selection (compare Chapter 6, B III on allometric growth). In large types the more progressive brain parts are relatively larger, the neurons are absolutely larger, and the number of dendrites is absolutely greater; hence, more complicated associations may be rendered possible.

Thus, the difference of histological structures represents several advantages in favor of the larger animals. The most important of such histological improvements seems to be the increase of the absolute number of sense cells.

4. We have found that in large poikilothermic animals the number of eggs and young is greater than in related smaller species (Chapter 6, B III). Hence, phyletic increase of body size involves an increased number and not an increased size of ovarian cells. It is quite obvious that an increased number of offspring is usually an advantage in selection. (This statement does not exclude the fact that in large types it may sometimes be more advantageous to increase care of the young and to reduce the number of offspring.)

5. We have seen that in large animals the processes of metabolism are more economical than in smaller types, due mainly to the relatively smaller loss of body heat of the former (see also Krumbiegel, 1933). This fact is a reasonable explanation of Bergmann's Rule applying to races of a species or a large 'Rassenkreis' (and partly to related species), the range of which extends from warm into cooler regions. Larger races of such species inhabit the cooler zones, which was proved to be the result of climatic selection (by minimum winter temperatures: B. Rensch, 1939). Quite a long time ago (B. Rensch, 1924; also 1952) I tried to call attention to the possibility that the most impressive examples of Cope's Rule, i.e. the increase of body size in mammals, might be partly explained by the principle of climatic selection. This is suggested by the fact that the increase of body size in certain phyletic lines was correlated with a lowering of the average temperatures of the Tertiary. These lines include horses, elephants, rhinoceroses, carnivores, and others. It would be possible to say that these examples show, over a period of time, the effect of Bergmann's Rule. As proved by the Titanotheriidae, the body size of which was enormously increased during the Eocene, climatic selection was certainly not the only causal factor increasing phyletic body size of mammals, but it may have played an important part in its causation.

This is illustrated especially by the widespread evolution of large or even giant animal types in nearly all orders of mammals during the Glacial Age. The best known of these large types of the northern hemisphere are the giant mammoths (*Elephas primigenius* and *E. imperator*), the woolly-haired rhinoceros (*Rhinoceros tichorhinus*), the giant hippopotamus (*Hippopotamus major*), the ancient bison (*Bison priscus*), giant elks and deer (*Alces latifrons*, species of *Megaceros*), the camel-sized *Macrauchenia* (order Litopterna), giant beavers about bear's size (*Castoroides*), cave lions (*Felis spelaeus*) and cave bears (*Ursus spelaeus*) about one-third larger than their Recent successors, the giant sloth (*Megalonyx jeffersoni*) of North America, and the large mole *Talpa magna* of Europe. In South America, the contemporary giants included large types of Xenarthra, like *Chlamydotherium*, an armadillo of about the size of a rhinoceros, or the related *Glyptodon* (2 m. long), and *Doedicurus* (about 4 m. long). Other contemporaries were the giant sloths (order Gravigrada),

Megatherium, Mylodon, Grypotherium, and *Scelidotherium* of about elephant's size, and the notungulate *Toxodon* (about the size of a hippopotamus). In the West Indies the giant beavers *Amblyrhiza* and *Elasmodontomys* and the sloth *Megalocnus* (Gravigrada) evolved during the period. During the Pleistocene the Australian fauna produced giant wombats (Phascolarctidae) of about the size of a rhinoceros, such as *Diprotodon, Nototherium,* and *Koalemus,* and similarly large relatives like *Phascolonus* and *Sceparnodon,* the giant kangaroos (*Sthenurus* and *Palorchestes*), and a giant emu (*Genyornis*) of about 30 cm. skull length. In New Zealand the giant *Dinornis maximus* reached 3·5 m. body height. The Prosimiae of Madagascar also attained large body size at this time (*Palaeopropithecus* and *Megaladapis* of about man's size), and the giant ostriches of the genus *Aepyornis,* also from Madagascar, were contemporary with these large Lemuroidea. Though large animals have evolved during almost every geological epoch (except the Cambrian), the accumulation of so many large and giant types during the cold Glacial Period is certainly not accidental, as these large types usually were the largest forms of their kinship and definitely larger than their Recent successors. It seems probable that the relatively small body surface of warm-blooded animals of large body size was a decisive character in the process of selection by extremely cold temperatures.

6. As may be seen from Table 20 (p. 163), large homoiotherms usually live longer than related smaller types. Apparently, this rule also applies to some groups of poikilotherms. Among reptiles, for example, crocodiles and giant turtles live considerably longer than their smaller relatives. The female of the tiny *Dinophilus gyrociliatus* (Archiannelidae) of about 0·7–1·3 mm. body length lives about two to three months, its male about eight to ten days, but the larger polychaete *Pectinaria koreni* has a life span of about one year, and *P. auricoma, P. belgica,* species of *Spirorbis,* and related types will live for several years (compare Remane, 1932). Increased longevity may be regarded as a favorable character, as animals living longer than their smaller relatives have more time to find the best habitat, to gather individual experience, to accumulate antibodies, and to become more versatile and plastic in reacting to various situations arising in the struggle for life. This is especially applicable to higher animal groups, of course.

7. There is one more factor which may possibly be considered to contribute to the causation of successive phyletic increase of body size in all animal (and some plant) groups, as far as Cope's Rule is applicable. This factor is heterosis (luxurious development) in hybrids. With advancing speciation, individual variability, and the formation of ecological and geographic races, heterozygosity also increases, and the effects of heterosis (Dobzhansky, 1951, p. 121) are more likely to appear. One of the results may be an increase of body size. An interpretation of Cope's Rule from this point of view is especially suggested by all those cases to which the selection advantages listed under 1–6 do not apply. Increasing body size in the numerous lines of Foraminifera, for instance (compare Hiltermann, 1952; Hiltermann and Koch,

1950; Newell, 1949), can hardly be understood on the basis of competition and selection for larger body size as the decisive character, but may become intelligible when the principles of heterosis and hybrid superiority are applied.

Summing up, we may state that phyletic increase of body size involves a series of distinct and more or less general selective advantages. Hence, Cope's Rule may well be interpreted by the principles of random mutation and infra- and transspecific selection of larger variants or species. These statements apply especially to mammals, which provide the most numerous and impressive examples of all animal groups, but do not exclude the possibility that in some other groups a decreased body size must be regarded as a favorable character. This is easily demonstrated by the fact that smaller animal types will be superior to their larger relatives when small biotopes are to be occupied, such as the dense leaves of a thicket, the mold humus of the soil, the sandy stretches of shallow waters, and so forth. Small body size is also advantageous in numerous aerial types, as it involves a more efficient aerodynamic mechanism of wing size and body weight.

One should not forget that the more intense metabolism, the shorter life cycle, and the quicker succession of the generations of small animals may represent favorable characters in certain types of selection. This may be why Cope's Rule does not apply to the bulk of the enormous multitude of insects. Regarding the other animal lines, however, which show a successive increase of body size, we may rightly assume that the factors discussed above provide a possible, though perhaps not fully sufficient, explanation of the phenomenon. At any rate, there is no need to suppose that Cope's Rule proves the existence of unknown autonomous forces of evolution; random mutation and natural selection provide a sufficient interpretation of this kind of orthogenesis.

Orthogenetic Evolution of Organs. The interpretation of Cope's Rule by general selection of larger forms applies also to numerous cases of special orthogenetic evolution of single organs. In Chapter 6, B III, on allometric growth, we have found that numerous organs are correlated to the whole body by positive or negative allometry. The teeth, for instance, are structures with positive allometric growth as, after the loss of the milk teeth, the permanent teeth, of course, must grow more quickly than the body as a whole. With successively increasing body size in the course of phylogeny, the permanent teeth will not only grow proportionately larger but will become disproportionately large and even excessively large. Teeth, then, seem to be one of those structures most likely to be affected by special orthogenetic over-development, and hence there are numerous examples of orthogenetic evolution and excessive development of teeth in the lines leading to large animal types such as elephants, walruses, hippopotamuses, large species of pigs, and the like. On the other hand, since the shape and structure of teeth are now strictly bound to and defined by their function, and teeth may not be altered *ad libitum*, too-excessive variants will be wiped out by special selection. This applies especially to the premolars and molars, which may not be altered too much without disturbing the proper articulation of the upper and lower jaws. In the case of the incisors and canines the situation is not so critical, and these teeth are often very much altered by allometric growth in consequence of increasing body size.

Figures 76 and 77 demonstrate some extreme examples of orthogenetic lines. The Proboscidea, for example, evolved from types which were related to *Moeritherium* and were about pig's or tapir's size. They had a complete set of teeth, of which the second incisors of the upper and lower jaw were already a little enlarged. With increasing body size the relative size of these teeth became larger and larger until finally four tusks had developed, as in the Oligocene *Palaeomastodon* and *Phiomia*, which were about as large as a small elephant. From these types evolved the larger Miocene and Pleistocene mastodonts, in which the lower incisors were lacking. From such types the elephants originated during the Pliocene. The largest elephants, the Eurasiatic *Elephas primigenius* and the North American *E. imperator*, had the relatively longest tusks. There is another branch of the Proboscidean ancestry in which the second incisors of the upper jaw were lost during the Oligocene, and this branch produced the Dinotheria of about mammoth's size, with huge tusks of the lower jaw pointing downward (compare Osborn, 1922, 1935). In the family of cats, the Middle Eocene *Hoplophoneus* of about lynx's size, which had relatively long upper canines, evolved to the Miocene *Machairodus* and the Pleistocene *Smilodon*, which was about the size of a lion and had enormously long upper canines (Figure 77). A smaller genus, *Eusmilus*, possibly belonging to an evolutionary offshoot, evolved excessively long canines during the Early Oligocene. A corresponding example is provided by prosimians (Lemuroidea), in which the small Eocene Adapidae of cat's size evolved into the giant type of the Pleistocene *Megaladapis* which was about chimpanzee's size but with a larger head, and which had extremely strong upper canines. The lower canines are lacking, which may be the result of material compensation. For further examples see the next section (Figures 83 and 84).

A similar predisposition to orthogenetic excessive development is met with in the bills of birds. In young birds, the bills are relatively short, but they grow with positive allometry (Figure 38). Hence, there are extremely large bills in the large types of various bird groups, as in the largest hornbills (*Ceratogymna, Dichoceros*, and the like), toucans (*Rhamphastus*), kingfishers (*Dacelo*), parrots (*Ara*), pelicans, snipes (*Numenius, Limosa*: Figure 38, *Recurvirostra*), herons (*Balaeniceps*), and others.

There are other bony structures of the skull, like antlers and horns, which become excessively large as a result of the correlative effects of increased body size. In very small types of Miocene deer (such as *Merycodus*), the antlers are but a little longer than the skull, but with increasing body size these antlers reach the enormous size of several times the length of the skull, which is to be seen in the Pleistocene giants, such as *Megaceros*. There is a line of descent in Titanotheriidae showing a gradually increasing body size and a correlated 'orthogenetic' and at last excessive growth of two nasal horns, which begins with the Middle Eocene *Manteoceras*, leads to *Protitanotherium* of the Late

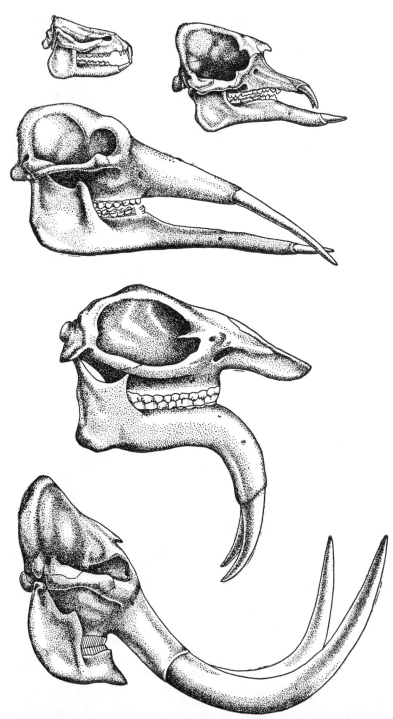

FIGURE 76. Excessive growth in proboscidians. Top to bottom: *Moeritherium*, *Phiomia*, *Trilophodon*, *Dinotherium*, woolly-haired mammoth. (After Romer, 1933.)

Eocene, and ends with the giant *Brontotherium* of the Lower Oligocene (Figure 82). Compare also *Triceratops, Pteranodon, Stegosaurus,* and *Dimetrodon,* which will be referred to in our discussion on 'overspecialization' in the next section.

In consequence of increasing body size, the proportions of the skull are also altered characteristically. In our consideration of allometric growth (in Chapter 6, B III) we have found that in large animals the head is relatively smaller and the facial part grows with positive allometry. Young mammals have a more rounded skull than old ones and, correspondingly, in larger species the facial skull is relatively longer than in related smaller species (Figure 35). This alteration and prolongation of the facial bones is especially obvious, as we have seen above (Figure 32), in the horse line. Automatically there is a relatively stronger development of the teeth. The front teeth generally become larger and the molars broader, and the premolars tend to develop into true grinding teeth ('molarization'). This intensified development of some teeth causes – apparently in consequence of material compensation – a strong reduction of the first premolars and a less strong reduction of the last

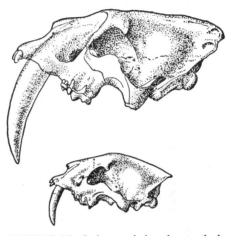

FIGURE 77. Orthogenesis in saber-toothed cats. Above: *Smilodon californicus* (Pleistocene of North America); below: *Hoplophoneus primaevus*(Middle Eocene of North America). (After Matthew from Abel.)

(third) molar. Hence, at least some of the basic orthogenetic trends of the equine skull so often debated upon are caused by the correlative effects of increased body size, and larger body size very probably was a favorable character in selection, as we have tried to show above. Besides this, a special selection by special environmental factors affects all phylogenetic stages of the lines of descent. One of these factors is the type of food provided by the habitat (compare Stirton's studies on the evolution of hypsodont teeth, 1947).

One can understand that some paleontologists who do not recognize the principles of allometry have again and again tried to use the orthogenetic trends of the equine skull as a proof of autonomous evolutionary trends, especially as such trends are evident when even short periods of the phylogeny are studied. Naturally, in most cases the paleontologist cannot properly evaluate the allometric growth of ontogeny. Consequently, the paleontologist can form only a tentative judgment of a certain orthogenetic development, and in most cases, if he does not know the correlated effects of phyletic increase of body size, he will be unable to discover that a certain organ has developed orthogenetically owing to the effects of selection.

We must not, of course, overlook the fact that in numerous cases selection affects the special growth gradients of single organs. This fact has been referred to several times in Chapter 6, B III. One last example, however, may be given as an illustration. In hoofed animals, living in the steppe will invariably cause a selection of the quick (i.e. long-legged) variants, while the slower types will be wiped out. Similarly, those specimens will be destroyed which do not recognize a possible enemy from a sufficient distance, which are unable to stand the dry seasons because they lack efficient water economy, and so forth. As selection by the environment of the steppe remains more or less constant, and as all stages of the phylogenetic transformation are subject to this selection, an orthogenetic evolution will result, which may be in contrast to some trends of allometric growth. All hoofed animals, for example, originated from small omnivorous or insectivorous Cretaceous ancestors, the legs of which very probably grew with positive allometry (as in Recent insectivores and carnivores) and the young of which therefore had relatively short legs. Because in the line of hoofed animals of the steppe it proved favorable to have relatively long legs and to be able to run more quickly immediately after birth, the prenatal growth gradient of the legs was affected by selection, and a negative allometry of the legs after birth evolved, as Recent Artiodactyla and Perissodactyla show. The evolution of the relatively longer intestine of these large forms probably took place despite the normal ontogenetic trend of allometry (compare p. 133).

Reduction of Organs. Just as positive allometry and an increase of body size may lead to the evolution of excessively large organs, so may negative allometry cause a reduction of structures and organs to vestigial structures. Striking examples of such vestigial organs are provided by the limbs of lizards. According to Schuster's findings (1950), the legs of lizards grow with negative allometry in relation to the trunk. Hence, the legs will become relatively shorter with increasing body size, and sometimes they will no longer have sufficient length to play an essential part in locomotion. Hence, locomotion will be brought about mainly by winding serpentine movements of the trunk, and finally the legs will become totally useless.

There are numerous species of lizards demonstrating this process of reduction until the various final stages are reached. One may establish anatomical lines of species ranging from types with normal legs (*Lygosoma, Scincus, Eumeces* of the family Scincidae, for example) to forms with more or less vestigial limbs (*Seps, Ophiodes, Ophisaurus, Anguis*). Sewertzoff (1931), studying such lines, stated that the number of presacral vertebrae was successively increased so that the body became more slender and serpentine. In this case curious intermediate types originated, such as, for example, the Mediterranean *Chalcides ocellatus* (Figure 78). In this lizard the legs are of normal shape, but so short that their locomotory function is very insufficient, and the whole construction seems rather odd. An animal of such construction is more or less bound to rely on a serpentine type of locomotion, which again causes or at least allows a further reduction of the legs. The next stage of such a process is seen in *Seps tridactylus*, for example, which has only very short legs, of no use at all in locomotion. In a final stage the forelegs are completely lacking, as in *Ophiodes striatus*, because the anterio-posterior differentiation gradient of ontogenetic development affected the forelegs first, so that the body material possibly 'meant' for the development of the forelegs was consumed by the trunk. As no special selection was at work to preserve the legs,

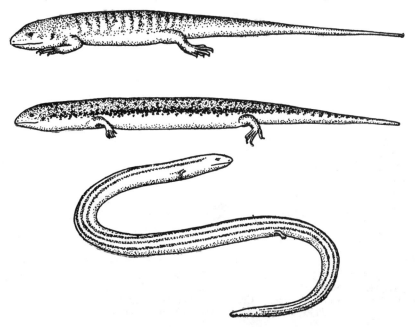

FIGURE 78. Successive reduction of limbs in lizards. Top to bottom: *Eumeces algeriensis, Chalcides ocellatus, Ophiomorus tridactylus.*

they became more and more reduced. In the Mediterranean *Ophisaurus apus*, although very tiny hind legs (about 2 mm. long) remain, they cannot serve any locomotory function, and the forelegs are totally missing. The final stage is represented by *Anguis fragilis*, the slow-worm, completely lacking external legs, and having only a vestigial cartilaginous pelvis with a remnant of a femur (Figure 79). In some cases, the evolution of vestigial organs took a different course, since the legs no longer capable of running became a means of burrowing, and a special selection remained in effect, which was demonstrated in several species of lizards by Steiner and Anders (1946). In snakes, too, it was the hind limbs that were reduced last, as may be inferred from the vestigial remnants in *Python* and *Ilysia*.

In most cases in both reptiles and whales, the distal parts of the limbs, i.e. the phalanges, were reduced first, while the pelvic girdle and the femur persisted longer. This principle was known to Fürbringer (1870). Similarly, it is

223

the distal parts and veins of insect wings that first become reduced (Figure 80), that is to say, these are the parts which differentiate last, and which are the first to be affected when the body material available is consumed partly or totally by those parts developed before. This is even more true in vertebrates, because in ontogeny the body as a whole grows rapidly and consumes the bulk of material available, so that the more distal parts go short. Landauer's findings (1931, 1934) on the growth of 'creepers', a race of domestic fowl, closely agree with this concept. He discovered that the short legs of this race

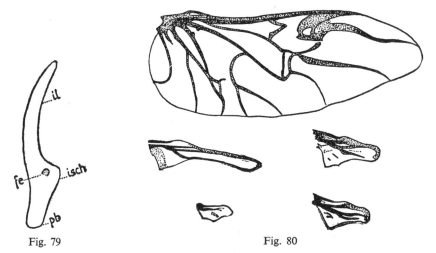

Fig. 79   Fig. 80

FIGURE 79. Pelvis of *Anguis fragilis* showing the remainder of the femur (fe). (After Sewertzoff, 1931.)

FIGURE 80. Normal hind wing and four reduced hind wings in five species of the ground beetle *Poecilus*. Above: *P. cupreus*. Center: left: *P. gebleri*, right: *P. marginalis*. Below, left: *P. expansus*, right: *P. lepidus*. (After Mařan, 1927.)

are due to the fact that there is a retardation of general growth in the embryos between the thirty-sixth and seventy-second hours of development. In other races with normal legs this period is characterized by a marked growth of the legs. Hence it is understandable that in 'creepers' the legs remain short, and that the most distal parts and the largest bone (the femur) are most affected (compare p. 147).

In some species, the vestigial organs or normal rudiments appear during the ontogeny only and become successively absorbed until they completely disappear. This process was called *aphanisia* by Sewertzoff (1931). In *Ophisaurus apus*, for example, at first there arise ontogenic buds of the fore limbs, but they develop no cartilaginous skeleton and soon become totally reduced (except for the scapula and the coracoid bone). Apparently the intense growth of the trunk consumes so much material that the anlagen of the limbs are also consumed by way of material compensation. A similar process can be studied in the hind limbs of the horse, where, during the earlier part of ontogeny, toe-like joints of the third and fourth digits develop, only to be subsequently reduced and disappear completely, while the metatarsalia are preserved as the well-known splint bones (Krölling, 1934). Hence, the only difference between 'aphanisia' and 'rudimentation' is that in the former the anlagen of the structure in question begin to develop, only to be totally consumed later on, while in the latter the anlagen continue their development, later suffering only a partial reduction, so that vestigial structures remain.

The interpretation of the origin of vestigial structures by the principles of negative allometry and compensation of body material also provides a possible answer to the question of why 'useless' vestigial organs persist in numerous animal groups through long periods of evolution (as the vestigial hind limbs of whales have done since the Lower Tertiary): the hereditary character is still preserved and hind limbs would develop if increased growth of other parts of the body did not cause a retardation in their development by way of material compensation. Selection directly affects the hind limbs only slightly if at all, because otherwise they would completely disappear.

The increased variability known from so many examples of vestigial organs (e.g. the hind limbs of *Ophisaurus*: Sewertzoff, 1931, and Figure 18; the hind wings of the ground beetle *Carabus nemoralis*: Krumbiegel, 1936) also indicates that the ontogenetic development of the organs concerned was interrupted or seriously reduced by compensation of body material. Hence, a balanced correlated system is disturbed by a secondary nonconstructive process. This is also evident in the various types of reduced eyes of cave-dwelling isopods and fishes, where the left eye is often found to be less strongly reduced than the right or vice versa (C. Kosswig, 1937, 1949; C. and L. Kosswig, 1940; De Lattin, 1939).

Finally, the reduction of many organs in one sex only seems to be a strong argument in favor of our hypothesis that competition for body materials, as well as the loss of genes, is important in the process of reduction. The two sexes have the same genes, except for the sex-linked ones. Hence the reduction of wings in many female butterflies, the reduction of the mammae in male mammals, and so forth may be mainly the product of such compensatory effects.

The reduction of eyes in cave-dwelling animals may, in my opinion, best be explained by the principle of material compensation, which is applicable to all growing parts of the body, as we have seen above (p. 180). Hereditary variants, with eyes reduced by an increased growth of other organs in a process of material compensation, would be wiped out in above-ground habitats. In caves, however, such variants remain alive, and the increased growth ratios of other organs, by which elongated legs and antennae, for example, are produced, must be regarded as favorable characters in this subterranean habitat. Besides this, progressive evolution in cave dwellers may be favored by other factors: the increasing homozygosity of characters due to the small populations (C. and L. Kosswig, 1936, 1937; De Lattin, 1939); genetic linkage of processes of reduction with photophobia (Ludwig, 1942; experimentally analyzed by Janzer, 1950), and finally, orthoselection, for example, by a steady selection of variants with long legs and antennae and with a tender body more sensitive to tactile stimuli than a hard-shelled one. It is not possible here to go into further details concerning this problem.

Hence, all orthogenetic lines may be considered to be the result of random mutation, selection, and loss of alleles by fluctuations of the populations. In addition, we should remember that the range of phylogenetic variability is reduced by the special structure of the body (compare p. 59) which is apparent, for example, from predetermined patterns like those studied by Vogt and his school, referred to as 'eunomias'. There is no reason to assume autonomous evolutionary forces with unknown physiological bases. As we have seen, it is necessary when judging processes of orthogenetic development to bear in mind the complex effects of allometric growth, compensatory consumption of body material, and many other mutual correlations. As long as we evaluate orthogenetic transformations of organs as a process *per se*, without consideration of their correlated effects, we are likely to pass a wrong judgment, because in numerous cases selection affects not the organ proper, but a structure or an organ more or less closely correlated with that under consideration.

Nor is there any reason to maintain the existence of a 'Law of Biological Inertia', which Abel (1928) formulated in order to reduce the phenomena of irreversible evolution, orthogenetic development, and progressive reduction of variability to a common denominator. Ehrenberg (1932) added the 'Biogenetic Rule' to this more general 'Law of Biological Inertia'. However, all these rules of evolution may be understood on the basis of the known principles of mutation, selection, and the mutual interrelations between single characters and the organism as a whole; the common factor rightly recognized by Abel and Ehrenberg is nothing more than the general constancy of hereditary elements.

C. OVERSPECIALIZATION, DEGENERATION, AND EXTINCTION

I. Excessive Growth and 'Overspecialization'

Positive allometry, and sometimes negative allometry, as well, causes the development of excessive structures, some of which are referred to as luxuriant or overspecialized structures, because their special function is unknown or does not seem to correspond to the amount of material needed

FIGURE 81. Excessive growth in various families of beetles. Small, non-excessive species of the same families are shown for comparison. 1. *Deliathus incana* (Cerambycidae); 2. *Macropus longimanus* (Cerambycidae); 3. *Carabus serratus* (Carabidae); 4. *Mormolyce phyllodes* (Carabidae); 5. *Dynastes tityus* (Dynastidae); 6. *Strategus titanus* (Dynastidae); 7. *Dynastes hercules* ♂ (Dynastidae); 8. *Chalcosoma atlas* (Dynastidae); 9. *Mecynodera coxalgica* (Chrysomelidae); 10. *Sagra buqueti* ♂ (Chrysomelidae); 11. *Rhynchophorus ferrugineus* (Curculionidae); 12. *Cyrtotrachelus dux* (Curculionidae); 13. *Eutrachelus temmincki* (Curculionidae); 14. *Euchirus longimanus* (Cetoniidae or Euchiridae); 15. *Pelidnota aeruginosa* (Cetoniidae); 16. *Chiasognathus granti* ♂ (Lucanidae); 17. *Cladognathus giraffa* ♂ (Lucanidae); 18. *Lissapterus howittanus* (Lucanidae). (After Heyne and Taschenberg, 1908.)

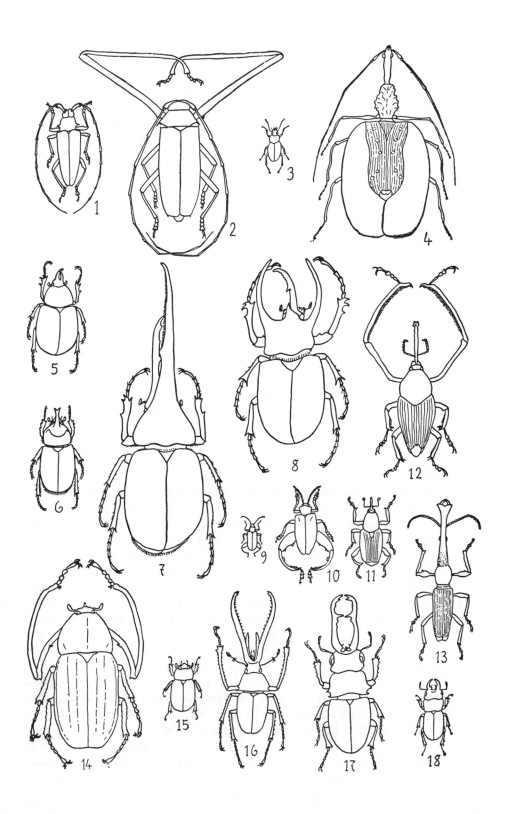

for their construction. As we have seen above, organs with positive allometry, such as tusks, canines, antlers, bills, and the tail feathers of some birds, tend to excessive growth when body size is increased in the course of phylogeny. Thus, the final stages of evolution represent the most advanced stages of excessive growth. Among the Recent species it is usually the largest types of their groups that provide striking examples of excessive growth. Figure 81 shows some of the most grotesque types of beetles of seven different families, all of which are giant types among their relations, as can be seen by comparing them with representatives of the same families that have medium body sizes. As will be seen in the figure, the mandibles, thoracic spines, and forelegs are especially affected by excessive growth, but in Carabidae and Curculionidae the heads, in *Mormolyce* the antennae, legs, and elytra, and in *Sagra* the hind legs are also disproportionately altered. Excessive growth is not strictly confined to large types, of course, and extreme structures may be developed in some smaller species (for example, *Macrocrates femoratus* among the Lucanidae and *Ludovix attenuatus* among the Curculionidae: compare the tables by Heyne and Taschenberg, 1908), although these cases are much rarer and such development almost never occurs in the smallest species of a family. But even some of these examples may be considered as large final stages of phylogenetic side branches. Moreover, in some cases a specialized selection may have caused the development of the excessive organs because they served an essential biological function. This has probably happened in both the cases just mentioned, as the mandibles of the lucanid *Macrocrates* are used in the sexual fights similar to those of our well-known stag beetle *Lucanus cervus*, and in the curculionid *Ludovix* the elongated head capsule is probably an efficient means of getting at the appropriate type of food.

A similar situation exists in other animal groups. The largest races of the drongo *Dicrurus paradiseus* (see p. 144) have excessive crests and tails, and it was the largest species of dinotheres, mastodonts, and elephants (see p. 219) that developed the relatively longest tusks. In the palearctic mammoth, this tendency to excessive growth produced an absurd structure, as in the adult the tusks pointed backwards in a huge curve. In the Pleistocene North American *Elephas columbi*, the spiral trend produced tusks the spire of which pointed inwards and backwards or downwards. The most excessive type of antlers in deer are found in the largest species, such as *Megaceros* of the Ice Age or the Recent *Cervus canadensis* (the wapiti). The longest nasal horns evolved in the largest types of titanotheres such as *Brontotherium* (Figure 82), and the longest canines of Notungulata appeared in the largest type of the ancestry, in *Astrapotherium giganteum*, of more than the size of a rhinoceros (Figure 83). In Recent walruses the canines are enormously long, while in the small ancestors of the Late Miocene (*Prorosmarus*) they were relatively smaller. *Babirussa*, a peculiar type of pig from Celebes and Buru, has canines which can hardly be used in digging and fighting, as their pointed ends are rendered less effective by being strongly bent (Figure 84). *Babirussa*, although not the largest pig, is probably the largest type of its line of descent. The Recent giraffe, which is the largest type among its relatives, has the relatively longest neck, while in the smaller okapi the neck is relatively short and resembles that of the Pliocene Giraffidae. Excessive structures are most likely to be met with in the largest types of *Sauria*. The huge *Stegosaurus* (body length about 9 m., Lower Cretaceous) had two rows of giant vertical bone plates along the vertebral column, and the giant *Triceratops* (body length about 8 m., Upper Cretaceous) had an enormous horn on the forehead and a huge neck plate. In *Pteranodon* of the Late Cretaceous the wings spanned about 6–7 m., which indicates that this type was one of the largest flying saurians, and in consequence of this large body size excessively long and pointed jaws developed, lacking any teeth, and an equally excessive supra-occipital crest appeared. Excessive dimensions of bills in some birds must also be rated as overspecialization (see above on p. 145).

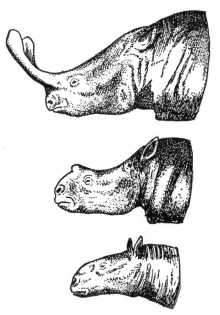

All such hypertelic structures must be regarded as allometric byproducts of increased body size, which were tolerable as long as they did not reach harmful dimensions or prove unfavorable. Selection often caused the increase of body size, as this was advantageous (compare the last section). According

FIGURE 82. Orthogenetic evolution of excessive horns in a phyletic series of Titanotheria. Above: *Brontotherium platyceras* (Lower Oligocene); center: *Protitanotherium emarginatum* (Late Eocene); below: *Manteoceras manteoceras* (Middle Eocene); (After Osborn.)

to Krieg (1937), the luxuriant (i.e. excessive) growth is 'a character that has exceeded the highest possible (not only the required and necessary) stage of adaptation, but has not reached a disadvantageous stage'. This opinion agrees with my own statements, though Krieg did not interpret excessive growth as a result of allometric growth, but thought it to be a 'consumption of surplus nutritive material'. He tried to support this opinion of his by stressing that it is the males, so little burdened by the physiology of sexual reproduction, that develop luxuriant organs or 'luxuriant motor patterns' (such as the howling orgies of monkeys, or the screeching of parrots and cuckoos during nonreproductive seasons). In my opinion, however, one may well agree with both views, provided one regards the allometric causation of excessive growth as the primary factor (this also applies to castrated females) and a compensatory reduction of the excessive parts in the process of sexual reproduction, affecting the female more than the male, as the secondary process.

Until now we have confined our attention to single excessive parts arising by allometric growth correlated to that of the whole body. We should not

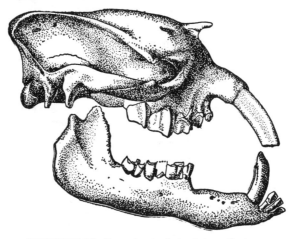

FIGURE 83. Excessive teeth in *Astrapotherium*.
(After Weber, 1927.)

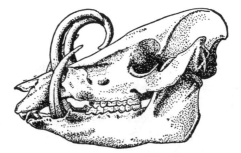

FIGURE 84. Excessive canines in *Babirussa alifurus*. (After Plate.)

forget, however, that there are numerous parts of the body with a less pronounced allometry, which is not strong enough to bring about excessive structures but which causes appreciable structural alterations with increasing body size. As some of these altered proportions become subject to new special processes of selection, the whole system of correlations is loosened by increasing body size and the structural type may then gradually be altered. This process of disorganizing the formerly well-balanced systems of correlations is commonly referred to as 'degeneration', but it may also be the starting point of progressive and constructive alterations. In Proboscidea, for example, the positive allometry of the facial parts of the skull and the second incisor has not only caused the compensatory disappearance of the remaining incisors, canines, and premolars, but has also favored the development of the trunk as a prehensile organ, as the lips, because the tusks became too long, became more and more incapable of taking up food. The negative allometry of the brain and of the cranial part caused a disproportionate shape of the skull, which would not have provided sufficient insertion areas for the neck and tusk muscles without the evolution of a peculiar lamellar inflation and enlargement of certain skull bones which is unique in the whole mammalian world. The necessity of evolving disproportionately bulky bones – the strength of bones grows by the square of their cross-section, as we have seen on p. 178 – and the heavy weight of the large body required the development of an especially soft and elastic tissue, formed by sole pads, which served as a cushion. Further alterations necessitated by increased body size include the anatomy of the nasal opening, the formation of motor areas in the brain to innervate the trunk by a special pyramidal tract, and the relative breadth of the iliac bones. Of course, we may only speculate about these phylogenetic processes, but surely it is a reasonable assumption that all these alterations were initiated by increasing body size; the altered structures represent characters which differ profoundly from those of the small Oligocene Moeritheriidae, the close relatives of elephants' ancestors. If no Recent and fossil types of large proboscidians existed, we would probably deny the possibility of structural types with such extremely long incisors and prehensile noses! The increase of body size also initiated the breakdown of the normal system of proportions in the hippopotamus and in some other Recent and fossil giant types, in which numerous cases of excessive 'degeneration' of the teeth are also encountered.

Finally, we must not forget that 'excessive' organs may arise as alterations independent of the body size and other correlations, simply as the result of undirected mutation, and that they may persist unless they are disadvantageous. Such structures seem to be excessive only when compared to those of related animal types. This applies to the apparently excessive shape and intricacy of some insect genitalia, as such organs of copulation may vary enormously, provided that their function remains that of lock and key. The extremely long penes of some snails – in *Limax redii* and *L. gerhardti* about 25–50 cm., i.e. three to eight times the length of the whole body – must be rated under this heading (Peyer and Kuhn, 1928; Gerhardt, 1944). The extreme contortions of body parts and feathers in the sexual display of some birds (bustard *Otis tarda*, ostrich *Struthio camelus*, birds of paradise, and so forth) may also be considered to be examples of some of many possible types of construction, as their effect after all is not fundamentally different from that of the less 'excessive' sexual display patterns of related species. That we so often find a phyletic intensification of sexual characters, sexual display, and of some other instinct patterns may be explained by the fact that 'supranormal' characters may evoke stronger reactions than do the natural releasers (experiments by Tinbergen and Perdeck, 1950, and others).

II. Phylogenetic Aging and 'Degeneration'

The evolution of excessive structures has often been regarded as indicating the aging of a line of descent. Haeckel (1866), who first recognized the 'Verblühzeit' (overdevelopment) of the lines as a special phylogenetic phase, thought that the degeneration of a group consisted of 'a limitation and decrease of its physiological capability' (p. 322). Later, senile degeneration and gerontism were referred to as typical symptoms of this phase. Hyatt (1894) thought that increased coiling in ammonites, a richer sculpture, and more complicated sutural lines of their shells were symptoms of phyletic old age.

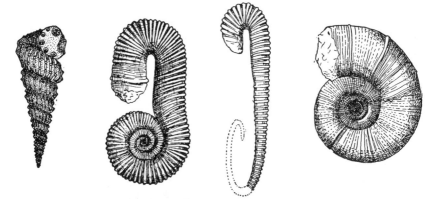

FIGURE 85. Excessive 'Nebenformen' in the ammonid family Lytoceratidae. Left to right: *Turrulites catenatus* (Gault), *Macroscaphites iranii* (Neocomian), *Hamites rotundatus* (Gault), and the normal *Lytoceras liebigi* (Tithon). (After Zittel.)

When considering the evolution of Crinoidea, Wachsmuth and Springer (1897) stated that 'extravagance of form and rank development in any group is the signal for its speedy extinction'. Beecher (1898, 1901) maintained that in Rhizopoda, Corallia, Brachiopoda, Trilobita, Ammonoidea, and Vertebrata the development of spiny types signifies the onset of old age in the lines of descent. Lull (1925) recognized that such a 'spinescence' in Mollusca, Brachiopoda, and Crustacea is brought about by a deterioration of the system of correlations, and explained that 'in general, such excrescences seem like growth force run riot, as though with the lessening vitality incident to racial old age, it is no longer adequately controlled'. In Recent times, phyletic morphological degeneration was thought to be essentially due to a process of 'dissociation from the typical pattern' (Beurlen, 1937, p. 86), and Hennig (1932, p. 33) expressed a similar opinion: 'A certain process of dissociation and a loss of the whole constitution so long preserved through all stages of transformation cannot be denied; they precede phyletic extinction.' In this context it is useful to remember the excessive types signifying the end of certain lines of descent which have been referred to above, such as *Brontotherium*, *Triceratops*, *Stegosaurus*, *Pteranodon*, and so forth. In his study on the evolution of Echinodermata, Gislén (1938) stressed the fact that types with excessively long and calcified spines were bound for speedy extinction.

232

Some examples often quoted in literature are not as typical as they are thought to be and should not be considered germane to the problem with which we are concerned here. This applies to some unusual forms, so-called 'Nebenformen', of Lytoceratidae, a family of Ammonoidea, in which the normal spiral shell seems to be 'degenerate', as the last whorl is preceded by a straight part, or the spiral is open so that the succeeding whorls do not touch the earlier ones. The spiral may be transformed into a hook or a straight secondary rod, or may even be tightly coiled in a snail-like way. Such excessive types, however, have appeared not only at the end but at intervals during the

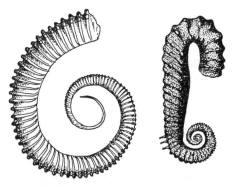

FIGURE 86. Excessive 'Nebenformen' in the ammonid family Kosmoceratidae. Left: *Spiroceras bifurcatum* (Dogger); right: *Crioceras* (*Ancyloceras*) *matheronianum* (Neocomian). (After Zittel.)

whole period of existence of this family, which lasted from the Early Jurassic to the Late Cretaceous. The rod-shaped *Baculina* evolved during the Middle Jurassic (Upper Dogger), *Pictetia* with an open spiral emerged during the Early and Middle Liassic, the hook-shaped *Hamites* and *Macroscaphites*, with its last whorl preceded by a straight part of the shell, appeared during the Early Cretaceous, and the screw-like *Turrulites* is found in the Gault and Cenoman deposits (Figure 85). Thus it is obvious that a striking lability of spiral coiling is part of the normal genetic background of this family. Moreover, the Lytoceratidae can hardly be regarded as a degenerate final stage of ammonite evolution, because this family persisted for about 115 million years, i.e. for more than one-third of the total ammonite existence (which lasted about 300 million years, from the Early Devonian to the Late Cretaceous). Furthermore, there are younger families of Ammonoidea which have a normal spiral (e.g. Desmoceratidae: Upper Cretaceous only; Prionotropidae· Gault-Senon). Similar examples may be found in the 'Nebenformen' of Kosmoceratidae, a family which existed from the Dogger period to the Late Cretaceous (Senon), as *Spiroceras* with an open spiral emerged during the Brown Jurassic, and *Crioceras* of the Lower Cretaceous had an equally open spiral with the latest coil preceded by a straight part of the shell (Figure 86).

233

In some cases, however, *pathological* degeneration has been proved immediately to precede phyletic extinction. Such proof exists in the rich fossil material of cave bears from the Mixnitz caves (Styria, Austria) studied by Abel, who found that in the latest types a great percentage of abnormal bones and degenerated dwarfs and an unusually wide range of variability developed. Abel (1929) tried to account for this fact by assuming that it was the optimum environment which lessened the effects of selection, so that many degenerate variants were preserved. He was inclined to interpret the extinction of other Pleistocene large mammals and of large Sauria on the grounds of the same assumption. Later, Fenton (1931) studied a large collection of material of *Spirifer*, a brachiopod genus of the Hackberry deposits (Upper Devonian), and tried to prove that in some older types of the species and races (according to his classification) the formation of wrinkles and transverse ridges was increased and the physiological capacity of repairing shell lesions was lessened in comparison with earlier ancestral types. Though the study is based on more than 10,000 specimens, its conclusions are not too convincing, as the phyletic lines put forth as evidence persisted together through several geological horizons, thus not too well proving the process of phyletic aging. In his study, Fenton quoted Child (1915), who from his studies on animal regeneration concluded that phyletic aging might be due to a progressive deterioration of plasma metabolism, a process which would be similar to ontogenetic aging.

Whether such physiological factors of phyletic aging may be regarded as genuine causes of extinction will have to be proved by future research (see below: Ottow, 1950). So far, however, we may reasonably assume that 'physiological' deterioration is only one of the numerous ways of phyletic aging, and a strictly morphological degeneration in its literal meaning, that is to say, a deviation from the typical shape of a taxonomic group, seems to be the more common type of phyletic senility. As there are numerous different ways of speciation and transspecific evolution, there may be many, or at least several, various ways of so-called 'phyletic aging' and of extinction.

However, it is essential to note that numerous lines become extinct without any warning symptoms of morphological or pathological degeneration. This statement applies to all those extinct groups which did not produce excessive or giant types, as, for example, Tarsioidea, Adapidae, and Archaeolemuroidea of the primate group; Phenacodontidae, Meniscotheriidae, and Mioclaenidae of the protoungulate types; and the Miacidae of the creodont relatives (compare also Remane, 1951).

III. Phyletic Extinction

The fact that phyletic degeneration and extinction of the structural type are not necessarily correlated is further emphasized when one studies various causes of phyletic death (compare also Osborn, 1906). The following causes may be distinguished:

1. Competition can lead to total replacement by superior types and hence to extinction.

2. Giant growth and consequent excessive development of special organs or structures or one-sided specialization may render further (selective) adaptation of the structural type impossible, so that the types are unable to survive in a changing environment (or to compete successfully with superior forms: see no. 1).

3. More radical changes of the environment due to climatic alterations, the appearance of new enemies, and so forth, may wipe out even such types as have not yet reached the stage of excessive growth. Single species may be destroyed by parasites or infectious diseases.

4. The types may die out in consequence of their own evolution, i.e. because of infraspecific competition.

Finally, the forms may persist more or less unchanged up to the present time.

Let us consider these possibilities in more detail.

1. In our discussion of the explosive phase of evolution (p. 111), we have concluded that animal groups may quickly be replaced by newly developed or newly immigrated superior types. The original carnivores (Creodonta) were rather quickly replaced by the Felidae and Canidae (especially Cynodictinae) of the Late Eocene. The Tarsioidea, nearly all types of which vanished during the Upper Eocene, were followed by the Catarrhini of the Oligocene. The primitive Palaeodictyoptera disappeared as soon as a stronger evolution of higher insect orders began in the Late Carboniferous, the Mesammonoidea of the Late Triassic were superseded by the Neoammonoidea, and so forth. In most of the cases enumerated here no symptoms of degeneration can be traced, but in numerous animals of these types one or more characters were found that rendered their bearers inferior to the newly evolved contemporaries. In other cases, 'mistaken' adaptations (Abel's 'fehlgeschlagene Anpassungen', such as the dentition of Titanotheriidae: p. 115) must be considered to have played an important role in phyletic extinction. Among vertebrates doomed to phyletic death, the relatively small size of the brain was a frequent cause of competitive inferiority and extinction in the old types (Amblypoda, Titanotheriidae, Creodonta, and the like).

The South American fauna, for example, was exposed to transspecific competition, as the Pliocene connection with the North American continent enabled the immigration of Artiodactyla, Perissodactyla, Carnivora, and others, which were superior to the South American species of Xenarthra and Taxodonta. Further examples of transspecific competition and extinction are provided by the well-known effects of the introduction of holarctic animals to Australia and New Zealand, in consequence of which many marsupials became extinct in large parts of their areas (for example, the marsupial wolf *Thylacinus* and the 'devil' *Sarcophilus* were displaced by the dingo: compare Palmer, 1898). None of these now extinct types revealed any symptoms of degeneration. However, it seems unnecessary to analyze these well-known cases in more detail.

2. More interesting from a theoretical point of view are those groups or single species which were not wiped out by superior competitors but became doomed to phyletic death in consequence of their own evolution. In this case we have to consider those lines of descent which might suggest the existence of autonomous trends and forces of evolution.

The study of many lines of descent comprising a sufficient number of different types will reveal that there is a general increase of specialization in the course of time. The much-discussed development of the horse family, for example, originated from small omnivorous protoungulates with several toes and led to specialized one-toed running animals with grinding teeth adapted to feeding on plants. The more or less omnivorous Creodonta evolved into the specialized Felidae, Canidae, and Mustelidae with carnassial teeth and improved sense organs. From an early offshoot of primitive carnivores originated the relatively small primeval whales (Archaeoceti), which had fin-shaped forelimbs and a tail well adapted to swimming while their dentition still resembled that of carnivores. These original whales were transformed into the giant right whales so perfectly specialized to feeding on small planktonic animals. Such transformations were necessitated, as the variants with better adaptations were superior in the infraspecific struggle for life. But the advantages of a present were to become the dangers of the future, as such specialized types were no longer able to adapt themselves to more essential changes of the environment. In particular, the organs once lost in the process of evolutionary specialization (such as toes and canine teeth) could not be developed a second time in their original shape and function (compare the discussion of irreversibility on p. 123). When the Yoldia Sea, for example, was transformed into the Ancylus Lake, all stenohaline animals inhabiting it were destined to die. During the Glacial Age nearly the whole forest fauna of central Asia was wiped out because the preglacial forests were destroyed. In the course of the postglacial devastation of East Africa, the animals of the thick forests retreated to the remaining forest islands, where, when even these forests vanished, the animals also died out. Only a few of them, such as the squirrels of the genus *Xerus*, which became adapted to a fossorial way of life (Lönnberg, 1929) managed to escape extinction. Thus,the process of specialization necessitated by selection often leads into a blind alley from which there is no way back (see also Osborn, 1906, and Decugis, 1941).

Occasionally, however, phylogeny can escape from such blind alleys when a certain reversal of the specialization is brought about by a prolongation of the juvenile phase and an abbreviation of the more specialized adult phases. We shall refer later to such cases in our discussion on the formation of new structural types (compare Chapter 6, E) but, on the whole, such reversals of evolutionary direction seems to be fairly rare. Sometimes extinction has been avoided by a functional change of organs and structures.

Adaptation to changing environmental situations, including new competing animal types, new enemies, new infectious diseases, and the like, is especially difficult in excessive types which – as we have seen above – often are giant forms. Giant hoofed animals, giant sloths, and giant anteaters of tapir's or elephant's size could not evolve into arboreal, fossorial, or aerial types, as was pointed out by Wallace (1891, p. 607). Hence, the development of new types of construction invariably started from small poorly specialized forms, like omnivorous species, which usually had not yet split into numerous lines. Haeckel (1866) recognized the evolutionary importance of such types and referred to them as 'generalists'. Cope stressed the importance of their status in his 'Law of the Unspecialized' as follows (p. 172):

> The validity of this law is due to the fact that the specialized types of all periods have been generally incapable of adaptation to the changed conditions which characterized the advent of new periods. . . . Such changes have been often specially severe in their effects on species of large size. . . . On the other hand, plants and animals of unspecialized habits have survived. . . . Animals of omnivorous food habits would survive where those with special food requirements would die. Species of small size would survive a scarcity of food, while large ones would perish. . . .

This highly important rule has been proved over and over again, and may also serve as a means of finding out about the probable ancestors of certain lineages.

The extinction of giant types was also furthered by the existence of excessive structures and organs which developed as by-products of allometric growth and which had reached the critical stage of biologically overspecialized or even harmful structures. Each minor change of environmental conditions and any advent of new transspecific competitors may extinguish such exaggerated types. It is for this reason, apparently, that giant species and giant structural types did not persist through longer geological periods.

Wallace (1891, p. 607) and, later on, Devaux (1928) thought that the retardation of sexual maturity and the correspondingly small number of young so often met with in giant animal types contributed to the causation of phyletic death. As the period of juvenile growth in the ontogeny of large animals is usually prolonged, the descendants were more exposed to possible dangers, and for a longer period, than their smaller ancestors. One should not forget, however, that the 'disadvantage' of a small number of young in large mammals is usually balanced by an improved care of the young, and poikilothermic animals with brood-care behavior, which tend to have an essentially decreased number of young, have not been wiped out so far. One more argument disproving Wallace's and Devaux's opinion is that large egg-laying poikilothermic animals lay a large number of eggs per clutch, as has already been emphasized above.

Finally we have to take into consideration that the absolute number of giant animals is often very small, especially in those cases in which each individual needs a rather large area for its own. Thus, random extinction may happen rather easily. (Compare Simpson, 1953, and Hutchinson and Ripley's statements about the Indian rhinoceros, 1954.)

3. Sometimes extinction may have been caused by a sudden destruction of a whole species. Though precise data are scarce, in several cases there are indications that this happened. The Pliocene immigration of Canidae, Felidae, and especially Machairodontinae (saber-toothed cats) to South America probably caused the extinction of numerous Xenarthra which were less heavily armored and had less well developed brains than their new enemies. The rapid spreading of the offspring of nine Indian mongooses (*Herpestes mungo*) introduced into Jamaica in 1872 caused the extinction of many species: the rodent *Capromys*, almost all rats, numerous snakes, lizards, turtles, terrestrial crabs, two species of doves nesting on the ground (*Columbigallina passerina* and *Geotrygon montana*), and the petrel *Aestrelata caribboea* (Palmer, 1898).

Occasionally, diseases and epidemics may have annihilated some species. By an extensive study of the bones of *Pezophaps solitarius*, the flightless giant dove which died out in historical times, Ottow (1950) could show that almost any individual of this species suffered from an apparently inherited bone disease similar to osteogenesis dysplastico-exostotica. Bone diseases seem also to have been fairly common among cave bears, which apparently suffered from bone tuberculosis and arthritis deformans (Tavani, 1951).

Such interpretations cannot be applied, of course, to the extinction of widely distributed animal groups or of whole faunas. The disappearance of the Late Cretaceous fauna and the appearance of the Early Tertiary types happened so quickly that a direct extinction especially of the marine animal groups (M) seems probable: Dinosauria, Plesiosauria (M), Mesosauria (M), Ichthyosauria (M), Pterosauria (M), Ammonoidea (M), and Belemmoidea (M) (except for a few surviving types), Rudistae (M), species of *Inoceramus* (M), Nerineidae (M), and other groups disappeared. Perhaps this general change of whole faunas was caused by the advent of new superior types, as in the Early Tertiary the mammals and birds and the teleost fishes and large sharks became dominant species of their habitats. It is true that mammals had been existing since the Jurassic, but at first their homoiothermy was not a particularly superior character, for the climate of the Jurassic and Paleozoic was warm throughout the terrestrial zones (Schwarzbach, 1950), and poikilothermic animals really were more or less 'homoiotherms' (compare B. Rensch, 1952). But the mammals had gradually developed better working brains, and in this way they became superior in competition.

Here we may also consider cosmic processes as causal factors. Giant types of Sauria may have received too little or too much ultraviolet radiation. Fluctuations of the amount of ultraviolet light penetrating to the terrestrial surface are likely to have occurred, as even today the ozone layer at about 40 to 50 km. altitude, which absorbs an essential part of the radiation, fluctuates in density and dimensions (Götz, 1931). The same applies to other high-energy rays affecting the mutation rate. Perhaps mutation pressure sometimes became more effective than selection pressure. However, no precise data concerning this problem have been brought forward as yet.

Audova (1929) thought that due to fluctuating temperatures the eggs of giant saurians no longer hatched, but this hypothesis can hardly be verified, as we do not even know whether giant saurians laid eggs, and, moreover, there are large egg-laying types, like the crocodiles, that survived these periods Finally, it is highly probable that the climate of the Early Tertiary was fairly warm. But it seems quite possible that thermophilic types were extinguished in consequence of the appreciably cooler climate of the Pliocene and Pleistocene. Ekman (1935), for example, proved that numerous typical marine groups of the eastern Atlantic disappeared during this period.

4. We need not enter into the details of phyletic extinction by infraspecific competition, as the superiority of some hereditary variants in the course of normal evolution has been referred to several times, and it is quite evident that by this process the species will gradually change and one species may be replaced by another in the course of time.

Finally, we may ask which of the types of phyletic death is most frequently met with in the study of evolution. It is quite obvious from a great number of lines with sufficiently rich fossil records, that extinction by superior transspecific competition was by far the most frequent type of phyletic death. The next most frequent type (connected with the first one) was dying out in consequence of highly increased specialization. The dangers of excessive specialization could not be evaded; these specializations were useful when they were being formed and established, but they became harmful later on, because of environmental changes. Conspicuous as the extinction of giant and excessive types may seem to have been, it was not so much caused by 'immanent' factors as by an inferiority in competing with less excessive groups. One may reasonably assume that all such excessive types would still be alive had they been preserved in suitable habitats by man.

At any rate, it should be acknowledged that there are various types of phyletic extinction which may be met with in combination or singly. Although phyletic death may result from giant and excessive growth necessitated by selection (Cope's Rule), there is no reason to assume unknown autonomous forces of evolution. Finally, one should not forget that the extinction of animal types cannot be regarded as a regular phase of an evolutionary cycle, because all the antecedents of the Recent animal types have not experienced this phase, and those groups that were not extinguished until the later epochs of terrestrial history were again and again subject to transformation and remodelling, so that no evolutionary cycles with phyletic death as a regular phase can be traced.

D. THE EFFECTS OF PHYLOGENETIC ALTERATIONS IN VARIOUS STAGES OF THE INDIVIDUAL CYCLE

I. Causation and Classification of Phylogenetic Alterations of Ontogeny

The fundamental source of evolution is mutation, which may affect genes, chromosomes, the whole set of chromosomes, or the cytoplasm of the germ cells. Thus there is a fundamental alteration of the individual cycle at least as to genic or chromosomal material. The morphological or physiological effects of this alteration may appear in various stages of the ontogenetic cycle, from the first cleavage of the egg to the final stages of development. Usually, however, such effects will become manifest in the more advanced stages of ontogenetic development. The effects of mutation on the ontogenetic cycle may be quite various, ranging from control of a single reaction (e.g. the definitive pigment formation in insect eyes or the synthesis of arginine in *Neurospora*) to an effect on some parts or the whole of the ontogeny. In addition to qualitative changes, quantitative alterations often occur by accelerations or retardations of certain phases in the development of organs or structures (heterochronisms). In his studies on the evolution of Ammonoidea, H. Schmidt (1925) stated that acceleration (tachymorphism) and retardation (bradymorphism) of the development are parts of the normal variability in these animals. In *Eumorphoceras reticulatum* of the Upper Carboniferous, for example, some specimens reached the so-called 'dorsalis' stage quite early and the adults then had a narrow umbilicus, while in other individuals from the same desposit (near Gevelsberg), which were slower in their ontogenetic development, the umbilicus remained relatively wider. Sometimes accelerations and retardations of single gradients may cause essential alterations of the normal process of differentiation, and important alterations in construction may arise. Some of these have been referred to in Chapter 6, B III. In addition, ontogenetic development of the whole body or of single gradients may be abbreviated when some phases are omitted, or a prolongation may arise from new phases added to the final stages.

Due to such alterations of single developmental gradients, either the whole system of reactions during ontogeny or the formation of single structures and organs and their proportions may be changed (Chapter 6, B III). Consequently, some hereditary characters, though left untouched, are prevented either from producing their normal effects or from causing any reactions at all. Thus, numerous characters may remain potential or become manifest only on rare occasions when certain favorable gene combinations or special modifications arise (as in occasional 'atavisms'). From this point of view the wide range of normal modificability of most animal types (the extreme of which is heteromorphism) becomes comprehensible.

This wide range of modificability also has some connection with the fact that certain structures are definitely shaped only by the onset of their function. The amphibian lungs, for example, will be definitely shaped by their respiratory function, i.e. the typical lung tissue will not develop in animals prevented from aerial breathing (Maschkowzeff, 1936). In higher vertebrate types, however, the typical respiratory tissues develop independently of any function. One more example may illustrate the relations between inheritable characters and environment: the visual centers in the brain of dogs will not develop properly in dogs which have been prevented from seeing by having their eyelids sewn shut immediately after birth (Berger, 1901).

We thus see that each alteration of ontogenetic phases may have extremely diverse consequences. It is not easy to state which ontogenetic alterations have occured most frequently in the course of evolution, because the fossil records suitable for such studies are few. The ontogenetic development of some characters has become known from but a few lines of descent. Material favorable for study is provided by the shells of molusks, as in these the succession of ontogenetic alterations can be traced even in the shells of adult specimens. This applies to the increasing complexity of sculptures of snail shells and to the suture lines of Ammonoidea. We also know the proportions of the bony skeleton of some fossil juvenile vertebrates, although we almost totally lack knowledge of their early ontogeny.

Thus, it is necessary to rely on the study of related species or related genera and families of Recent animal types, provided that their common ancestry is probable. In such cases we can compare the whole ontogenetic development of two related species and find out to what extent it is identical and in which stage alterations of development begin. The importance of such phenogenetic studies was stressed by Haecker (1918, 1925). In vertebrates a great deal of such work has been done by Sewertzoff (1931) and his school, and numerous other authors have since taken up this kind of research. The results obtained so far indicate that phylogenetic alterations may appear first in very different stages of the ontogenetic development. In some cases, the two ontogenies will be different from the first stages of cleavage or from the formation of the first buds of organs. Such cases were referred to as 'archallaxis' by Sewertzoff. In other cases, the differences will appear at a later phase of development: deviations of various kinds will arise. Sometimes the alteration will affect only the early or intermediate phases of the ontogeny and will be lessened successively in later periods. Additional phases may be added to the final stages of the ontogeny (anabolies in the sense of Sewertzoff). Furthermore, intermediate phases may be omitted (abbreviations), and accelerations and retardations may complicate the differences between the lines compared. Sometimes one finds that characters typical of the final stages of the ontogeny are gradually shifted toward more juvenile phases (proterogenesis).

We shall consider some examples typical of each of the categories mentioned and try to determine the regularity of these phylogenetic phenomena in the course of ontogeny. We shall also determine which of these alterations is most frequently met with and which of them is most likely to initiate the evolution of new organs, new structures, and new types of organization.

II. Archallaxis

Evolutionary alterations occurring in the early phases of the ontogeny were referred to as 'neogenesis' by Garstang (1922). Sewertzoff (1931) coined the term 'archallaxis', meaning that 'the very first embryonic stages of an organ become altered in a certain direction'. In such cases a recapitulation in the sense of the biogenetic rule (see below) takes place only in stages before the appearance of the organ in question. In my opinion, one should distinguish between such an organ archallaxis and a total archallaxis, which shows differ-ences in the stages of cleavage, in gastrulation, and so forth.

The genetic basis of archallactic alterations can be found only in cases of infraspecific differences. In *Drosophila melanogaster* the mutant minute 1, for example, can be recognized from form and color differences of the egg (Li, 1927). In most trans-specific alterations an adequate analysis of the genetic background is im-possible, because hybrids of species and genera usually are inviable or in-fertile. Thus, we must rely on comparative studies of ontogenetic development. Transplantation and ex-plantation experiments, however, provide sound clues towards the proper evaluation of the poten-cies, affinities, and in-duction effects of single cells and organ buds, from which one may draw some conclusions regarding the genetic causes of archal-lactic alterations.

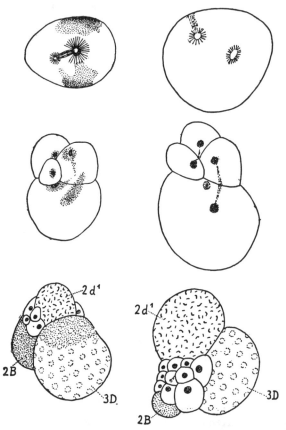

Total archallaxis can morphologically be studied most easily in species with eggs of the mosaic type, as in these one can often distinguish definite parts of eggs and cleavage stages and trace their subsequent development. This applies,

FIGURE 87. Differences in the type of cleavage in two species of Oligochaeta. One-, four-, and seventeen-cell stage of *Tubifex rivulorum* (left) and *Pachydrilus lineatus* (right). Polarplasm dotted. Homologous blastomeres shown alike. (After Penners, 1922 and 1930.)

for example, to *Pachydrilus lineatus*, an oligochaete worm of the Enchy-traeidae, and the species *Tubifex rivulorum* of the related family of Tubificidae. According to Penners' studies (1930), ontogenetic develop-ment of both species can be distinguished morphologically from the egg stage onward. The egg of *Pachydrilus* lacks the polar plasma so typical of the *Tubifex* egg (Figure 87), and contains more yolk material. This causes an unusual enlargement of both somatoblasts in *Pachydrilus*, and the entodermic cells remain relatively small, while in *Tubifex*, with small somatoblasts, the main bulk of the early embryo is formed by entodermic cells. Therefore, in *Tubifex* the germ band will form a distinct spiral, while in *Pachydrilus* only a slight ventral curvature develops. In the latter, the two original cords of the central nervous system will fuse later in ontogeny than those of *Tubifex*. In *Pachydrilus* the complete development of the intestine precedes that of the ventral nerve cord; in *Tubifex* the development is in the opposite order. In the latter, the closing of the blastopore begins at the posterior end, and in *Pachydrilus* it starts from the anterior part. Thus, almost any stage of development is different in the two related species.

In Echinodermata, the development of many specific and generic traits has been found to occur at early phases of ontogeny. This was established by

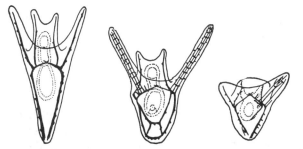

FIGURE 88. Early ontogenetic origin of specific differences in Echinodermata. Left to right: pluteus of *Parechinus miliaris*, *Echinocyamus pusillus*, and chimera of *Par. miliaris* with additional *Echinocyamus* micromeres. (After Von Ubisch, 1933.)

hybridization, experiments on isolated blastomeres, and the study of chimeras (see, for instance, Von Ubisch's studies, 1933). Normally, the four micromeres of the sixteen-cell stage of the sea urchin will migrate into the blastocele where, after further cell divisions, they form two separate groups of cells from which the skeletal spicules will arise. These four micromeres have been removed and replaced by four micromeres taken from a different species of sea urchins. Similarly, four foreign micromeres may be grafted into the isolated animal half of an eight-cell blastula. Micromeres taken from *Parechinus miliaris* and grafted into a half blastula of *Echinocyamus pusillus* developed into typical *Parechinus* spicules, and micromeres of *Echinocyamus* implanted in *Parechinus* blastulae produced spicules of the *Echinocyamus* type. Micromeres taken from *Echinocardium cordatum* and implanted in complete blastulae of *Parechinus microtuberculatus*, that is to say, into a species with a very different larval skeleton, developed into a mixture of both skeletal types. This proves that the characters typical of the species were determined at early ontogenetic stages (Figure 88). In a hybrid of the two species the shape of the skeleton was similarly intermediate between those in the parent species. If the parent spicules were of the same type, as in *Parechinus miliaris* and *Paracentrotus lividus*, the chimeras had normal skeletons, but the additional micromeres produced additional skeletal spicules in about 25 percent of the material. The early ontogenetic differentiation of generic traits could be proved in this manner, whereas it would have been impossible to establish such proof by making a mere morphological study of normal ontogeny.

Species hybrids of *Triturus alpestris* and *Tr. palmatus* are not viable, but chimeras can be produced by uniting two specific halves of the gastrula stage (Rutz, 1948). In such specimens, the anterior half of which was specifically different from the posterior one, the further development of the material was specific, but when presumptive dorsal and ventral material of two species was combined, the determination of body size, color patterns, and the onset of metamorphosis were caused by the dorsal material.

A total archallaxis is also exemplified by related species in which the tempo of ontogenetic development is different from the early stages. Riddle, Charles, and Cauthen (1932) found in four races of pigeons differing in body size that although there are only slight differences in the first postnatal stages of ontogeny (Figure 51) the rate of development differs greatly among the races from the twentieth to the forty-fifth day after birth. Four species of the snail *Crepidula* which differ considerably in adult body size differ only slightly in egg size, and this difference is not proportionate to body size. In geographic races of the moth *Lymantria dispar* differing in body size, the size of the eggs is about equal, but the differences of size and color patterns develop during the various larval stages (R. Goldschmidt, 1933). The embryonic stages of the two flatfish, *Pleuronectes platessa* and *Pl. flesus*, are extremely similar except in egg and chromosome size, but their ontogenetic growth rates are essentially different (Von Ubisch, 1950). At birth, the three cyprinodonts *Xiphophorus helleri*, *Lebistes reticulatus*, and *Heterandria formosa* are of about equal body size, but the subsequent rates of ontogenetic development are distinctly different. At first the growth gradients of males and females are more or less identical, but after sexual maturity the males grow at an essentially slower rate than the females (Wellensiek, 1953). Thus we see that archallactic alterations by different growth ratios are rather common.

Similarly, organ archallaxis may begin with differences or be caused by different growth rates and differentiation during ontogeny. This is especially well exemplified by the formation of certain specific numbers of body segments in lizards and snakes, the numbers being fixed by the faster or slower rate of differentiation of the primordial mesodermal segments (Sewertzoff, 1931). In heterometabolic insects, the differentiation of certain specific structures begins rather early and increases until the final stage is reached. This implies that the specific differences of the structures considered must be fixed at embryonic or early larval stages. Spett (1931) could show this in two species of Orthoptera, *Chortippus albomarginatus* and *Ch. parallelus*, which show strong specific differences in the first nymphal stage. These are mainly differences in relative size and in the growth rates. A similar situation will probably prevail in most heterometabolic and holometabolic insects, as the differentiation of many, if not most, specific and generic characters seems to commence prenatally. R. Goldschmidt (1924–33) has also stated that infra-specific differences of a moth (races of *Lymantria dispar*) begin at very early stages of ontogeny (growth rates, cell size, but not egg size).

For a proper evaluation of organ archallaxis, transplantations and explantations again proved very useful. Heteroplastic transplantation between *Triturus taeniatus*, *Tr. alpestris*, and *Tr. cristatus* proved that many organs are specifically determined in early stages of ontogeny, so that they continue their specific development after they have been transplanted to other host

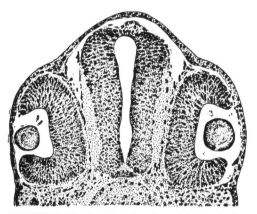

FIGURE 89. Horizontal section of the head of a larval *Triturus cristatus* in which the lens and cornea anlagen of the smaller species *T. taeniatus* were implanted in the left side ten days after the operation. The lens maintains its specific develop-ment, i.e. it is too small for the larger eye cup of *T. cristatus*. (After Rotmann, 1939.)

organisms. Epidermal parts transplanted during the early phase of the gastrula stage did not develop altogether specifically at first, but later became specific, and remained specific even after metamorphosis. The glands of the epidermis also seem to remain specific (Rotmann and MacDougald, 1936). Heteroplastic transplantations of adhesive organs performed on early gastrula stages produced specific or chimeric structures (Rotmann, 1935). Limb buds transplanted heteroplastically in the tail-bud stage also developed specifically, and when a heteroplastic exchange of presumptive epidermal or mesodermal material of limb buds was made, chimeric limbs resulted (Rotmann, 1933). Such development also occurs in the eye lenses of *Triturus cristatus* and *Tr. taeniatus*. The size of these lenses seems to be specifically determined quite early, as after the transplantation of skin material of the small species *Tr. taeniatus* in front of the optic cup of an embryo of the larger species *Tr. cristatus*, the lens developed a size typical of *taeniatus* lenses (Figure 89, after Rotmann, 1939). Hybrids of *Triturus taeniatus* and *Tr. cristatus* showed typical characters of both species at early stages of ontogeny (Popoff, 1935).

On the other hand, endodermal material transplanted at the late gastrula or even the neurula stage proved to be not yet determined specifically, and became harmoniously incorporated into the intestine and allied organs after heteroplastic transplantation (Balinsky, 1938).

In the axolotls *Ambystoma punctatum* and *A. tigrinum*, numerous organs are determined specifically in early ontogenetic stages, which was proved by transplantations of prospective limbs, hearts, auditory vesicles, lenses, and the like. The transplanted organs developed more or less specifically, but there was also a certain modification of the specific features by the host organism. The small auditory vesicles of *A. punctatum*, for example, tended to grow larger when transplanted to *A. tigrinum* (Richardson, 1931, 1932), and the limb buds of the larger *A. tigrinum*, when transplanted to an embryo of the smaller *A. punctatum*, caused a hypertrophy of the shoulder girdle in the latter (Schwind, 1932). Replacing the sixth, seventh, and eighth somites of *A. punctatum* by those of the larger *A. tigrinum* induced the development of more muscles in the host organism (Detwiler, 1938). On the other hand, the substituted brachial region of the spinal cord did not develop specifically, but was transformed into spinal cord typical of the host organism (Detwiler, 1931).

Early determination of organ buds has also been proved by heteroplastic transplantations in species of frogs and toads. In these genera various modes of determination of the lens have been found. In *Rana temporaria*, for example, the formation of the eye lens is induced by the optic cup, so that by auto- or homoioplastic transplantation eye lenses can be evolved in various parts of the body. In *Rana arvalis*, however, the lens will be formed independently, not requiring the presence of an optic cup, and the skin is no longer predetermined to develop eye lenses once the tail-bud stage has been passed. Hybrids of *R. arvalis* and *R. temporaria* react like *R. arvalis*, as far as transplantation of optic cups and lens formation are concerned (Popoff, 1935). In *Bufo viridis*, the inducing power of the optic cup is less effective than in *B. vulgaris*, as a *vulgaris* cup will induce the formation of a lens in the epithelium of *viridis*, but a *viridis* cup will not do so when transplanted to a *vulgaris* embryo (Lasareff, 1938). Further results were reported by Baltzer (1950 a,b), who studied the effects of xenoplastic transplantations in species of *Triturus* and *Rana*, *Triturus* and *Salamandra*, and *Triturus* and *Bombina*.

The more detailed studies on the physiology of development in amphibians and echinoderms proved that organ archallaxis is quite a common type of evolution, and one may well infer that this phenomenon is of considerable general importance in the whole animal world.

III. Permanent Deviations and Heterochronies

Sewertzoff (1931), following Franz (1924), coined the term 'deviation' for an 'aberration' of development beginning at intermediate stages of the ontogeny, a type of development already known by F. Müller and Naef. But Sewertzoff did not give a more precise definition of such 'deviations', and it is not quite clear what was meant by 'early, intermediate, and later' stages. From the examples quoted one may infer that an attempt was made to divide not the whole ontogeny but only the special morphogenesis of single organs into three parts. In the process of deviation, differences arise at intermediate stages of ontogeny and proceed in a direction which cannot be regarded as 'additions' to the preceding stages of ontogeny (see below), but which effects a distinct aberration from the characters typical of the original group. A similar aberration at a later stage of ontogeny was referred to as 'anabolia' by Sewertzoff and, hence, there is no clear-cut difference between the two types. Therefore it seems useful to restrict the term 'anabolia' to the prolongation of morphogenesis (by additional stages, not by a temporal prolongation). Correspondingly, all phylogenetic alterations of intermediate and late phases of the ontogeny will be referred to as deviations. Naturally, no line can be drawn between the two categories, as in many cases additions and deviations of single structures will be mixed.

Deviations may be permanent or temporary. Permanent deviations occur when morphogenesis is altered in a certain phase and when the differences become more and more distinct with increasing ontogenetic age, while in cases of temporary deviations the differences will be confined to early or intermediate stages and the later development will converge and lead to more or less similar adult stages. Permanent deviations seem to be very common in the evolution of animals and plants, and I shall first discuss some examples demonstrating the effect of this type of phylogenetic alteration.

In vertebrates the phenomenon is exemplified by the development of nasal grooves in fishes (Sewertzoff). At first epithelial folds will arise on the bottom of these paired grooves. Later on, the development may lead in different directions, forming a radiate pattern (as in Salmonidae) or a simple pedunculate ridge (as in *Belone, Scombresox*, and the like). Furthermore, the grooves of the former type are covered by a bridge-like structure, while those of the latter remain open.

A similar example is provided by the differentiation of the mucous glands and the dart sac of central European banded snails. *Cepaea nemoralis* and *C. hortensis* differ in the coloration of the lip and in the size of the dart sac, which is definitely smaller in *hortensis*, whereas the mucous glands of the latter are relatively longer and spindle- or club-shaped rather than cylindrical (Figure 90). In *nemoralis*, the number of glandular tubes varies from three to five, while in *hortensis* only two to four will be found on each side. In *hortensis* the darts are considerably smaller and more strongly curved than in *nemoralis* and each of the four longitudinal ridges of the darts is separated into two parts by a central groove. These differences appear in intermediate stages of development, but cannot be distinguished in the early stages of mucous glands or dart sacs (Figure 90). Similar differences arise in the structure of jaw and radula. As the first whorls of the juvenile *Cepaea* shell are very similar and sometimes identical, one can hardly distinguish the species at early onto-genetic stages (which does not mean, of course, that a very detailed study would not reveal certain differences). A typical example of deviation is also provided by the relatively late differentiation of visual rods and cones from more cone-like prestages in amphibians (Birukow, 1949). Many further examples can be quoted (for instance, very late differentiation of sense cells and hair cells in Trichoptera: Rönsch, 1954).

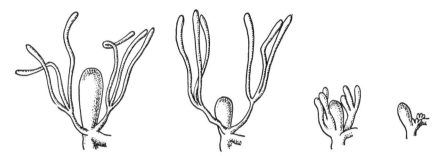

FIGURE 90. Dart sac and mucous glands of the snails *Cepaea nemoralis* (left) and *C. hortensis* (second from left). The drawings on the right show the same organs from the same two species at two much earlier ontogenetic stages where no specific differences can be seen.

All these considerations refer to purely morphological differences only, and one should not forget that at early ontogenetic stages most organs have a more or less spherical or oval structure which cannot be much altered in their external morphology. Apparently it is for this reason that early ontogenetic stages of related groups are so similar. Closer examination will, however, reveal certain differences, especially between groups which are not too closely related. For example, Krabbe's thorough studies (1942) on the morphogenesis of the brain in three related mammalian orders (Insectivora, Chiroptera, Dermoptera) revealed striking differences in the proportions of the brain parts in the three orders (Figure 91), which would hardly have been expected in such early stages of ontogeny.

In those cases in which no real morphological differences can be found before the deviation begins, one may always try to detect chemical differences of the tissues, or different affinities and capacities of induction; this may be done by histological staining or by transplantation or explanation (see the experiments by Baltzer, Rutz, Von Ubisch, and others, quoted above). This more detailed physiological analysis of development is likely to reveal numerous intermediate types of deviation and archallaxis. Hence, the above classification of phylogenetic alterations is not intended to be more than a general scheme facilitating the discussion of the various cases.

Finally, there are numerous deviations represented by accelerated or retarded development of single organs or structures, which are referred to as *heterochronies*. Similar effects may arise from an alteration of body size (see the detailed accounts of this process in Chapters 6, B III and IV), as the correlations of organs and structures with the body as a whole will be shifted in a similar manner in both cases. The examples of deviation quoted in this chapter also involve more or less essential effects of heterochrony. Thus it may be sufficient to state that heterochrony is likely to be met with often in evolutionary studies. In the study of the causal factors underlying the process of deviation, experiments on modifications also prove useful if the physiological mechanisms producing modifications correspond with those involved in the causation of hereditary differences: see the studies on the origin of different densities of cell layers in the wing of *Drosophila* by Henke (1950, 1951).

Insofar as phylogenetic alterations arise as pure deviations and not as archallaxes, the first parts of morphogenesis in related species, genera, and larger systematic units will remain identical or similar. This is the basis of Von Baer's 'Law of Individual Development' (1828), stating that ontogenetically the special characters develop later than the general traits, which means that the characters typical of the class will appear earlier than those of the order, those of the order will precede those of the family, and so forth. Darwin (1859) explained this 'Law of Embryonic Similarity' by the fact that in most cases phylogenetic alterations affect later ontogenetic stages. He did not omit some exceptions to this rule, such as the formation of adaptive organs in larvae or successive alterations occurring at very early stages of ontogeny, nor did he neglect the fact that sometimes the alterations tend to be successive shifts towards more juvenile phases of ontogeny. On the whole, however, his statements on this problem are hardly more than mere suggestions, and one cannot help thinking that he did not see the whole problem

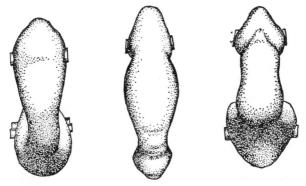

FIGURE 91. Three comparable stages of early brain development, shown from above. Left to right: *Talpa europaea* from an embryo of 2 mm. body length, *Vespertilio* species of 2·2 mm., and *Galeopithecus* of 4 mm. length. (After Krabbe, 1942.)

of the evolutionary differentiation of the major structural types; this involves the question, in which manner late deviations could produce the differences between different orders and classes observed in early stages.

Von Baer's Rule is generally applicable to the higher systematic categories. In the lark family (Alaudidae), for example, the characters typical of larks, i.e, special type of scales on the legs and special palatine bones, will not develop until the characters typical of the passerine order have appeared, which occurs after the formation of the traits typical of the bird class. To lower categories and to infraspecific differentiation, at least, the inversion of Von Baer's Rule is applicable, stating that genes common to numerous animal types may be considered to be relatively old. The dominant factor 'bobbed', for example, was found to exist in the genetic composition of six species of *Drosophila*, and hence it should be credited with a relatively old phylogenetic age, which is also suggested by the fact that this factor exerts a certain influence on meiosis (Lüers, 1937).

IV. Deviations confined to Early and Intermediate Stages of Ontogeny (Coenogeneses)

In numerous cases, at early or intermediate stages of ontogeny, more or less pronounced alterations arise, which are partly or completely balanced in the

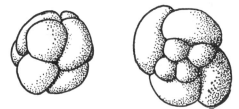

FIGURE 92. Two different types of eight-cell stage in the turbellarian *Prorhynchus stagnatilis*. Left: almost equal and radial cleavage; right: inequal and spiral type. (After Steinböck and Ausserhofer.)

course of subsequent development, so that hardly any difference is to be found in the adults. Haeckel referred to such cases as 'coenogeneses'. In my opinion, coenogeneses may be divided into early ontogenetic deviations and intermediate deviations. Early ontogenetic deviations may result from various causes. It may be that different types of cleavage or gastrulation are possible within a group, or these early stages were adapted to various environmental conditions in the course of phylogeny. Intermediate deviations are normally brought about by such hereditary adaptations to the environment as, for example, larval organisms, which need special structures that may be useless in later stages. Such cases were referred to as 'convergent poecilogony' by Verrier (1950).

A most conspicuous example of such early ontogenetic deviations was reported recently by Steinböck and Ausserhofer (1950). In the Turbellarian *Prorhynchus stagnatilis* two different types of cleavage may be found (possibly depending on environmental temperatures). Radial cleavage of the normal type will yield an eight-cell stage with blastomeres of about equal size, while the other possible type, spiral cleavage, will produce four macro- and four micromeres (Figure 92). The embryo may grow inside the yolk material, developing into a typical pleroblast, or it may grow on the surface of the yolk, forming a hollow koiloblast. In consequence of these two types of cleavage temporary differences will arise as regards the sequence of the formation of organs. The adult animals, however, will be obsolutely identical.

Deviations on early ontogenetic levels are, of course, more frequent in the differentiation of species, genera, families, and orders. In the calcareous sponge *Sycon*, for example, an amphiblastula develops and the slender flagellate micromeres will invaginate into the spherical body made up of macromeres, while in *Clathrina*, also a calcareous sponge, groups of flagellate cells will migrate into the macromere layers and will be united here to flagellate chambers. In *Aurelia aurita* gastrulation may occur by invagination only or by multipolar immigration. In Semaeostomae, gastrulation is generally brought about by invagination, but in Stauromedusae the entoderm arises by a multipolar immigration and no archenteron is formed. In Cubomedusae, gastrulation is the result of multipolar immigration and subsequent delamination, and in the gastrulation of the Coronatae invagination and multipolar immigration are combined (compare Thiel, 1936–8; B. Rensch, 1953, Figure 2; 1954, Figure 3). In spite of such basic differences of ontogeny, the adult stages are similar.

Corresponding situations are met with in Polychaeta. In *Eupomatus*, for example, gastrulation is the result of invagination, while in *Nereis* (with sterroblastula) the formation of the entoderm is brought about by epiboly and in *Ctenodrilus* by multipolar immigration. Similar differences of gastrulation are found in Prosobranchia, where eggs poor in yolk material will develop a coeloblastula and an invagination gastrula (*Vivipara*, *Planorbis*, *Limax*, etc.), while eggs with a rich content of yolk will develop a sterroblastula, and gastrulation will arise by epiboly (*Crepidula*, *Purpura*, and so forth). In *Patella* the formation of the entoderm layer is brought about by multipolar immigration.

In Anura, normally gastrulating by epiboly and invagination, the early ontogenetic stages may also be quite different. The eggs of *Eleutherodactylus* from Brazil, for example, are very rich in yolk material, and their cleavage is probably of the discoid type. As in reptiles, the embryo grows on the surface of a spherical body consisting of yolk material overspun by blood vessels. No external gills are developed, as there are no free larvae. In spite of this essentially different ontogeny, the adults do not differ from the normal anuran type (Lynn, 1942; Lynn and Lutz, 1946; B. Rensch, 1953, Figure 3).

In forms with metagenesis, which may be looked at as a prolonged individual cycle (see below), one will sometimes find that the first generation of two related species, for example, the polyps of hydrozoans, differs more conspicuously than the subsequent sexual generation, i.e. the medusae. As in many cases we do not yet know which polyp belongs to which medusa, separate taxonomic keys are applied to the generations, one to the polyps and one to the medusae. Now, some species of medusae belonging to one family are regarded as belonging to separate polyp families and vice versa (see Broch, 1924).

Intermediate deviations, that is to say, cases with similar early development but with differences arising at larval stages and disappearing again at later

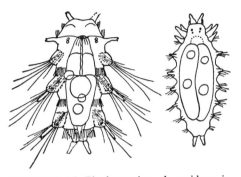

FIGURE 93. Planktogenic and nereidogenic larva of the polychaete *Nereis dumerili*. (After Wilson, Hempelmann, and Friedrich.)

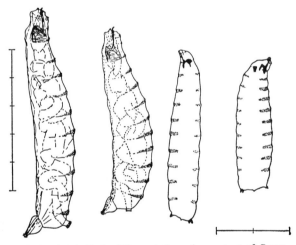

FIGURE 94. Early differentiation of a mutant of *Drosophila melanogaster*. Left to right: newly born larva of the normal type and of the mutant chubby (scale: 0·1 mm.), final larval stage of normal *D. melanogaster* and of chubby (scale 1 mm.). (After Dobzhansky and Duncan, 1933.)

phases of ontogeny, are even more numerous than early ontogenetic alterations. In Prosobranchia, for example, fully developed larvae of the veliger type may arise, while in other related types no larvae are developed at all, but the adult animals of these types do not differ very much, and their differences cannot be attributed to these differences of the larval phase. The marine snail

Littorina littorea, for instance, develops a normal veliger. *L. obtusata* lays eggs from which young snails will immediately hatch, and *L. rudis* is viviparous (Linke, 1933).

Nor are strong differences of the larval organisms likely to affect the structural type of the adult animal. This is exemplified by the various morphological and physiological differences of mosquito larvae (Figure 16): *Culex* with tracheal respiration, *Corethra* with cutaneous respiration, *Tendipes* with skin and gill respiration, and so forth. Different types of larvae may even arise in one and the same species, as in *Nereis dumerili*. This polychaete produces planktonic larvae when certain segments become sexually mature, but nereidogenic larvae will develop when such segments have not been developed (Figure 93). Finally, one example of intermediate deviation may be mentioned, which was analyzed genetically by Dobzhansky and Duncan (1933). The recessive gene 'chubby' on the second chromosome of *Drosophila melanogaster* causes a considerably chubbier shape of the larvae and pupae, but only a slight difference in the adults (Figure 94). This alteration occurs at embryonic stages, as in newly hatched larvae the difference is already clearly noticeable.

Summarizing the more essential findings, we may say that early ontogenetic and intermediate deviations are quite commonly met with, but that even quite striking differences of juvenile morphology do not necessarily affect the shape and structure of the adults.

V. Alterations by Additional Phases of Development (Anabolies)

Quite often, phylogenetic transformation seems to be brought about by the addition of new phases to the ontogenetic development; this has been referred to as anabolies by Sewertzoff (1931) and as hypermorphosis by De Beer (1940). The frequency of this phenomenon seems to be due to the fact that alterations of early and intermediate stages of ontogeny will more readily be wiped out by selection; more developmental reactions are disturbed by such alterations than by mutants adding new phases to the final stage of the morphogenesis. A characteristic example referred to by Sewertzoff is the development of jaws of the small fish *Belone acus*. In larvae of about 10 mm. body length, the head is similar to that of the related Percesoces, such as *Atherina hepsetus*. In subsequent stages an excessive growth of the lower and then of the upper jaw begins, which results in the formation of an enormously long snout, shaped like a forceps (Figure 95). Corresponding examples are provided by the development of the pectoral fin of *Lophius*, which is turned horizontally, and by the dorsoventral flattening of the body of this fish. Later Sewertzoff referred to the shape of the head of the sea horse (*Hippocampus*) and to the three free marginal rays in the pectoral fin of *Trigla* species as further examples of the anaboly. It is evident that these cases are partly the result of positive allometry, which we have already discussed in Chapter 6, B III and V. Further examples would be represented by the long bills of some birds growing with positive allometry after birth (Figure 38) or by the development of the final shape of the facial skull and of excessive teeth, horns, antlers, and similar structures in mammals, and so forth.

Some of the above-mentioned examples also refer to fossil types. Sewertzoff certainly was right in emphasizing that conclusions drawn from the study

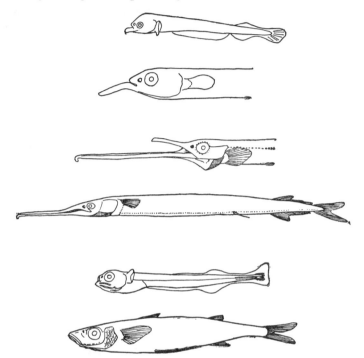

FIGURE 95. Addition of growth phases to the final stage of ontogeny demonstrated in growth of jaws in *Belone acus*. Top to bottom: stages of 10 mm., 21 mm., and 91 mm. body length, and the adult fish. For comparison: the related species *Atherina hepsetus* of 12 mm. body length, and in the adult stage, lacking the excessive growth of the jaws. (After Sewertzoff, 1927.)

of fossil material cannot be safe from possible errors arising from wrong evaluations of phyletic offshoots, and that they will sometimes be as hypothetical as the conclusions drawn from anatomical series of Recent animals. Nevertheless, it is quite obvious that paleontological material will be more important for the problem under consideration. This is proved by the following examples. The antlers of Miocene deer, for instance, did not exceed the stage of six points, while in Pliocene and Quaternary forms the juvenile type with two, four, or six points was exceeded and developed into stages with eight, ten, fourteen, and more points. The ontogenetic stages of the cave bear (*Ursus spelaeus*) correspond surprisingly well to phylogenetic stages of its ancestors (Ehrenberg, 1932), though there are some special alterations (e.g. the lacking premolar of the cave bear) which should not be overlooked. Certain ontogenetic stages of the cave bear are named after corresponding stages of phylogenetic ancestors: the *U. arctos* stage (resembling the similar *Ursus etruscus*), an early phase of postnatal ontogeny, is followed by the *U. deningeri* stage, which finally is replaced by the *U. spelaeus* stage. The close parallelism of ontogenetic and phylogenetic stages is well demonstrated by the proportions of limbs and, especially, of the skull. The skull of a one-year-old cave bear is similar to that of a three-year-old brown bear, and the skull of a cave bear of two to three years will resemble that of an adult *Ursus deningeri*. Finally, the growth of the cave bear exceeds that of *U. deningeri*. The forehead becomes steeper and the first premolars disappear (possibly by material compensation), and hence, the typical head of *U. spelaeus* arises (Figure 96). One should not forget that such a phylogenetic increase of body size may have been caused not only by a prolongation but also by an acceleration of the growth process, as the growth of Recent races of small and large body size or

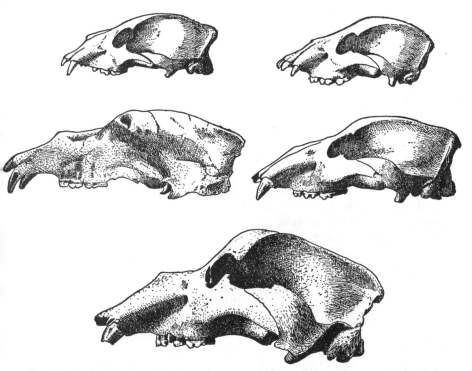

FIGURE 96. Upper left: skull of a three-year-old brown bear *Ursus arctos*; middle left: adult *Ursus deningeri*. Upper right: cave bear *Ursus spelaeus* at about one year (*arctos* stage); middle right: cave bear at about two to three years (*deningeri* stage); below: full-grown cave bear (*spelaeus* stage). (After Ehrenberg.)

related large and small species shows. This acceleration of growth often begins in rather early stages of ontogeny. It is quite possible that such an acceleration of growth affected the ontogeny of the cave bear, too, as the genital passage of an adult *U. spelaeus* is hardly any wider than that of the Recent brown bear, and thus the embryo of the cave bear cannot have been larger than that of the brown bear (Ehrenberg, 1925).

Infraspecific differentiation by phases added to the final stages of ontogeny is exemplified by the development of relative brain size in large races of domestic fowl referred to on p. 142. The relative brain size of juvenile specimens of giant races is identical with that of adult dwarf hens of equal body size (which is about the same as that of the wild ancestral types). The growth ratios of single brain parts, however, are different, and a young specimen of a giant shows a histological structure different from that of an adult dwarf. The same applies to the eyes of vertebrates, and as phylogenetic increase of body size involves less dense layers of retina cells and a thinner layer of granular cells, the phylogenetic addition of further phases of development to the final stages continues the ontogenetic tendency.

Sewertzoff (1931, p. 274) pointed out that such additions do not always mean a continuation of the trends developed so far (an 'overstepping': De Beer), but may equally well mean the beginning of a new tendency of development at the end of 'normal' morphogenesis. This is exemplified by the loss of a premolar in the cave bear to which we have referred above.

One might object that these differences by addition to the final stages certainly could be traced at earlier ontogenetic levels provided one applied a sufficiently subtle method of detection. Indeed, embryological studies done from this point of view are rather rare and there are many cases still lacking a clear statement of the bifurcation of ontogenetic divergence in closely related forms, though the importance of such studies was emphasized by Haecker's phenogenetic analyses (1918, 1925). Studies of this kind are rendered fairly difficult because the descriptions of ontogenetic stages by various authors are hardly comparable. However, such studies can well be done, as is proved by Penners' (1922, 1923, 1929) comparison of the ontogenetic development of *Tubifex rivulorum* and *Peloscolex benedeni*, i.e. of two oligochaetes belonging to two closely related families (Tubificidae and Lumbriculidae). In these two worms the first cleavages, the relative size of micro- and macromeres, the position of certain cells, the type of formation of the dermal layers, and the differentiation of the primitive germ cells and of the intestine are alike or very similar. The characters typical of *Peloscolex*, i.e. the cuticular papillae, the shape of the ventral bristles, and so on are not preceded by recognizable differences on the early and intermediate levels of ontogeny.

As far as genuine additions to the final stages of ontogeny are concerned, it is clear that von Baer's Rule is applicable to them because the development of specific and generic characters will be preceded by that of properties typical of families and orders. Another consequence is the ontogenetic recapitulation of characters typical of phylogenetic ancestors. As the ontogeny of the cave bear shows, sometimes a real ontogenetic recapitulation of adult ancestors happens (characters of the ancestors *Ursus etruscus* and *U. deningeri*). The reality of such a type of recapitulation has long been doubted. Besides anabolies in the narrower sense, that is to say, relatively short additional phases of development restricted to closely related types, there are 'recapitulations of the type as a whole', which means that structural characters of major taxonomic groups are recapitulated in a more general manner, but in the sequence of their phylogenetic origin (Lebedkin, 1937). In these cases the recapitulation begins during early phases of ontogeny, in which the organs in question correspond to the same organs in adult ancestors, without showing, however, the histological differentiation of the adult organs. The embryonic structures of the descendants correspond to those structures of ancestors which were embryonic as well as to adult structures, that is to say, to structures which are 'definite' characters (compare Jeschikov, 1933, 1937). In the ontogeny of the mammalian cochlea, for instance, a certain stage resembles that of reptiles in which the basilar papilla is not yet coiled spirally, but represents a curved tube not yet capable of any physiological function. 'Recapitulations of the type as a whole' are even more evident in cases of regressive development. In right whales, numerous embryonic teeth develop, showing a quite normal process of calcification, but shortly before the birth of the embryo the teeth are absorbed again and disappear. Similarly, the larvae of sturgeons (*Acipenser*) develop normal teeth, functioning for some time, but afterwards they are resorbed, so that in young sturgeons they are wanting.

Such additions, then, more or less affecting only the final stage of ontogeny, provide a basis for comprehending the highly disputed process of ontogenetic recapitulation of phylogenetic characters. Indeed, anaboly and recapitulation are firmly linked, as was especially stressed by Sewertzoff. Inversely, all cases of recapitulation, as far as they refer to final embryonic stages, may be regarded as examples of anaboly (in a wider sense) or of a succession of anabolies. The numerous cases of ontogenetic recapitulation which have been studied provide suitable material for a proper estimation of the frequency and importance of this type of evolution. Usually, however, those examples of recapitulation have been studied which show 'unnecessarily crooked ways' of development (*Entwicklungsumwege*), comprehensible only as remainders of the paths of development of former phylogenetic stages.

From the rich material demonstrating such ontogenetic by-paths, a few examples will be sufficient. In birds and anteaters (Myrmecophagidae), anlagen of teeth are developed, but suffer subsequent resorption. The same holds good in right whales (Mystacoceti) and in ground hogs (Tubulidentata), showing teeth buds in the anterior part of the jaw. In armadillos (*Dasypus*) a similar development and subsequent resorption of five small calcified embryonic teeth hidden in the gums of the lower jaw can be traced. Other peculiarities of ontogenetic development are exemplified by certain bones arising from two or more separate embryonic parts and joining at later phases of ontogeny to form one solid unit. In Bovidae the second and fourth metacarpalia are separated at early embryonic stages and later the second metacarpale will unite with the canon; similarly, the carpal os centrale of the elephant is at first separate and fuses with the scaphoid bone in later ontogeny.

The embryonic development of a chorda dorsalis and of gill pouches in terrestrial vertebrates cannot but be regarded as such by-paths or detours of development, and the same holds for the embryonic development of pro- and mesonephros in amniotes, the gills of amphibians with terrestrial ontogeny (gymnophions) or viviparity (*Salamandra atra*), the bipolar embryonic stages of unipolar ganglion cells in higher vertebrates, and so on. It is true that many such stages of development serve a special function – gills may be a means of nutrition in the embryo, pro- and mesonephros may function as excretory organs, and the chorda dorsalis exerts an effect on histological induction (compare Lehmann, 1938, on the morphogenetic importance of primordial anlagen) – but this has no bearing in this context. I want only to point out that such prestages represent detours in development, which should have been replaced by straight-line developing prestages if they had not been preserved in consequence of additional phases of ontogeny.

Even more numerous are those cases where the characters typical of species, genera, and families are the result of a development of late stages in a new direction without any detours. In these cases, too, the early and medium stages of ontogeny are recapitulated. This is exemplified by the various types of antlers and horns of artiodactyls originating from similar embryonic or juvenile structures and developing various final stages as regards their length, curvature, and surface sculptures. Further examples of this kind are provided by the molar teeth of vertebrates varying in number, size, and form of cusps and ridges, or by the development of feathers from an embryonic stage that is identical with that of scales (compare Matveiev, 1932). All these and similar cases must be accounted for in estimating the frequency of phylogenetic alterations brought about by addition to final stages. Besides this, the general evolutionary importance of anabolies is also suggested by the fact that numerous types of larvae, typical of whole phyla or classes (such as cysticerci, trochophorae, nauplii, or the larvae of Pantopoda) are to some extent representations of the heads of the adult stages. The three pairs of legs in a nauplius, for example, represent the first and second antennae and the mandibles, and in a protonymphon (the larval stage of Pantopoda) they become the chelicerae, palpi, and the third pair of extremities, which does not as yet belong to the walking legs. In a trochophora, the body proper of the worm is 'added' to the larval body at the stage of the metatrochophora. In all these cases the typical structure of the adult body is an 'addition' to the phylogenetic prestages, which were more or less similar to the larvae. Though many more phenogenetic comparisons of ontogenetic development in related types are urgently wanted, one may even now state that evolution by anaboly is quite a common and frequent process.

VI. Palingenesis, Proterogenesis, and Neoteny

In the discussion of ontogenetic alterations, I have not yet considered the question of whether the alterations appearing in the successive species of a line of descent diverged quite at random or whether they pointed to the same direction. I shall have to refer to such directed alterations of ontogeny in more detail, as they have been the basis of certain hypotheses on the origin of new structural types. Orthogenetic alterations of ontogeny may be caused by various factors.

1. Ontogeny may be gradually prolonged or abbreviated, and new stages may be added to the final phases or some ontogenetic phases may be omitted entirely.

2. On the other hand, the duration of ontogeny may remain constant, but the development of the whole body or of single organs may be accelerated or retarded, which will also result in the formation or the disappearance of additional structures.

3. Both the intensity and the duration of ontogenetic development may be affected by successive alterations.

4. Selection may cause the disappearance of certain stages of ontogeny only (e.g. the veliger larvae, which are lacking in some fresh-water proso-branchs). Characters typical of early ontogenetic phases may gradually be shifted towards later phases of ontogeny in the course of evolution. As it is almost impossible clearly to separate the effects of alterations of the length of ontogeny from those of the rate of ontogenetic development, we shall restrict our considerations to (A) abbreviations and (B) successive shifts of juvenile characters towards later ontogenetic phases (proterogenesis).

(A) The gradual abbreviation of ontogeny by the omission of certain stages has been referred to by various terms, the most general and neutral one of which is 'abbreviation'. Abbreviation corresponds to Haeckel's palingenesis, Buckman's (1900) lipopalingenesis, T. Nevill George's (1933) tachygenesis, and the 'Law of Juvenile Preponderance' of Wedekind (1927), which means the shift of adult characters towards more juvenile phases (compare also Schindewolf, 1946). Beurlen's term 'neomorphosis' (1937), based on Wede-kind's 'juvenile preponderance' will be dealt with in the following chapter, in the discussion on the origin of new structural types.

The omission of final ontogenetic stages is exemplified by the disappearance of the third and fourth digits in the horse line. All horses from the Early Tertiary to the Miocene *Merychippus* have fully developed digits with three joints each. In *Phiohippus lullianus* of the Lower Pliocene, the second and fourth digits consist of one bone only, which cannot be seen without special preparation. In Recent horses, the separate anlagen of the joints of the vesti-gial digits can be seen in juvenile phases only (Abel, 1928).

There are numerous additional cases of vestigial *physiological* characters. Hinsche (1932) and Krumbiegel (1938, 1941) found fighting reactions in Amphibia which probably are behavior relics of their stegocephalian ances-tors. Insects and birds lacking the ability to fly will still react to certain passive movements by postural reflexes of wings and tail, and stump-tailed monkeys and cats will 'use' the tail stump as though it were still a means of maintaining their balance in climbing, and so forth.

Further examples of abbreviated ontogeny are provided by numerous cases of regressive development, especially frequent in parasites. The ancestors of parasites had more complicated organs of perception and locomotion than their descendants.

When intermediate ontogenetic stages are omitted or abbreviated, characters typical of the adult stage may successively be shifted towards earlier phases of development. The forebrain of *Equus*, for example, will begin to develop cortical folds at a juvenile stage in which the brain size corresponds to that of the adult fossil *Eohippus* (*Hyracotherium*), which did not develop folds at all (Edinger, 1948). The juvenile stages of some Mesozoic Ammonoidea have quite intricate sutural lines, while in their phyletic ancestors these lines were still very simple. On the whole, however, it is remarkable that such shifts of adult characters towards more juvenile phases of ontogeny are rather rare, and it is significant that regressive animal evolution is caused mostly by the omission of final ontogenetic stages (and simultaneous deviations).

(B) Schindewolf (1925, 1936) must be credited with first stressing the evolutionary importance of shifts of juvenile characters towards adult stages. He also pointed out that orthogenetic effects might arise from this process. Proterogenesis, as this process is called, was first described in Cretaceous ammonites by Neumayr at the end of the last century. Pavlov studied the *Simbirskites* of the Lower Cretaceous and referred to their juvenile phase as the 'phase prophétique', but this statement was not appropriately acknowledged by later authors (compare Schindewolf, 1940, p. 377). As early as 1914, Wedekind stated that in *Goniatites* of the Late Carboniferous 'new characters may be acquired in juvenile stages and successively shifted towards adult phases'. Proterogenesis, then, is opposite to palingenesis, since in the latter it is the adult stages that recapitulate the phylogenetic stages. Proterogenesis in its simplest form may be caused by a general retardation of growth, so that the whole organism is still in a rather juvenile stage when its sexual maturity or its definitive size is reached. A retardation of growth might have been brought about by selection if it was advantageous. In Primates and in some other large mammals, for example, a prolonged growth period extended the time that could be devoted to the care of the young or to the acquisition of more complicated associations of the central nervous system. Such a rejuvenation was called 'fetalization' by Bolk (1926), 'paedomorphosis' by De Beer (1930), and 'diametagenesis' by Mijsberg (1931); and Wedekind (1927) referred to this process in his 'Law of Adult Preponderance'. (In 1897 Eimer coined the terms 'epistasis' and 'genepistasis', denoting a standstill of development at earlier or later phases of ontogeny from an evolutionary point of view. Apparently, however, he did not think of retarded development and rejuvenation, but wished to denote cases with additional stages added to the final phase of ontogeny.)

In some cases, fetalization causes the onset of sexual maturity in early juvenile or larval stages. This is known from tiger salamanders (*Ambystoma*), and may be supposed to be true in other urodeles with persisting gills. Hence, this would be a phylogenetic neoteny, which in some cases is likely to be a favorable character, as Margalef (1949) could show in fresh-water Crustacea (more efficient means of defense against predators, shorter cycles of development, and so forth).

Proterogenesis may also be caused without absolute retardation of morphogenesis by successively increasing positive allometry of organs. This is possible because growth gradients, and not only definite 'characters', are inherited and selected. Thus, organs originally restricted to juvenile phases

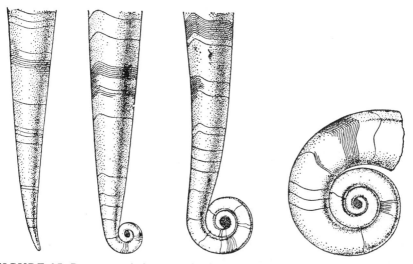

FIGURE 97. Proterogenetic increase of coiling in Nautiloidea from the Lower Ordovician. Left to right: *Rhynchorthoceras, Ancistroceras, Lituites, Cyclolituites.* (After Schindewolf, 1936.)

will persist longer and longer, until at last they will have reached the final stages of ontogeny. This seems quite applicable to one of the model examples quoted by Schindewolf and belonging to the *Lituites* line. These types of Early Ordovician Nautiloidea (Endoceraceae) have a straight shell (*Rhynchorthoceras*) in the oldest deposits (vaginate lime horizon B 3 b). In subsequent horizons (*Platyurus* lime, C 1 a), the same type of shell is met with, but a new shape of shell also appears, in which the first part is slightly but distinctly coiled, forming one or two whorls (*Ancistroceras, Lituites*). Still later (in the layers of the *Chiron* lime horizons, C 1 b), fully coiled types also appear (*Cyclolituites*: Figure 97). Apart from possible doubts regarding the assumed phylogenetic relationship of the types mentioned above, one might think that coiled shells were favored by orthoselection, as this type means reduced resistance in locomotion. Hence, the growth gradient causing the asymmetrical coiling of the shell became more and more efficient and the effects of symmetrical growth were more and more shifted to later stages of ontogeny. Schindewolf did not distinguish the accelerated development of the single character from the retarded development of the whole body, but it seems appropriate to refer to the former process as proterogenesis in the strict sense and to the latter as pedomorphosis or fetalization. Both processes may, of course, occur together. An example is also provided by two parallel lines of Jurassic Ammonites: *Macrocephalites – Cadoceras – Quenstedticeras* and *Macrocephalites – Kepplerites – Kosmoceras*. The opening of the whorls of these ammonites, when viewed in a transverse section, is similar to a broad, flat kidney in the original forms, while in subsequent types it resembles a heart. This transformation might also suggest a process of orthoselection by which shells with crested whorls (i.e. with heart-shaped openings) were favored because their shape was more efficient in swimming. In *Kosmoceras* lines, also, Brinkmann (1929) found that it is the juvenile shells that show more phylogenetic progress.

Proterogenesis (in its broader sense) was thought by De Beer and Schindewolf (1936) to have a special bearing upon the origin of new structural types. We shall refer to this problem later (6 E). In the context of this chapter it may suffice to state that proterogenesis has been proved in several cases, but that it is certainly not as frequent as archallaxis, deviation, and addition to final stages.

VII. Alternation of Generations

Ontogeny may also be altered by the development of alternating generations. The simplest type of alternation is metagenesis, in which sexual reproduction is followed by asexual reproduction. All generations are produced by germ cells, but normal bisexual reproduction and parthenogenesis alternate regularly or in a more complicated succession. It is not necessary in this context to give a detailed account of all the possible cases, and we shall restrict our considerations to an evaluation of the evolutionary importance and implications of these processes.

Metagenesis is not a difficult problem from an evolutionary point of view, as it may be regarded as a normal individual cycle with asexual reproduction occurring at a certain ontogenetic stage, e.g. at the larval stage. Because of this reproduction, we are forced to speak of 'generations'. Apart from this formation of asexual offspring, there is nothing unusual in the further 'ontogeny' of such metagenetic species, as the further 'ontogenetic' development is continued by the second (or even third) asexual generation. This is exemplified by *Echinococcus granulosus*, the cysticerci of which will propagate by budding, but the general development of this worm from egg to sexual maturity is not essentially different from that of the related species of *Taenia*. Such a metagenesis, which is an exception in tapeworms, has become the rule in other animal groups, like Hydrozoa and Scyphozoa, and its secondary abolition (as in *Hydra*) must be regarded as an adaptation to living in fresh water, where osmoregulation in the medusae would be difficult.

Evaluating and understanding the various types of *heterogony* is definitely more difficult, as each generation originates from germ cells, and the question of how very different individuals can arise from the same genome in rhythmical alternation is a difficult one. At first one might argue that heterogony should be regarded as an autonomous course of developing alternately differing generations. If the alternation of generation coincides with the alternation of diplophase and haplophase, one might think that the latter differences partly or totally cause the morphological alternation. But there are numerous other examples showing alternation of generations for which we have as yet no explanation of how a reversible change of the genome (the chromosomes or the cytoplasm of the germ-cells) could be produced. Hence, it seems probable that such remarkable changes of morphology and physiology in subsequent generations are caused by external factors affecting certain labile phases of ontogeny. This is suggested especially by numerous species in which the alternation of generation coincides with a change of the habitat (or of the host organism in parasites), or by types which under optimum conditions will develop parthenogenetic generations and sexual generations when environmental conditions (especially temperature) become worse. In Cladocera, Rotatoria, and Aphidae, for example, there is parthenogenetic reproduction in summer and sexual reproduction in autumn. The alternation of generation of numerous plant lice and gallwasps (Cynipidae) coincides with their change of host plants. How complicated the causal factors of alternation of generation may be, can be inferred from Luntz's studies on the rotifer *Pterodina elliptica*. This rotifer will be parthenogenetic if it is constantly kept at pH=6·8–7·6 in a culture medium of 0·01 percent, and fed on a constant diet of the same species of flagellate (*Chlamydomonas*). If there is a change of the diet, the following generation will develop male-producing eggs and resting eggs, provided the pH is kept constant. If transferred to a culture medium of 0·1 percent concentration, mictic females will arise, but only from the late-developing eggs. However, this will not happen if the animal lives in another range of pH values. Thus, the genetic basis of the phenomenon is the faculty of developing parthenogenetic and male- and female-producing eggs; but to produce an alternation of generation quite distinct environmental factors are required. In numerous other cases, especially in insects, these internal and external conditions of heterogony are far from sufficiently known, and a proper evaluation of the evolution of such processes of alternations is thus not possible at present (compare also the synopsis of heterogony by Hartmann, 1947).

VIII. Frequency of the Different Types of Ontogenetic Alterations and their Bearing on Evolution

Reviewing the various types of ontogenetic alterations arising in the course of phylogeny, we have to state that the primary undirected evolution, initiated by random mutation, is quite obvious also in the transspecific differentiation of ontogenetic development, as in many groups of animals almost any possible alteration of ontogeny can develop, provided that it remains biologically tolerable (Chapter 4). This means that once more there is no reason to assume special autonomous factors of evolution causing a certain direction of ontogenetic alterations, and that mutation and selection provide sufficient

263

explanations of the phenomena in question. Also palingenetic and protero-genetic alterations of ontogeny, altering successively whole lines of descent, may often have an adaptive value. We cannot completely deny the possible existence of further directing forces, but so far we are not obliged to assume such factors, and we should not introduce unknown and unanalyzed factors to veil our ignorance of some evolutionary happenings.

We cannot yet successfully try properly to evaluate the relative frequency of the various types of phylogenetic alterations of ontogenetic development, as careful phenogenetic studies are required. What we may safely say is that archallaxis and early deviation have been proved in numerous cases, but that late deviations and additions to the final stages of ontogeny are more frequent than the former two types.

Nor should we forget that the classification of the various types of phylo-genetic alterations of ontogeny attempted in the previous sections is intended only to facilitate our analysis of the various ways of phylogeny. In reality, almost any example of transspecific alteration involves the effects of two or more types, as many characters are usually altered, although sometimes the effects of one type may be especially conspicuous.

A very important question is whether we may ever arrive at a safe inter-pretation of the evolutionary types of ontogenetic alteration, as we have to restrict our studies to the divergence of 'characters' and cannot analyze the divergence of genic actions, which often will be of pleiotropic nature. Let us consider, for example, the effects of the recessive allele a, which causes red eyes in the normally black-eyed moth *Ephestia kühniella* (Becker, 1938). The red-eye pigment does not develop until the pupal stage is reached, because a certain agent is lacking which in the wild form (with the allele a present) is developed by the optic imaginal discs, the brain, and the gonads. Red-eyedness might then be regarded as an alteration of late phases of ontogeny or even as an additional phase added to the final stage. The lack of this agent in the red-eyed race, however, affects all larval stages right from hatching from the egg, as the skin is white (instead of reddish), the eyes of the larvae are slightly pigmented (instead of strongly pigmented), and the testes are whitish (instead of red-brown). With regard to these characters, the mutant a might be regarded as belonging to the category of alterations of inter-mediate ontogenetic stages, i.e. of deviations. Finally, the allele a causes a general retardation of development, and thus a very early onset of the alterations, and this might well be regarded as a case of archallaxis. A similar situation is met with in the *Drosophila* mutants vermilion and cinnabar, but in these the onset of eye pigmentation is a still clearer case of addition to a final stage, as the agent causing the pigmentation is produced in a larger and more efficient quantity only one to three hours after pupation, though small quantities are produced at earlier stages.

In this context some studies on larval salamanders by O. Kuhn (1933) should be noted, showing that sensitivity to thyroid hormones is gradually developed by certain organs. Even in larvae with a clearly developed yolk sac and filiform gills, some thyroid follicles produce the colloid, a process which increases rapidly later on. Now, gills, epidermal cornification, and Leydig's cells are already sensitive to thyroxin at the yolk sac stage, but the mesodermal pigmentation will not be affected by thyroxin until shortly before birth or after a long period of free aquatic life. Cutaneous glands will react to the hormone only during their phase of development, while the process of epidermal cornification remains constantly sensitive to thyroxin. As there are so many hormones and genes affecting various stages of the ontogeny, one should not overrate the importance of classifications of evolutionary alterations of ontogeny. Nevertheless, the practical value of the terms 'archallaxis', 'deviation', and 'anabolic addition to the final stages of ontogeny' has been proved.

One might assume that the major processes of transspecific evolution, and especially the origin of new structural types and new organs, might be the results of archallaxis altering the whole ontogeny. Therefore it is essential to note that many far-reaching deviations arising on early ontogenetic levels are compensated for in subsequent ontogenetic phases, so that the final stages are quite normal again. On the other hand, numerous radical alterations of the structural type and many new organs are the result of late deviations and additions to the final stages. This is exemplified by the evolution of eyes of the vesicular type with cornea, lens, retina, and pigment insulation of the visual cells in *Mollusca*. These highly specialized organs may be regarded as the result of mutational steps gradually improved by adaptive development and selection. At first they were hardly more than an accumulation of photosensitive cells; then the groove eye developed, followed by the eye of the vesicular type; and finally the lens and cornea were developed. This is suggested by anatomical series established by a comparison of the eyes of Recent species (compare Chapter 6, E, and Figure 101). All these alterations arise at late stages of ontogeny, and the eyes of adult Helicidae will regenerate even after the antennae have been lost. Feathers, so essential in the development of flight and in the structural organization of the bird organism, must also be regarded as the result of late deviation and additions to the final stages, as these highly differentiated structures arise from the embryonic buds which are homologous to the buds of reptilian scales. The increase of relative size of the forebrain, so essential in the progressive evolution of vertebrates, begins only at intermediate stages of embryonic development.

Temporary alterations of early ontogeny, balanced in the course of subsequent stages, and numerous examples of archallaxis imply that Von Baer's (1828) Rule of Individual Development (compare p. 256) is not generally valid. In the majority of cases, though, the early ontogenetic stages of related forms and groups are more similar than the later stages, but in some animals the characters typical of the higher category will develop after those typical of the lower category. This is the case in the frog *Eleutherodactylus nasutus*, in which the first phases of ontogeny are by no means typically amphibian but rather reptilian (yolk sac). In birds, the feathers, a character typical of the class, will not develop until many characters typical of the order, family, or genus have differentiated (e.g. fixed number of neck vertebrae; constellation of palatine bones).

The validity of the highly disputed Biogenetic Rule (Law of Recapitulation) should not be overrated, either, though it applies to numerous animal onto-genies. It is partly based on Von Baer's Rule, as far as it applies to the 'Law of Conservative Prestages' (Naef, 1917). A genuine recapitulation of phylo-geny (not only the recapitulation of certain ontogenetic prestages) is probably met with in but a few cases of additions to the final stages of ontogeny, as in the evolution of the cave bear skull. The Biogenetic Rule does not apply to archallactic alterations.

E. EVOLUTION OF NEW STRUCTURAL TYPES AND NEW ORGANS

I. The Origin of New Structural Types

From our considerations on the problem of alterations of structural types, on orthogenetic evolution, and on overspecialization (Chapter 6, C), we could conclude that in many cases genera, families, and even higher categories may be regarded as the results of continued mutation and selection, and that there is no need to assume the existence of additional autonomous factors of evolution. However, this conclusion cannot be considered to prove that trans-specific evolution is generally similar to or identical with the process of infra-specific differentiation. Thus, we are left to examine whether random muta-tion and the 'accidental' character of selection may lead to the origin of new organs and the formation of completely new structural types. During the last decades this problem has been tackled several times, especially by paleon-tologists, members of a branch of science which has been rightly skeptical of mutation-selection theories, because the intermediate stages of animal phyla dating back into pre-Cambrian times are not known, and because many classes and orders are not sufficiently linked by intermediate evolutionary stages proved by the fossil record obtained so far. Moreover, the origin of new structural types, as inferred from numerous fossil records, seems to have happened very quickly or even by 'saltations' (see Chapter 6, A). Hence, there is an obvious possibility that the origin of new structural types – like those of mammals, insects, and gastropods, or of smaller groups such as bats, beetles, and mussels – might be due to factors fundamentally different from those governing the adaptation of such structural types to the requirements of various habitats. Consequently, the origin of new structural types was distinguished as aromorphosis (Sewertzoff, 1931), phylogeny proper (Parr, 1926), or typostrophism (Schindewolf, 1950) from the process of adaptive evolution or idioadaptation (Sewertzoff), or adaptiogenesis (Parr). From the geneticist's point of view, Plate (1913) distinguished a more or less constant 'Erbstock' from the normal genes following Mendel's rules; R. Goldschmidt (1940, 1948) distinguished mutations altering the structural type from nor-mal gene-mutations, and Remane (1939) drew a distinction between the

266

experimentally studied 'real mutations' and other types of alterations not yet analyzed.

For a proper discussion of these problems, it is necessary first to define what is meant by 'structural type' or simply 'type' (compare the criticism by Gross, 1943). In this context I shall use this expression only to denote the characters typical of a higher systematic unit, and shall give a more precise formulation of the problem by trying to answer the questions of whether the evolution of the characters of orders, classes, phyla, and their subordinate categories is the result of a process fundamentally different from that of infraspecific evolution and whether special autonomous factors are required for a reasonable interpretation of the process.

Paleontologists and zoologists, doubting that random mutation and selection are sufficient causes for the origin of new structural types and new organs, are inclined to believe that especially large mutational steps occurred – possibly a series of them pointing in the same direction – by which the new types originated. As we have seen in Chapter 6 of this book, such macromutations will usually be lethal. Apparently, speciation by macromutation, for example, by duplication of certain organs, will be an exceptional event. In the discussion on the alteration of structural types (Chapter 6, B III), however, I have stated that certain 'constructive genes' exist, which may affect many characters by their pleiotropic and more or less harmonious effects. Moreover, ontogeny may phylogenetically be altered from its initial stages on, as we have seen in the preceding chapter. All these evolutionary processes may well cause a relatively quick and radical alteration of a structural type, and therefore such an alteration must not be considered as the result of one major step of mutation.

We may not even maintain that such constructive genes are typical of alterations of structural types, as they are equally or even more often met with in adaptive evolution of the structural types as, for instance, in cases of allometric growth. Early ontogenetic alterations are not characteristic exclusively of changes of the structural type, but are also met with in species of a genus, like *Triturus* or *Rana*. Finally, late deviations and anabolic additions to the ontogeny may produce very radical alterations of the structural type. On the whole, however, archallactic alterations and constructive genes seem to be of major importance in the formation of new structural types.

Paleontologists have been accustomed to assume, and probably will continue to assume, major evolutionary saltations when a gap appears in an otherwise complete series of successive fossil stages. I do not think it necessary to consider this problem in more detail, as it has so often been discussed (compare Weigelt, 1943). The gaps in the fossil records have become narrower and narrower year after year as new paleontological evidence has been found. We must, however, remember the highly important discovery of the crossopterygian *Latimeria chalumnae* in 1938 (compare Gross, 1939), a genus not until then represented in fossil records in the whole Tertiary, and we must bear in mind such huge gaps in the fossil record as that between the Recent cyclostomes and their Paleozoic ancestors, the ostracoderms. Hence, we see that there are large gaps which are caused only by the lack of fossil findings and not by saltatory evolution.

On the other hand, we should not forget that gaps are especially frequent in the earliest stages of new structural types as, with the origin of a new type, evolution usually proceeds at a faster rate (compare our findings on 'explosive phases' in Chapter 6, A). It is for this reason that in many lines with a rich fossil record proving gradual phylogenetic transformations the very first stages of the structural type are missing. This is the case, for example, in the well-known lines of horses, proboscidians, cetacea, sirenia, amphibians, and so forth. Simpson (1944) proved that in nearly all mammalian orders the duration of the lines known so far was longer than that of the original lines not yet known to science. The quicker evolutionary rates following the formation of new types were due partly to the fact that the new types conquered new habitats by which selection was highly intensified (whales, amphibians), or that they conquered the habitats of more primitive types bound to rapid extinction when competing with a new superior animal. Simpson (1944) rightly points out that probably the new types represented at first small populations, from which not only a faster rate of evolution, but also a poor fossil record, resulted.

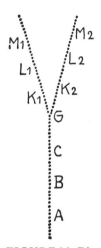

FIGURE 98. Diagram of phyletic bifurcation, demonstrating the rarity of intermediate forms.

Moreover, it is evident that morphologically intermediate stages of various constructional types cannot be frequent, as the period of phyletic divergence was relatively short and – in a diagrammatic representation – will correspond to a single point at the bifurcation of the types (Figure 98). Evidently, we are far more likely to find a fossil type of the periods before the branching (A, B, C) or after the bifurcation (K_1, L_1, M_1, or K_2, L_2, M_2) than just the type denoting the branching point (G). Moreover, the gaps in paleontogical series may be looked at as proving saltations of evolution only in the groups in which many individuals are preserved as fossils (as, for example, in mollusks or mammals). It is significant that in such groups with rich fossil evidence the morphological gaps are not usually wide, though they do exist. One need not assume saltatory evolution of new structural types because of these fairly small gaps. This is convincingly demonstrated by certain lines of Ammonoidea, of hoofed animals (with regard to the origin of the horse type from small, omnivore, many-toed protoungulates similar to *Hyopsodus*) and of carnivores (the various families like Felidae, Mustelidae, Ursidae, Canidae, etc., which descended from creodont ancestors). Hence, there is no reason to accept such skeptic hypotheses as Przibram's apogenesis, Westenhöfer's 'phyletic shrub', or Dacqué's far-reaching doubts.

Above all, one should not underestimate the numerous intermediate links of structural types that really have been found. We can hardly expect that a phylogenetic link should precisely represent the point of phyletic bifurcation (see above), but we will often recognize types with intermediate morphological structure, with corresponding paleontological age, and with close relationship to a younger convergent line of ascent. There is quite a reasonable number of such links. Of course, fossil evidence linking the animal phyla cannot be expected, as the intermediate types must have lived in pre-Cambrian epochs, and not many fossils unaffected by secondary deformation have come to light. Many animal classes and orders, however, are linked by numerous well-known intermediate stages. I shall mention only the best-known structural types of vertebrates. Among mammals, the distinct structural type of carnivores is closely linked to that of the ancestral Cretaceous insectivores by the Early Tertiary creodonts, and similarly artiodactyls are linked by protoungulates, and Primates by tarsioids and lemurs to the same insectivorous ancestors of the Cretaceous. It is sometimes difficult properly to classify Paleocene fossils of these groups. Also closely linked to the creodonts are the whales, as in the intermediate Eocene Archaeoceti (such as *Protocetus*) the teeth were of a more or less normal type, the vertebrae of the neck region were not yet fused, and the nasal openings were still close to the front of the head. The Recent sirenians approach the primitive hoofed animals by the Eocene *Eotherium*, and the same applies to proboscidians, which are linked by the Oligocene *Moeritherium*. The whole class of mammals is so closely linked to the reptiles by the Late Triassic theromorphs that it is not clear whether the oldest mammalian genera should be classified as mammals or as reptiles. *Tritylodon*, for example, is mammal-like, but the os articulare is still fastened to the lower jaw in a typically reptilian manner (W. G. Kühne, 1943; Hennig, 1947). The mammalian characters of the intermediate saurian ancestors, the theromorphs, include the more or less mammalian-like dentition, the reduced os quadratum, the long external auditory meatus, the double condylus occipitalis, the probable existence of cutaneous glands and hairs, and the like. On the other hand, the reptiles are relatively well linked to the birds by the Archaeopterygiformes, and the distinct order of turtles is linked to the primitive cotylosaurs by the Permian *Eunotosaurus*. The cotylosaurs, as the ancestors of reptiles, are linked to the amphibian stegocephalians. Between these and the crossopterygians as the nearest relatives among fish there was a gap in the fossil record for quite a long time. Some years ago, *Elpistostege watsoni* was discovered, a stegocephalian of the group of Ichthyostegalia (Figure 99), which is such a perfect link between the two groups that Westoll (1938) is certainly right in his statement that there is a 'perfect transition between the crossopterygian and ichthyostegid patterns of dermal bones'. As the transition to terrestrial life involved an intense selection, the tetrapods seem to have evolved very rapidly during the Late Devonian. Moreover, Jarvik (1942) proved that the skull of osteolepiform crossopterygians is similar to that of stegocephalians and anurans, and the skull of porolepiform crossopterygians resembles that of urodeles, so that the two classes, fishes and amphibians, are linked at least morphologically. A similar morphological approach cannot be denied between the phylum (or sub-phylum) of Acrania and that of Vertebrata, as the jawless and limbless agnatha of the Silurian had a branchial intestine, one central nasal opening, an incomplete labyrinth, and other prevertebrate characters.

Moreover, the gaps between several phyla and classes are bridged by a number of Recent links, at least morphologically. Protozoans and metazoans are linked by the Cnidosporidia with their multicellular spore, by the Myze-tozoa (Myxomycetes) with their multicellular sporangia, and by the Mesozoa,

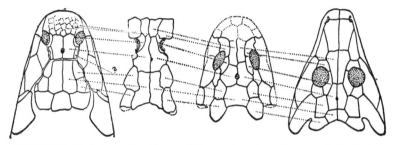

FIGURE 99. Intermediate forms of Crossopterygia and Stegocephalia. Left to right: *Dipiplopterax* (Crossopterygia, Middle Devonian); intermediate form *Elpistostege* (Lower Late Devonian); *Ichthyostegopsis* (Upper Late Devonian or Early Carboniferous); *Actinodon* (Early Permian). Homologous bones linked by dotted lines. (After Westoll, 1938.)

which have only one epidermal layer of cells (somatoderm) besides the gonads. The structural types of Coelenterata and Turbellaria, looking so extremely different, are linked morphologically by forms like the ctenophores *Coelo-plana* and *Ctenoplana*, which have a ciliary sole for locomotion, a ventral mouth with an ectodermal pharynx, and excretory organs with flame cells, and which develop micromeres and show a slight spiral cleavage. The Tur-bellaria, again, may have some relationships with higher worms, as forms like *Leptoteredra* have 'invented' a genuine anus. The gap between the two classes of Turbellaria and Trematoda is slightly reduced by parasitic tur-bellarians and ectoparasitic trematodes. The cestodes show an approach to the nearest class, the trematodes, by certain forms (like *Nesolecithus* and *Amphilina*) with a totally reticulate excretory system and a terminal excretory pore, and by the lack of a differentiation of the scolex and proglottids. Scolecida and Annelida are morphologically linked by various characters in the intermediate types of Nemertina (lacking a coelom, but having a system of blood vessels, intermediate larvae, and so on). Annelids and arthro-pods, on the other hand, are linked by the peculiar Onychophora (*Peripatus*), in which annelid characters (legs not jointed, cutaneous layers not chitinized, nephridia, equal cleavage) are mixed with arthropod properties (tarsal claws, dorsal vessel with ostia, tracheae lacking chitinous spirals). The linkage of annelids and mollusks, finally, is brought about by the worm like, shell-less

Solenogastres and the Placophora, as the larvae of the latter are similar to a trochophore and as some segmentation of the body can still be seen in the serial arrangement of the gills. Still more primitive and worm-like are the newly discovered Monoplacophora (*Neopilina*) with segmental muscles, nephridia and gills, and with the anus directed backward.

These examples may do to prove that a certain morphological and onto-genetical linkage of several phyla and classes is not impossible, even though there are no fossil records of common ancestors.

In discussions on the evolutionary bearing of such phylogenetic links and ancestors, the argument is sometimes raised that presumed prestages are too specialized 'in general' (compare Fuhrmann's discussion of the origin of cestodes and Bresslau's study on the descent of Turbellaria in Kükenthal-Krumbach's *Handbuch der Zoologie*, 1928–33). In my opinion, however, it is sufficient when *one* single genus or *one* single species may be considered as the link or as similar to a link, because all other forms must, by the nature of things, belong to some specialized group or side branch of the hypothetical ancestral type. Of course, linking types will be found especially among small and poorly differentiated animal forms. For example, the relatively complicated shape of the gonads in larger turbellarians is due only to the fact that in a large body lacking a circulation system the transport of nutritional material is difficult. Hence, we must not expect to find phylogenetic links among such specialized forms, but should look for them among the small Rhabdocoela.

Summarizing our findings, we may state that, even in cases in which paleontological evidence is lacking, it is highly probable that intermediate phylogenetic links have existed, bridging the gaps between the structural type of two orders, two classes, or two phyla. Gaps in the fossil record must be ascribed to rarity of fossilization, to accelerated evolution after a new structural type had evolved, or after a totally distinct habitat had been conquered (water – land), and (partly) to the small size of the populations.

Contrary to the above statements, Schindewolf (1936, 1950) regards such gaps as proofs of saltatory evolution of new structural types, and R. Goldschmidt (1940, 1948, 1953) thinks that they prove the effects of macromutations. Though such macromutations are usually lethal, Goldschmidt thinks that occasionally some 'hopeful monsters' may have found a suitable habitat to which the new structural type was perfectly adapted.

A special hypothesis by Schindewolf holds that new structural types (like that of *Archaeopteryx*, for instance) originated 'all of a sudden and by saltation' as the result of proterogenesis, by which the alteration of the type at first affected the embryonic stages and was successively shifted towards later stages of ontogeny. Therefore, if one compares only the adult stages, there should be a 'saltation' of the structural type. Consequently, Schindewolf concludes:

> The search for a series of successive transitional stages between two types, demonstrating the gradual formation of the new structural type, will be in vain, because such stages have never existed

(p. 22). At first sight, this opinion may seem tenable, but one should remember that by proterogenesis only single or – at best – a group of characters can be altered, and that Schindewolf's examples refer to specific and generic differences only. The origin of a complete structural type, however, requires so many new characters that their simultaneous formation by proterogenesis is extremely unlikely. Hardly any zoologist will agree with Schindewolf's statement that 'the first bird hatched from the egg of a reptile'. Moreover, we have seen from the preceding chapter that proterogenesis is not too frequent among the processes of evolution.

Even more speculative, as well as more disputable, is the theory of 'neomorphosis' advanced by Beurlen (1937), partly based on ideas of De Beer (1930). This theory holds that occasionally a rejuvenation of a type is brought about by a successive shortening of the specific life cycle. 'The whole differentiation as the result of the preceding somatic development is reduced, and the juvenile, plastic, and adaptable organism may commence a new process of differentiation' (p. 204). This theory, then, combines two well-known phenomena (discussed in former chapters): that essentially new types are likely to originate from poorly specialized forms (Cope's 'Law of the Unspecialized'), and that phylogenetic abbreviations of ontogeny are quite common. This combination, however, does not account for the fact that phylogenetic abbreviations of certain ontogenetic phases usually involve simultaneous additions of new stages and further differentiations of other phases, because the steady process of selection requires continuous adaptations. Moreover, so far as I know, no paleontological proof has as yet been given demonstrating such a secondary rejuvenation of a specialized structural type and the subsequent redifferentiation of a completely new one. This statement does not deny that a 'rejuvenation' may have occurred in some lines, especially in those with successively decreasing body size. As we have seen in Chapter 6, B III, small mammals have a relatively large cranial and a relatively small facial skull, the bony crests of the skull are flat, the relative length of ears and legs is different from that of larger relatives, their general metabolism is more intense, and their relative sexual differences tend to be less marked than those of related larger types. After a phylogenetic reduction of body size, then, it seems quite possible that a new line of differentiation could begin, especially as small animals are more easily adaptable to new ways of living (e.g. to a fossorial or arboreal life) than large types. In some cases, and in a restricted sense, 'neomorphosis' may well be regarded as a suitable interpretation. (Compare also the contrary opinion expounded by W. Gross in his critical review of Beurlen's hypotheses, 1943.)

The hypotheses on the origin of new structural types advanced by Schindewolf, Beurlen, and R. Goldschmidt are based on the assumption that specialization and origin of new types are two essentially different processes, and that the latter cannot be explained by the principles underlying normal evolution. It is true that new structural types have often evolved relatively fast and that in subsequent epochs only minor adaptations and alterations of the type occurred (compare Chapter 6, A, B, on explosive evolution and the phase of specialization), but there is ample proof that some structural types have evolved relatively slowly, governed by modes of evolution not at all different from those of normal racial, specific, and generic differentiation.

Supposing that only the Eocene *Eohippus* (*Hyracotherium*) and the Recent Equidae (wild horses, wild asses, zebras) were known from the horse line, we would certainly classify them as two different structural types (orders), because the Eocene forms have several toes and a skull shape differing from that of Recent horses; the forebrain is relatively small, lacks cortical fissures, and does not cover the posterior brain parts; and the dentition of these primeval horses, because of their bunodont teeth, was also quite different from that of our Recent horses. Hence, it is because of the numerous successive phyletic links known from fossil evidence that *Eohippus* and *Equus* are classified as one family. This classification, then, is based on historical rather than on morphological facts. However, the rate of transformation from one stage to the next was so uniform and gradual that Mathew (1914) thought that the progressive morphological alteration of certain characters might be used as a time scale. Though in subsequent studies this did not prove to be a workable hypothesis, as Abel (1918) pointed out, the type of evolution in this line does not differ from the usual evolution by speciation and was by no means a 'neomorphosis'.

Similarly, we would certainly classify the Eocene and Oligocene Moeritheriidae as totally different types from Recent elephants, had intermediate links not become known from the fossil record. Elephants differ very remarkably from the only slightly specialized Moeritheriidae by their unique tusks, by their complex molars, by the peculiar holes in their skull bones, by their unique trunk, etc. But many intermediate types of the Tertiary show that *Moeritherium* and *Elephas* may be grouped in this order.

Numerous orders of birds seem to have evolved by normal speciation, as various combinations of certain constant characters are met with in certain groups, and some of them must be regarded as typically adaptive traits. In the dove order (Columbae) we find tubular nostrils, no nasal glands, quill feathers without aftershafts, ten primaries, a crop producing in some cases a secretion for feeding the young, and no gall bladder. In the Galli we find stouter legs with the fourth toe inserting more proximally, quill feathers with aftershafts, ten primaries again, a crop not producing any crop milk, and a gall bladder. The divers (Colymbi), finally, have long lobate toes for swimming, short wings with eleven primaries, a very short tail, feathers with aftershafts, nasal glands, and a gall bladder. Such examples, proving a gradually advancing adaptation with some characters remaining constant, could easily be extended.

In this context it should be noted once more that the variety of structural types displayed by all Recent and fossil animal phyla, classes, and orders strongly suggests that most biologically tolerable possibilities have been realized by evolution and there is no reason to assume directed transformation processes of an autonomous nature. In any organism the basic biological functions are alike: feeding and breathing, resorption and gaining energy, storage and excretion, stimulus perception and reaction, reproduction and ontogenetical transformation. But what a wealth of diverse structural types has evolved to perform these functions! All processes mentioned may happen in a single cell (protozoans), or they may be divided among several groups of cells specially differentiated (metazoans). In some cases the food is passed into the body by a mouth, while in others it passes through the whole integument (sporozoans, trematodes, cestodes, and the like). Sometimes the food is actively seized, and sometimes the particles are transported to the mouth opening by whirling and rotating mechanisms, as in ciliates, corals, or tentaculates. The digestion of food may be of the intracellular type, as in protozoans or coelenterates, or of the extracellular type; there may be intra-intestinal or extraintestinal digestion (as in spiders), there may be a sac-like intestine (as in turbellarians or trematodes) or a special anal opening may allow a continuous passage of the food. The acquisition of energy may be dependent upon or independent of oxygen. Oxygen may enter the organism by the integument, by gills, by lungs, or by tracheae, and within the body it may be transported by a system of blood vessels or may be distributed without this system. Various chemical compounds such as proteins, fats, and carbohydrates, may be stored, and the mechanisms of storage are also quite various. Extremely diversified, too, are the patterns of reproduction of the various structural types, which range from asexual to sexual reproduction, including metagenesis and more or less complicated heterogony, automixis, conjugation, and copulation by the most diversiform mechanisms. An equally rich diversity is also met with in the shape of the body of the various animal types: there are spherical bodies in sea urchins, filiform ones in nematodes, and radiate forms in coelenterates and echinoderms, and there is the more or less bilateral symmetrical type of the vertebrates. The body may be jointed or may lack any segmentation at all; it may lack limbs or may consist of little more than limbs (as in Pantopoda, where parts of the intestine, of the gonads, and of the excretory organs are situated in the legs). There are sessile, creeping, digging, swimming, floating, climbing, running, jumping, and flying types. It is this diversity of all possible morphological and physiological specializations that strongly suggests that the diversity is the result of undirected processes which resemble random mutation and selection. In consequence of this randomness, the characters used in taxonomic keys are often of no importance for the morphogenesis of the type concerned; characters which seem rather irrelevant, judged from the standpoint of phylogeny, are often used only for convenience in key construction.

II. The Origin of New Organs

The arguments brought forward by defenders of autonomous forces of evolution usually run as follows: mutations generally cause disturbances of characteristic development and will thus decrease viability or fertility or even be lethal. The few mutations not causing any deterioration of biological qualities and those producing positive effects may affect any morphological or physiological character. However, it seems extremely unlikely that a succession of random mutations and selective processes by random factors of the habitat should bring about the formation of a purposeful organ of a completely new type, such as a kidney, an eye, or even a brain. The improbability seems even greater when we consider that such organs require the differentiation of highly specialized histological structures (e.g. the lens, the vitreous body, the cornea, the retina, with visual cells and pigment layers of the eye) and the harmonious coordination of various systems (e.g. blood vessels, lacrimal glands, protective structures arising from neighboring bones and connective tissue, neurons allowing central nervous connections, reflexes, instincts, and voluntary actions). These arguments, however, do not account for the fact that new organs usually do not arise as something absolutely new but result from differentiations and organ systems that evolved long ago (such as neurons, blood vessels, and inducing agents), and that the evolution of new organs is not directed towards a certain aim. This is quite obvious from the numerous 'detours' in development, in which certain structures had more or less completely to change their function once or several times (compare Dohrn, 1875; Kleinenberg, 1886).

For a proper consideration of this problem we shall turn to the well-known evolution of the auditory ossicles of the mammalian ear. The malleus developed from the articular bone of the lower jaw, which in fishes, amphibians, and reptiles joins the lower to the upper jaw. (In mammals the goniale – developing, like the malleus, from Meckel's cartilage – is included.) The incus originated from the quadrate which in fishes, amphibians, and reptiles is part of the jaw articulation. The stapes, finally, is the result of a transformation which started from the hyomandibular of the fishes, joining the jaw to the brain case; but in amphibians it has already become reduced to the tiny rod-shaped columella which is a true auditory ossicle – though the only one possessed by amphibians – transmitting sound waves from the drum to the internal ear. Later, with advancing evolution, this ossicle moved still deeper into the ear, and finally it became the innermost part of the three auditory ossicles typical of the mammalian ear. Hence, the auditory ossicles developed from bones with quite different functions, and the functional changes occurred at quite different stages of their evolutionary transformation. In this process, the negative allometry of the bones later to be used in hearing seems to have played an important part, which may be inferred from their relatively large size in mammalian fetuses (Figure 100). The tympanic cavity, the site of the auditory ossicles, is derived from the first gill slit, which first developed into the spiracle of Elasmobranchia and was thus a part of the breathing apparatus. The evolutionary course of all these accessory organs of the future mammalian ear, then, was by no means directed right from the first steps of transformation, and it may well be interpreted by the assumption that several mutational steps occurred, some of which were favored by selection, as hyomandibulare, quadratum, and articulare became unimportant and grew with negative allometry when the skull became reinforced. But as these ossicles were situated in the neighborhood of the ear developing in land animals, they were now favored by selection because they could develop an efficient mechanism for perceiving aerial sound waves. Without knowing the consecutive steps of this phylogenetic development, some paleontologists and zoologists would certainly speak of 'macro-evolution' (Watson, 1952).

Similar statements apply to the phylogenetic development of other organs, the first stages of which often served functions quite different from those of the final structure, or were constructions of minor importance. However,

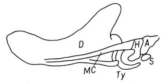

FIGURE 100. Lower jaw of fetal kangaroo *Macropus*, showing the relatively large hammer (H), anvil (A), and stirrup (S). D=dentale; Ty=tympanicum; MC=Meckel's cartilage. (After Broom from Hilzheimer and Haempel, 1913.)

after a new habitat (e.g. the terrestrial habitat) was conquered and a new mode of living became necessary, they were favored by selection in new directions. This is exemplified by the evolution of lungs starting from the paired air bladder of fishes, which in some types was used as an accessory organ of respiration, though this function was as yet of no vital importance. Some divergent types of fishes, however, possessing this accessory breathing mechanism, were enabled to live in poorly oxygenated water or could spend the dry periods of hot seasons buried in the mud (as the Recent *Protopterus*). With the evolution of terrestrial animals this organ gained vital importance, and in higher reptiles it developed into a spongy organ providing an extremely large surface for a more efficient oxygen intake, which was an essential prerequisite for the development of homoiothermy. It is evident that the evolution of lungs is sufficiently interpreted by the principles of random mutation and natural selection, by which variants with a more efficient breathing mechanism (i.e. a larger respiratory surface) were preserved.

Employment of pectoral fins as organs supporting the body was of some (though not of essential) importance in fishes resting on the bottom. The rays of the pectoral fins gained considerable importance as soon as certain crossopterygian types evolved into the terrestrial stegocephalians, and the vital importance of the forelimbs has increased ever since up to the Primates. The 'fins' of penguins so perfectly adapted to swimming evolved from wings like those of some guillemots and auks (Alcidae), not fully efficient for flight, but occasionally used as accessory organs in diving and swimming. Finally, the wings of birds, so characteristic of the structural type of this class, undoubtedly did not serve exclusively for flying right from the beginning, but seem to have been used mainly for climbing and occasional gliding, which is suggested by the fossil record of the Archaeopterygiformes.

In all these examples, we cannot state with certainty in which stages of development the various steps of transformation occurred, but in numerous cases the assumption seems justified that the transformation took place by successive alterations of final morphogenetic stages rather than by archallaxis. This is especially suggested by the evolution of eyes of the vesicle type. As we have already seen in Chapter 4, these highly specialized organs have evolved independently but convergently in various groups of animals. In coelenterates, annelids, gastropods, and echinoderms, we meet with photosensitive spots and photoreceptors of the groove type, which may be arranged in an 'anatomical line' of descent representing the evolution of the eye from the primitive stigma to the highly developed camera-type eye with lens, pigment layer, and retina (compare Hesse, 1908, and Plate, 1924). For example, photosensitive plates consisting of but a few sensory cells separated by tall pigmented epidermal cells are found in the polychaete *Eunice*. The cuticle in front is only a little thickened like a very flat lens. In *Ranzania* the sensory cells are arranged in a sort of groove, and a cuticular cone projects into this groove from above. The protecting pigment is part of the visual cells. In *Scyllis* and – even more conspicuously – in *Nereis* a vesicular cavity has evolved, containing a true vitreous body and a real retina consisting of visual and pigment cells. In *Vanadis*, finally, the number of visual cells has markedly increased and a spherical light-gathering lens under a genuine cornea has developed (Figure 101). Corresponding anatomical lines can be established in gastropods, as there are slightly grooved sensory plates in *Patella*, and eyes of a deeper and more vesicular type in *Haliotis*, *Trochus*, and others, while eyes with lenses are met with in *Murex* and many other Prosobranchia and Pulmonata. Similar lines can be perceived in coelenterates (e.g. *Catablema–Sarsia–Charybdaea*) or in echinoderms (exemplified by *Astropecten mülleri – A. aurantiacus – Asterias glacialis*). In all four groups mentioned, the eyes do not develop until more advanced stages of ontogeny, and there are no photoreceptors, or different ones, in the larval forms. Hence, it is probable that the evolution of such perfect organs as eyes of the camera type was brought about by additions to final ontogenetic stages or by late deviations.

Moreover, the fact that structures so different as the various types of eyes evolved independently and convergently, each representing a 'special structure' on the common basis of the transformation of epidermal nerve cells into photoreceptors, suggests that the evolution of organs was not brought about by autonomous forces, but by the directing influences of selection. An increased number of visual cells may evolve by mutation at almost any stage of phylogeny, and it certainly is a favorable character. A slight grooving of the

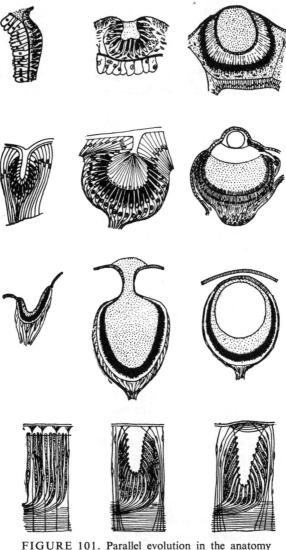

FIGURE 101. Parallel evolution in the anatomy of photoreceptors. Left to right, row 1: Coelenterata: *Catablema, Sarsia,* and *Charybdaea*; row 2: Polychaeta: *Ranzania, Scyllis,* and *Vanadis*; row 3: Gastropoda: *Patella, Haliotis,* and *Murex*; row 4: Echinodermata: *Astropecten mülleri, Astropecten aurantiacus,* and *Asterias glacialis.* (After Hesse, Schewiakoff, Pfeffer, and Plate.)

sensory plate is often seen in consequence of the increased number of photosensitive elements, and this is the beginning of perception of the direction from which the light comes – again a character likely to be favored by selection. The development of a transparent tissue in front of the photoreceptors certainly meant an essential advantage, and it is quite comprehensible that from this stage the origin of a genuine light-concentrating body (i.e. a lens) was not far away. Mutation and selection, then, provide sufficient explanations of the evolution of the spherical eye of the camera type.

In this context it will be useful to remember our findings on developmental influences exerted by certain environmental factors (see Chapter 4), and causing convergent development, and even the formation of new organs (such as myzetomes, stridulating organs, webbed feet, and the like). This explains the fact that similar or identical organs may arise from quite different anatomical substrata. The vasa malpighii, for instance, will originate from the ectoderm in insects and from the endoderm in spiders. Stridulating ridges and spines of insects may develop on various parts of legs, wings, thorax, or abdomen, and will be favored by selection, once organs of sound perception have evolved, and the production of sound has become a means of attracting the sexual partner. The lack of a unique 'plan' in the evolution of organs is also exemplified by the fact that a certain organ is present in one species and is lacking in a closely related type: in *Mustelus laevis*, a shark, a placenta will develop from parts of the yolk sac, while in the closely related *M. vulgaris* and *M. antarcticus* no placental layers will be found to connect the embryo and the uterine walls.

III. Summary

Summing up, we may state that the evolution of new structural types and of new organs needs no other explanation than specific and generic differentiation, i.e. the combined effect of mutation and selection. Possibly, mutations causing harmonious alterations of early ontogeny may sometimes be especially important, but late deviations and additions to the final stages of ontogeny may also lead to the formation of totally new organs. A certain lack of directedness is obvious when the origin of various structural types is compared. Sometimes, certain limits of evolution are established by the structural type. The first stages in the evolution of a new organ are often of minor importance to its possessor; a new organ may change its function once or several times, giving rise to detours in the development which suggest a randomness in the process of origin of the organ. Fundamentally similar to this biological process is the development following the invention of an engine. After the internal combustion engine had been invented, an array of most diversified 'structural types' was developed in adaptation to various kinds of 'habitats': motor cars, tractors, motor ships, agricultural machines, and aeroplanes. In much the same way various structural types, such as bats, hoofed animals, carnivores, whales, and so forth, evolved, once the principle of homoiothermy and more complicated central nervous systems had been 'invented'. In both cases, the evolution of quite different structural types was a consequence not only of invention of a superior and more efficient mechanism, but also of the presence of various types of habitats that could be filled in by types progressively adapting themselves by mutation and selection. The formation of new organs and new structural types is not restricted, however, to the phases of flourishing radiation, though there will often be an accumulation of newly arising types during such periods.

'Neomorphosis,' i.e. phyletic rejuvenation preceding a period of new transformation, may sometimes occur, though it is not a necessary prerequisite of the origin of new structural types and new organs. There is no reason to assume the existence of autonomous unknown principles of evolution. The gaps between the structural types are better bridged – at least morphologically – than many discussions might lead us to believe.

These statements are by no means intended to maintain that all causes and processes of organic evolution and the origin of new structural types have been sufficiently 'explained' or 'cleared up'. The opposite is true in many cases. What is urgently wanted in the further analysis of phylogeny is the completion of more detailed studies on comparative ontogeny, on phenogenetics, and on the complex effects of mutations, especially of selection. (Compare the examples quoted in Chapter 6, B III.)

7

Anagenesis (Progressive Evolution)

In the considerations of the processes underlying infra- and transspecific evolution and the formation of major animal groups, I have repeatedly emphasized that mutation and selection and their complex interaction effects in animal transformation are sufficient clues to lead us to a reasonable interpretation of the phenomena. But I have not yet referred to the question of why the phyletic tree is usually shown in an 'upright' position, indicating that there is a progressive evolution towards higher levels, and it must now be asked whether this process might suggest the existence of autonomous principles. It is a well-known fact that many fossil lines begin with 'primitive' types and end at 'higher' levels, and the phyletic tree as a whole represents 'lower' types and 'higher' forms in different branches. Though it has rightly been argued that the 'lower' species may be 'perfectly' adapted to their special habitats, the existence of a 'lower' and a 'higher' has not been denied. There is no doubt that most of our Recent types of mammals are 'higher' than their smaller Jurassic and Cretaceous ancestors, and that Recent insects represent a higher form of organization than their annelid ancestors. However, as a 'higher', level may be caused by increased complexity, improved rationalization, or greater versatility in reacting to environmental stimuli, the problem of progressive evolution, observed in so many phylogenetic branches, needs a consideration of its own. I have proposed referring to it as anagenesis, whereas the normal branching of lines of descent should be distinguished as kladogenesis.

(Sylvester-Bradley, 1951, advised me that the term 'anagenesis' was used in a different sense by Hyatt, 1875, and proposed using 'phylogenesis' for steadily progressive evolution and 'phylogeny' for the description of the special lines of descent and their ramification. However, such a differentiation between these two terms is likely to cause confusion, as in almost every language the two words have been used as synonyms. Moreover, the literal meaning of 'phylogenesis' would not imply that emphasis should be laid on the progressive element of the process. Hence, I would suggest that 'anagenesis' be retained, as this term has already been adopted by several authors, and as there is no other word so clearly emphasizing that the progressive element of evolution is especially referred to.)

Elementary knowledge of various levels of animal organization dates back to extremely ancient times. Later, Anaximander (610–547 B.C.) thought that from a primeval slime certain fish-like creatures had evolved, followed by terrestrial animals and, finally, by man. A further and more precise manifestation of such 'evolutionary' thought is the Aristotelian classification of all living animals into nine categories. In the modern science of natural history it was Cuvier who first recognized the progress to 'higher' levels in subsequent faunas. At about the same time, Erasmus Darwin (1794–8) developed a clear idea of the gradual improvement and perfection of the animal world in his *Zoonomia*, though his statements were somewhat speculative. In 1809 Lamarck advanced his theory of speciation and progressive evolution from 'incomplete' to most complete animal types and to man, using as a basis a fairly rich body of material drawn from various scientific fields. He was the first biologist who characterized 'progressive' development by increasing complication and greater versatility. He thought that progressive evolution was brought about by a 'constant law of nature'. He did not analyze this 'law' any further, and one may assume that he had in mind certain autonomous developmental forces. Not long before, in his considerations on the principles of morphology, Goethe (Introduction: 1807) had already advanced the idea of progressive improvement. He characterized this progression by an increasing 'dissimilarity' and 'subordination of the parts of an organism', i.e. an increasing differentiation and centralization. Thus, the two essential principles of anagenesis were discovered rather early, and have again and again been emphasized by later authors, such as Milne-Edwards (1851) and Bronn (1853), the latter also stressing the importance of the division of labor, of a decrease in the number of single parts of an organism, and of the progressive shifting of external organs and structures towards the interior of the body. (For the history of these and similar ideas of phylogenetic improvement, the reader is referred to Franz, 1920, 1924, and Uschmann, 1939.) In Chapter 4 of his book *On the Origin of Species*, Charles Darwin (1859) also considered the problem of progressive evolution. Referring to Von Baer's statements, he regarded increasing differentiation and division of physiological labor as the essential characters of anagenesis, but did not fail to stress the importance of 'psychic' development in the lines of descent and the increasing similarity of vertebrates to man. We should note that Darwin did not assume the existence of autonomous principles underlying the processes mentioned, as he considered the evolution of higher types the result of natural selection. That 'lower' types are still existing today is explained by the assumption that these forms evaded competition with 'higher' types (as fishes and mammals; marsupials, edentates, rodents, and monkeys in South America). In Chapter 11 of his famous book, Darwin stated once more that some forms had to remain 'adapted to more primitive conditions of life'. Haeckel, too (1879, pp. 253–4), listed the division of labor, the reduction of identical parts, and centralization as essential characters of progressive evolution, but distinguished progressive adaptation as a phenomenon of its own, which does not or need not lead to higher levels of evolution.

This precise characterization has sometimes been overlooked by subsequent authors, and there has even been doubt as to whether natural selection is a cause of progressive development. Von Nägeli (1884) assumed a 'drive to improvement' which, however, according to him, was basically caused by 'molecular forces'. Gaudry (1896), trying to demonstrate the improvement of sensory and nervous performances in phyletic lines, wrote: 'Il me semble que l'activité divine s'est manifestée d'un manière continue' (p. 209). Korschinsky (1899) thought that the 'tendency to achieve progress' is 'a fundamental internal quality not depending on external factors'.

During more recent times, Franz (1920, 1924, 1935) and – inspired by him – Plate (1925, 1928) have worked on this problem. Plate regarded the following traits as characteristic of progressive evolution: increase of size and number of elements, formation of new elements, division of labor, general improvement of ontogenetic processes, more efficient means of dispersal, and increase of fertility. Nevertheless, higher types need not be represented by more species than lower forms (compare apes and monkeys). Plate (1928, p. 757) stated: 'Biological perfection arises from a harmonious increase of the number, complexity, and efficiency of adaptations.' 'Perfection, then, is identical with reaching a higher level of adaptedness' (p. 758). The latter sentence implies that perfection was regarded as the result of natural selection. But this view is not satisfactory, as each progressive adaptation and each specialization would mean a 'perfection' in the sense of anagenesis. Hence, we may see that the problem of anagenesis will not be solved by assuming that anagenesis and improvement are identical phenomena. Nor was the problem solved by Franz's more elaborate statements (1935), which run as follows:

> The degree of perfection in a type of organism (species, genus, order, and so on) will depend upon the efficiency of the mechanisms serving to maintain the existence of the type (p. 36).

The mechanisms that are here referred to serve to increase, reinforce, and accelerate all functions of the organism. Franz's reference to such increased efficiency suggests that he thought of natural selection as the essential principle of anagenesis. Sewertzoff (1931, Chapter VIII), on the other hand, thought that the acquisition of new characters was the essential trait of 'morphological and physiological progressive evolution or aromorphosis', and that such new characters would increase 'the energy of the vital processes' of the organisms and 'would be favorable in many disadvantageous alterations of environmental factors'.

In spite of these numerous correct statements, skeptical views have been expressed by various authors during more recent decades. Hennig (1927), a well-known paleontologist, stated:

> All progressive development from the simplest cell to the vertebrate animal and the highest type of plant can only be regarded as a pure progress

in a certain field, and it definitely is not a process towards biological perfection. The trend of evolution results in enormous complication; no biological sense can be traced in it with certainty (p. 27).

Dacqué (1935) did not deny the trend to perfection, but thought that it was the result of autonomous forces of development rather than of natural selection. Beurlen (1937) thought that the 'perfection of types' was essentially brought about by a wider range of reactions to environmental stimuli ('Umwelterweiterung') and an 'increased autonomy, i.e. a reinforcement of the autonomous organic structural principle' (p. 221), basically caused by a 'will to own formation' ('Wille zur Eigengestaltung', p. 222). And W. Zimmermann (1931) thinks that 'nobody has precisely demonstrated so far what are the real differences between recent "higher" and "lower" beings' (p. 141). This series of skeptical views by recent authors could easily be extended.

A more satisfying aspect of the problem of anagenesis appeared in J. S. Huxley's brilliant book *Evolution, the Modern Synthesis* (1942) (which was not known to me when I finished the first German edition of this book at the end of the Second World War). Huxley, not discussing the above-mentioned opinions, listed the following criteria as typical of 'higher' organization: increasing complication, evolution of dominant types capable of producing many new forms dominating over and not depending upon factors of environment, and generally improved efficiency of all vital processes. Later on, Huxley (1954) defined biological progress as an unrestricted type of improvement permitting further improvement. Similarly Simpson (1949) emphasized that 'progress that broadens the chances of further progress' and 'progress in adaptability' are the most important characters of evolutionary progress. Huxley does not think it necessary to assume the existence of unknown autonomous forces in evolution, and is convinced that anagenesis is comprehensible as a result of natural selection.

Thus it will be necessary to analyze the various factors referred to by the above-mentioned authors (and by myself in the first German edition of this book) and to find out whether these factors provide a reasonable explanation of the important phenomenon of progressive evolution. A more detailed study of the principles of anagenesis seems worthwhile, because – in my opinion – two important factors have not been sufficiently accounted for as yet; the increasing knowledge of animals and their relevant environment, and the improving plasticity of structures and functions.

B. FACTS OF ANAGENESIS

The average figures of oxygen consumption per kilogram body weight in an hour are about 2,310 cm.[3] in *Paramaecium*, 51 cm.[3] in an earthworm, 319 cm.[3] in the snail *Deroceras agrestis*, 78 cm.[3] in the perch, 4,130 cm.[3] in the mouse, and 220 cm.[3] in the horse (after Von Buddenbrock, 1924–8). From these figures we may infer that in protozoans the metabolic rates may be surprisingly high, and that the metabolic processes of vertebrates are not more intense in general than those of invertebrates, so that no differences of this kind of 'perfection' exist between higher and lower types. When we accept Charles Darwin's statement (1859, Chapter 11) that all 'recent species have proved their superiority over their extinct ancestors by their survival', we should remember that this superiority does not necessarily mean a result of progressive evolution. From our consideration of the process of specialization (p. 236) we have seen that selection usually causes progressive adaptation to special conditions of the environment and that this adaptedness may prove a fatal disadvantage in future development, because radical environmental changes can no longer be coped with. It is for this lack of general versatility that lines with progressive specialization are so often doomed to extinction in spite of the 'superiority' of each subsequent type over the ancestral form. Hence, the major lines of ascent so typically exemplifying progressive evolution originated from unspecialized types ('Law of the Unspecialized'). Nevertheless, it has become general usage to refer to higher and lower forms in lines of progressive specialization, also.

From these considerations we may infer that no clear-cut definitions have been established as yet, and that it will be useful to distinguish the process of improvement and perfection from that of anagenesis or progressive evolution in the narrower sense. To do this, we shall at first refer to some cases exemplifying progressive evolution and try to find out the principles which have a more general bearing. As Franz, Plate, and J. S. Huxley have devoted detailed studies to the problem, we shall confine our analysis to the most essential questions.

At first it seems essential to list briefly the more general 'perfections' of the animal phyletic tree. We shall take into account only those alterations of the structural type which, once they had been 'invented', were maintained by all subsequent species and genera.

The first of such 'inventions' seems to have been the formation of molecules of nucleoprotein capable of reduplication and thus providing a possible basis of continuous reproduction and the origin of living substance (compare Chapter 8). If such 'living matter' were to evolve into living beings, an increasing complication of its components was required, and variants must be developed that were capable of adapting themselves to various environmental conditions so that some of them could persist. To achieve persistence, some molecules or molecular groups had to remain unaltered in the varying processes of living showing only a dynamic balance of the living substance. Genes in all living beings are such persisting molecular structures. The union of certain genes to form the first chromosomes was important, as by chromosomes the simultaneous and equal distribution of genic material in the process of cell division and reproduction was guaranteed. Chromosomes became even more important, as the absolute constancy of the genic material became subject to gene mutations producing variants of differing adaptability to environmental alterations. The production of variants was also facilitated by sexual differentiation. All these 'inventions' of nucleoproteins, genes, chromosomes, and sexuality apparently proved so favorable that they became common traits of 'higher' types of being. The invention of the cell as the basic element of the organisms proved favorable, because it guaranteed the persistence of hereditary characters during ontogenetic development as most cytoplasmic effects of the genes are more or less confined to a short range of controlled reactions. Needless to say, the cellular architecture led to division of labor and to specialization within the multicellular organism. Nerve cells, one result of such specialization, were a favorable character, as they enabled their possessor to perceive danger signals and vital stimuli at a distance, so that the possession of nerve cells and receptors often meant the avoidance of possible death. The invention of a central nervous system proved to be favorable, as the coordination of several reactions, the formation of inheritable reflexes and instincts, and the origin of memory were rendered possible by it.

Correspondingly, there was a series of favorable inventions that led to the evolution of the organs of metabolism and their improving efficiency (compare Hesse, 1929). Here, too, most basic 'inventions' have been maintained ever since.

In contrast to the structural organization of plant organisms, taking their nutrients directly from the environment and hence developing a large surface for intake, animals had to develop internal cavities for the chemical decomposition of food. The first type of such cavities was that of coelenterates and flatworms, forming a sac with an epithelial lining. As in this type the food and waste material passed into and left the body cavity by one opening, the 'invention' of the first anus, by which a more continuous exploitation of the food and a division of labor was rendered possible, was an important event. This type of intestinal passage proved so efficient that it has persisted ever since, and is met with in the structural types of all subsequent forms, such as the Recent vertebrates, arthropods, and mollusks. In nemertines, the intestine is imbedded in the parenchymatous layers of the body and movement of the food in the intestinal tract is brought about accidentally when the body is moved in general locomotion. The invention of special muscles for peristaltic movements of the intestine and the formation of a coelom at the phyletic level of the jointed worms (Annelida), then, were also favorable events, as a more efficient and rational type of digestion was rendered possible. As soon as the metameric segmentation of the coelom was abandoned, the intestine was no longer tied to each single segment of the body, but could be coiled so that its length was increased and the resorption surface was enlarged. This again meant an improved digestion of food, and this improved mechanism, like the others mentioned above, was maintained in subsequent structural types (Sipuncoloidea, Mollusca, Arthropoda, Vertebrata).

Closely parallel with the above-mentioned alterations, a functional differentiation of the digestive tract developed; numerous mechanisms of cutting and mastication originated (mandibles of insects, radula of mollusks, teeth of vertebrates, the muscular stomach of many birds, and so forth);

secretion of enzymes was more and more confined to the anterior and intermediate parts of the intestinal tract; resorption of the broken-down food material was assigned to the intermediate and final parts, and the preparations for getting rid of the waste material were taken over by the distal part. It seems unnecessary to refer to these facts and to their chemical and physiological implications (improved system of enzymes, increased efficiency of hematochromes, and so forth) in more detail.

All these and numerous similar examples from other fields of biology, e.g. improving histological structure, suggest that a general perfection and improvement of structure and function can be traced in the major lines of descent. In the introductory sentences of this chapter it has been stated that the evolution of 'primitive' to more advanced types does not necessarily mean 'perfection', but this limiting statement does not apply to the general improvement of subsequent structural types of the phyletic tree, which was designed on the basis of the results of studies on comparative morphology, embryology, physiology, and paleontology. As each of the 'inventions' mentioned above and each subsequent general improvement was a favorable character, anagenesis in this more general meaning may well be comprehended by the principle of natural selection.

Before we enter a more detailed examination of the major characters of perfection, it seems important to stress that a general improvement is met with in all major animal phyla and in numerous animal classes so far as we may judge their phylogeny. Only a few further arguments are necessary to prove this.

As has been stated above, an increasing complication and a progressive functional improvement of the digestive system are met with in most diverse animal groups, and certain favorable properties will persist once they have been 'invented'. We have already referred to the general improvement of the intestinal tract of worms and arthropods. A similar, if not more conspicuous, series of subsequent improvements can be studied in vertebrates. There is only a poor differentiation of the intestinal tract in Agnatha, and the internal surface of the intestinal tube is enlarged only by a simple longitudinal fold. The length of the intestine and its differentiation are increased in the higher types of fishes and amphibians. In reptiles, the internal surface of the digestive tract is considerably enlarged by fine folds, and in mammals this enlargement is brought about by the still more efficient villi.

A parallel evolution of improved eyes – from the primitive photosensitive grooves to the eyes of the camera type – happened in Coelenterata, Polychaeta, Gastropoda, and Lamellibranchiata. In vertebrates, the subsequent general improvement of photoreceptors is again quite conspicuous. Accommodation is brought about by moving the lens by muscles in a rather primitive way. A much more improved type of accommodation is seen in reptiles and higher vertebrates, especially mammals, working on the principle of focusing by changing the curvature of the elastic lens.

The relative size of the central nervous system is gradually increasing, and a gradually progressing differentiation of nervous structures is apparent in the lines of arthropods (Phyllopoda to Insecta), mollusks (Amphineura to Cephalopoda), and vertebrates (Agnatha to Mammalia). Parallel improvements of general importance which evolved in various classes and orders of these phyla were traced by Lartet (1868) and – especially – by Marsh (1874). The brain of *Archaeopteryx* was relatively small, and resembled that of reptiles in its proportions. The forebrain of this ancestral bird was relatively smaller than that of higher bird types (Edinger, 1929). The brains of early mammals were relatively smaller and of a more reptilian type than those of their subsequent descendants, which is exemplified by fossil horses from *Eohippus* to *Equus* (Edinger, 1948), and by the ancestral creodonts and their carnivorous successors (Edinger, 1950), and also by the brains of lemurs and monkeys, or of anthropoids and hominids. In the horse line, for example, the successive increase of relative brain size is accompanied by a relative increase of the size and fissurization of the forebrain, which was proved by endocranial casts (Figure 102).

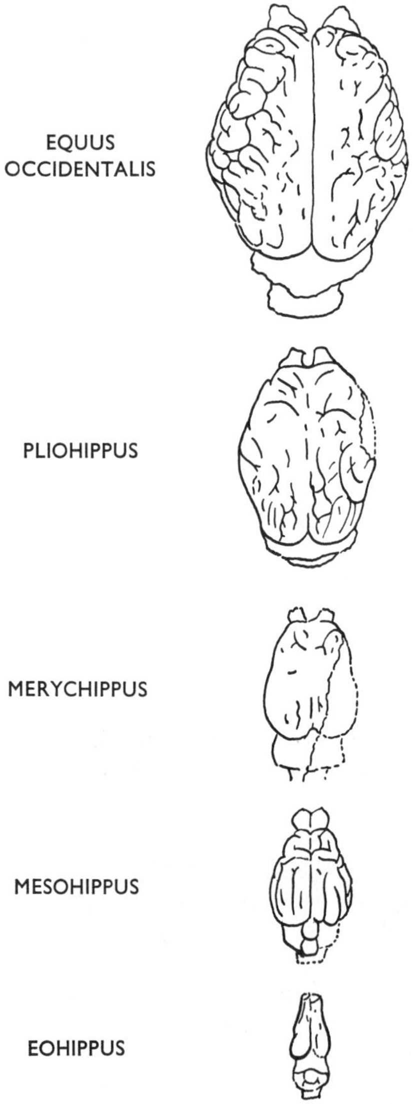

These examples may do to prove that a parallel anagenesis has happened in numerous phyla, classes, and orders and to demonstrate the various characters typical of a more general improvement. We should not forget, however, that sometimes even such subsequent phylogenetic types are referred to as 'higher' forms, in which no immediate improvement is to be seen in

EQUUS OCCIDENTALIS

PLIOHIPPUS

MERYCHIPPUS

MESOHIPPUS

EOHIPPUS

FIGURE 102. Endocranial casts of the horse line (on same scale). (After T. Edinger.)

comparison with their ancestors. This is exemplified by the first terrestrial types of arthropods or vertebrates, which, although they certainly were not more 'perfect' in a general sense than the highest aquatic types of their category, yet are regarded as 'higher'.

C. ANALYSIS OF ANAGENESIS

From our historical review of progressive evolution we have had to conclude that various authors have passed various judgments concerning the distinguishing characteristics of anagenesis. According to my opinion, the following characters may be listed as typical of anagenesis: (1) increased complexity; (2) rationalization of structures and functions (including increasing centralization); (3) special complexity and rationalization of central nervous systems (implying progressive evolution of parallel psychic phenomena); (4) increased plasticity of structures and functions; (5) improvement permitting further improvement (partly identical with point 4); (6) increased independence of the environment and increasing command of environmental factors (progression of autonomy). These factors, including some additional characters typical of anagenesis and referred to under different headings by other authors, will now be analyzed in more detail.

I. Increase of Complexity

Since Lamarck, all authors who have discussed the problem of progressive evolution and improvement have agreed that increased differentiation should be regarded as a main character. How is this differentiation achieved, and what is its evolutionary value?

The size of a protozoan cell cannot be enlarged *ad libitum*, because the sphere of action of the nucleus is limited. A protozoan, then, cannot be enlarged unless several nuclei are developed (polyenergid cells). Essentially increased complexity of individual organization requires increased body size, and this will not be brought about successfully and efficiently unless the numerous nuclei are separated from each other by cell walls, thus preserving the proper sphere of action of each nucleus. In such multicellular organisms, i.e. metazoans, various layers of cells will be exposed to various environmental factors – for example, internal cells will be different from those of external layers – which implies the origin of the differentiation of various structures and functions. This differentiation will increase with foldings and invaginations of the cell layers (for example, with the formation of ectoderm and endoderm in Coelenterata), and the selective value of genes controlling such differentiations will change. Hence, mutations affecting strictly quantitative alterations of the number and size of cells may produce further differentiations.

As such an increase of body size may be brought about at almost any stage of phylogeny, increased differentiation and formation of new organs are rendered possible on each evolutionary level.

Mutations causing positively allometric growth of certain tissues will be especially important, as an increase of cell number may mean the first step of differentiation and the formation of new organs. One need only consider the origin of parapodial appendages of polychaetes, the formation of wings in insects, the fins of primeval fishes, feathers and hairs, the neural groove, the optic cups, the semicircular ducts of the otic labyrinth, the basilar papilla of the internal ear, the neopallium of the vertebrate brain, and many others. This principle – the origin of new organs as a consequence of a strictly quantitative increase of certain tissues causing new growth gradients – is applicable to many animal types and has been too much neglected in evolutionary thought. It is especially important to notice that such an increase of a tissue may not serve any *special* function at first. Later on, however, the 'superfluous' tissue may be 'employed' by a new function in the course of subsequent evolution.

The following examples may illustrate this process. In vertebrates many functions previously located in the midbrain were shifted towards the forebrain as soon as its relative size was enlarged by positive allometry which originated on the amphibian-reptilian level of phylogeny. In amphibians the forebrain is already relatively large, but is still fairly unimportant, as proved by extirpation experiments. The positive allometry of the pars dorsalis medialis of the amphibian brain may possibly have initiated the origin of the cerebral cortex of higher vertebrates (Nolte, 1953), although as yet no important functions seem to be localized here. After centers for all senses had developed in the brains of mammals, additional regions originated and could now be used as centers of association. The same holds good for the corpora pedunculata of the polychaetes and arthropods, as well as for the frontal and temporal lobes of mammals. In the human forebrain, one of these temporal regions, Broca's region, could also assume the motor functions of speech, thus contributing to the rapid development of human culture (compare Chapter 7, D of this book and B. Rensch, 1953).

It is natural that periods of increasing complexity may sometimes be interrupted by processes of secondary simplification, but anagenetic development will not be interrupted. As we shall see below, secondary simplifications are quite frequently met with, because more economic and more rationalized functions are thus rendered possible.

Morphological complexity is often accompanied by an increasing chemical differentiation which arises from the huge multitude of possible protein compounds, as well as other possible organic compounds. Hence, even the formation of new species and new races often involves corresponding alterations of the chemical background of the body, which has been proved by precipitin tests and paper chromatography. In the evolution of mammals from lower to higher types, a distinct increase of new subunits of proteins (so-called 'proteals') is obvious, while only a few of the old types of compounds will be abandoned (Mollison, 1938). Hence, it is again a more quantitative alteration that underlies the process of differentiation. This quantitative principle applies also to the successive increase of blood and hemoglobin contents in invertebrates, which is so important for the intensification of functions and ultimately for the origin of homoiothermy, as may be illustrated by a few data taken from Von Buddenbrock (1924–8, p. 304): in fishes 100 cm.^{3} of blood will contain 5·7 gm. of hemoglobin, while the corresponding figure is 6·3 in amphibians, 6·9 in reptiles, 12·4 in birds, and 15 in mammals.

II. Increased Rationalization

There is a second character typical of anagenesis, which has been referred to under various headings by various authors. Goethe's term was 'increased subordination of parts'; Lamarck spoke of 'decreasing complexity of organs', and Bronn of division of labor, and centralization and shifting of organs towards the interior of the body ('interiorization'). Von Baer, Milne-Edwards, and Darwin emphasized especially the division of function; Haeckel, Franz, and Plate stressed centralization and improvement of performance as the important traits of anagenesis, and Franz (1935) regarded an improved working efficiency as an essential character of evolutionary progress. In my opinion, all the above-mentioned phenomena may be grouped under the heading of 'rationalization of structures and functions'. The division of labor is essential, as it enables a more efficient performance. Serial organs of equal form and function will therefore be transformed into organs of different form and performance (primeval polyisomerism evolving into anisomerism). This is especially well exemplified by the ever-increasing heterodont dentition of vertebrates, especially of fishes, theromorphs, and mammals, and the improving division of labor in crustacean limbs and insect mouth parts (compare also Gregory, 1935).

On the other hand, rationalization is often, but not always, accompanied by a process of simplification, which does not mean a regressive development, but an improvement of efficiency by eliminating and abolishing superfluous parts which originated as a sort of by-product in the process of random mutation. Haeckel recognized this process of the reduction of identical parts as early as 1879. A good example of this process of reduction is provided by the phylogenetic reduction of the number of skull bones from fishes to mammals (Williston's Rule). In the crossopterygian skull this number is about 150, while in man it is only 28 (Figure 103).

This process of reduction often involves an increased centralization, which may also be seen from the following examples. A reduced number of cranial elements gave rise to a more compact shape of the skull, and the numerous 'hearts' (elastic contractile parts of the major blood vessels) of annelids and chordates were replaced by one central pump, the proper heart. Similarly, numerous decentralized breathing organs were abolished in favor of one central organ of respiration, and the nervous system was also centralized. Just the same kind of processes of rationalization and centralization have occurred in man's technical development of tools and machines, which is evidenced by modern locks and keys in comparison with their medieval predecessors, or by modern and old aeroplanes, modern and old vessels, and so forth.

It is essential to notice that rationalization may sometimes be initiated by a quantitative mutation, i.e. by an increased number of certain cells. This is

291

exemplified by the visual cells of the retina, which will be capable of a better analysis of complex stimuli if their number is increased. In more primitive eyes a simple quantitative alteration may thus mean the important advance from perceiving only the direction of light to genuine visual perception of forms and images. The same quantitative aspect applies to the compound eyes of arthropods. Even a tooth, say the canine, acquires its special function by slight quantitative alterations, i.e. the division of labor may arise as a

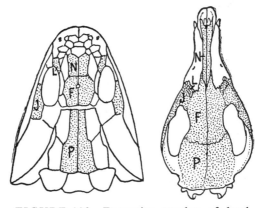

FIGURE 103. Decreasing number of head bones in the course of evolution. Left: *Osteolepis* (Devonian Crossopterygia), right: dog. N=Nasale F=Frontale, P=Parietale, J=Jugale, L=Lacrymale or Praefrontale. (After Watson and Wiedersheim.)

consequence of quantitative changes. It is also a quantitative alteration that finally causes the complete division of the ventricles in the reptilian heart and the corresponding improvement of oxygenation of the arterial blood. On the other hand, all cases of rationalization are advantageous and will therefore always be favored by selection.

III. Special Complexity and Rationalization of Nervous Systems

Increased mental development and the approach to the human level were regarded as typical characters of evolutionary progress, especially in the vertebrates, by Charles Darwin (1859). This view was supported by Lartet's (1868) and Marsh's (1874) studies on the brain size of fossil vertebrates. Gaudry (1896) had tried to prove a 'progrès de l'intelligence' and a 'progrès de la sensibilité' in correspondence with phylogenetic levels of brain and sensory organs in vertebrates and invertebrates. As we have seen in Section B of the present chapter, an increased differentiation of nervous structures and a general improvement of the nervous system as a whole are indeed typical characters of anagenesis. Such nervous improvement is met with in the evolution of mollusks, especially of cephalopods, of arthropods from polychaetes, of annelids from more primitive worms, of scolecids from possibly ctenophore-like ancestors, of coelenterates from more primitive, nerveless metazoans, and – more conspicuously – in the evolution of vertebrates. It is significant that in phylogenies which are doubtful as to their anagenetic character – like that of the descent of the Pulmonata from the Prosobranchia – no essential improvement of the nervous system and sensory organs can be traced. Hence we may rightly regard the improvement of nervous and sensory system – working so economically with regard to metabolism – as a typical characteristic of anagenesis. It is for this reason that we once more refer to them in detail to supplement the general treatment of differentiation and rationalization.

Of course, it is difficult to evaluate the special selective value of nervous and sensory improvements, as they are usually accompanied by other progressive features of organization. In vertebrate evolution, however, the selective value must have been great, as the fertility of primeval reptiles was hardly greater than that of contemporary amphibians, and the Mesozoic mammals were hardly more fertile than their reptilian contemporaries. As the newly evolving types were of rather small body size (Cope's Rule; compare Chapter 6, B IV), they can hardly have been superior in body strength. Due to their improved nervous systems and sense organs, especially to the extraordinary refinement of dermal and muscular innervation, they were, however, superior in their greater versatility, and in their greater ability to make use of favorable environmental conditions and to evade adverse factors of environment. These increased faculties were favorable in the competition for nesting sites, feeding grounds, and so on. This is why so many South American pouched mammals and Xenarthra were rapidly wiped out by the advent of North American carnivores, hoofed mammals, and monkeys, after the two American continents were linked in the Later Tertiary. A similar 'brain victory' was gained by the higher placental mammals (Monodelphia) after they had been introduced to Australia, where the more primitive endemic types were rapidly wiped out in many places. It seems probable that in higher animal types the young in general are especially protected, as in these the relative brain weight at birth is higher than in lower animals, and the postnatal development of the brain is relatively slower. In apes and men the relative brain weights at birth will be distinctly higher than in the marsupials, in which the brain grows with less negative allometry after birth.

Improved nervous performance may arise from a purely quantitative increase of nerve cells brought about by one or more cell divisions. Dubois (1930) advanced the interesting hypothesis that vertebrate brain size increased by doubling the number of neurons as a whole, rather than by gradual growth of different brain parts (see p. 144 of this book). With regard to this problem, one must also account for the fact that within a species the size of the neurons will differ in large and small specimens. The same is true in closely related species differing in body size, but belonging to the same level of cephalization (i.e. having an identical number of neurons). Regarding brain and body, the exponential formula $E = k \cdot P^{5/9}$ is applicable, in which E stands for brain weight, k for cephalization coefficient, and P for body weight. When the data used in this kind of calculation refer to the weight of the whole brain, one should also account for the fact that the cerebellum and the medulla oblongata differ quite considerably in various mammalian groups.

Dubois' idea that phyletic increase of brain size, i.e. reaching a higher level of cephalization, comes about by doubling, tripling, etc., the number of neurons of the forebrain, was supported by Brummelkamp (1939 a,b,c) who worked on a large number of rodents, ungulates, amphibians, and fishes. According to him, 'jumps' of cephalization do exist, but they are smaller than might be expected. The cephalization coefficient of the above-mentioned formula increases as a geometrical series with the modulus $\sqrt{2}$, that is to say, $(\sqrt{2})^0 - (\sqrt{2})^1 - (\sqrt{2})^2 - (\sqrt{2})^3$, etc., which is equal to the series $1 - 1{\cdot}42 - 2 - 2{\cdot}83 - 4 - 5{\cdot}67$, etc. (Brummelkamp established this by plotting the logarithms of the brain and body weights on a system of coordinates: the values then fall on approximately straight lines, the gradients of which correspond with the $5/9th$ power [0·56] while the distances between them correspond to the above-mentioned 'jumps' in cephalization.) Jerison (1955), however, calculated an allometric coefficient of 0·67 (163 measurements).

Though Brummelkamp's results are based on a rather large body of material – 32 rodents and 56 ungulates were studied – they must be considered skeptically, because mostly species with rather large differences in body size were chosen and those of intermediate sizes were not studied. Studies of this kind should be extended in the future to forms gradually increasing in body size, including geographic races and borderline cases of race and species. Brummelkamp's results are not supported by the findings of Bok and Van Erp Taalman Kip (1939), who found an equal number of cortical brain cells in homologous sections of the mouse, the rat, and the rabbit, though – according to Brummelkamp – all three species should be classified in different categories of cephalization. Above all, however, as has already been stated, different cerebral areas grow with different and sometimes changing positive or negative allometry during postnatal development (see Harde, 1949, and Chapter 6, B III of this book). In the phylogeny of closely related groups such cerebral growth gradients can also be traced (C. Schulz, 1951; B. Rensch, 1953), and it is evident that they are the result of independent inherited characters. The hereditary basis of each cerebral region is also suggested by the considerable range of variation in the formation of certain brain parts in man. Eminent musicians, for example, will have relatively large temporal lobes, and in great speakers the motor regions of speech tend to be unusually large (see, for example, Kleist, 1934). Such variants probably are hereditary, as the hereditary character of parallel psychic qualities (musical talent, etc.) has been proved. Recently, Sholl (1948) has seriously criticized Dubois' concept of cephalization, and thus we cannot do more than state that, although cephalization probably advances sometimes by large steps, these steps are not reached by doubling or multiplying the total number of neurons of the forebrain.

Dubois thought that increased cephalization was the result of autonomous improvement not primarily involving any selective processes, and Versluys (1939, p. 15) adhered to this view, because the increased number of neurons had not been made use of until the development of late animal types (he referred especially to ancestors of man). On the other hand, both authors agree on the selective advantage of an increased number of neurons. In my opinion, we cannot give any precise estimation of the selectional value, and should restrict our judgment to the general statement that an increased number of neurons represents an essential advantage in selection. Consequently, all mutants with an increase of relative brain size would be favored, and a general phyletic increase of brain size would result. Hence, there is no reason to assume the existence of autonomous factors underlying the process of brain evolution, and this is highly important for the estimation of the evolution of parallel psychic phenomena (see Chapter 10).

IV. Increased Plasticity of Structures and Functions

In my opinion, discussions on the diagnostic characters of anagenesis have given insufficient attention to an important trait common to many of these cases: increased plasticity. Very often higher animals differ from lower ones by the development of organs, structures, and physiological characters which allow a greater variety of reactions, by which the individual specimen is rendered more autonomous and independent of its surroundings, and by which it is enabled to 'choose' the most favorable kind of environment. A few examples will demonstrate this. We shall take them from the vertebrate phylum, as here the fossil record is well-established.

Progressive improvement and increased versatility are evident in the evolution of the metabolic systems. There is a continuous increase in the relative size and versatility of secretion of the salivary glands and of the pancreas in the vertebrate lines of descent. Moreover, the regulation of the amount and kind of secretion by nervous or hormonal stimulation is continuously improved. The highly developed pancreas of mammals will start to secrete its digestive juice a few minutes after the taste buds of the mouth have been stimulated. Pancreas secretion may also be evoked, however, by a hormone (secretin) acting upon the gland via the blood as soon as acid food enters the intestine. The kind of enzyme secreted can be regulated to a certain degree, as the protease will remain inactive when the food consists mainly of carbohydrates and will be activated when flesh is to be digested. The regulating faculty of the liver so essential in the metabolism of glycogen and fat did not develop until the vertebrate phylogeny had reached the fish level, and it is wanting in the Acrania (which are probably similar to their ancestral types) and poorly developed in Agnatha. Above all, it is the mechanisms of homoiothermy that show a highly developed plasticity. The decomposition and resynthesis of glycogen are regulated by different quantities of a hormone; other regulatory mechanisms bring about shivering from the cold and induce hibernation; regulations controlled by nervous mechanisms bring about cooling off by perspiration, accelerated respiration, and increased capillary circulation; also involved are migrations to warmer regions, instinctive avoidance of harsh temperatures by hiding in caves, inherited seasonal molting of hair or feathers, growing longer hairs with decreasing environmental temperatures, and seasonal changes of the subcutaneous layers of insulating fat. The faculty of increasing the number of red blood corpuscles in adaptation to higher altitudes and decreased oxygen supply must also be rated as a mechanism of plasticity.

Some parts of the vertebrate body have acquired a greater and freer mobility in the course of phylogeny. This applies especially to the neck region, which after the reptilian stage could be moved more freely and easily and could thus bring the sensory organs of the head and the mouth in the direction of the sources of environmental stimuli. At the same time the tongue became more versatile and capable of more complicated movements, which improved the chewing process and were an essential prerequisite of vocalization. A similar development was the increased mobility of vertebrate limbs, especially of the hands of Primates, which are movable in all directions.

The improvement of locomotion obviously made the more freely moving organisms superior to the heavily armored previous types, and thus the latter were more or less completely replaced by forms with a lighter type of armament or even by organisms lacking any armor at all. This is exemplified by the extinction of the ostracoderms, so abundant during the Silurian, and of the stegocephalians and large Mesozoic stegosaurs, and by the successive evolution of slightly plated bony fishes, of soft-skinned amphibians and of freely mobile types of lizards and snakes. Correspondingly, the numerous Mesozoic Ammonoidea protected by a rather heavy shell were replaced by the naked Cephalopoda of the Cenozoic, of which about 600 species are known today (compare also Gaudry, 1896; Stromer, 1944).

The general improvement of vertebrate photoreceptors referred to previously, i.e. the improvement of accommodation, adaptation, and similar accessory mechanisms of vision, must also be listed as exemplifying improved regulation. This is especially applicable to the growing complexity of the central nervous systems. More complicated spinal cords and brains enabled their possessors to acquire highly complicated reflexes, instincts, and actions guided by control mechanisms (for example, feed-back mechanisms), by experience or even by insight. Thus, environmental stimuli were not reacted to simply by a rigid inherited behavior pattern, but by responses to each special situation. Actions guided by insight in most Primates, and probably also in some carnivores, may be regarded as the highest type of plasticity. They are brought about by a series of nervous 'experiments' in the brain, trying to discover the most favorable type of reaction to a complex stimulus situation, in which the actual performance is postponed until a favorable 'solution' is reached. This intracerebral 'testing' of possible reactions – a parallel psychic situation: 'Spiel der Motive', competition of motivations – certainly is an extremely advanced type of plastic behavior. High types of regulation are also evidenced by animals actively searching for places of optimum temperatures (compare the experiments by Herter). Almost any complex instinct involves elements of such regulation, which is well demonstrated by the seasonal migrations of birds, fishes, and numerous mammals (reindeer, animals of the steppe, bats). Such migrations certainly are an efficient means of avoiding adverse environmental conditions.

It seems superfluous to compile examples proving increased capability of regulation in the realm of invertebrates. In these, at least, the increased complexity of sensory and nervous systems also meant an improvement of regulation and a more favorable reaction to environmental stimuli. On the other hand, however, one must not forget that some important faculties of regulation have been lost with advancing phylogeny. Examples of such lost regulations include the ability to encyst oneself, shown by many protozoans, cercariae, and the like, the faculty of regeneration so common in amphibians and so radically reduced in reptiles and homoiotherms, and the great modificability of some lower organisms. It seems that most such losses happened by increased specialization when characters suited to many types of environment were replaced by more special adaptations.

A very important character of general regulation, developed even in the lowest types of organisms, is sexuality. It was the process of fertilization that allowed the continuous production of new combinations of genes capable of adaptive radiation in various kinds of habitats. This was emphasized by Weismann (1913, vol. 2, p. 29). Sexual reproduction therefore persisted in all classes and orders, though in numerous cases asexual processes were also possible.

Besides such trends of increasing plasticity, which I had already mentioned in the first German edition of this book (compare also Waddington, 1948; Schmalhausen, 1949) a genetic plasticity also exists, called genetic homeostasis by Lerner (1954). In his inspiring book this author could show that natural selection favored genetic variability and balanced polymorphism to protect the population from changes in the environment, and that heterozygosity also provides a basis for developmental self-regulating properties. At about the same time, J. S. Huxley (1955) compiled a number of examples proving the bearing of polymorphism on evolution.

V. Improvement Permitting Further Improvement

In Chapter 6, C III, I tried to show that natural selection normally causes increasing adaptation, which may nevertheless lead finally to the extinction of a line of descent. Hence, definite evolutionary progress is possible only by a progress that broadens, rather than narrows, the chances of further progress. Simpson (1949) therefore emphasized that 'progress in adaptability' is one of the characteristics of evolutionary progress, and in this context J. S. Huxley (1954) spoke of 'nonrestrictive improvement' permitting further improvement. We could also mention this type of improvement in the previous section, as we are dealing here with a special type of plasticity, that is to say, with growing evolutionary plasticity. It may, however, be better to emphasize this character by a special heading. The line of descent leading to man may illustrate this nonrestrictive improvement if we consider the changes involving the forelimbs, the brain, and homoiothermy.

VI. Increased Independence of the Environment and Increased Autonomy

First H. Spencer and later Plate, Sewertzoff, Beurlen, J. S. Huxley, and Simpson regarded an increased independence of and a better control over the environment as typical characters of evolutionary progress, enabling the progressive types to extend their ranges of distribution and to produce more diversified forms. It is obvious that a superior structural type could displace inferior forms from their habitats and occupy the conquered niches, whereupon a period of specialization was likely to begin (compare Chapter 6, A). As was stressed by Plate, there are numerous exceptions to this statement. Recent apes, for instance, inhabit a smaller range at present than do the lower monkeys, and the number of species is by far greater in the latter than in the former. Correspondingly, the number of holocephalian species is smaller than that of the more primitive selachians.

In many cases, such increased autonomy is the result of improved sensory and nervous systems. In man this autonomy finally led to a control of the factors of environment. Another essential means of increasing the autonomy was the establishment of homoiothermy, by which the higher groups of vertebrates became more or less independent of the environment. Consequently, only homoiothermic animals were able to conquer successfully extreme habitats like the polar regions. General characters of increased autonomy, then, are a growing independence from environmental factors, and an increase of plasticity, of internal, or internally caused, physiological processes.

VII. Regressions

In consequence of the randomness of all mutations, processes of regressive development may be initiated at almost any level of phylogeny. In sessile or parasitic types, especially, secondary reductions of characters typical of higher organization were likely to develop, as with abundant and easily obtained food it was no longer necessary to have highly efficient organs of locomotion, perception, and reaction. The number of biotopes providing such facilities of life is not too large. Nevertheless, in German animals, for example, about 25 percent of all species rely on parasitism (Arndt, 1940), 75 percent of these parasites are insects which are parasites during their larval development only.

Regressive development may also affect single organs like extremities, which originated in the course of a clearly anagenetic process, but, however, had no further value after the mode of life had been altered. Such regressions and their causation can be studied especially well in certain groups of lizards, such as Anguidae, Pygopodidae, and Amphisbaenidae, which originated from types with normally developed legs. Anatomical lines of Recent types (Figure 104) suggest that with increasing length of the trunk the legs could no longer support the body, so that the belly rested on the ground and had to be dragged along in locomotion. Hence, the limbs so successfully developed in the ascent to the amphibian-reptilian level became more and more useless, as this type of organism had to rely on sinuous movements rather than on the tetrapod type of locomotion. The relatively long-legged Agamidae, on the other hand, could keep the belly free from the ground in locomotion, and, consequently, no regressive development is met with in any of their lines.

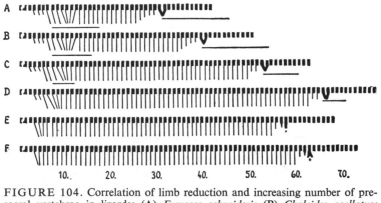

FIGURE 104. Correlation of limb reduction and increasing number of presacral vertebrae in lizards: (A) *Eumeces schneideri;* (B) *Chalcides ocellatus;* (C) *Ophiomorus tridactylus;* (D) *Ophiodes striatus;* (E) *Ophisaurus apus*; (F) *Anguis fragilis*. Relative length of limbs indicated by length of horizontal bars. (After Sewertzoff, 1931.)

From such regressions and from the existence of persistent genera and the persistence of primitive structural types, it is quite evident that there is no general trend of progressive evolution. Anagenesis, then, is only one of the possible types of evolution, and it was caused by characters representing selective advantages of a more general importance.

VIII. Summary

We may summarize our findings by stating that anagenesis is characterized by several phenomena, the most essential of which are increased complexity and rationalization of form and function of organs and structures. The other characters referred to in the previous considerations – improvement of central nervous systems and performances, improvement of adaptability, growing independence of environmental factors, and advancing autonomy – are more or less special cases, consequences of increased complexity and rationalization, though any of these factors may initiate anagenesis. It is especially important that essential factors of anagenesis may arise from strictly quantitative mutations.

As by an increased complexity of a structure a more efficient division of labor becomes possible, this results in a selective advantage. All other characters of evolutionary progress which have been mentioned, such as the increase of rationalization, plasticity, and autonomy are also favored in selection, and they are possible at any evolutionary level. Hence, anagenesis will result automatically and necessarily. This concept was advanced by Charles Darwin, who thought that natural selection must inevitably lead to a gradual improvement in numerous lines. Thus, there is no need to assume a 'drive towards perfection', a 'faculty of self-improvement', or any similar principles postulated by Lamarck, Naegeli, Korschinsky, Dacqué, Woltereck, Beurlen, and others.

[An interesting philosophical aspect opens up if, in the sense of a general causality, the temporal sequence of the phylogeny is regarded as a totally determined process. Then, the evolutionary past as well as the evolutionary future would have to be considered as a 'Gestalt' completely determined in every detail, and besides the *causae efficientes* one could think of *causae formales* in the sense of Aristotle. (This problem is independent of the question whether it is justifiable to introduce the time factor as a fourth dimension to the three spatial ones. As is well known, this is a justified and logically correct method of mathematics, since in metageometrical constructions all four dimensions have equal qualities. This is not so when the reality experienced by us is considered from the viewpoint of the theory of cognition, as then the temporal dimension differs essentially from the three spatial dimensions. The differences of the temporal dimensions consist in the continuously advancing point of the present, in the impossibility of altering the direction of advance, and in the causal connections between elements of a temporal sequence which do not exist in a spatial series: also compare Ziehen, 1934.)]

As at any phylogenetic level anagenesis is only one of several evolutionary possibilities, it is comprehensible that sometimes regressions will occur and that sometimes the anagenetic level will remain unaltered.

Even the initial stages of lines leading to the formation of new classes do not always manifest anagenetic characters, as is exemplified by primitive mollusks, like *Chiton* or *Neopilina*, which certainly cannot be regarded as higher forms than Polychaeta, or the first Devonian Collembola, which probably did not represent a higher type of organization than their worm-like predecessors. All these types were adapted to certain conditions of new habitats, and the anagenetic process proper did not commence until later phases of shaping and improving of the structural type, so that – to refer to our examples – finally the higher cephalopods and winged insect orders came into being.

The chief evolutionary force initiating and maintaining the process of anagenesis was selection. At all times there have been habitats suited not to higher types, but exclusively to primitive, simple, structural types, and it is because of the existence of such habitats that protozoans, lower worms, primitive insects, and similar groups could evade extinction, not being forced to enter into competition with higher animals. There were also large habitats, like the open seas, providing so much food and space that no competition arose in obtaining food and defending territory. Consequently, the most diversified types persisted side by side, as is evidenced by pelagic Coelenterata and Pteropoda, Cephalopoda and Crustacea, Tunicata and fishes.

In this context we have to ask why certain animal groups evolve anagenetically and why others do not. As we have found in Chapter 6, C, natural selection normally effects an increasing adaptation to the environment, so that the phylogenetic development may often end in a blind alley, because the once highly adapted types may be extinguished when the environmental factors become altered. An escape from such culs-de-sac may be possible by a change of function or by occasional phyletic rejuvenations (neomorphosis), but often there is no escape at all. Anagenesis, then, requires a certain nonspecialization in the sense of Cope's 'Law of the Unspecialized', but this nonspecialization may be accompanied by an increase of complexity or of rationalization. That is to say: there must be an improvement of adaptability or at least an improvement not restricting further improvements. This may come about by random mutation at any evolutionary level. If such improvements appear, simultaneously, anagenesis may be initiated.

Seen from an ecological point of view, the problem is quite similar. Normally, selection will cause a growing adaptation to the respective habitat, i.e. a stenoecia will develop. Under certain conditions, however, a general euryoecia may be maintained, which must be considered a 'general improvement', allowing further improvement similar to the nonspecialization of morphological structures.

D. REMARKS ON HUMAN DESCENT

Aristotle recognized many morphological characters common to monkeys and man. Ever since, the taxonomic relationship of monkeys, apes and men has been regarded as a scientific fact, and Linné expressively defended his inclusion of apes – the 'cousins of man', according to his view (H. Schmidt, 1918) – in the genus *Homo*. Goethe searched for the intermaxillary bone in man, because he was convinced that higher mammals were of the same structural type as man. It was Lamarck (1809) who first tried to establish the direct phyletic descent of man from apes, but it was not until 1863 that this idea was proved to be true by T. H. Huxley. See also Vogt (1863), Haeckel (1865, 1874), and C. Darwin (1871). Since then, the human descent has been convincingly illuminated by a large series of fossil material. Though numerous details are still open to further scientific discussion, the major trends of the human descent are quite clear, and we may rightly state that an at least plausible morphologically intermediate stage between apes and man has already been found (levels of Neanderthal man, *Pithecanthropus*, *Australopithecus*, and intermediate stages). For details the reader is referred to such authors as Abel (1931), Weinert and Lechter (1936), Heberer (1940, 1951, 1952), Von Krogh (1943), Gieseler (1943), Broom and Schepers (1946), Weinert (1947), and Le Gros Clark (1950). Some authors (Cope, Osborn, Westenhöfer) have tried to eliminate the apes from the line of human descent by linking the types of Recent man directly to primitive mammals, but these ideas seem to have been influenced by the opinion, so frequently met with, that the simian ancestry must be considered as a degradation of man. However, the fact that apes and monkeys are commonly regarded as caricatures of human beings actually proves the close morphological and biological relations between these two primate groups.

We should rather ask why out of so many higher structural types it was the apes that gave rise to man, and which phylogenetic processes were the essential elements of this last step of anagenesis. As the literature pertaining to this subject is vast (compare J. S. Huxley, 1940; La Barre, 1954), we may restrict our considerations to a few points.

One essential fact of human anagenesis was the lack of specialization in the ancestral apes ('Law of the Unspecialized'). Dentition and digestive system were of the omnivorous type and did not require very special habitats and modes of life, and arboreal life had preserved the 'primitive' prehensile type of hand, which had not undergone a process of special adaptation, but had maintained its versatile mobility. In addition, its nerves were immediately connected with the forebrain by the pyramidal tract. Climbing and grasping in arboreal life caused the erect posture of the trunk which was such an essential prerequisite to erect walking on the ground.

As regards the phylogeny of the human foot, one cannot overrate the importance of the human type of foot in the mountain gorilla (and its similarity to that of the human embryo). It is quite evident from this type of foot (Figure 105) that the prehensile foot of the other apes could be transformed into a strictly running organ quite readily and quickly. The change from an arboreal mode of life to walking on the ground, and using the hands for many performances, such an essential process of human phylogeny, seems to have been necessitated by this special condition. In accordance with Cope's Rule, the human line evidences a steady increase of phyletic body size since the prosimian stage, and it is obvious that larger animals are not and cannot be so successful and versatile in arboreal life as their smaller relatives. This is convincingly demonstrated by smaller monkeys in comparison with the large gorilla, orang-utan, and chimpanzee. Quite similarly, the australopithecine-like ancestors of *Pithecanthropus* were probably forced to stay on the ground by their relatively heavy bodies. Thus, it becomes comprehensible that the legs were 'hominized' earlier than the skull. Life on the ground favored the human character of the anterio-ventral curvature of the caudal end of the trunk axis, which is so typical of the human structural type.

High importance must also be given to the increase of the absolute and relative brain size, to the relative increase in size and the steadily improving division of labor in the forebrain, and to the selective advantage of the freely movable and improving plasticity of the nervous system. We have already referred to the more general characters of progressive evolution common to several animal groups.

The forebrain of apes is distinctly larger than that of monkeys, and the

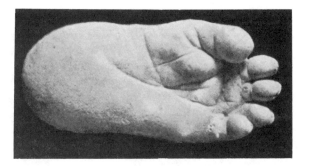

FIGURE 105. The human-like foot of the mountain gorilla (*Gorilla beringeri*). (After Gregory.)

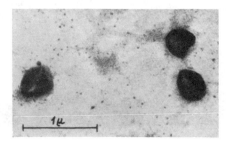

FIGURE 107. *Cystidium gönnertianum,* or bronchopneumonia virus of mice. (After Ruska.)

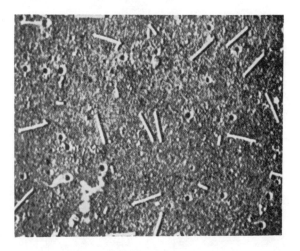

FIGURE 108. Tobacco mosaic virus. Enlarged 27,000 times (After Schramm and Wiedemann, 1951.)

FIGURE 109. Tobacco necrosis virus, showing the regular molecular structure. Enlarged 68,000 times. (After Markham, Smith, and Wyckoff.)

localization in the cortical areas is more pronounced and differentiated; for example, the motor area of the anterior gyrus centralis is strikingly similar to that of man (Von Brücke, 1934). The transformation of Late Tertiary apes into the human type was characterized not only by an increase of the frontal regions of the forebrain (Brummelkamp, 1938), but also, and mainly, by the development of the speech center (Broca's region; Figure 106), which in the simian brain is still lacking or functions in a different manner (regions homologous to Broca's field have been found in *Hapale* by Peden and Von Bonin, 1947). We do not know whether this essential transformation had

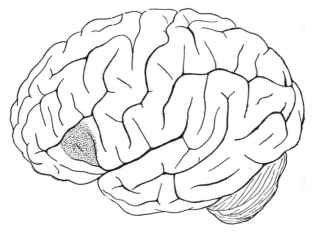

FIGURE 106. Left hemisphere of human brain showing Broca's motor area of speech. (After Rauber and Kopsch.)

already happened at the *Pithecanthropus* level, or whether it did not commence until the Neanderthal stage, nor do we know whether it proceeded gradually or by jumps. At any rate, further inquiries will have to account for the striking fact that during the rise of the *Pithecanthropus* stage to the level of primitive Neanderthal man, covering several hundred thousand years, the cultural progress was very slow (judging from the development of the crude stone and bone tools of *Sinanthropus* to the slightly better chipped tools of early Neanderthal man). This slow development may possibly be due to the fact that no concepts and no abstract meanings or words had yet been 'invented'. In our evaluation of such problems we shall have to confine ourselves to the ancient tools of prehistoric man as the typical characteristics of mental progress; Verworn (1910) has rightly referred to tools as 'prehistoric petrified ideas'.

On the other hand, we should not overlook the fact that Recent apes, especially chimpanzees, are able to perform numerous kinds of actions, involving reasoning, insight, and the use of remarkably complicated tools (such as bicycles). Moreover, they will spontaneously 'invent' the uses of certain objects as 'tools', and have repeatedly given ample proof of their faculty of using instruments (Köhler, 1921; Yerkes, 1945; Hediger, 1952; Hayes, 1952, and others).

The increasing 'fetalization', which has an essential bearing on human descent, also deserves brief mention. Bolk (1926) has convincingly demonstrated that the retardation of ontogenetic development and the retarded onset of sexual maturity in man caused a prolonged juvenile phase, which proved essential because it provided a prolonged and intensified period of learning, during which numerous nervous associations could be established and favorable plastic reactions could be acquired. At the same time, this juvenile phase required the evolution of parental behavior towards the helpless young, and this pattern is, no doubt, the evolutionary basis of human society. According to Bolk, this retardation was caused by a successive decrease of hormone production (especially that of the anterior pituitary), which commenced at the ape level. (One could, of course, assume that there was first a retardation of cell division, followed by a successive decrease of hormone production.)

Such a phylogenetic rejuvenation, then, would correspond to the process of neomorphosis referred to in Chapter 6, E. Retarded sexual maturity and a prolonged growth period involved the reduction of the offspring to one or two young per season, a characteristic already acquired by the monkey ancestors. The retarded growth of body hair in man, corresponding to that of apes at birth, and the relatively small jaws of man may well be regarded as adaptive disadvantages. This was stressed by Versluys (1939), who tried to connect Bolk's hypothesis with Dubois' assumption of the doubling of neurons in the phyletic growth of the forebrain (which has already been criticized in a previous chapter). The disadvantage of retarded growth could be compensated, however, by the plasticity of the juvenile behavior and by the improved performance resulting from the increased size of the forebrain.

This plasticity also applies to the genetic basis. Man has apparently formed only one species with many geographic races (a 'Rassenkreis') ever since the *Pithecanthropus* stage (compare Mayr, 1950), and certain races arose by climatic selection (inherited skin color, more or less pronounced fat deposits in certain parts of the body, large or small body size partly according to Bergmann's Rule; compare B. Rensch, 1935), or by mutation pressure, especially in small populations (random mutation of ear, nose, and hair structure). These races, however, were again and again mixed by migrations, so that at present man is the most polymorphic being on earth (compare Haldane, 1948) and hence it is he who is most gifted in the colonization of all extreme habitats.

It is expecially interesting that many characters essential in the process of human phylogeny (increased brain size, phyletic rejuvenation, and so forth) were caused by strictly quantitative effects of mutation, i.e. by random mutations that could happen at any evolutionary level.

We now come to the basic question whether this highest type of being has originated by the well-known factors of evolution: mutation, selection, isolation, and population effects. Bolk, Versluys, and numerous other authors, who did not make this assumption, postulated directing autonomous factors. In my opinion, however, this skepticism is not justified. In the human line of descent it is quite reasonable to regard the increase of the brain, the development of a motor area of speech, the prolongation of the juvenile phase, the erect posture in walking, and the tool-using hand as essential selective advantages. Human descent, then, was largely necessitated, and is no exception to the general processes of evolution and evolutionary progress which have affected other animal lines. In human phylogeny the typical characteristics of anagenesis are exemplified perfectly well: increased complexity and rationalization, improved plasticity of structures and functions, independence of the surroundings, and autonomy. After all, it was a brain victory that brought man into the ruling position of the world, as it was a brain victory when the creodonts were defeated by the carnivores and the prosimians by the monkeys and apes.

Will man, this relatively young product of evolution – the Recent races are not older than about 30,000–40,000 years, equalling 1,000–3,000 generations – continue his evolution along the lines followed so far? Will the brain proceed to differentiate? Can problems like these be discussed without leaving the firm ground of scientific knowledge? It is certain that predictions pertaining to such problems can never be free from doubt, and we shall restrict ourselves to a few statements.

Natural selection, such an essential factor in the process of anagenesis, has been increasingly restricted ever since human cultures developed, as houses, clothes, improved food, medical care, and similar factors made man almost completely independent of the environment and its selective effects. This statement does not deny that even at present the weaker and more sensitive persons are more easily affected or extinguished by diseases than the more resistant individuals, and that weak and mentally inferior peoples will more readily be wiped out than vigorous and superior peoples. It is true that a certain sexual selection is still at work as crippled or ugly persons will not find a marriage partner so readily as those with an intact body and a handsome appearance ('ugly' and 'handsome' denoting qualities not corresponding to the innate releasing pattern; also compare Lorenz, 1943). On the whole, however, the effects of natural selection have been strongly reduced, and in times of war a counterselection will reduce in the male sex the number of individuals with valuable characteristics, such as courage, strength, and the like (for a mathematical treatment of counterselection, see Von Hofsten, 1951). In addition to this sad fact, persons of higher cultural levels will often limit their offspring. The reproduction rate of the races varies quite radically, and it is an important mathematical consideration to estimate the racial components of mankind in about 100, 200, or 500 years from now.

One should notice, however, that man has reached a unique evolutionary position in the realm of organisms. Innate behavior patterns, so essential as a guidance in animal life, have been strongly reduced. The chasing instinct has often been overcome by the aversion to hurting, torturing, or killing animals, and the hoarding instinct has been reduced to the 'vacuum activity' of collecting stamps, coins, copper engravings, and the like. Only the instinctive patterns of reproduction, especially with regard to choosing a 'handsome' partner, seem to have remained unaltered. Even the number of favorable and purposeful innate movements has become very small, as pointed out by Ziehen (1924, p. 527).

In consequence of the enormous complexity of his brain, man has acquired a fundamentally new evolutionary faculty. He is capable of reasoning in a much more effective manner than any animal and can perform 'free' actions, that is to say, in a given situation he can choose the most favorable type of motor reaction, as many possible reactions are tested in the brain by a complex process of nervous facilitations and inhibitions, until the most suitable type of reaction has been discovered. (The physiological side of this process may be traced by recording action potentials from the organs before they are actually moved.) Compared to apes, man has gained the essential advantage of possessing the motor area of speech (Broca's region) by which he can label visual and auditory perceptions, and also abstractions with spoken (or speakable) words. The evolution of speech and of a speech center in the forebrain – possibly happening at the *Pithecanthropus* stage – rendered possible the rise of a quite new type of general concepts (ideas), because similar sensations or ideas were associated with the same acoustic and motor components of speech. Hence, man began to think in words rather than in unnamed images (as in dreaming), and this is why he was enabled to do more rapid, more abstract, and more complex thinking and to advance complicated concepts, ideas, and speculations of a typically human character. It is quite reasonable to attribute the essential part of this evolutionary process to natural selection, as by concepts and reasoning man could plan for the future, prepare himself for success, and avoid disaster (also compare Ziehen, 1924).

One more simian character proved essential in human anagenesis: social life. Man, too, is a 'zóon politikón'. Aided by his speech, he could convey not only excitations, moods, and general nervous dispositions (as most animals can), but also his experiences to other individuals of his group. These, then, could prepare themselves for coming events and meet such events more successfully than if they had been unprepared. Moreover, they could produce further and more complicated abstractions. Thus, speech became the basis of tradition, which increased from generation to generation when the young were educated during the prolonged plastic juvenile phase. This process was fundamentally different from any animal evolution, as it did not require new mutations and selections. Human cultures could not develop until 'biological heredity' was aided by this type of 'social heredity' (Dobzhansky). Then, the division of labor could be handled according to a new scheme, as with the ever-increasing complexity of social communities the leader and follower, the inventor and imitator, the white-collar worker and the man in the factory could mutually assist each other, so that a new type of cooperation arose.

Highly advanced cultures, however, could not develop until planned education of children was introduced or until writing had been invented. The development of writing and, later, of printing had an essential bearing on human culture, as books could store a far greater number of experiences than could be remembered by an individual. Books represented 'extracerebral chains of associations', and the individual could at any time connect his cerebral chains of associations to the ever-growing and eventually extremely complicated knowledge stored in books. Thus, aided by the experience of generations communicated to him by books, the individual is able to achieve superindividual effects. It is by these extracerebral chains of associations that we can perform processes of thinking which otherwise could not be mastered by the mental capacity of the individual. This written and printed tradition provided a basis for the universal culture and the world-embracing knowledge common to mankind, and the union of human beings all around the globe by this common knowledge is stronger than any bond between individual animals could be. As this universal knowledge has grown, the individual human being has become more and more dominated by superindividual tradition. Customs, laws, ideas, religions, methods of administration, offices, periodicals, and other human institutions are more persistent in time than the individual human being involved in them.

On the other hand, the directing force of environmental factors guiding animal evolution was replaced by human freedom arising from the non-inheritable accumulation of mental abilities, from education and literature. We cannot discuss here the problem of whether this freedom is absolute or only apparent, because this involves the enormous complications of determined processes of association (also compare Seidenberg, 1950).

At any rate, this mental 'freedom' permitted man to give life an ultimate 'sense', although the attainment of this sense could be only a human goal. An 'absolute' sense of life *a priori* cannot be confirmed by biology. To human beings of some ten thousand years ago life was certainly a given fact, as it is to any animal. All aims and purposes of life are imagined and established by man, by man above a certain cultural and mental level. This is also evidenced by the fact that primitive races of man lacking an advanced cultural level do not bother at all about an *a priori* sense of life.

However, man should always be aware of the fact that a high level of mental capacities may easily prove unstable, as it is not maintained by inherited characters. Suppression or extinction of the material basis of this human level, of literature and other means of mental culture, may mean a total interruption of tradition and a reduction to the level of primitive beings. There have been some examples indicating the possibility of such regressions during the last few decades. Man's task, then, will be to guide his own phylogeny in future (compare J. S. Huxley, 1940) and to prevent phyletic death arising from – as it has in many lines of animals – an excessively growing organ – in this case, the human brain. Future human evolution will also depend largely on the dangers and still more on the blessings of the development of technological sciences. This development will become all the more necessary, as the more rationalized and improved inventions will replace the less efficient implements, and thus technical progress is made an inevitable and essential element of future human evolution.

8

The Evolution of Life

In the last chapters I have tried to show that no autonomous evolutionary forces need be assumed for a reasonable explanation of evolutionary progress and human descent. We shall have to examine the question of whether the very origin of phylogeny, the origin of life in its primordial form, is also comprehensible by the principles of evolution which have been mentioned, or whether special forces must be assumed for a proper interpretation of this process. Can a gradual transformation of inorganic matter into living substance possibly have occurred at all, and is it possible to discuss this question, which is also very important from a philosophical point of view? We are far from having solved this problem, although hopeful approaches have been opened up in very recent times. However, numerous hypothetical ideas are involved in the following considerations, and hence the discussion will be restricted to a few essential problems. For a historical review of former hypotheses on spontaneous generation, the reader is referred to Pasteur's classical study (1862) and to the works by H. Schmidt (1918) and Lippmann (1933).

Bearing in mind that a consideration of the problem of the origin of life must be highly speculative, one might be inclined to answer the question by a simple 'Ignoramus'. In 1929 the brilliant physiologist Jordan referred to the hypothesis of the origin of life as 'the most absurd hypothesis ever advanced', because he thought that the production of amino acids and proteins from inorganic material would be practically impossible and because life 'is not the quality of a homogeneous substance, but the performance of a system'. During the last three decades, however, rickettsias, viruses, and bacteriophages have been studied intensively, and we cannot deny that these organisms and organism-like beings may be arranged in progressing series representing a model of the possible evolution of living organisms from nonliving material.

The classical statement 'omne vivum e vivo' has not been been disproved so far, and no living organism that did not originate from living ancestors has as yet become known. Hence, we are justified in asking the question from which kind of living substance the lowest protists, such as the smallest and simplest bacteria, may have originated.

When I tentatively wrote this chapter for the first German edition of the book in 1945 and 1946, I felt that I was engaging in a questionable enterprise. But during the last few years quite a number of similar publications have appeared.

One could evade this question by assuming that the formation of primordial bacteria occurred on celestial bodies which cannot be properly judged as to their environmental conditions. Such theories were first advanced by the physician Richter (1865), followed by the physicist Thomson (Lord Kelvin), and by Helmholtz. Finally, the astronomer Arrhenius formulated these ideas more precisely, rendering them as a panspermia hypothesis (also compare H. Schmidt, 1918; Lundmark, 1930), which holds that bacterial spores may be transported from one celestial body to another by cosmic particles or by radiation pressure. (Transport by radiation pressure would take twenty days from Mars to the earth, eighty days from Jupiter, and 9,000 years from Alpha Centauri.) The low interstellar temperatures would be survived by the spores, but the extremely dry conditions of space and the strong short-wave radiation would be very dangerous. Even if the latter difficulty did not exist, the panspermia hypothesis would not solve the problem and the quest for protobacterial beings and their origin would remain necessary, so that we may as well discuss it from a terrestrial point of view.

B. CHARACTERISTICS OF LIFE

If we wish to trace the phylogeny of life, we shall first have to enumerate the characters distinguishing living organisms from nonliving material. Generally the following characters are named (more or less following Roux, 1915): metabolism (especially assimilation), growth, development, reproduction, excitability, and faculties of regeneration, restitution, or general regulation (compare, for example, Sierp, 1939). Besides these characters regarded as typical of life, Driesch (1909, 1932) postulated adaptability, instinct, and intelligence as further vital symptoms. We may rightly omit, however, the last two characters mentioned, as instinct and intelligence cannot be traced with certainty in the lowest animals nor at all in plants. As very little is known of bacterial regeneration and restitution, we may also safely exclude these characters from our consideration. The same applies to the faculty of regulation, unless this term is meant to denote the general process of adaptation of an organism to a particular environment, but this would be covered by the term 'adaptability'. The remainder of the above-mentioned symptoms may be grouped together under the headings of metabolism and energy transformations, reproduction, excitability, and change of form (compare also Hartmann, 1927).

It has often been emphasized that these characters are not exclusively symptomatic of life, and O. Lehmann (1906, 1907) proved that the so-called 'liquid crystals' (e.g. of paraazooxycinnamicethylester) will also show growth (by intussusception), change of form (lengthening and coiling), 'reproduction' (autonomous division into smaller parts when a certain size is reached), and autonomous regeneration of destroyed parts. Johnsen (1930) also demonstrated restitution and reproduction faculties in ordinary crystals, and the phenomenon of 'excitability' (in the form of heliotropism and galvano-tropism) is also displayed by iron chloride in a solution of water glass. Even the faculty of selective adaptation is not restricted to living organisms, as crystals tend to grow more intensely with their pyroelectrically analogous 'weaker' pole, which is more easily affected by solvents. Hence, there is hardly any character exclusively reserved to living organisms, and life cannot be characterized by single criteria, but must be regarded as the performance of a system, as has been stressed by H. J. Jordan, Von Bertalanffy, and others. One should note, however, that this system must be made up of elements containing mainly proteins and nucleoproteins. In my opinion, then, one should give a more precise rendering of the term 'living system', by defining it thus: Living organisms are individualized systems containing proteins (especially nucleoproteins) and capable of metabolism, energy transformation, reproduction, excitability, and change of form, maintaining a specific constancy of the systems through very many generations in spite of the alterations affected by the above-mentioned processes. This definition would be applicable to the 'classical' organisms, i.e. animals and plants with cellular organization, including the simplest types of bacteria and spirochetes. (The character of cellular organization might also have been added to the above definition, but I have rejected it because this concept is difficult to define. We know for instance, that viruses and phages may have a membrane.)

Tracing the above-mentioned characters in the evolutionary line leading from the simplest bacteria to even simpler protobacterial types and to non-living matter, one might expect that the single characters mentioned in our definition would become indistinct and difficult to delimit, and that the number of such characters would decrease with the approach to the lower levels. We do not actually know any line of ascent of this type, but there are numerous organisms and 'beings' resembling organisms that are characterized by a decreased number of the 'vital' traits mentioned in the above definition. These primitive organisms and 'beings' may even be arranged to form a sort of model series linking nonliving matter with living substance and representing a possible evolutionary line of life. Hence, we can no longer evade the problem of the origin of life or do away with it by a simple 'Ignoramus'.

C. SERIES OF ORGANISMS WITH SUCCESSIVE REDUCTION OF VITAL CHARACTERS

As is well known, bacteria and Cyanophyceae are distinguished from the remainder of protists by the lack of a nucleus. In bacteria, however, distinct small particles are found resembling a nucleus, and these nucleoids will divide in two before each division of the cell. The nucleoids contain not normal chromosomes but thymonucleic acid and, as many mutations of bacteria have been reported, one may assume that genes are contained in the nucleoids. In this context it is important to remember that essential modifications of virulent bacteria can be produced by having the respective strain pass an organism which is quite different from its normal host type. Such modifications will persist through several passages and generations. It is difficult, however, to judge whether the 'modification' is genuine or whether a certain strain of different quality has survived the passage through the foreign host in a process of selection. In such processes, plasmatic genes may also be subject to selection. All bacteria have a cellular membrane. The usual size of bacteria ranges from 1 to 10 μ, but there are considerably smaller forms, too, like *Micrococcus prodigiosus* of about 0·5 μ.

There are numerous organism-like beings quite diverse in structure, simpler than bacteria, and referred to as viruses, rickettsias, anaplasmids, and the like. Some of these beings may rightly be regarded as genuine organisms, even though some typical characters of life are lacking in them.

One step down from the level of bacteria is that of rickettsias (compare, for example, Gildemeister, Haagen, and Waldmann, 1939; Holmes, 1948; Ruska, 1950; Schramm, 1952; Friedrich-Freksa, 1954). They are known to produce spotted fever and similar diseases. Their spherical or rod-shaped body is covered by a cellular membrane. Their chemical components are of complexity similar to those of bacteria. Their reproduction is a division. Unlike bacteria, they can be cultured only on live substrata. Apparently, then, they lack an agent necessary for growth, which might be an enzyme produced only by living tissues. Possibly, their organization has become simplified secondarily by their parasitic way of life, and certain agents required in the process of reproduction may have been supplied by the host organism. We seem justified, then, in regarding rickettsias as living organisms, closely related to bacteria.

The same statement seems to apply to several forms of similar size (0·2–0·7 μ) grouped under the common heading of 'cysticetes' by Ruska and Poppe (1947) (also compare Ruska, 1950). The shape of these bodies is quite diverse, often like vesicles with longitudinal folds or appendages (Figure 107). There is no genuine cell membrane, but an external 'cover' of varying consistency. The body contains carbohydrates, lipoids, enzymes, and – of course – nucleoproteins. Reproduction processes range from simple division and the formation of gonidia to multiple division of 'inclusion bodies'. The group of cysticetes includes the infectious agents of bronchopneumonia in mice, of pleuropneumonia in cattle, and of psittacosis, and also the organisms isolated from compost soil by Seiffert, and from sewage water by Laidlaw and Elford. The latter types, as well as the producers of bovine pleuropneumonia, can be grown on nonliving media.

There is a third intermediate group between bacteria and viruses, which is commonly referred to as 'Anaplasmidae'. This group comprises the causal agents of *Bartonella* diseases of men and animals (*Bartonella*, *Anaplasma*, and the like). *Bartonella bacilliformis* will enter the red blood corpuscles and form minute bodies shaped like rods, dumbbells, or balls, but it will not generate any spores (compare Kikuth, in Gildemeister, Haagen, and Waldmann, 1939). This species of *Bartonella* will also grow on nonliving culture media.

Rickettsias, cysticetes, and anaplasmids may be regarded as representing real organisms, though these are of a simpler type than bacteria. This is justified by their chemical complexity, requiring the assimilation of special components, by their type of reproduction, and by their change of form. This statement cannot safely be applied to lower forms, which have been described as 'organized types of viruses' by Ruska (1950). These pathogenic viruses contain nucleoproteins, fats, and enzymes, and are of different size. The virus of cowpox, for example, is about 210×260 mμ and contains fat, biotin, catalase, phosphatase, and polynucleotidase, and its cube-shaped body is covered by a sheath resistant to pepsin. The influenza virus consists of spherical and filiform bodies of about 80–100 mμ and contains fat, besides the other constituents mentioned above. Fat and nucleoprotein are also found in the spherical virus causing equine encephalomyelitis. This virus is below the limit of visibility in the light microscope (size: 40–50 mμ). All these types, including the chicken plague virus (100 mμ), the mumps virus (180 mμ), and others, can be grown on living tissues only. The reproduction of these types is hardly known. Hence, these 'organized viruses' share only a few characters with the living organisms in the proper sense: their contents of nucleoprotein, their specific constancy of shape and pathogenic effects, a certain 'assimilation' of agents and compounds required to build up the body, and hence a certain change of form.

Bacteriophages, 'parasites' on bacteria, are small spherical bodies of about 12–40 mμ in diameter or rod-like particles of about 140 mμ in length, and should be placed on a similar level of organization. Sometimes the rods have a spherical appendage of about 35 mμ on one end, which is chemically different from the rest of the body. Head and appendage are covered by a common membrane. Recent studies have revealed that the phages enter the bacterial cells only by their nucleoprotein component. The formation of new phages inside the bacterial cell is brought about by multiplication connected with a cyclic change of form (compare Penso, 1955). If one bacterium is entered by two 'species' of phages, new forms of phages may arise combining the characters of both 'ancestral' types (see Delbrück and Balley, 1947; Hershey, 1947). Similarly, one may assume that in some cases other viruses do not simply reproduce themselves identically, but cause the host cell to form virus particles. But it is not likely that all smaller viruses may be looked at as 'deteriorated parts of cells' only, since after the infection of different hosts the virus normally remains constant.

Finally, not all 'macromolecular viruses' can be rated as living substances, because they consist of only one type of nucleoprotein and may be crystallized without losing their faculty of reproduction. Viruses of this type include the famous tobacco mosaic virus (Figure 108) discovered and first crystallized by W. M. Stanley (1935), the tobacco necrosis virus, the potato-X-virus, the virus of foot-and-mouth disease, that of poliomyelitis, and many others. These viruses consist of spherical bodies of 19–26 mμ in diameter, which may join to form octahedra, cubes (Figure 109) or dodecahedral accumulations (tomato bushy stunt). Other types of these viruses, such as the tobacco mosaic virus (15×280 mμ), and the potato-X-virus (15×430 mμ), will form rod-shaped bodies with trilateral symmetry or very narrow spirals. The smallest macromolecular virus probably is that of foot-and-mouth disease, which consists of spherical particles about 12 mμ in diameter.

A thorough analysis of the tobacco mosaic virus by Schramm (1943) suggested that this virus consists of a protein 'skeleton' made up of serial subunits arranged like coins one after the other (possibly slightly spiraled). Each subunit is combined with one molecule of nucleic acid. The molecular weight was calculated as 46,000,000.

It is highly important that these macromolecular viruses are also subject to mutation. Such mutations, traceable in strains derived from a single individual, have, for instance, been observed in the tobacco mosaic virus by Schramm (1948). After the mutation a certain part of the nucleic acid seems to be more firmly tied to the protein, so that at pH$=6$ these parts will not be able to dissociate.

From the above findings we may infer that these macromolecular viruses consist of nucleoproteins which may even be crystallized, and that they can hardly be referred to as beings endowed with 'life' in its normal sense. We should not forget, however, that these viruses display four essential characters which are typical of living beings: (1) their chemical structure, with prevailing protein and nucleoacid; (2) their faculty of identical reproduction (though possibly not always direct), permitting a type constancy through long chains of 'generations'; (3) their individual cycle (compare especially Penso, 1955); and (4) their mutability. One vital character is still lacking, however: energy metabolism. Considered strictly morphologically and physiologically – apart from any phylogenetic point of view – we may refer to all the above-mentioned groups of organisms and organism-like beings as a graded series with successively decreasing characters of life, ranging from the size of bacteria to that of large protein molecules: (1) bacteria; (2) rickettsias, anaplasmids, cysticetes; (3) organized viruses and bacteriophages consisting of various chemical compounds; (4) macromolecular viruses. It is within this series that we have to look for the borderline between living and nonliving substances, and whether we want to incorporate only bacteria, rickettsias, and cysticetes, or to include organized viruses or even all types of viruses in the category of living beings depends on our definition of 'life'. As the definition of life is of high importance to our philosophical view of the universe, the lack of a fixed borderline between living and nonliving substances will have far-reaching consequences. It should once again be stressed, however, that the above-mentioned series cannot be claimed to represent the actual phylogeny of life. It is only a model intended to aid in a discussion of the origin of life. We do not know whether the most primitive proto-organisms similar to macromolecular viruses still exist or whether they existed only during the initial phases of terrestrial life, after which they were wiped out because they could not exist beside the bacteria except in the form of their parasites. However, the various levels of organization represented by the various protoorganisms and organisms mentioned above provide a possible clue to understanding the origin of life as a series of successive steps, the first and most important of which was the formation of nucleoproteins capable of reproduction.

It is of great importance that macromolecular viruses are very similar to genes, as they are nothing more than nucleoproteins capable of identical reproduction. Single genes, especially the cytoplasmatic genes, certainly are not organisms in themselves, but they are the most essential elements of an organism, because their capability of identical reproduction warrants the constancy of successive generations and their mutability is the starting point of evolution.

Almost nothing is known as yet of the reproductive processes of macromolecular viruses and genes. We must assume that reproduction is brought about by identical reduplication. Some authors have developed a hypothesis to explain this process. As the specific structure of a protein molecule is characterized by the pattern of positive charges of their hexone bases, Friedrich-Freksa (1940) supposed that the negatively charged parts of nucleic acid would be tied to this positive pattern. The negative parts, in turn, would again attract and tie positively charged protein molecules or parts of them, and thus reproduce the virus. In contrast to this view, P. Jordan (1944) thought that identical reduplication was caused by resonance attraction, i.e. by mutual attraction and interaction of thermically rotating molecular groups, a concept advanced by quantum physics (compare also Von Bertalanffy, 1944; Dehlinger and Wertz, 1942). We cannot yet decide whether such a reduplication really does occur in viruses and genes. As stated above, larger viruses and phages do not reproduce in this way, but in the process of reproduction the nucleic acid component of the virus is again the essential part. At any rate, there is no reason to believe that the reproduction of viruses and genes should be a special 'vital' process fundamentally differing from all other processes of microphysics. Hence, it is evident that one need not assume unknown autonomous evolutionary forces as the causes of the successive evolution of the characters typical of life.

These statements are not meant to imply that we know the origin of life. The study of viruses has only shown that a model series may be arranged bridging the gap between organisms proper and nonliving molecules by various levels of organization and vital characters.

D. THE EVOLUTION OF NUCLEOPROTEINS

Quite different and even more difficult is the problem of how proteins capable of reproduction, especially nucleoproteins, may have originated under terrestrial conditions. The formation of primordial peptides, their accumulation to form proteins, the origin of purine and pyrimidine bases and of pentoses, and the formation of nucleic acid, which was to 'meet' proteins so that the vital nucleoproteins arose: our knowledge concerning all these problems and many more of this kind has not yet passed a highly speculative phase. This is why we shall refer to them only briefly.

315

The principal purpose of the following consideration is to show that it is not too absurd to try to solve the above-mentioned problems. Pflüger (1875), Oparin (1936), Beutner (1938), Bernal (1951), and S. L. Miller (1953) and others have advanced reasonable theories on these fundamental problems. According to these authors, carbon compounds may have reacted with water vapor to form various types of alcohol, sugar, and organic acids. These, in turn, would react with ammonia which had been formed simultaneously, and amino acids could arise. The polymerization and condensation of carbon compounds containing nitrogen was probably favored by the intense ultraviolet radiation strongly penetrating to the terrestrial surface because of the lack of an ozone layer. More complex organic compounds may have originated before the existence of terrestrial life under the influence of numerous electric discharges. This possibility has been proved by the very important experiments of S. L. Miller (1953), who obtained amino acids like glutamic acid and alanine (and probably also asparagin acid and amino butyric acid) by electrical discharges in a mixture of methane, ammonia, hydrogen gas, and water vapor (types of gas which are supposed to have existed in the primordial atmosphere of the earth). Such amino acids were very stable compounds, lasting probably for millions of years, as they have recently been isolated from Devonian fossils (compare Blum, 1955, p. 178 C). According to Bernal, the first protein molecules may have originated from an adsorption of amino acids to clay or quartz molecules. Sugar, amino acids, and the simplest proteins would persist in these early days because there were as yet no microorganisms. In spite of all these quite reasonable theories, the real causes underlying the formation of the highly complex nucleoproteins are still unknown.

Even more important is the problem of how nucleoproteins capable of reproduction could originate. Perhaps the formation of adenosine triphosphate was decisive in the origin of the first systems capable of self-reproduction (Haldane, 1954). P. Jordan (1945) suggested that among the proteins developed in the above-mentioned way a few molecules might – by chance – have possessed the faculty of reproduction. Because of this important ability this type of protein would soon be dominant. According to this concept, the initial stage in the evolution of life would have a remarkable resemblance to the phenomenon of mutation, because both are the result of a single process. This assumption implies the idea of mutation and selection, as the nucleoproteins could only persist and reproduce themselves in those 'habitats' in which suitable and sufficient material could be obtained. Hence, an adaptation to certain 'habitats' would develop, and the differences of environmental conditions would cause the differentiation of various 'species'. The basic factors of all evolutionary processes may thus have been the essential causes in the early development of the prestages of life. Let us hope that the highly speculative character of the above considerations on the origin of life may soon be replaced by positive knowledge in the corresponding borderline fields of astronomy, geology, geochemistry, and biology.

9

Autogenesis, Ectogenesis, and Bionomogenesis

In the preceding chapters of this book we have repeatedly considered the question of whether mutation, selection, fluctuation of populations, and isolation – the essential factors of race formation and speciation – would also provide a sufficient explanation of transspecific evolution, or whether special autonomous forces should be assumed guiding the major trends of evolution. So far, we have not had to assume such special autonomous factors, either in the processes of normal phylogenetic ramification (kladogenesis), the special processes of orthogenesis, parallelisms, 'harmonious' alterations of organ systems, and of structural types of phylogenetic 'cycles', or in progressive evolution (anagenesis), including the descent of man and the origin of life. This negative statement cannot, of course, prove the nonexistence of autonomous evolutionary forces. There is no reason at all, however, to assume the existence of obscure and unknown factors as long as other factors proved by scientific analysis provide a sufficient explanation of the phenomena in question. This attitude cannot be altered by the fact that we do not know the causes of spontaneous mutation and that we are far from a sufficient knowledge of the complex 'systemic relations' involved in major alterations of structural types, because all such analyses done so far have revealed that mutation, selection, and isolation were the only causal processes. In our opinion, then, evolution was not a process of autogenesis, but rather of ectogenesis, as the path of evolution was continuously guided by environmental factors. Until now, only spontaneous mutation might be referred to as an autogenetic element of the evolutionary process. Plate (1913), who introduced the above-mentioned terms, referred to autogenesis as the vitalistic, and to ectogenesis as the 'mechanistic' (we might better say: causalistic) point of view.

To many people such an ectogenetic aspect of evolution may seem unsatisfactory, as it implies the idea that the wonderfully purposeful structures of animals and plants, the evolution and descent of man, and perhaps also the origin of life, are nothing but the result of random mutation and equally 'randomized' selection and isolation. This evolutionary aspect is likely to collide with philosophical and, perhaps, religious ideas and maxims. It must be stressed, however, that in the realm of science an aversion to a scientifically founded opinion should be based on strictly scientific arguments only. We shall therefore evaluate the meaning of 'purposeful' characters and the 'randomness' of evolution from this point of view, and then we shall refer to causality and its bearing on aspects of evolution. Some other problems of natural philosophy connected with the science of evolution will be dealt with in the last two chapters of this book.

It was, above all, the surprisingly complete usefulness of nearly all characters, organs, and structures that gave rise to grave doubts concerning random mutation and selection as the sole causal factors (compare, for example, the recent publications by Cuénot, 1941; Lillie, 1945; Vandel, 1949). In the present book I have tried to dispel such doubts by deliberately demonstrating that even extremely complex structures and functions can successfully be approached by causal analysis. Besides this, one should remember that the numerous examples of 'purposeful' adaptations are simply a matter of course and not surprising at all because it would be absurd and incomprehensible if unfavorable and disadvantageous characters had persisted in large numbers (this was recognized by Empedocles). With respect to the continuous processes of selection, it is much more surprising that some unsuitable, disadvantageous organs, 'luxuriant' structures, and unnecessary vestigial structures still exist.

In addition, it is to the terms 'randomness' and 'accidental processes' that unjustifiable objections and opposition are often raised. 'Randomness' of mutation and selection in its proper meaning need not imply a negative judgment in the sense of 'not according to terrestrial regularity', but means only that we do not know the complicated interrelations of causal factors underlying so many phenomena in the world of organisms. Numerous physiological processes have not yet been causally analyzed because of the systemic coordination of a great many single events, and yet it is certain that such processes are causally determined, as they can be precisely predicted and circumscribed by 'balance formulas'. Von Bertalanffy (1932, 1942) has tried to give a mathematical expression of such processes in his inspiring *Theoretische Biologie*. The evolutionary processes, especially the directedness of many processes, dealt with in the present book can also be interpreted as causally determined. It is the task of natural science to search for causal factors, a statement especially emphasized by Hartmann, (1927). Consequently, the vitalists have not been able to prove a special vital regularity or lawfulness as a category of its own, though they have invented several terms to the not-yet-analyzable residue (vital forces, entelechy, psychoids). It is significant for the vagueness of these terms and ideas that the field covered by these alleged autonomous vital forces becomes smaller and smaller with advancing and improving causal analysis (compare especially developmental physiology). On the other hand, the vitalists (especially Driesch) must be credited with having emphasized again and again the systemic correlations and effects playing such an essential role in the processes of life. The prevalence of such systemic correlations, however, is a consequence of the vast complexity already shown by each protein molecule, and not a typical symptom of life. 'Systemic regularity' ('Gefügegesetzlichkeit') may be seen also in inorganic compounds, as exemplified by the completely new qualities of sodium chloride (NaCl), so different from those of metallic sodium and of poisonous chlorine gas.

There are, however, two strong arguments which seem to contradict a causalistic interpretation of the evolutionary processes and which should not be overlooked. One of these objections refers to the fact that possibly all living beings are characterized by phenomena of consciousness, which we have not mentioned in our previous deliberations. We know these phenomena with certainty only from our own experience, although – by analogy – we may attribute them to higher animals and, because of the proposed uninterrupted phylogenetic lines, to lower animals, too. Consequently, all organisms may be credited with possessing some sort of consciousness. Consciousness is based on processes known as parallelisms accompanying certain physiological phenomena of nerves and sensory organs, and no causal interference of these parallel processes with the functions of nerves and sense organs has been proved so far. The parallel processes are referred to as 'epiphenomena' and governed by laws of their own (the connection of certain parallel processes with certain physiological processes). As the present chapter is devoted to a causalistic interpretation of evolution, I shall postpone the discussion on parallel processes, but in the final chapter of this book I shall refer to the parallel evolution of the phenomena of consciousness in more detail.

The second argument against a strictly causalistic aspect of evolution derives from the fact that in the realm of natural sciences doubt has recently been cast upon the general validity of the laws of causality. As is well known microphysical processes cannot be predicted according to the laws of causality, and it is from this point of view that Bohr and P. Jordan derived the conclusion that noncausal processes of microphysical dimensions might be amplified to macrophysical effects in the living organisms. This view is partly based on Heisenberg's uncertainty principle, which holds that microphysical processes cannot be predicted because the instruments for measuring the details of minute processes will invariably cause a disturbance, so that the act of measurement and observation cannot be kept apart from the process proper without causing serious disturbing reactions. Heisenberg's equation shows that the product of observational errors inherent in the microphysical data, such as place and speed of an electron, will never be smaller than Planck's constant. Hence, it is obvious that quantum mechanics is based on statistics, and consequently most microphysicists conclude that there is no sense in searching for the cause of such micro-processes, because one cannot even observe or control all factors of the process. Such a noncausalistic point of view, which can be comprehended only in the language of mathematics, led Bohr and P. Jordan to apply it to the phenomena of life and to postulate an 'interior freedom of life' ('innere Freiheit des Lebendigen'). Jordan, in particular (1932, 1945) expounded the view that in spite of all analyses there might remain a certain part of unconquerable territory – which would mean that also in the realm of biology a complementarity might exist of which one would not doubt that it is fundamentally connected with the complementarity of quantum physics. . . .

He advanced a hypothesis according to which the effects of microphysical control mechanisms can be amplified into macrophysical processes. This amplifying principle applies – according to him – to the effects of genes in the physiology of development, and to the induction of a gene mutation by a single ionization due to a 'hit' by a hard radiation. He also gives examples of the sterilization of bacteria by a few quanta of light hitting a certain control system of minute dimensions and causing death, though these bacteria may have been exposed to millions of quanta before the lethal hit occurred. It seems possible to him that amplifying principles of this kind underlie the processes of nerve actions.

These hypotheses, ingenious and inspiring as they may be, have not been accepted unanimously, and Bünning (1935, 1943) and Zilsel (1935), especially, have pointed out the fact that most important phenomena of life – cellular respiration and metabolism, effects of stimuli, etc. – have been analyzed on the basis of macrophysics rather than by the concept of microphysical indeterminism. Life as such is most probably a process of macrophysics in which the quantum-mechanical indeterminism is more or less balanced and does not manifest itself. All biological structures of molecular dimensions known so far are found in large numbers in all cells, so that their effects are a problem not of microphysical uniqueness but rather of balanced statistics. The same holds good for all stimulating agents, and the effects of gene actions also can very well be understood from a strictly causalistic point of view. Bünning (1943) rightly emphasized that it seems rather hopeless to think that the problems of life might be 'solved' by trying to analyze certain centers of 'increased' activity, because life certainly is a problem of the interaction of a great many organized parts. With regard to nervous processes, Zilsel (1935) pointed out that they are governed by Talbot's Law, which holds that in a stimulus situation of rapidly changing intensity the biological reaction is proportionate to the average intensity rather than to certain high or low stimuli:

> Most probably each kind of consciousness is correlated with, or the result of, the statistical average of a vast number of atomic elementary processes, so that the individual microphysical process is not really very important.

(For further details, see also Hartmann, 1937.)

Finally, one is justified in doubting that Heisenberg's uncertainty equation leads, of necessity, to abandoning the concept of causality in the realm of microphysics, because in Heisenberg's theory causality is definitely an essential element and the basic assumption of the formula as 'it concerns and takes into account the inevitable causal interference of the object and the means of observation'. Moreover, microphysical processes are by no means indeterminate, but are restricted to certain limits which can be calculated by the methods of quantum physics: these processes

> cannot be predicted with regard to the single micro-process but only from a statistical point of view which, however, yields quite predictable statements. Quantum mechanics does not maintain that microprocesses can be compared to the 'arbitrary' actions of 'animated' humans, but compares them to a die which when cast can assume only a limited number of final positions.

Above all, however, we do not seem justified in thinking that predictability is a necessary criterion of causality. Macrophysical processes can be predicted, and this fact led to the belief that physical processes could be comprehended without accounting for the possible interference of process and means of observation. As stated by Hermann (1935), the law of causality means only that nothing happens which is not caused, in its physically comprehensible characters, by former processes. 'From this point of view the concept of continuous causality is not only in accord with quantum mechanics but an essential prerequisite of it.' Thus, the validity of the law of causality has not been contradicted, but, rather, clarified by quantum mechanics, and characters not necessarily part of the definition of causality have been abandoned. Though a reference to the causality of quantum-physical processes may seem to have 'no sense', this cannot, of course, be considered a proof of a microphysical noncausality (see also Lindsay, 1948).

Thus, there is no reason to assume noncausal or autonomous processes in evolution, though certain important events, such as radiation-induced gene mutation, are based on microphysical phenomena. The typical feature of all phylogeny, then, is not some sort of autonomous independence but strict causality. Hence, we cannot speak of autogenesis. But ectogenesis is not a fully sufficient term, as it emphasizes too strongly the 'randomness' of environmental effects. In order to imply that all regularities ('laws') of evolution referred to in the above chapters are the result of vastly complicated causal reactions I propose to introduce the term 'bionomogenesis'. This term means that evolution is a process governed by the development of laws (or regularities) of the living matter, which are true laws (mostly systemic laws) because of the complication of interrelations, which are based, however, on causal processes.

10

Evolution of Phenomena of Consciousness

A. INTRODUCTION

As mentioned in the previous chapter, we have to examine a last group of facts which might indicate an autonomous course of evolution: these facts – purposely not dealt with so far – belong to the phenomena of consciousness. In the following discussion 'consciousness' will mean not only the acts of voluntary thinking occasionally referred to as 'conscious experience' or self-consciousness, but all the components of consciousness, such as sensations and conceptions, the formation of conceptions, conclusions and judgments, feelings and volitions. At first these facts of consciousness can be experienced only as phenomena, i.e. as immediate reality, by one's own self, but by analogy one feels sure that they are likewise experienced by other human beings, and it is only a matter of gradation to assume that they are experienced also by higher vertebrates and, in successively simpler forms, by lower animal groups, also.

Four main questions have to be answered in the course of our discussion:

1. Is it probable that these phenomena of consciousness were subject to some sort of evolution?

2. Down to which evolutionary level of the animal kingdom may we assume sensations, ideas, concepts, feelings, and processes of volition?

3. May somatic evolution have been influenced by such phenomena of consciousness?

4. How does the idea of the evolution of consciousness fit into our philosophy in general?

The material for an analysis of these problems is furnished by human psychology, by human and animal behavior, and by the theory of cognition. It is obvious, then, that in discussing certain aspects of these fields we shall have to leave the realm of strictly natural sciences and enter fields of scientific research where the pertinent literature is so vast that in spite of his best efforts the biologist cannot possibly cover all relevant papers. On the other hand, however, one must mention the sad fact that most trends of present philosophy do not sufficiently take into account the results of modern biology. This may

322

be due partly to the splitting of our old faculties of philosophy into two faculties (humanities and natural sciences), so that philosophy 'officially' belongs to the faculty of the humanities. (Consequently, the holders of the philosophical chairs have come from the side of the humanities and have an inadequate background in the natural sciences.) This very fact makes it urgent that problems of natural philosophy be tackled also by natural scientists. This is all the more so because – as we have seen – the results have an important bearing on fundamental problems of biology.

Modern philosophy, unfortunately, shows many divergent tendencies, and only the smaller part of it is based on the findings of the natural sciences (see the periodical *Philosophia naturalis* and the former philosophical journal *Erkenntnis*, in which attempts have been made to establish a 'nonmetaphysical' philosophy of science). Hence, a special statement concerning the author's own point of view as to the theory of cognition precedes the chapters dealing with the above-mentioned problems. In these remarks I have had to combine concepts of the natural sciences, of psychology, of epistomology, and of logic. My own point of view developed along a line which may be characterized by the names of Descartes, Spinoza, Locke, Berkeley, Hume, Condillac, Kant, Spencer, Mach, Wundt, and Ziehen. Above all, the extremely critical works and often quite new results of Ziehen (1898, 1913, 1924, 1934, 1939) – a former psychiatrist – seem to be a sound basis for discussions on the theory of cognition for the natural scientist. Of course, there are many other quite different aspects of the cognition theory, but we need not deal with most of them in this book, because the major part of the problem of the evolution of consciousness is independent of the kind of philosophy followed.

To facilitate our discussion we shall first refer to some psychological facts necessary for an understanding of the phylogeny of consciousness. Additional necessary remarks on the cognition theory will be given along with the problems proper (mainly in Section E).

B. GENERAL REVIEW OF THE PHENOMENA OF CONSCIOUSNESS

As many biologists (unfortunately) are not too well acquainted with cognition theory, we shall first presume that – in the sense of a naïve realism – there is a fundamental difference between 'psychic' and 'material' components. (Later we shall refer to them more precisely as parallel components and 'reduction components' of our phenomena.) What are the components to be detected in our own psychic experiences, and in which way are they correlated with each other? Not until we have answered these questions may we try to find out about the possible phylogeny of the various psychic factors.

When we are awake, our psychic experiences are a complex of continuously changing phenomena, and it is difficult to isolate single factors from this complex. A more thorough analysis, however, shows that there exist only two groups of fundamental phenomena: sensations and ideas. Groups of ideas accompanied by intense feelings are sometimes referred to as 'feelings', and some directed processes of representations are considered processes of volition.

The basic experiences upon which all other phenomena depend are the sensations. As in all phenomena, their properties cannot be defined, and one can only characterize them by describing them as quality, intensity, locality, temporality, and of pleasant or unpleasant character. The qualities (red, sweet, the sound C, etc.) certainly are not inherent characteristics of 'material objects', as they arise only after the parallel physiological processes in the relevant sense organs have started (e.g. after the retinal cones have been stimulated by certain light waves causing certain chemical reactions). In some cases these qualities can be arranged to form a series of gradations in which no zero point can be established (for instance, the range of colors). Intensity, on the other hand, can be characterized as a continuity with a fixed zero point, but the various grades of intensity cannot be measured geometrically, though one can estimate them by comparing them, and this gives them a slight similarity to the grades of quality. Three-dimensional locality shows continuous steadiness and continuity, but the various steps can be related to zero points chosen at will. Most probably this also applies in a very similar manner to the 'material objects' (more precisely: to all causal components). Temporality is one-dimensional and shows the same continuous duration and continuity and may also apply to the 'material objects' in a similar manner. In contrast to locality, however, temporality as a phenomenon is typically experienced by us through the ever-progressing points of the immediate presence. Whereas the above-mentioned symptoms of a sensation are obligatory, this is not so with regard to a positive (pleasant) or negative (unpleasant) feeling accent accompanying most, although not all, sensations. We do not know as yet whether we have to refer also to the impressiveness ('Eindringlichkeit', 'Empfindungspotential', Weber, 1941) of a sensation as a proper characteristic of sensation, because possibly this impressiveness and significance are nothing but the result of fusions of sensations and subsequent perceptions and their intensity and character. The same applies to Hornbostel's (1931) intermodal parallelism of 'light' and 'dark' qualities, e.g. in the realm of odors. As it has been found that 'light' odors increase and 'dark' smells decrease the body tonus in general, 'light and dark' characters may be brought about by the effects of parallel intensities and positive or negative accents. It is interesting to note that Von Schiller (1933) has proved that such parallelisms also appear in vertebrates: fishes were trained to choose the brightly colored dish of a pair, one bright and one dark. When offered a 'light' and a 'dark' odor, these fishes would also choose the 'light' one.

It is generally impossible to describe or judge properly pure sensations, because they are immediately followed by mental images, without which the sensations cannot become part of our consciousness, i.e. we do not become aware of our sensations without subsequent conceptions. Moreover, we do not experience single sensations, but complexes, so-called 'Gestalten'. Thus it is justifiable to refer to these complex sensations modified by immediately subsequent concepts as perceptions.

All mental images are derived from residual images of sensations. The primary images not yet transformed by abstractions differ from the underlying sensations only by a less vivid intensity. The causal components (i.e. the physiological processes of the nerves concerned) of sensations differ from those of mental images by the fact that sensations are caused by external stimuli, while mental images arise from central, internal excitations (usually by association).

It is not yet certain whether the engrams of the various sensory performances are strictly localized in certain regions of the brain (Von Economo and Koskinas, 1925; Von Brücke, 1934; Kleist, 1934; Dusser de Barenne, 1937; Kornmüller, 1935, 1937) or whether they should be considered as the result of a more widespread activity of major parts of the central nervous system (see, for instance, Von Kries, 1898; Bumke, 1923; Von Brücke, 1934; Goldstein, 1927; Brown, 1927; Franz, 1926). Whether one adheres to the former or latter concept depends on whether one thinks sensations and mental images have a parallel occurrence in the same histological structures or whether one assumes that different histological elements underlie these two groups of phenomena. Ziehen (1924) furnished good evidence for the latter view. I myself am inclined to a quite different opinion as to where the physiological processes parallel to sensations and mental images are localized: psychic components arise parallel to processes in the sense organs; the brain is only a sort of central switch mechanism, and enlarged or reinforced nerve fibers are the basic element of the engram (see B. Rensch, 1952). These psychophysical problems, however, do not necessarily interfere with the discussion of the phylogeny of sensations and mental images.

Intermediate phenomena between sensations and mental images are represented by hallucinations, which are caused by central excitations and are as vivid as sensations, or more so. In some respects the subjective images of eidetics represent very similar intermediate phenomena.

If several equal or similar processes of sensation and perception are repeated, primary and secondary abstractions are formed by omitting certain temporal and spatial relations in the sequence of events. Thus, experience may lead to the formation of secondary conceptions. At first, primary individual conceptions will arise (for instance, of a certain yellow rose); several similar experiences (with roses of other colors) may result in a secondary conception (rose) or even in a general concept (flower).

In the formation of such abstractions or concepts, the highly discussed functions of differentiation are an important element. They, too, are special phenomena, that is to say, they are a part of direct psychic reality. They comprise: (1) the analytical function which we have just mentioned, by which the abstraction of the single experience from the temporal and spatial relations is performed; (2) the categorical function, permitting the recognition of the categories of equality, inequality, similarity, constancy, and change; and (3) the synthetic function. At the same time these functions are essential elements of the processes leading to the association of ideas and the formation of judgments. Here, the coincidence or some other fixed relation of the spatial and temporal coefficients of the phenomena connected in the judgment are essential (e.g. 'This rose is yellow': 'rose' and 'yellow' are experienced at the same spot and at the same time). Physiologically, then, two complexes of excitation apparently coincide or happen at the same time. An accumulation of coinciding abstractions leads to the combination of judgments to form conclusions (e.g. flies have six legs; beetles, butterflies, grasshoppers, etc., have six legs; hence: insects have six legs). Judgments and conclusions do not involve any new psychic (parallel) components, but are due only to the effect of the functions of differentiation upon the conceptions.

For the following discussion it is also important to note that there is no *a priori* 'ego' (self) (compare especially Ziehen, 1915, § 14–20), but that the self is nothing but the result of the connection of psychic processes by one central nervous system. The special character of the concept of an ego is due mainly to the following facts. Only the sensations of my own body are omnipresent, whereas sensations caused by external stimuli are less permanent. Furthermore, it is only my own body which causes kinesthetical sensations, and the sensations arising from my own body are those most frequently accompanied by intense positive or negative accents (e.g. pain, hunger, sexual feelings, etc.). Thus, there is only a gradual formation of a primary concept of the self during the early time of childhood and an equally gradual development of the 'secondary idea of the ego', i.e. of the concept of my own self as a personality by the synthesis of my own experiences and by the formation of complexes of ideas concerning my individual preferences and principles. A primitive concept of matter and soul leads to the idea of a soul of the ego. This concept was termed 'tertiary idea of the self' by Ziehen. (Many philosophers think it to be proof of a special 'self-intuition' which, however, does not seem to be a justifiable conclusion.) It is the very fact of realizing that my own self is not an *a priori* entity which justifies my conclusion that there are other egos, and this fully justified concept is an essential basis of our conclusions that there might also be some less complex types of 'egos' lower down in the animal line of ascent.

As yet sufficient studies have not been done to solve the problem of whether attention is a primary phenomenon. Whether a certain sensation out of several simultaneous ones is selected, and whether our attention is focused on this special sensation, depends on its intensity, its emotional accent (positive or negative), and its assimilability (i.e. its equality or similarity to sensations experienced previously). The same applies to the 'selection' of a certain concept out of a series where the vividness, accent, and constellation of that particular concept in relation to others play the most important role. In sensations the psychologically specific character of attention is caused only by the various movements of intention and accommodation of the sense organs concerned.

In the processes of association, dominant, intense, and positively accented mental images will gain supremacy and repress other less intense or less pleasant ones, so that a sequence of associations proceeding in a certain direction results. 'We cannot think the way we would like to, but our thinking is determined by the kind and constellation of associations at any moment' (Ziehen,

1924, p. 498). This determination of processes of volition is even more evident in animal experiments, e.g. in maze trials, where the types of motivation are not so complex (see below). Processes of volition, then, must be considered as specific phenomena, but their specific character is only a result of the constellation or relation of concepts. (Contrary to this view the voluntarists think that volition is a primary phenomenon, to which they – e.g. Schopenhauer – even ascribe a primacy in spiritual processes.) Apparently the phenomenon of the personal 'resolution' ('Entschluss'), so typical of processes of volition, arises from the antecedent sensations caused by kinesthetic sensations of preparations and intentions of muscles and sense organs. A final reason for the belief that volition is an independent and primary process is the fact that our actions are partly determined by the effects of engrams which are not accompanied by parallel components (i.e. of which we do not become 'aware').

Finally, we have to ask whether feelings, moods, and emotions should be regarded as primary processes. As we have seen, positive or negative feelings of various intensity may and usually do accompany the sensations (sensory feelings) and mental images (intellectual feelings). Thus, they are a property of sensations and mental images, and not a category of their own. Coherent concepts may cause a transfer of these feelings to whole complexes of associations, so that 'moods' and 'emotions' will result. It is possible that the parallel physiological process is due to chemical alterations of the blood. There is no general agreement yet as to whether one has to acknowledge further differences in feelings beyond pleasantness and unpleasantness. (One may, for instance, compare the various feelings evoked by a chord in major and one in minor, or the 'dark' and 'bright' qualities of various modalities mentioned above.) At any rate, we cannot refer to the complex and often disputed 'feelings' such as love, envy, etc., as pure feelings because they certainly are complex concepts accompanied by various positive or negative feeling accents.

The psychological and epistomological analysis briefly expounded here leads to a point of view from which we recognize that only the sensations, with their various properties, and the mental images are the basic elements of consciousness, that is to say, of the indubitable reality of our experience. The course of concept formation involves the functions of differentiation, which we have mentioned above. Attention, volition, feelings, moods, and emotions do not represent or contain psychic elements different from those of sensation and imagination.

As stated at the end of the introductory remarks, it is not necessary for a proper understanding of the following pages to accept the kind of cognition theory advanced here. Only the classification of psychic phenomena and their interrelations as given above should be acknowledged.

C. GENERAL PHYLOGENY OF THE PHENOMENA OF CONSCIOUSNESS

Almost anyone will agree that most animals must be credited with some processes of consciousness, be it in the form of sensations, feelings, or a more or less well-developed power of retention. This concept, as the result of the wide and general experience of man, remains valid even from the critical point of view of the philosopher. Thus, one can trace this idea in a more or less well-established form throughout the history of philosophy. As early as in ancient Indian philosophy one finds the direction that one should respect all living beings and not kill any of them. This idea is especially clearly expressed by the famous '*Tat tvam asi!*' ('That is you!') of the Upanishads (about 800 B.C.). Likewise, the great Indian philosopher Shankara, in the eighth century A.D., stated that animals differ from men only by degrees and not fundamentally, as they possess only a lower capability of recognition.

The Greek physician Alcmaeon of Croton (about 520 B.C.) thought that animals – in contrast to men – can perceive but not comprehend. Similar ideas were advanced by Anaxagoras and Aristotle, and the latter ascribed sensation, memory, anger, drives, happiness, and pain, but no reasoning, to all animals (*Works*, 1847 ed., vol. 2). In his *Confessions* (about A.D. 317) St. Augustine wrote: 'It is true that beasts and birds possess memory, for how could they find their way back to their nests or holes if not by memory.' One of the most important sentences of Spinoza's *Ethics* (II, 7th corollary, footnote) reads as follows: 'Omnia, quamvis diversis gradibus, animata sunt.' In his famous *Essay Concerning Human Understanding*, Locke (1690) stated that animals not only possess the capability of perception and retention, but that (p. 276) '. . . some of them in certain instances reason'.

Hence, it is natural that as early as the beginning and middle of the nineteenth century discussions on the theory of evolution extended also to the phenomena of consciousness. Lamarck (1809) emphasized the increasing mental capabilities of animals in the course of evolution, and Charles Darwin, in Chapter 8 of his *Origin of Species* (1859), referred to the evolution of psychic capacities. H. Spencer (1880) gave a very precise statement of the problem (pp. 291–2):

> . . . if the developed nervous systems of such creatures have gained their complex structures and functions little by little; then, necessarily, the involved forms of consciousness which are the correlatives of these complex structures and functions must have arisen by degrees.

Not all authors, however, have succeeded in dealing with the problem from the necessary critical point of view, and anthropomorphic interpretations of animal psychic performances are not always avoided. Thanks to behaviorism, premature psychological interpretations were eliminated and more precise studies on nerve and sense physiology and ethology were begun. Although these certainly did not approach the problem of the animal psyche, they emphasized how questionable it is to draw analogies between human sensations, conceptions, and emotions and those of animals. One should not forget, however, that it is equally inadequate to consider the problem from the standpoint of pure ethology only, as one may easily underestimate the importance of some well-founded analogies, so that the fundamental problem of animal consciousness remains undiscussed.

328

If we now want to deal with the phylogeny of psychic components (parallel components), we shall first have to find out which kinds of judgment by analogy there are and how weighty they are. Unquestionable reality in the sense of phenomena experienced by our own selves exists only for each ego. But numerous experiences concerning the reactions of other human beings soon convince us that our fellow men must also be credited with sensations, imaginations, feelings, etc., essentially the same as ours. This conclusion, clearly based on an analogy and derived only from acquaintance with other people, that is to say, from my corresponding auditive sensations and subsequent mental images, is well-established; I rightly consider that it indicates a very high degree of probability – truth, in popular parlance – and without such conclusions, no scientific problem can be approached. Similar conclusions regarding the sensations, imaginations, feelings, emotions, and volitions of their fellow men also apply to the more primitive races of mankind, to *Homo neanderthalensis, Sinanthropus*, or *Pithecanthropus*, though the certainty of our judgment here is a little weaker. There is no obvious reason, then, to limit these conclusions to a part of the evolutionary scale, and of necessity one has to include the highest types of mammals among the group of beings possessing the above-mentioned psychic phenomena. The lower we descend in the animal kingdom, the less human are the type of behavior and general appearance of the animals concerned – e.g. reptiles and fishes, though they must be considered as belonging to the direct line of human descent – so that we have to give a more precise statement of the criteria suggesting the assumption of psychic[1] phenomena in animals (see also Ziehen, 1921).

The purposive character of some animal reaction movements, so often referred to in the older literature, certainly does not prove the existence of psychic components, since it can easily be understood on the basis of causal laws as the result of selection, and since often these reactions are not accompanied by any parallel components (e.g. opening of the pylorus, or peristaltic movements of the intestines).

More reliable criteria of psychic processes, used by many authors, are furnished by the phenomena of memory (retention) and conditioning. But these criteria are not so safe as they may seem at first sight, for there are processes of a similar kind which are apparently not accompanied by psychic components, e.g. the permanent modifications of Protozoa representing a 'conditioning' to environmental factors. Hering (1870) and Semon (1905) even thought that memory and heredity could be interpreted – in terms of Lamarckism – as very similar processes, and this opinion has been shared by some later authors. There are certain phenomena of becoming conditioned to chemical substances – 'sensitization' of the animal body by the injection of certain substances, i.e. increasing reactivity to produce antibodies, or produce immunity to certain stimuli – which also reveal some qualities similar to

---

[1] In the following discussion the term 'psychic' is always used in its common meaning. One should keep in mind, however, that, according to the cognition theory, expounded further below, the causal components are also considered to be 'psychic'.

'memory', and the same applies even to some processes of the inorganic world, e.g. to the slow improvement of working of new machines, gears, engines, etc., and to the physical processes involved in disc and tape recording.

It might seem, then, that associative memory and the capacity of choosing would provide better criteria of psychic processes, but there are similar inorganic processes also. This can best be shown by a machine which will start running according to certain stimuli and 'learn' to react to one out of two simultaneous stimuli. We shall purposely select a very simple machine of this kind, though in reality there are robots of far more elegant design. Let

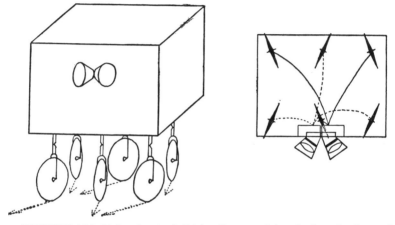

FIGURE 110. Robot (see text). Right: diagram of the wheel mechanism and the 'selenium eyes'

us imagine a small box on six wheels. On the front of this box there is a small hole with a lens and two plates of metallic selenium (Figure 110). When exposed to light, the electrical conductivity of this metal is strongly increased. Each of these selenium plates is connected to a small engine working three out of the six wheels. Three wheels point 45 degrees to the left and three 45 degrees to the right. If a beam of light comes from the left side it will be 'perceived' by the selenium plate on the right, which works the engine of the three wheels pointed 45 degrees to the left, so that the box will move forward and left, i.e. towards the source of light. Correspondingly, the box would react to a beam of light coming from forward and right. The movements of the machine, then, may well be compared to the well-known 'positive phototaxis'. It is easy to provide this machine with a device preventing the braking effect of the three wheels not used in one of the two movements (by flapping sideways).

If all six wheels are now furnished with brakes of a material which easily wears out, we shall soon find that the oftener we 'exercise' the phototaxic movement mentioned above, the more quickly our machine tends to react: our machine 'learns'. If we 'train' the machine mainly to a movement forward and left by offering the appropriate light stimulus in the majority of trials, the brakes of the wheels serving the movement forward and right will hardly wear out, while those of the wheels producing the movement forward and left will become practically ineffective. After a sufficient period of 'learning', the machine is offered two simultaneous and equally strong stimuli of light from the two sides. All six wheels will start turning, but the three wheels pointing forward and left (where the brakes have become almost ineffective, due to long 'exercise') will work more quickly and more efficiently than the other ones, and the result will be movement forward and left by the box: our machine 'chooses' the forward left direction, reacting according to the direction it has been 'taught'.

Finally, we can also imagine a model experiment to imitate the process of association. We furnish the box with a thermostat which shuts an electrical circuit as soon as a certain environmental temperature is reached. The circuit makes all six wheels move, but this movement is much slower and much less vigorous than that caused by the mechanism controlled by the photo-cells. We now 'train' the machine by offering a beam of light from forward and left together with a general temperature stimulus. Although the weak effect of the thermostat makes all six wheels move, the quicker and more vigorous movement caused by the light-controlled mechanism will be superior to the weaker effect of the thermically controlled mechanism, so that the box moves forward and to the left. After a certain 'learning' period (i.e. after the brakes controlling movement forward and left have been fairly well worn out) only the temperature stimulus is offered. The result will be a turning of all six wheels, but the ones which move forward and to the left will turn more quickly and vigorously – no or almost no brakes being effective there – and the movement forward and left will result. Hence, our robot has 'associated' both stimuli and 'learned' to 'choose' the movement forward and left, though only the general thermic stimulus is offered.[1]

I hope that this model experiment will not be taken in the sense of a 'l'homme machine', as it is only meant to demonstrate that associative memory and the capability of learning and choosing are not sufficiently reliable criteria in the search for processes of animal consciousness. Similar analogies are furnished by the performances of electronic calculating machines which have a 'storage' representing a 'memory', and a control center which starts certain actions automatically as soon as certain data in the process of calculating coincide. This is quite similar to the animal ability to choose one out of several stimuli. (For further analogies, see Wiener, 1948.)

Hence, in our search for animal consciousness, we can use only certain criteria from the fields of morphology, anatomy, and brain and sense physiology, i.e. we may try to infer from human phenomena similar animal

[1] After designing this experimental robot in 1946, I came across a reference by Buyten-dijk, mentioning a machine constructed by Hammond, which had 'photoelectrical eyes' and was capable of phototactic movements like an animal. A more detailed description of this machine is not accessible to me, and hence I do not know whether Hammond's and my model have any similar features. Very much more elegant types of robots have been constructed by Walter (1953).

phenomena only by the approximate similarity of the material basis. From human physiology, and especially pathology, we know that sense organs, nerve systems, and brains are necessary prerequisites of sensations and conceptions. Moreover, it has been well established that for the formation of ideas and judgments certain regions of the brain cortex are required, and that feelings and emotions are essentially correlated with the action of the diencephalon (Hess, 1948, etc.). As histology (Von Economo and Koskinas, 1925), electrophysiology (Kornmüller, 1937; Schäfer, 1942), and pathology have proved that the cerebral cortex can be divided into a great number of regions and areas of different functional significance, there has been an attempt to correlate these regions with certain psychic performances (see Kleist, 1934, for instance). It is doubtful, however, whether such strict correlation is justified (see Brown, 1927; Goldstein, 1927), and we must confess that we do not even know whether sensations and mental images are based on the physiological processes of the same or of different cortical regions. More precisely: psychic phenomena are correlated with processes in sense and nerve cells, but we do not know which events should be regarded as the psychophysical processes proper. I have tried to overcome these difficulties by a new psychophysical hypothesis (B. Rensch, 1952), which assumes that sensations and mental images are the parallel phenomena of physiological processes in the sense organs, while the brain, especially the cerebral cortex, serves as a vastly complicated switch mechanism connecting the various excitations and bringing about the coherence of consciousness. This hypothesis agrees with our knowledge of the phylogeny of sense organs and nerve systems, and is also well in accord with the findings of animal psychology – but we need not here discuss it in detail.

If we want to find out to what extent conclusions by analogy are justified regarding the problem of animal consciousness, we shall first have to trace the morphological structures essential for our own psychic parallel processes in the human line of descent. The brain structure of monkeys, especially of apes, in morphological, histological, and electrophysiological respects, is very similar to that of man (and even more similar to that of man's ancestors of the *Australopithecus* group). Only the frontal and temporal lobes are less differentiated and the motor area of speech is lacking. The sense organs of apes and monkeys are fundamentally equal to those of man. These facts, and the rather well-established and theoretically continuous line of human descent, strongly suggest that apes and monkeys must be credited with psychic processes. Naïve men have always been convinced of this. The conclusions thus reached by analogy may well be extended to include other mammals having a brain structure which is more primitive, but essentially similar to that of man. This is all the more so as in these animals we find a number of almost human facial or general expressions (such as the appearance of anger, joy, or fear in dogs) and the capacity of learning. Obviously, more complicated processes of association are lacking, but for the time being we want only to trace and discuss the possible existence of psychic phenomena in animals in general.

Mammals originated from theromorphs during the Late Triassic. The transition, especially with regard to brain structure, was apparently a very gradual process. This is proved by skull casts of mammals of the Early Tertiary, such as *Eohippus* (Figure 102), or of certain creodonts which show a relatively small and barely fissurized forebrain and a serial position of the various brain parts resembling that of reptiles (see Edinger, 1948, 1950). In modern reptiles the cerebral cortex is only three-layered (not partly seven- or five-layered as in mammals). The five brain parts and the sense organs of reptiles, however, are fundamentally similar to those of mammals. This applies also to amphibians and fishes, in which there is no multi-layered cerebral cortex. Their forebrains are relatively small and mainly serve the sense of smell, while optical excitations are associated with the midbrain. But fishes also show a fairly well developed learning ability; like men, they are subject to visual delusions (Herter, 1953); in situations of conflict, they give evidence of 'irresolution' and of neurotic behavior. Hence, there is no reason to deny that fishes are capable of (at least very simple) psychic performances such as sensations, simple images (memory), emotions, etc., though they lack a cerebral cortex. Moreover, as we have seen above, the emotions and feelings of man are also closely connected with the diencephalon, rather than with the forebrain.

Below the vertebrate level we can no longer base our considerations on a well-established phylogenetic record, and therefore we have to rely mainly on the evidence of comparative anatomy. Some of the morphological and physiological criteria referred to so far no longer apply, as in invertebrates the central nervous systems are totally different from those of vertebrates. Hence, it would be possible to conclude that no invertebrates can be credited with psychic components. Such an opinion, however, cannot be accepted unconditionally, because there are numerous facts strongly suggesting that invertebrates, especially insects, spiders, cephalopods, etc., are capable of psychic performances such as sensations, simple images (memory), feelings, and volitions. These facts include the following:

1. The histological substratum of the parallel physiological causal components, i.e. neurons and sense cells, in invertebrates is very similar to that of vertebrates. The central nervous system of cephalopods, hymenoptera, or beetles is of a highly complicated order, and the eyes of *Octopus*, for example, are very similar to those of vertebrates.

2. With regard to the plastic behavior of higher insects and cephalopods and to their learning abilities (e.g. in bees) it would not be justifiable to deny psychic processes of the kind which are commonly assumed in the lowest vertebrates (e.g. in fishes or *Petromyzon*).

3. The phylogenetic tree shows continuous lines of descent. Below the level of Agnatha the phylogenetic series may be continued by animals like Acrania (at present represented only by the specialized type of *Branchiostoma*) down to more primitive invertebrate groups. Hence, it seems improbable that

'psychic' phenomena, so fundamentally different from 'material' ones, should have originated as a completely new event at a certain level of the uninterrupted and gradually transforming series of evolution.

The gradual evolutionary transformation traceable in the phylogenetic tree, in particular, implies that one has to attribute at least primitive and not highly differentiated sensations accompanied by positive or negative feelings and also simple images (retention ability, memory) to all those invertebrates which are provided with the 'psychophysical structures' of nerve and sense cells. This is suggested by the primitive learning abilities of lower worms and actinians (see Pieron, 1908; Bohn, 1903), though it is not a fully sufficient criterion (see above). Hence, one might think that 'psychic' components (parallel processes) originated during the evolutionary period in the coelenterates. This would mean that sponges and meso- and protozoans should not be credited with sensations.

This view, however, is likewise not in accord with the gapless series of phylogenetic transformations ('synechological argument': Ziehen, 1921). It is not very probable that in the continuous process of transformation entirely new laws of psychic parallelism should have suddenly emerged, because the differentiation of certain cellular systems into nerves and sense cells represents only an improvement, not a quite new creation, of qualities inherent in all cells, i.e. of irritability and the capability to conduct electrical phenomena of excitation, etc. Moreover, even in protozoans we find structures like neuronemata and sense organelles specialized for the perception of stimuli and conduction of excitations (see, for instance, Von Gelei, 1936, 1939a,b). Ciliates have also been credited with the possession of learning ability (see Alverdes, 1937; Bramstedt, 1935), though this has been justifiably doubted (Koehler, 1939; Grabowsky, 1939; Alverdes and Bramstedt, 1939). Any observer will receive the impression that the plastic behavior of certain ciliates – e.g. *Dileptus* and *Lacrymaria* searching and probing with their 'proboscis' – shows some close analogies with the behavior of higher worms (*Stylaria*, Archiannelida). But even assuming that memory and conditioning are improbable in protozoans, the possibility of sensations cannot be denied. This view is supported by the fact that all concepts are essentially based on sensations. So one would expect that the evolutionary origin of psychic (parallel) components would be represented by sensations.

If we agree that the synechological argument is valid right down to the protists – where our conclusions by analogy, so 'safe' on the vertebrate level, have become more and more vague – we have to ask whether the origin of psychic phenomena (of laws governing parallel 'psychic' components) coincided with the origin of life. Before we try to answer this question, it will be necessary to analyze the manner in which the number of psychic phenomena decreases as we go from higher to lower levels of the phylogenetic tree, and to establish which phenomena should probably be regarded as the oldest in phylogeny. To summarize our findings so far, we may state that it is not improbable that all living beings should be credited with some sort of psychic processes, though we can conclude this only on the basis of analogy.

D. PHYLOGENY OF THE SPECIAL PHENOMENA

The emergence of psychic phenomena or of the 'soul' (according to our terminology the parallel components, because all objects are only experienced as 'psychic' ones) has been discussed ever since the time of Aristotle and in more detail since Lamarck. We cannot give a summary or critical review of all the numerous treatises on the topic in the present chapter (the reader is referred to Von Uexküll, 1909; Kafka, 1914; Hobhouse, 1915). We shall deal only in brief with the evolution of those psychic components which can be precisely defined in terms of the theory of cognition, and we shall confine the discussion to the following questions:

1. In following backward the branches of the phylogenetic tree, how far down may we think that sensations, feelings, and emotions are probable?
2. Down to which evolutionary levels may we expect primary concepts (memory)?
3. To what extent may we assume secondary concepts, i.e. abstractions and generalizations, and how far are the categorical functions involved?
4. To what extent may we assume rudiments of animal judgments, conclusions, and 'insight' (reasoning)? After we have answered these questions relating to animal psychology proper, we shall try to fit our findings into a philosophy based on the theory of cognition.

I. Phylogeny of Sensations

Out of the multitude of stimuli to which each organism is continually exposed, only relatively few – those of biological importance for the organism cause an excitation. Man, for example, lacks any sense organs for the perception of electric or magnetic stimuli, of radio and light waves shorter than 395 mμ, and of sound waves higher than 20 kilocycles. In spite of this, the number of different qualities of stimuli perceived by man and – most probably – all homoiotherms is extraordinarily great. We can differentiate among more than a hundred colors, numerous odors, sounds, tones, and qualities of touch, pressure, pain, temperature, etc. In the perception of some of these qualities man and the higher vertebrates have reached the optimum sensitivity: if our retinal sensitivity were slightly increased, we would perceive the irregularities of light emission and thus get a less adequate image of the environment. The same applies to our acoustic sense, in which we would perceive the Brownian movement of the atmospheric molecules.

If we consider the phylogenetic ladder from the homoiotherms to the poikilothermic vertebrates, such as fishes, the number of sense qualities does not diminish considerably (in contrast to the decrease of retention ability and the parallel decreasing complexity of the brain structures). It is not until we reach the level of invertebrates that a clear diminution of the variety of sensual qualities, which is parallel to a poorer development of sense organs in these animal types, becomes obvious. Many of these animals lack fine discrimination of acoustic and static perceptions; the ability to taste and smell is replaced by a common 'chemical' sense, and the other senses discriminate far fewer qualities than in higher vertebrates, for instance.

This diminishing ability of discrimination suggests the interesting question of whether the various abilities originated one after the other in the course of phylogeny or whether the abilities and modalities evolved by gradual differentiation from ancestral 'basic' qualities. Democritus thought that all sensations were modifications of our sense of touch, and a similar opinion was favored by H. Spencer (1855, 1882) and a number of more recent authors. Correspondingly, Wundt, Leydig, O. and R. Hertwig, Nagel, and others thought that in lower animals there is only one type of universal sense organ capable of perceiving various sensory stimuli equally well, without implying a plurimodal universal sensation. This latter concept was clearly defined by Eimer (1888), and Plate (1924) followed him in stating: 'Originally, visual, acoustic, olfactory, and taste stimuli are not perceived as such, but as stimuli of touch.'

We should not overlook, however, that these conclusions are by no means borne out by facts. If an organism reacts to different stimuli in the same way (e.g. by a positive or negative taxis) this does not imply that the various stimuli evoke only one kind of sensation. Men likewise run away to avoid the effects of quite different stimuli. In the majority of cases, we do not know enough about the correlation or connection of sensations with the underlying physiological processes or morphological structures in lower animals. Above all, we should not forget that a homogeneous sensation does not imply causation by homogeneous stimuli: the sensation 'green', for example, can arise from a stimulus by light of 520 mμ or from a simultaneous stimulation by light of 585 and 470 mμ, which – if offered singly – would cause the sensation of 'yellow' or 'blue' respectively. By simultaneous stimulation by light of various wavelengths, the uniform sensation of 'white' can be produced. Correspondingly, most of our sensations of taste, experienced as homogeneous phenomena, consist of fusions of olfactory and gustatory sensations with those of touch. On the other hand, each new quality arising in the course of phylogeny represents an essentially new psychic phenomenon parallel to the fundamentally new process of transforming the new stimulus into nervous excitation. This is also the reason why we are unable to define the sensory qualities experienced by us, and can only state that there are certain differences between them. (It is for this reason that it is impossible to explain the sensations of color perception to a color-blind person.)

Nevertheless, there are a number of hints suggesting that some sensations evolved from common 'ancestors' phylogenetically. Proof of this view can, of course, be furnished only by human introspection. In the visual sector, for instance, one can easily trace the way in which a light-dark sensation changes into the sensation of color. Each spectral light passes the sensory threshold without causing a color sensation, and it is only after a certain increase of intensity (this 'colorless interval' differing with the various wavelengths) that the color sensation arises (see Ziehen, 1924, p. 190). Correspondingly, the retinal rods and cones differentiate in the course of ontogeny from intermediate rudiments (like those in the retina of adult cyclostomes).

In the realm of olfactory sensations, evidence of similar differentiations is furnished by the highly interesting introspective findings of the physiologist Hofmann (1926). He reported that after an inflammation of the nasal cavities he had completely lost his sense of smell, and the olfactory sensations returned one after the other, at first being quite abnormal (p. 251):

> Diluted acetic acid and correspondingly diluted butyric acid smelt completely alike. Later on my sense of smell differentiated between the two in a progressively improving manner. . . . The same applied to benzene, toluene, and xylene, which at first smelt alike, and only later on could they be differentiated olfactorily.

Thus, it is at least conceivable that a gradual evolution of the qualities of the chemical sense has taken place and that lower animals experience only few of them or even only one basic quality. This is also suggested by the findings of Mangold (1943) in earthworms. These animals react similarly to all four basic qualities of taste (sweet, bitter, sour, and salty) even if only weak concentrations have been offered. If different taste stimuli are combined the reactions will accumulate. Hence, it is possible to conclude that earthworms experience only the positive or negative side of a single uniform quality of taste. On the other hand, we must remember that too much should not be inferred from the fairly simple reactions of earthworms, and it is quite possible that earthworms distinguish all four qualities of taste, though their apparent reactions to all of them are alike. With regard to the intermediate stages of sensory qualities experienced by man (see above), it is probable that these are phenomena parallel to the correspondingly changing chemical processes in the sense cells.

Here one should also consider the possible development of auditory sensations from those of vibration and touch, which was first assumed by Mach (1865). Regarding sensations of vibration, Von Frey (1926, p. 107) emphasized:

> It is hardly possible to decide by introspective observation whether a certain sensation should be considered as a formal modification of an already known quality or of a new and special quality or modality.

In regard to all these conclusions and deductions one has to keep in mind, however, that there is a great difference in the structure of auditory and tactile sense cells. In insects, for instance, sound is perceived by tympanic organs, but vibration is perceived by different organs (see Autrum, 1941).

Summing up, we must conclude that it is not possible to decide whether or to what extent a phylogenetic differentiation of sensory qualities and modalities evolved in close correlation with new chemical and physical processes in the sense organs, or whether we have to account for alterations due also to the laws of parallel psychic components. The former assumption could mean a gradual but salient 'mutative' increase of psychic components, whereas the latter concept would be possible because of the origin of new qualities (e.g. 'green' from the simultaneous effect of two stimuli which produce the qualities of 'blue' and 'yellow' when not applied simultaneously).

If we agree that the possession of sense organs and nerve cells and the evidence of corresponding animal reactions suggest the presence of parallel psychic qualities, and if we apply this conclusion throughout the gapless phylogenetic tree right down to the protozoans (which may be justified: see Chapter 10, C), we certainly cannot expect that all the different sensory qualities will finally merge into one. At the low level of unicellular beings we have to expect a considerable number of basic sensations accompanying the various psychophysical processes of this one cell: sensations of light, touch, temperature, and chemical stimuli, possibly even sensations of pain and hunger. As even in larger bacteria one can trace photo-, chemo-, thigmo-, and thermotaxis – where Weber's Law is applicable (proved by experiments applying chemical stimuli) – we may possibly assume that various modalities are experienced even by these lowest true organisms. This aspect leads us to the problem of the origin of psychic components, but this will be dealt with in a later section on a more general view of cognition theory (Chapter 10, E).

So far we have considered only the most striking features of sensations: their qualities. We shall have to deal also with three more properties: locality, temporality, and the (pleasant or unpleasant) feelings accompanying a sensation (we need not go into details concerning the intensity of sensation).

Spatiality is an inherent attribute of all sensations, though in olfactory and gustatory sensation it need not be very clear and conspicuous. Hence, parallel with the phylogenetic development of the senses, there has been the development of a visual space, a tactile space, an auditory space, a space perceived by means of the vestibular organs, etc. All these spatial properties of different sensations can merge into a uniform sensation of space, as experienced by man. This fusion of various sensations is facilitated by the fact that all modalities prove that space is a homogeneous phenomenon (contrary to the inhomogeneous qualities of the sensations). Apparently space experienced by man corresponds closely to space as a physical reality (in contrast to the qualities of sensations). We may thus assume that a quite similar homogeneous space is experienced by the higher animals. In lower organisms such as worms, lacking a vestibular organ, ears, and (well-developed) eyes, this experience of space will, of course, be much 'narrower', because their 'space' can be perceived only as the fusion of the stimuli from the tango- and chemoreceptors which are scattered over the body surface in a mosaic manner. The existence of a right-left localization in earthworms has been proved by conditioning. In coelenterates, on the other hand, which lack a central nervous system, one may well assume that the localities of the single sensations do not merge into a homogeneous experience of space, so that on this evolutionary level the reality of physical space is not too well represented in the 'psychic' experience of space (see Section E on space concept by reduction of sense properties; also compare Ziehen, 1934, p. 133).

Relatively little evolutionary development need be assumed for the temporality of sensations, since the temporal organization of sensations corresponds fairly well with physical time. And yet it is possible that in lower animals the temporalities of various sensations are superimposed on each other, so that there is no fusion into a homogeneous experience of time.

Positive and negative feelings accompanying sensations are relatively labile and not obligatory traits; they not only depend on the kind and intensity of the stimulus, but are also influenced by the properties of previous sensations, concepts, and associations (irradiation, agreement, etc.). Hence, they have been more readily subjected to often rapid evolutionary changes. The sight of an owl's head may excite a sensation with inherited positive feelings in another owl during mating time, but with negative feelings in small birds. As positive or negative feelings can accompany all sensory qualities, one may assume that they arise from very general physiological processes. Intense feelings are often connected with engrams of former (pleasant or unpleasant) experiences, and, above all, with many innate behavior patterns of flight from enemies, of feeding, breeding, etc., i.e. those activities which are of high importance in securing the existence of the individual of the species. Usually, the most intense feeling decides which activity – out of a variety of possible behavior patterns – will begin: e.g. in case of danger from an enemy, the animal flees in an opposite direction instead of continuing its previous path; sexual activity often dominates over feeding, etc. The biological importance of such feelings as a means of ensuring vital processes strongly suggests that sensations are experienced by all animals. It is significant that we can speak of 'shock' reaction, thermo-'phily' and the like even in protists.

Summarizing, we may postulate a gradual evolution of the qualities and accompanying feelings of sensations, but this most probably applies in only a limited way to the locality and temporality of sensations.

II. Phylogeny of Conceptions

As already expounded in sections B and C, all mental images arise from engrams of previous sensations. One may regard them as weakened sensations activated by central excitations. Their simplest form, the primary image, is an unaltered or incomplete memory of an individual sensation. If we credit animals with sensations, we may equally well credit them with primary mental images. Like those of man, images probably will be markedly less vivid than sensations. This, of course, applies only to animals in which a retention ability (memory) has been proved, but such proof has been established as far down as the coelenterates. We cannot yet decide to what extent the net of ciliate neuronemata is capable of retention (see the Alverdes-Koehler discussion, and refer also to the fourfold reactions of *Stentor* to various stimuli discovered by Jennings, and Buytendijk's opinion that *Stentor* is capable of 'a simplest kind of habit formation').

The origin of all the simpler concepts probably involves a process of selection. If an animal has to react to several qualitatively equal or similar stimuli occurring successively, it certainly is advantageous for the animal not to react to each of them separately, but to perceive them as the same object. If, for example, the eye of an octopus perceives five similar stimuli one after the other, each of them indicating the presence of an object of prey, it will be advantageous for the octopus to 'summarize' these several stimuli as a homogeneous concept of 'an object of prey'. Hence, this octopus will react as though the five similar stimuli came from the 'same' object. Thus, it seems probable that in animals a concept of a constant 'carrier' of stimuli is formed (in the sense of a simple reductive concept; see below, Section E). This simple kind of concept could be improved and broadened by man with the help of speech, and resulted in the concept of 'substance' in the philosophical rather than the chemical sense. (In spite of this deep phylogenetic root, the idea of such a 'carrier' of the qualities of sensation is not a necessary prerequisite to cognition theory; see Ziehen, 1934, § 20.)

As animals usually experience certain sensations several times in succession, so that the same object is perceived under varying conditions of illumination, distance, position, etc., normally secondary concepts will also have evolved, in the sense of abstractions. Thus, a sort of concept or idea arises comprising only those components which are common to all sensations caused by the object under consideration, while the casual and singular components of the single sensations are suppressed. The essential difference between the animal kind of concept and the human one usually lies in the absence of verbal associations. Man usually 'thinks in words', and this is the basis of his highly developed ability of abstraction. Besides these verbal ideas, there are also averbal concepts, e.g. that of a plant which we have repeatedly seen, but which we do not know by name. Animals possess only most primitive averbal concepts derived from and developed by experience. The number of such averbal concepts in higher vertebrates, especially, seems to be considerable. (The 'releaser mechanisms' causing certain instinctive reactions, like those caused by perception of the red ventral side of the male stickleback [see Lorenz, Tinbergen] probably can be regarded as similar in some respects to these averbal concepts.)

It is difficult to decide how far down in the line of descent one may expect such averbal ideas in animals. Probably some arthropods must be credited with possessing them (e.g. visual patterns under varying illumination near the entrance of a beehive serve as a guide for the insect). Besides, we may conclude that all animals capable of learning must also possess the ability of abstraction, as the stimulus situations are hardly ever identical. On the other hand, such averbal concepts are not very essential in invertebrates, especially in insects, as false reactions due to a lack of experience or ability, which cause the death of the individual, are rather easily compensated for by the enormous overproduction of offspring.

It is surprising that among vertebrates the formation of very abstract averbal concepts has been demonstrated. This is most impressive in birds and mammals, in which Koehler and his school could trace a remarkable 'averbal counting ability' (see especially Koehler, 1941, 1943, 1949; Arndt, 1939; Marold, 1939; Braun, 1952; Hassmann, 1952). These experiments proved that pigeons, budgerigars, parrots, jackdaws, magpies, ravens, and squirrels can learn to discriminate between certain numbers of food grains, even though the arrangement of the food objects varies continually. Besides this, they can learn to take only a certain number of grains out of a greater number offered successively at irregular intervals, varying from one second to one minute, so that no clue could be derived from a 'rhythm' or temporal sequence. The greatest number of discriminations mastered in these two ways was seven. Some animals were even able to learn double and triple tasks of this kind, i.e. they would stop feeding after they had taken two, three, four, five, or six grains, if they were offered two or three different colors or acoustic signals indicating the number of food grains 'allowed' to be taken. Marold (1939) emphasized that it is not necessary to conclude that the birds really 'count', but is sufficient to assume that the birds 'are able to retain a series of stimuli or activities accompanied by a decreasing positive accent so that this decrease indicates the end of the series' (of 'allowed' food grains). This kind of decrease (i.e. of the length of the series and of the number of grains to be taken) can fairly easily be taught to the animal by conditioning. Koehler added in 1941 that in none of these cases can an animal understanding of numerical series in the human sense be assumed. Nevertheless it seems justifiable to refer to a prelinguistic concept of numbers in these animals. This is suggested by the remarkable results obtained from a raven and some large parrots by Koehler and his school. The raven was offered five food dishes covered by five cardboard discs on which from two to six pieces of Plasticine, varying in size and arrangement, were fastened. In the trial a sixth cardboard for training was shown, on which two, three, four, five, or six Plasticine pieces or colored spots indicated which of the five boxes would yield the food reward. After a long training (at first in the single tasks) the raven was able to solve this kind of task (Figure 111; Koehler, 1943). An African grey parrot learned to open one out of three food boxes covered by cardboard plates with two, three, or four dots, after the 'correct' box was indicated by a corresponding number of acoustic stimuli at the beginning of each trial (Braun, 1952). Recently, Koehler was able to demonstrate an averbal concept of the pattern of a maze in mice. The mice learned to master a rather complicated maze; when, after this, they were offered the inverted image, they were able to master the maze without new learning.

The problem of whether there is some kind of animal concept of the self (ego) is especially important. As I have tried to show in Section A, the concept of the 'ego' is not an *a priori* entity, but – contrary to the opinion of many philosophers – the result of ontogenetic (arising during the first months of a child's life) and phylogenetic developments. Following the analysis of Ziehen, we had previously accepted these criteria concerning the formation of a concept of self: (1) the omnipresence of sensations arising from one's own body (contrary to those caused by the environment); (2) the active kinesthetic and proprioceptive sensations (including reciprocal sensations when touching one's own body); and (3) the strong feelings accompanying sensations arising from one's own body (pain, hunger, thirst, sexual sensations). All these components can be assumed at relatively low phylogenetic levels in the animal

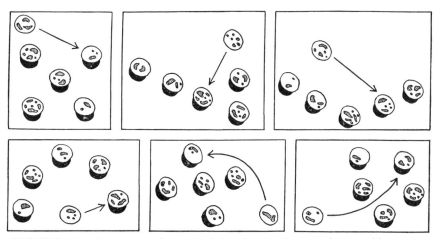

FIGURE 111. A sequence of correct choices in sampling experiments with a raven. Lids covering food dishes bearing two to six irregular spots of different size and position. Arrows connect the sample pattern with the 'correct' lid. (After O. Koehler.)

world, so that it seems possible that a primary concept of the ego may be formed as soon as animals are capable of some prelinguistic ideation. A bare rudiment of such a concept of self should possibly be attributed to the coelenterates, as some jelly fishes – especially *Tiaropsis indicans*, named after its ability to 'indicate' which tentacle has caught a prey, but also *Sarsia*, *Gonionemus*, etc. – are capable of exactly localizing on their velar periphery the spot where an animal has been caught by a tentacle (Figure 112). The movement of the gastral opening towards the spot where the prey has been caught is probably accompanied by a strongly positive and pleasant sensation of the body. It is true that coelenterates show certain evidence of a limited learning and retention ability, and must therefore be credited with some sort of primary concepts concerning their own bodies. They cannot, however, properly discriminate among the various engrams and, hence, they cannot be credited with 'awareness' of their own selves, though they may experience a sort of trace of this concept. The example of the coelenterates was selected only to show how far down in the lines of descent one may expect some essential components of the concept of ego which developed later.

Besides this, one may well assume the existence of rudiments of a secondary concept of the self among animals, as far as a memory of the experience of the body, and dominant complexes of individual properties may be supposed. This is suggested, for example, by experiments on fowl, where a fixed 'pecking order' – irrespective of physical strength – is maintained over long periods of time. Such a 'pecking order' implies that each hen has a sort of 'selfconsciousness', a concept of her self and the 'value' of her personality in relation to her fellow-hens. Katz and Toll (1923), who reported these findings, even referred to typical 'differences in the character' of these animals. Even clearer evidence of secondary ego-concepts was furnished by Diebschlag's (1940) studies on domestic pigeons. Here, also, each individual maintains its position within the 'pecking order', and also lays claim to its own sleeping place, as a special threatening behavior indicates that a certain

FIGURE 112. Movement of gastral opening to the prey caught by the peripheral tentacles in the jellyfish *Tiaropsis indicans*. (After Von Buddenbrock.)

pigeon owns and maintains a certain seat. Diebschlag took a very inferior pigeon out of the dovecot and strengthened her courage by letting her always win in mock fights against dummies, so that finally this particular pigeon would always fearlessly attack the dummy and feel herself to be the victor. After a certain period, this pigeon was brought back to the dovecot, where she promptly fought fearlessly and violently and became No. 1 in the 'pecking order'. Now, when a stuffed pigeon was brought into the dovecot, the new No. 1 pigeon became a little frightened, so that when fights for a new 'pecking order' among all the mates resulted, the new No. 1 was defeated due to her impaired 'self-consciousness'. Though one has to account for instinctive actions and reactions when evaluating such experiments (compare the increased 'courage' of the male stickleback in the vicinity of his nest), one may yet assume the essential elements of the concept of the ego among these animals.

Finally, the evolution of the concept of the self might also be furthered by the ability to imitate the actions of other animals, i.e. to transpose visual conceptions into locomotory conceptions of the animal's own body.

If, following Ziehen (1924), we define volition as the result of a series of conceptions in which one is considerably dominant, or as the result of a particular constellation of sensations and images, then we have to credit animals with volition. If the experimenter prevents a fish from reaching food by putting a transparent pane between the food and the fish, all the movements of the fish – including possible detours in trying to get to the food – are determined by the dominant, strongly positive accent accompanying the perception of the piece of food behind the glass. After all, it is only a matter of definition as to what extent similar processes in lower animals should be regarded as volition. Perhaps the borderline of the genuine processes of volition is reached when the dominant sensations and concepts are replaced by inherited behavior patterns. In such cases it is the 'mood' (e.g. the desire to feed, to copulate, etc.) which decides to which stimulus the animal will react (e.g. a butterfly in the mood to copulate will follow a moving dummy of the opposite sex rather than react to the flowers yielding honey). In the field of human drives there are similar examples of actions determined by an emotion (e.g. anger, sexual desire) rather than by 'free' will.

III. Judgments, Conclusions, and Actions of Insight

It seems possible to extend the meaning of 'judgment' beyond its range in human psychology, where it usually means the comprehension of psychic connections and relations described by words. The extension would also cover averbal concepts in which a certain correlation or relation of spatial or temporal coefficients is comprehended (see Section A). In this broadened sense, judgments must be attributed to all animals capable of learning in an experiment that 'red' or the tone 'C' 'means' food and that they have to select this special relation of 'red and food' out of a wide variety of other relations (e.g. 'purple and food', 'tone F and food', etc.). This aspect, of course, involves the assumption made above that by analogy animals may be credited with parallel psychic processes. The evolution of judgments, then, might have come about like this: the averbal conception or secondary conception of 'red' or of 'C' was connected to the drive to find food, so that gradually the 'judgment': 'red goes with food' was formed. To emphasize the essential lack of verbality in these animal rudiments of judgments I propose to refer to them as averbal judgments. Most probably this kind of judgment can also be met with in man, especially in 'uneducated' persons. The neurophysiological processes parallel to these psychic phenomena are represented by simplified associative connections (not connected to a speech or word center) dominating over other connections. The accompanying psychic (parallel) processes, i.e. the peculiar process of perceiving and recognizing similarity, equality, etc., cannot of course, be 'explained' from this point of view. These phenomena (functions of differentiation) cannot be defined, and we can characterize them only by referring to our own experience.

This kind of averbal judgment, then, could possibly be attributed not only to lower vertebrates but also to higher invertebrates, especially to insects and cephalopods. One has to keep in mind, however, that in these homoiotherms such judgments are not too important biologically, as in most of these animals activities and reactions are controlled by innate behavior patterns. According to the detailed studies by Lorenz (1940, 1943), Tinbergen (1942, 1951), and their school the perception of a releaser (e.g. the red ventral side of the stickleback) causes an innate reaction. The difference between an innate releasing process and an averbal judgment is that the former is not a result of repeated experience (e.g. red ventral side 'means' male stickleback), but simply an inherited 'knowledge'. (It may be mentioned here that it is by no means probable that instinctive actions lack consciousness.)

It seems obvious that in animals the formation of abstract ideas is not very common, and also that the number of concepts is smaller than that of sensations. Hence, in the animal psyche the 'external object' and the complex of corresponding sensations are identical. This also applies to a large extent to the more primitive type of man. Hence, the external world ('Umwelt') and internal world ('Innenwelt') merge into one, as was expressed by Buytendijk, who thought that in animals this fusion of external and internal world is 'almost as complete as the unity of their body' (see also Volkelt, 1914). To us human beings, the easiest way really to understand the animal psyche is to consider such complexes of our sensations as are followed by few or no subsequent conceptions: e.g. all our senses may be occupied by eating, by sexual desires, etc.

Averbal judgments became of major importance as soon as the associative psychological and physiological complexes had, in the course of phylogeny, become sufficiently numerous to serve as a means of grouping and labelling some types of the extremely complex and often rapidly varying stimulus situations of the environment. In this context one should also consider the enormous increase of associations rendered possible by the immense number of ganglion cells of the mammalian brain, which can total many millions (e.g. the human brain contains about 13 billions). Moreover, most of these neurons branch into many ramifications (see Figure 113). As soon as these neural bases of association had become sufficiently numerous, averbal judgments gained considerable importance, and due to their selective advantage, their material substratum was favored in evolution. The adaptation of the method of judgment to the realities of the external surroundings ('material' world subject to the laws of causality) automatically led to the evolution of the three basic functions of comparison, analysis, and synthesis, which make it possible to react favorably to the stimuli of the environment and to deal with them properly. Finally, all higher forms of judgment are based on these three elementary functions (compare Kant's 'categories', and the critical review by Ziehen, 1913).

The connection of such averbal judgments resulting in averbal conclusions may well be attributed to at least some birds and mammals. Conclusions of this kind are apparently reached by Koehler's birds and mammals capable of 'prelinguistic counting'. A raven, for instance (see Figure 111), reacted according to the 'signboard' indicating by the number of dots the respective number of grains 'allowed' to be taken. Transcribed into human speech, the prelinguistic conclusion of the raven would run approximately as follows: 'The signboard shows three dots; hence food will be found in that dish covered by cardboard with three dots on it independent of their size, shape, and arrangement.'

345

The capacity of drawing such averbal conclusions consists in a comprehension of the mutual relations of objects (Köhler: 'simple comprehension of meanings'), which in modern animal psychology is usually called insight or reasoning.

The essential feature of such processes seems to be the fact that the correlated nervous events of such reasoning are not based on inherited (instinctive) or learned connections, but that without actual 'trial and error' the

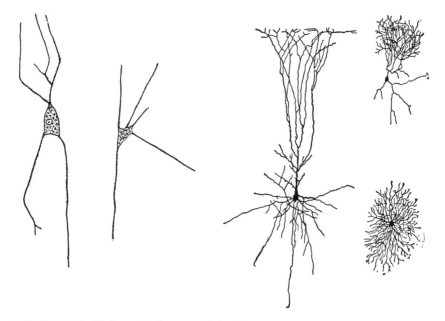

FIGURE 113. Phylogenetic increase of dendrites per neuron. Left: two multipolar ganglion cells from the ectodermal nerve plexus of the jellyfish *Rhizostoma*. (After Bozler from Hanström). Right: ganglion cells from the cortex of man, motor cell of III layer and two sensory cells (after Kölliher and Cajal from von Economo and Koskinas).

possible actions or reactions are 'experimentally' tested in the brain before the 'selected' motor action is performed (competition of motivations). The parallel psychic components of such cerebral processes are usually called 'deliberating', 'Spiel der Motive', or 'thinking it over', and I can see no reason why these terms should not be applicable in animal psychology also. One can very well imagine that the underlying physiological process is causally determined: the action currents pass through complicated pathways of associative connections and some of these pass more easily, due to previous facilitations, while others are inhibited. Thus, the action currents will not reach all motor areas in equal strength (some of them will be stronger, probably due to the 'facilitated' pathways they can travel, while others will reach the respective motor areas only as a very weak electrical signal: see Verworn, 1919). The concept is supported by findings of slight action currents or even intention movements previous to the actual motor performance (see Allers and

Sc̆reminsky, 1926; Berger, 1937, etc., and also Ziehen, 1924, p. 483). This process of reasoning, in which inhibitions may arise from intended motor reactions which would be unfavorable, contains the elements of future reactions which can be realized later. Hence, one can agree with McDougall's (1931) opinion: 'Insight is foresight.' It is obvious that the ability of reasoning represents an enormous selective advantage, and consequently animals capable of reasoning were favored in evolution.

Many examples proving simple degrees of insight are furnished by detour ('Umweg') experiments. If an animal tries to reach a goal (e.g. food) but is prevented from taking the direct way by an obstacle, there are two ways of solving this problem: learning the detour by trial and error performance or confining the trial and error to processes in the brain, that is to say, thinking these various possibilities over without actually trying them. The latter solution undoubtedly is the beginning of insight.

Further examples of insight are provided by cases of genuine imitation, i.e. of connecting visual or acoustic sensations and conceptions concerning other individuals with motor concepts of the animal's own body. Genuine imitation is the result not of random associations, but of direct transpositions of activities perceived in other individuals to the activities of its own 'person'. Hence, it is obvious that previsions of the near future (i.e. concerning the performances of its own body in the near future) determine the result of association processes and the behavior actually performed. Generally, animal imitation is much less frequent than is usually assumed. This applies even to monkeys and apes; and in the case of parrots, starlings, and other birds which are perfect 'imitators', very often only latent associations are revived. This certainly is so as far as instinctive releasers are concerned. If in a flock of newborn chickens one animal starts to drink, many others will 'imitate' this behavior, and if a 'pecking' sound is produced by knocking on the floor with a pencil the chickens will be stimulated to peck also (see Lloyd Morgan, 1909). Such cases prove only the release of an instinct or of a mood and not genuine imitation (which would require 'conclusions'). Nevertheless, genuine animal imitation and the association processes involved should be carefully studied. The highest degree of animal reasoning and conclusion has evolved in apes, and especially in chimpanzees (Köhler, 1921; R. M. and A. W. Yerkes, 1929, etc.), which are able to use tools without previous conditioning. Such cases of the use of implements include the use of sticks to get fruits which have been suspended to the ceiling, and as tools for digging and fighting, and the placing of several wooden boxes on top of each other so that a desired object can be reached. This way of handling boxes as tools has also been demonstrated in capuchin monkeys (Bierens de Haan, 1940). It is important to note that the solutions of such tasks often are found not after trial and error, but after spontaneous 'deliberation'. This implies that such experiments demonstrate first stages of creative imagination which are quite similar to the corresponding human reactions and which therefore suggest the assumption of similar psychic processes in animals.

Hence, the deep gap between primitive Recent man and the highest mammals seems to be fairly well bridged by such evidence of insight in animals. Nevertheless, one should not forget the fundamental difference and eminent advantage of human thinking: the ability to think in words, i.e. to connect most associations to the motor area of speech (Broca's region), which is lacking in animals. It was this fact that enabled modern man to discriminate between 'objects' and his sensations and conceptions. In this context it is interesting to note that this discrimination apparently is only poorly developed in primitive man, and that in the oldest Greek literature there is no term for 'matter' (Ziehen, 1937, p. 31). It is well known that the naïve type of modern man can hardly separate the qualities 'red', 'sweet', 'hard', etc., from the object which only produces these qualities in his sense organs and neurons, and neither does he analyze the other properties of his sensations and conceptions. For all animals, then, objects and the sensations arising from them seem to be identical to a large extent.

In spite of these facts, the higher animals *begin* to divide the phenomena experienced into varying 'psychic' (parallel) components and constant 'material' (causal) components. An ape comprehends causal connections when he uses various objects as tools or when he 'invents' the use of boxes to get a desired fruit which is beyond his reach. Parallel to the loss of so many inherited instincts and to the increasing hazards arising from man's mental freedom, there is evident in the higher animals an increasing adaptation to the realities of the external 'material' world (to the reality constructed by 'reductions': see below). This adaptation to reality led to man's attaining control of the world of matter. As this ability presented a definite selective advantage, it was favored in the process of evolution. Hence, we may assume that all evolution of thought and reasoning must of necessity be confined to certain directions, since wrong abstractions of facts were invariably followed by severe corrections from the 'external world', so that it was necessary for thinking to be in close correspondence with the material world. Verworn (1910, 1919) was altogether right in stating (1919, p. 39): 'If there are beings capable of thinking on Mars or any other cosmic body, their kind of thinking must be the same as ours.' According to my opinion it is very essential for a proper evaluation of phylogenetic processes to note that the evolutionary necessity – often referred to in the preceding chapters of this book (see especially Chapter 4) – also applies to the origin of human thinking.

The essential features of the evolutionary step from ape to man are man's ability to rely on certain insights and experiences of high predictability as truths, and the continual increase in range of such truths. It is easy to realize, however, that such truths can arise only from repeated experiences always giving the same result. Let us look at a photograph of the surface of the moon. If the light and shadows cast by the lunar mountains coincide with the light cast by the lamp under which we look at the picture, we shall have no doubt that the lunar mountains really are elevations above the surface and we shall be sure that this picture is 'true' with regard to the real lunar surface. If we now turn the photograph by 180 degrees, so that the light of our lamp and the shadows cast by the lunar mountains are opposite, we get the impression that the mountains are deep valleys. Quite similar 'true' impressions arise when we are in a train or plane which is taking a sharp turn, so that the vertical axis of our body no longer coincides with terrestrial gravity: in such moments buildings, trees, etc., seem to be in an oblique position. If we touch the upper row of our incisors with the lower surface of our tongue we get an equally wrong impression. Hence, to us those sensations which are in accord with common experience seem to represent the truth. Even the 'truth' of geometrical axioms (e.g. a straight line is the shortest distance between two points) is only the result of repeated and uncontradicted experience, though an essential part of such experiences is gained 'experimentally' from purely imaginative thinking (paradigmatic conclusions in the sense of Ziehen).

E. PHILOSOPHY AND THE EVOLUTION OF PSYCHIC PHENOMENA

In our attempt to trace a possible phylogeny of psychic phenomena, we have had to rely on conclusions derived only by analogy. As a matter of course, these analogies become more and more vague the farther we descend the evolutionary ladder. Nevertheless, we may distinguish the following gradations:

1. Insight and reasoning (at least to a limited extent) must be credited to apes, monkeys, and carnivores.

2. The formation of averbal conclusions from averbal judgments has been demonstrated among some birds (parrots, crows).

3. Judging from their behavior, one has to assume that many invertebrates are capable of averbal judgments, though we must not forget that their kind of 'judgment' is similar to that of man only with respect to the basic psychic components.

4. The same applies to animal averbal mental images.

5. We may rightly assume that mental images – at least in the restricted sense of primary images – are formed in all animals capable of retention, i.e. down to the lower invertebrates.

6. As there is no gap in the lines of descent ('omne vivum e vivo') we may credit with sensations all animals possessing sense organs or being capable of sensory reactions, i.e. 'down' to protozoans. Along with sensations we would also have to credit them with perception of the properties of sensations, hence with accompanying feelings.

Owing to the above conclusions, we have attributed the simplest psychic components to a group of lower organisms, comprising the flagellates and mycetozoans (myxomycetes) and sometimes considered as belonging to the realm of plants. As higher plants also show evidence of sensory performances and reactions (tropism, taxis), accompanied by electric action currents quite similar to those of animals, one may well ask whether plants should be credited with sensations. Of course, we cannot decisively answer this question – similar to that raised by the case of the protozoans – but with regard to the gapless evolutionary descent of animals and plants we are hardly justified in flatly denying the existence of sensations in plants, though we have no certain knowledge of any details concerning them. Most probably, however, these sensations would not be connected with each other, and therefore would be basic elements essentially different from ours, which are imbedded in a continuous 'stream of consciousness'.

On the other hand, the possible occurrence of protozoan sensations leads to the problem of the origin of psychic components in general. In view of the often vague conclusions regarding the evaluation of protozoan 'psychology', one might be disinclined to a further discussion of such problems. We shall see, however, that a treatment of these questions from the standpoint of cognition theory can support the more or less vague analogies as far down as the protozoan level. After all, the fundamental problem is whether or not there was a *de novo* formation of the 'psyche' (i.e. of the laws governing parallel components: see below). It seems advisable to delay dealing with this question until we have first considered a radically opposite view.

Hempelmann (1926, Chapter XIV), treating the problem from a point of view clearly based on cognition theory – as not too many biologists usually do – advanced the opinion that consciousness most probably originated spontaneously and suddenly in the course of phylogeny. To him it is certain that protozoans and plants lack any kind of psychic phenomena, and it is questionable whether even insects should be deemed capable of sensations, feelings, etc. 'It is quite possible that the majority of animals lack consciousness' (p. 620). Hempelmann enumerated the following reasons that led him to this conclusion:

1. From our personal experience we know that consciousness is a discontinuous process interrupted by sleep, faints, or narcosis.

2. The existence of thresholds of sensations shows that the phenomena of consciousness arise as discontinuous events.

3. Even complicated cerebral processes need not be accompanied by (psychic) parallel processes. (Here one could especially refer to certain 'automatic' actions, such as climbing a staircase, etc., which were not mentioned by Hempelmann.)

4. Life itself originated suddenly, as there are no 'semi-living' creatures.

However, these arguments are not so safe as they seem to be at first sight.

1. We can hardly prove that consciousness is a discontinuous process, as we can adduce only occasional interruptions of this stream of phenomena which we call the self. In determining the continuity of our consciousness, we have to rely on our retention ability, from which we derive the fact that we have experienced phenomena of consciousness in the past. Even if the normal continuity is actually interrupted (by decreased oxygen supply, etc.) there might be some psychic processes going on, and it is only our inability to

recall them, i.e. to keep them in memory, that leads to the assumption of discontinuity. Of course, one need not accept this assumption, but Hempelmann's first argument is not sufficient proof of the contrary.

2. The existence of sensory thresholds is due to strictly physiological facts because all stimuli not reaching a certain intensity are barred by physiological and anatomical mechanisms from being perceived, and only in the course of phylogeny have the sensitivity and complexity of sense organs been increased. This fact alone suggests that psychic processes need not be restricted by absolute thresholds, but that they can also accompany sensory processes of atomic dimensions (e.g. in the case of light perception). Moreover, one should note that the psychic process, after passing the threshold, does not commence in full strength but progresses with gradually increasing intensity, which suggests that there might be lower degrees of the phenomenon which are unable to effect an engram and to be connected with the stream of consciousness.

3. The fact that complicated cerebral processes (e.g. an 'automatized' action like climbing a staircase) are not necessarily accompanied by psychic processes does not furnish proof in contradiction of the gradual phylogeny of parallel processes. We do not know whether such actions really lack conscious phenomena or whether there are psychic processes of which we do not become aware. (We can, for instance, very well recall having performed an automatized action.)

4. As we have found in Chapter 8, the assumption of a spontaneous and sudden origin of life – quoted as a parallel example to the origin of the 'psyche' by Hempelmann – is not too well founded and must even be regarded as improbable.

However, Hempelmann's assumption of a sudden origin of sensations could be supported by a fact not mentioned by him: the fact that the qualities of sensations – contrary to their intensities – so far as they can be arranged in a linear series, do not show an approach to a zero point. This is apparently due to the specific structure and function of the sense organs. In the human eye only the visual purple and some other visual substances of the cones are capable of causing certain qualities when they are decomposed chemically, but this fact does not exclude the gradual phylogenetic evolution of sense qualities and the phylogenetic differentiation of basic qualities into more highly developed ones. Generally such arguments as Hempelmann's cannot deny that the components of consciousness have possibly been existing at all times, and that they are and always have been connected with specific physiological processes by certain laws. From this viewpoint it seems improbable that such a fundamental fact as the connection of 'psychic' and physiological processes by the laws of parallellism or identism should have originated all of a sudden in the course of a continuous phylogeny in which no other real gaps can be traced.

It is because of this lack of any serious evolutionary gap that – in my opinion – one cannot accept the view maintaining that only vertebrates show evidence of psychic processes (Edinger, Richet, and others), or that only metazoans possessing neurons can be credited with parallel processes. According to our previous findings and discussions we are justified in assuming – along with Fechner, Wundt, Ziehen, Plate, and others – psychic (parallel) processes of some kind in all living beings.[1] As stated above, there is no serious argument contradicting this view, though – on the other hand – conclusions by analogy favoring this assumption become more and more uncertain the farther we go down the phylogenetic ladder (see also Ziehen, 1921).

As we have seen in Chapter 8, the gap between living and nonliving systems is bridged to some extent by micrococci and rickettsia types, small and large viruses, and phages, forming a sort of model gradation leading down to the nonliving world, though they cannot be considered as the phylogenetic persistence of the evolutionary lines of descent. Here again it is difficult to assume a sudden origin of first psychic elements somewhere in this gradual ascent from nonliving to living systems. It would not be impossible to ascribe 'psychic' components to the realm of inorganic systems also, i.e. to credit nonliving matter with some basic and isolated kind of 'parallel' processes.

Such a 'hylopsychic' view will most probably not be accepted by very many biologists. However, quite a number of facts in addition to our findings in comparative animal psychology support this view. These facts belong to the field of cognition theory and even of microphysics.

As many modern biologists are not too well acquainted with philosophy, a broad review of the problems ahead is necessary. This will also facilitate the proper understanding of the concluding pages. If we want to analyze the relations between the so-called 'psychic' and the so-called 'material' phenomena, we have to use a cognition theory as a foundation. All cognitive considerations have to start from phenomenal facts whose real existence cannot be doubted, as they are most immediately experienced by ourselves (hence we cannot properly define them). These phenomenal facts are: sensations, mental images, judgments and conclusions, 'feelings', and volitions. Perceiving the image of a rose, I can state that I experience sensations of red and green color, accompanied by sensations of touch and smell, which cause pleasant feelings. All these sensations are connected with processes of thinking, i.e. of memory and judgment, such as 'this is a rose' or 'this is a flower'. All these phenomena actually experienced as reality are strictly psychic ('phenomena' in the sense of Kant or – more appropriately – 'gignomena' in the sense of Ziehen). I cannot maintain with the same certainty that there is a 'material' rose, irrespective of my sensations, though I can conclude from repeated experiences giving the same sensations and also from remarks about the sensations experienced by other egos that there might be a rose showing a 'constancy' beyond and independent of the range of myself. As a biologist, however, I shall admit that a rose beyond the range of my sensations most probably is

---

[1] Indian religions, in particular, presuming a transmigration of souls in all kinds of living beings, show that the conviction of a general consciousness (Brahma) is very old.

not 'red', as the sensation 'red' arises only after the visual cones of my retina have been stimulated by light of a certain wavelength. Similarly, I am convinced that such a rose *per se* does not 'smell', as the pleasant sensation of an agreeable odor arises only from the stimulation of my nasal sense organs by certain molecules emitted by the rose. Finally, the blossom of the rose certainly is not 'soft', because this sensation is only the result of molecules pushing against my touching hand. As to the 'material' basis of such a 'reduced' rose (lacking its qualities, i.e. being not-red, not-smelling, not-soft), I am told by modern physics that it is made up of complexes of energy having the characteristic of corpuscles and waves at the same time (hence the term 'Wellikel'), so that it cannot be described in terms of geometry nor comprehended adequately by any kind of imagination. Hence, a strictly scientific view yields a concept of such a 'material' rose that has in it nothing like the 'thing' or 'matter' in the sense of a naïve realism.

Now, according to the cognition theory there need not be any contrast between material and psychic phenomena, and Ziehen (1934) has convincingly proved that no fundamental difference can be found between these two complexes which at first sight seem to be quite unlike. It is not true, for instance – though it has been maintained – that only matter can be spatial, mobile, and divisible, as visual sensations are also arranged in a spatial manner, they are mobile, and one can distinguish parts of them. Hence, there is no primary psychophysical dualism at all. One can distinguish two components, however, in the phenomena, especially in sensations, as the essential bases of mental images: one (e.g. wavelength of the light perceived) which is subject to causal laws, and one (the qualities of the sensation, i.e. the color) which is connected by certain laws of parallelism. This aspect reveals the 'psychic' nature of the causal components, so that the term 'psychic' as a means of discriminating between two fields of phenomena may be abandoned in discussing the theory of cognition; it suggests primary differences between 'material' and 'psychic' which do not exist. All 'existence' is the existence of phenomena. This view was stressed by Berkeley in his identification: 'esse percipi'. Hence there is a neutral 'last something' (not in the sense of a material 'carrier' and thus not to be called by Kant's term 'Ding an sich') which changes according to causal laws (energetic changes in the language of the physicist) and at the same time according to laws of parallelism, and appears at first as sensations. (A third type of laws, that of logic, applies to both categories.)

It is an important philosophical question whether there is an 'extra-phenomenal' reality, i.e. whether objects not perceived by anyone really exist and, if so, what properties they have. Usually we assume that such objects do exist (i.e. a rose not seen by anybody would nevertheless exist). Our assumption is the result of eliminating the qualities and various intensities of our sensations (i.e. eliminating the redness, the pleasant smell, the softness of touch, etc.: see above). This process of elimination and reduction reveals the pure causal components of sensation, which are governed only by causal laws, referred to by Ziehen as 'Redukte'. In these reduced components the locality of sensations is left intact in its (Euclidean) three-dimensionality, while the qualities of the sensations are eliminated. The absolute 'subjectivity' of our concept of space, as assumed by Kant, cannot be proved. (One has to admit, however, that the indefinable peculiarity of our experience of spatiality, with its differentiation between above and below, right and left, and in front of and behind, and with its continuity of space as perceived by our visual organs, must be considered as a parallel component. Hence we would have to refer to the spatial qualities of the reduced components as 'locative' and to those of the sensations as local.) Similarly, the temporality of sensations remains unaltered when it is reduced to the aspect of causal regularity, as in the reduced components temporality also is one-dimensional, continuous, and lacking a zero point (though the over-progressing 'presence' in our sensation of time is a parallel component, so that in the sense of Ziehen we should rather refer to the 'temporativic' qualities of the reduced components). From such a consideration one may infer that the assumption of an extraphenomenal reality of such reduced components is justified. This extraphenomenal reality of the reduced components could be experienced, provided that a phylogenetic development should yield suitable sense organs. With this assumption we exceed the positivistic principle of immanence (i.e. that all statements based on cognition theory should remain within the realm of actual phenomena) but only in the sense of transgredient (not transcendent) conclusions. As such a 'reductive (transgredient) realism' (Ziehen) emphasizes the 'psychic' character of all phenomena and refuses any 'material' objects, it is still in the realm of idealism.

The typical feature of parallel 'psychic' components, i.e. of alterations of phenomena due to the law of parallelism, is that they are qualitative (they cannot be arranged serially or, if so, there is no zero point) or, at least, not purely quantitative like the alterations governed by causality (one need only compare the sensations of various colors and the series of corresponding wavelengths). Parallel 'psychic' components are not autonomous, but are always correlated to certain causal processes in sense or nerve cells in a completely parallel manner. It is impossible to show a connecting link between the last element of the causal chain and the origin of sensation. On the other hand, it is typical of causal reactions that they proceed along a spatial way (hence there is an element of truth in the otherwise false concept of 'matter' as being mainly characterized by spaciousness). For some nervous cerebral processes one can trace or assume the probability of the path from the primary stimulation of the respective sense organs via many stimulations of central ganglion cells to the excitation of the final and reacting organ, as Verworn (1912, 1920), in his *Mechanics of Mental Processes*, tried to show. Only certain parts of this causal reaction (such as certain sensations and concepts, competition of motivations, etc.) are accompanied by parallel processes. These parallel processes are fully coincident with the causal events and happen instantaneously. Hence, they are fundamentally different from causal components, i.e. they are fully acausal. Contrary to the opinion of vitalists, who assume acausality in complicated physiological processes, it is only here that we have reached the border of causality. Trying to coordinate parallel 'psychic' processes with causal events, we find it is only here that we meet with irrational phenomena, i.e. with phenomena which can be experienced but not explained.

Let us now continue our phylogenetic considerations. We have seen that it is possible to assume parallel 'psychic' components even in unicellular organisms and that the laws of parallel correspondence may possibly be as 'eternal' as causality. Hence, molecules and atoms should also be credited with basic parallel components of some kind. These parallel processes can be recognized as such only after the respective molecules have become part of the psychophysical substance (nerve and sense cells) of an organism, so that the parallel components form a complex of conscious phenomena which can be 'experienced'. Such a hylopsychic point of view is supported by the opinion of modern physicists that 'matter' is nothing but a condensation of energetic fields ('wave packages'), i.e. of something that is in the last analysis 'immaterial'.

To this hylopsychic opinion one might oppose the argument that along with the origin of new parallel components – e.g. of auditive sensations – new laws (governing these new auditive processes) might have originated, i.e. that parallel regularity might have been subjected to an evolutionary alteration and hence may have 'begun' in some remote time. This would be contrary to the laws of causality, which have always existed without any change, though their complexity has increased. Regarding these problems, Ziehen (1913, p. 249) summarized his opinion as follows:

Unalterable causality creates complicated complexes of reduced components, destroys them, and creates them anew, so that – correspondingly – the parallel components arise and disappear along with their new regularity.

One may seriously doubt, however, that both regularities should last for only such limited periods of time. This becomes evident from a consideration of the evolution of our terrestrial or solar systems, suggesting that the manifestation laws of both parallel and causal components evolved gradually. As long as our earth was a glowing, gaseous ball, there were no laws of osmotic pressure, of raised boiling and decreased freezing points, of lever action, of pendulum regularity, etc. Of course, these laws were already implicit parts of the eternal laws governing gravitation, diffusion, etc. One may well imagine that the laws connecting parallel 'psychic' components with certain causal processes evolved in a similar manner.

We shall not go into further detail regarding this problem of cognition theory, but we may state that a hylopsychic concept is well in accord with many findings and facts of the natural sciences, and that it is possibly the most suitable basis for a universal philosophy. Hence, Spinoza's famous sentence,

'Omnia, quamvis diversis gradibus, animata sunt' has been accepted (on the basis of identism or parallelism) by many philosophers (e.g. Mach, Wundt, Von Hartmann, Haeckel, Fechner, Ziehen). In this context the zoologist will be interested in the opinion of Spemann (1943), who – from the fact that a group of cells 'destined' to become a part of the skin can be transformed into cerebral cells if transplanted early enough – concluded that:

> the organism is animated (*beseelt*) in all its living parts, [and that] the parts of our body on which we are standing and walking might as well become parts for thinking if they had happened to be in a different place in the whole.

F. PSYCHIC PROCESSES AND SOMATIC EVOLUTION

In many cases hylopsychic concepts of modern biologists have been used to support a vitalistic opinion. This results from our present inability to fully analyze certain vital processes on the basis of causality. Hence, the action of psychic processes in the sense of psychophysical interactions is suggested. Often the apparent 'freedom' of human volition is taken as a model of such hypotheses, as in Driesch's 'entelechy', Pauly's 'cellular intelligence', or Bergson's 'élan vital'. As we have seen, however, it is probable that processes of volition do not contain any fundamentally different parallel 'psychic' components, as they can be understood as processes necessitated by the properties and constellation of conceptions. And especially in the animal world, 'will' certainly is not 'free', but depends on the intensity of excitations and emotions, as well as on inherited association patterns. Above all, there is no indication that the causality of nervous processes should not be as continuous and gapless as all other processes of physiology. One can neither expect theoretically nor prove empirically that causality is interrupted in some place where acausal parallel processes might become 'inserted', i.e. might act as a 'cause'. On the contrary, it is for their constant parallelism that parallel 'psychic' components have been given their name (therefore sometimes referred to as epiphenomena).

It is even more improbable that the parallel components of volition (e.g. 'élan vital') might interfere in a finalistic sense with non-nervous physiological processes, as has been assumed by some vitalists, especially in order to 'explain' certain processes of the physiology of development which have not yet been fully analyzed. By such assumptions, however, nothing is explained, as an unknown process is circumscribed only by a vague terminology not based on any facts. Woltereck's statement (1940) that 'an ontic center', a 'non-spatial interior', 'expresses itself materially' or 'belongs to other factual relations' can hardly be accepted as a serious explanation of any problem (not to mention the admitted fact that all sensations must be considered as having the property of locality, so that – from the point of view of the theory of cognition – Woltereck is wrong in suggesting a 'non-spatial interior', and also not to mention that a reference to an 'interior' and an 'exterior' – like all comparisons – cannot explain anything in this context).

There have also been occasional attempts to advance the hypothesis that voluntary processes might govern the major trends of evolution: 'It is the drive to realize an autonomous, own shape (*Gestaltung*), i.e. a will to individual perfection' (Beurlen, 1937; also see Schuh, 1937). One can understand, however, that such concepts were possible among paleontologists as long as there was no satisfactory causal explanation of orthogenetical series and of the phylogenetic 'cycles' in the lines of descent. In this book, however, I hope to have shown that when the methods and findings of modern zoology are applied to an analysis and explanation of these complex problems of evolution it is strongly suggested that we need not accept or assume the influence of psychic (acausal!) phenomena in the uninterrupted course of causal processes.

11

Conclusion

Reviewing the manifold problems of transspecific evolution and the results obtained from an enormous wealth of material and facts, we feel justified in stating that, while many fields certainly need further analysis, the general lines of thought in the field of evolution are, on the whole, well established. If there are some special problems to which we can only say 'Ignoramus', we need not add 'Ignorabimus', as Du Bois-Reymond did in his famous speech of 1872. Harvey's statement at the beginning of the seventeenth century, 'Omne vivum e vivo', has again and again been proved to be true and is a sound basis for the assumption that even in cases of evolutionary gaps we can be sure that the enormous multitude of forms is the product of one continuous phyletic tree. This giant tree has many branches, some of them dead, but on the tips of the ramifications life is carried on by an enormous number of individuals, proof of the coherent stream of life, which is diversified only by evolution. This wealth of forms is the result of continuous, undirected mutation and is patterned by the respective conditions of selection. As we have seen, this process of animal transformation is to a large extent necessitated.

In spite of this grand evolutionary aspect, many biologists will probably consider one essential feature unsatisfactory: that the 'randomness' of mutation and of natural selection should be the decisive factor of progressive evolution and even of the origin of man. This might seem an even more serious objection with regard to the fact that (compare the previous chapter) the origin of 'psychic' or parallel components, i.e. of our sensations, conceptions, volitions, and feelings, is indissolubly connected with somatic evolution.

In the course of our consideration of progressive evolution, we have already stressed the fact that the term 'randomness' denotes only our inability to analyze sufficiently extremely complicated but causally determined processes of evolution. And the unsatisfied feeling certainly is not a sufficient reason to deny or reject the above view, which, though it may seem to be incompatible with some philosophical opinions, should be considered as a stimulus for a deeper analysis of what is as yet unknown. In my opinion, such scruples are not necessary. Actually, one should realize that cosmic and terrestrial processes, including all somatic and 'psychic' evolution, are effected by eternal laws, the causal, the 'parallel' ('psychic'), and the logical laws. On our earth these basic laws have resulted in regularities of a higher order (systemic laws) parallel with the growing complication of causal and parallel ('psychic') components. This theory reveals a universal aspect of great sublimity. Let us emphasize once again that this aspect can also be based on an idealistic opinion. Since all phenomena are 'psychic', there is no contrast of subject and object or of matter and 'soul', and the natural scientist need not search for an abstract world of reduced matter ('Redukte') beyond the realm of general 'consciousness'. The concept of three types of laws and the idea that all beings are 'animated' although, of course, not in the human sense of a self-conscious-ness, certainly can also be fitted into the philosophy of scientists not working in the fields of natural science. It is a personal matter whether one makes such a hylopsychic view a part of a general theistic concept, or adheres to Spinoza's 'deus sive mundus' or Goethe's 'God-Nature', or contents oneself without these or any similar solutions of an old human problem, by considering one-self as a part of an incomprehensible cosmic regularity.

Postscript

Translation and production of this book took a rather long time. It was therefore impossible to discuss all those papers on evolution which appeared during the last three years. However, it may be useful to give references of some more important corresponding titles here. In most cases the investigations mentioned fit well in my conception and may be considered as good supplements.

Problems of mutation, selection, and speciation (Chapters 2 and 3) were treated by several authors. It may be sufficient to enumerate the following papers:

c. h. waddington. 1957. *The Strategy of the Genes*. London.

th. dobzhansky. 1957. 'Mendelian populations as genetic systems.' *Cold Spring Harbor Symp. Quant. Biol.*, **22**:385–93.

h. lüers and h. ulrich. 'Genetik und Evolutionsforschung bei Tieren.' In g. heberer. 1954. *Die Evolution der Organismen*, 2nd ed., 552–661. Stuttgart.

w. ludwig. 'Die Selektionstheorie.' In the same book, 662–712.

p. m. sheppard. 1958. *Natural Selection and Heredity*. London.

th. dobzhansky and o. pavlovsky. 1957. 'An experimental study of interaction between genetic drift and natural selection.' *Evolution* **11**: 311–19.

a. and v. grant. 1956. 'Genetic and taxonomic studies in *Gilia*.' *El Aliso*, **3**: 203–87.

t. m. sonneborn. 1957. 'Breeding systems, reproductive methods and species problems in Protozoa.' In e. mayr. *The Species Problem*, 155–324. Washington D. C.

c. g. sibley. 1957. 'The evolutionary and taxonomic significance of sexual dimorphism and hybridization in birds.' *Condor*, **59**:166–91.

A new method of evaluating evolutionary rates (Chapter 6 A) has been worked out by b. kurtén. 1958. A differentiation index, and a new measure of evolutionary rates. *Evolution*, **12**:146–57.

Questions of evolutionary speed and 'explosive' phases are also treated by:

b. schaeffer. 1952. 'Rates of evolution in the Coelacanth and Dipnoan fishes.' *Evolution*, **6**:111–11.

c. stern. 1954. 'Genes and developmental patterns.' *Caryologia*, **6**, Suppl.: 355–69.

m. f. glaessner. 1957. 'Evolutionary trends in Crustacea (Malacostraca).' *Evolution*, **11**:178–84.

Many papers were published on the evolutionary effect of allometric growth, which has been treated in some detail in our book (Chapter 6, B III). It may be sufficient to refer to the following titles:

L. VON BERTALANFFY. 1957. 'Wachstum.' In HELMCKE, VON LENGERKEN, STARCK. *Handb. d. Zool.*, vol. 8, pt. 4 (6). Berlin.

B. RENSCH. 1958. 'Die Abhängigkeit der Struktur und der Leistungen tierischer Gehirne von ihrer Größe.' *Naturwiss.* 45:145–54.

M. SHIMIZU. 1956. 'On the allomorphosis in the bones of the birds. I—IV.' *Bull. Fac. Educ. Shinshu Univ.* 6, 7, 9.

W. ROSE. 1956. 'Allometrische Änderungen bei der Ontogenese und Phylogenese des Eidechsengehirns.' *Zool. Jahrb., Abt. Anat.* 75:433–80.

K. W. HARDE. 1957. 'Die Verschiedenheiten der Körperproportionierung bei *Acanthocinus aedilis* L. (Col. Ceramb.).' *Z. Morph. u. Ökol. Tiere*, 46: 293–320.

R. MÜLLER. 1957. *Vergleichende Untersuchungen über die Grössenverhältnisse des Ohrlabyrinths einheimischer Säugetiere.* Diss. Erlangen. Leipzig.

L. DINNENDAHL and G. KRAMER. 1957. 'Über größenabhängige Änderungen von Körperproportionen bei Möven.' *J. F. Ornith.*, 98:282–312.

W. REETZ. 1958. 'Unterschiedliches visuelles Lernvermögen von Ratten und Mäusen'. *Z. f. Tierpsychol.* 14: 347–61.

R. NEDER. 1959. 'Allometrisches Wachstum von Hirnteilen bei drei verschieden großen Schabenarten.' *Zool. Jahrb., Abt. allg. Zool.*

On evolution of man and of human culture (Chapter 7, D) compare two books of Dobzhansky and the author's new book:

TH. DOBZHANSKY. 1956. *The Biological Basis of Human Freedom.* New York.

TH. DOBZHANSKY. 1957. *Evolution, Genetics, and Man.* New York and London.

B. RENSCH. 1959. *Homo sapiens. Vom Tier zum Halbgott.* Göttingen.

New summaries on human phylogeny have been published in G. HEBERER's *Die Evolution der Organismen*, 2nd edition. Stuttgart; W. GIESENER. 1957. *Die Phylogenese der Hominiden*, 951–1109; G. HEBERER. 1959. *Die subhumane Abstammungsgeschichte des Menschen*, 1110–42.

Compare further on:

H. J. MULLER. 1956. 'Genetic principles in human populations.' *Amer. J. Psychiatry*, 113:481–91.

J. T. ROBINSON and R. J. MASON. 1957. 'Occurrence of stone artefacts with *Australopithecus* at Sterkfontain.' *Nature*, 180:521–4.

The origin of life (Chapter 8) has been discussed in a Symposium at Moscow, summarized by N. W. PIRIE. 1957. 'The origin of life. Moscow Symposium.' *Nature*, 180:886–8.

Compare also:

M. CALVIN. 1956. 'Die chemische Evolution und der Ursprung des Lebens.' *Naturwiss.* 43:387–93.

Bibliography

ABEL, O. 1918. 'Das Entwicklungstempo der Wirbeltierstämme.' *Schrift. Ver. Verbreit. naturwiss. Kenntn.* Wien, **58**:91–120.
— 1920. *Lehrbuch der Paläozoologie.* Jena.
— 1928. 'Das biologische Trägheitsgesetz.' *Biol. gen.*, **4**:1–102.
— 1929. *Paläobiologie und Stammesgeschichte.* Jena.
— 1931. *Die Stellung des Menschen im Rahmen der Wirbeltiere.* Jena.
ADAM, W. 1941. 'Cephalopoda. In: Res. Sci. Croisières "Mercator",' **3**. *Mém. Mus. Hist. Nat. Belge*, 2nd sér., fasc. 21.
AEPPLI, E. 1952. 'Natürliche Polyploidie bei den Planarien Dendrocoelum lacteum (Müller) und D. infernale (Steinmann).' *Z. indukt. Abstamm. u. Vererbungsl.*, **84**:182–212.
ÅGRELL, I. 1948. 'Studies of the postembryonic development of Collemboles.' *Arkiv f. Zool.*, A 41, No. 12.
ALEXANDER, W. B. 1941. 'The index of heron population 1940.' *Brit. Birds*, **34**:189–94.
ALLEE, W. C., and K. P. SCHMIDT. 1951. *Ecological animal geography.* 2nd ed. New York and London.
ALLERS and SCŘEMINSKY. 1926. 'Über Aktionsströme der Muskeln bei motorischen Vorstellungen und verwandten Vorgängen.' *Pflügers Arch.*, **212**:169–82.
ALPATOV, W. W., and A. M. BOSCHKO-STEPANENKO. 1928. 'Variation and correlation in serially situated organs in insects, fishes and birds.' *Amer. Nat.*, **62**:409–24.
ALTEVOGT, R. 1951. 'Vergleichend-psychologische Untersuchungen an Hühnerrassen stark unterschiedener Körpergröße.' *Z. f. Tierpsychol.*, **8**:75–109.
ALVERDES, F. 1937. 'Das Lernvermögen der einzelligen Tiere.' *Z. f. Tierpsychol.*, **1**:35–8.
ALVERDES, F., and F. BRAMSTEDT. 1939. 'Erwiderung auf das Koehlersche Nach- und Schlußwort.' *Verh. Dtsch. Zool. Ges.*, **41**:470–3.
AMADON, D. 1950. 'The Hawaiian honeycreepers (Aves, Drepanididae).' *Bull. Am. Mus. Nat. Hist.*, **95**, Article 4.
ANCONA, U. D'. 1939. *Der Kampf ums Dasein. Eine biologisch-mathematische Darstellung der Lebensgemeinschaften und biologischen Gleichgewichte.* Berlin.
— 1939. 'Grandezza nucleare e poliploidismo nelle cellule somatiche.' *Monit. Zool. Ital.*, **50**:225–31.
— 1942. 'I Niphargus Italiani. Tentativo di valutazione critica delle minori unità sistematiche.' *Mem. Ist. Ital. Speleol.*, Ser. Biol., Mem. IV. Trieste.
ANDER, K. 1948. 'On the correlation between body length and ovipositor length in Ensifera (Salt.).' *Opuscula Entomologica*, 64–8.
ANDREWS, R. C. 1921. 'A remarkable case of external hindlimb in a humpback whale.' *Amer. Mus. Novit.*, **9**.
ARIENS-KAPPERS, C. U. 1934. 'Feinerer Bau und Bahnverbindungen des Zentralnervensystems. In: Bolk, and others.' *Handbuch vergleich. Anat. der Wirbeltiere*, **2**, 1. Hälfte, 319–486. Berlin and Vienna.

ARISTOTLE. 1902. Aristotle's *Psychology*. Trans. by W. A. Hammond. London.

ARMBRUSTER, L., H. NACHTSHEIM, and T. ROEMER. 1917. 'Die Hymenopteren als Studienobjekt azygoter Vererbungserscheinungen.' *Z. ind. Abstammungs- und Vererbungslehre*, **17**:273–355.

ARNDT, W. 1939. 'Abschließende Versuche zur Frage des "Zähl"-Vermögens der Haustaube.' *Z. f. Tierpsychol.*, **3**:88–142.

ARNDT, W., Sr. 1940. 'Der prozentuale Anteil der Parasiten auf und in Tieren im Rahmen des aus Deutschland bisher bekannten Tierartenbestandes.' *Z. f. Parasitenkunde*, **11**:684–89.

—— 1941. 'Die Anzahl der bisher in Deutschland (Altreich) nachgewiesenen rezenten Tierarten.' *Zoogeogr.*, **4**:28–92.

—— 1943. 'Das "philippinische Elefantenohr" Spongia thienemanni n. sp. Zugleich ein Überblick über unsere bisherige Kenntnis des Vorkommens geographischer Rassen bei Meeresschwämmen.' *Arch. f. Hydrobiol.*, **40**:381–442.

ARRHENIUS, S. 1907. *Das Werden der Welten*. Leipzig.

AUDOVA, A. 1929. 'Aussterben der mesozoischen Reptilien.' *Palaeobiologica*, **2**:365–401.

AUERBACH, C. 1949. 'Chemical mutagenesis.' *Biol. Rev.*, **24**:355–91.

AUTRUM, H. 1941. 'Über Gehör und Erschütterungssinn bei Locustiden.' *Z. f. vergl. Physiol.*, **28**:580–637.

BABAK, E. 1921. 'Die Mechanik und Innervation der Atmung.' In: Winterstein, *Handbuch d. vergleich. Physiol.* **1**, 2. Hälfte, 265–1027. Jena.

BACKMAN, G. 1943. 'Wachstum und organische Zeit.' *Bios*, 15. Leipzig.

BAER, K. E. VON. 1828. *Über die Entwicklungsgeschichte der Tiere*. Königsberg.

—— 1876. *Studien aus dem Gebiete der Naturwissenschaften*. St. Petersburg.

BAKER, H. G. 1953. 'Race formation and reproductive method in flowering plants.' *Symposia Soc. Exper. Biol.*, No. VII, *Evolution*, 114–45.

BALDI, E. 1949. 'Différents mécanismes d'isolement au sein de populations de crustacés planktiques.' In: *Symposium sui fattori ecologici e genetici della speciazione negli animali. La Ricerca Scientif.*, **19**, Suppl., 119–22.

—— 1950. 'Phénomène de microévolution dans les populations planktiques d'eau douce.' *Vierteljahresschr. Naturforsch. Ges. Zürich*, **95**: 89–114.

BALDWIN, J. M. 1902. *Development and evolution*. New York and London.

BALDWIN, P. H. 1953. 'Annual cycle, environment and evolution in the Hawaiian honeycreepers (Aves: Drepaniidae).' *Univ. Calif. Publ. Zool.*, **52**:285–398.

BALINSKY, B. J. 1938. 'On the determination of entodermal organs in Amphibia.' *Compt. Rend. Acad. Sci. Paris* (N.S.), **20**:215–17.

BALTZER, F. 1950a. 'Chimären und Merogone bei Amphibien.' *Revue Suisse Zool.*, **57**:93–114.

—— 1950b. 'Entwicklungsphysiologische Betrachtungen über Probleme der Homologie und Evolution.' *Rev. Suisse Zool.*, **57**:451–77.

BARTELS, E. D. 1941. 'Studies on hereditary dwarfism in mice. 3. Development of the adrenals in dwarf mice.' *Acta path. scand.* (Kobenh.), **18**:20–35.

BASLER, 1911. *Arch. ges. Physiol.*, 143.

BATES, M. 1939. 'Hybridization experiments with Anopheles maculipennis.' *Amer. J. Hyg.*, **29**:1–6.

BATES, R. W., T. LAANES, and O. RIDDLE. 1935. 'Evidence from dwarf mice against the individuality of growth hormone.' *Proc. Soc. Exper. Biol. Med.*, **33**:446–50.

BAUER, H. 1947. 'Karyologische Notizen I. Über generative Polyploidie bei Dermapteren.' *Z. f. Naturforsch.*, **2b**:63–6.

BAUER, H., and N. W. TIMOFÉEFF-RESSOVSKY. 1943.' Genetik und Evolutionsforschung bei Tieren.' In: Heberer, *Evolution der Organismen*, 335–429. Jena.

BEADLE, G. W. 1947. 'Genes and the chemistry of the organism.' *Science in Progress*, **5**:166–96.

BECKER, E. 1938. 'Die Gen-Wirkstoff-Systeme der Augenausfärbung bei Insekten.' *Naturwiss.*, **26**:433–40.

BEECHER, C. E. 1898. 'The origin and significance of spines. A study in evolution.' *Amer. J. Sci.* (ser. 4), **6**:1–20, 125–36, 249–68, 329–59.

1901. *Studies in evolution.* London.

BEER, G. R. DE. 1930. *Embryology and evolution.* Oxford.

1940. *Embryos and ancestors.* Oxford.

BEHEIM-SCHWARZBACH, D. 1943. 'Über den taxonomischen und isolierenden Wert der Forcepsvariationen einiger Caraboidea.' *Z. Morph. Ökol.*, **39**:21–46.

BEKLEMISHEV, N., and A. N. ZHELOCHOVTSEV. 1937. 'Die geographische Verbreitung von *Anopheles maculipennis* und seiner Unterarten in URSS.' *Med. Parasitol.*, **6**:819–35. (Russian, French summary.)

BENAZZI, M. 1945. 'Dendrocoelum lacteum verbanense: nuova razza del Lago Maggiore.' *Atti acad. d. Fisiocritici, Sez. Agr.*, **10**:31–3.

1949. 'Ricerche genetico-sistematiche sui tricladi.' *Ric. Scient.*, **19**, Suppl., 113–18.

BENICK, L. 1939. 'Die höhlenbewohnende Silphide Choleva holsatica, ein Beispiel für die Altersbestimmungsmöglichkeit rezenter Tierformen.' *Verh. 7. Intern. Kongr. Entom. Berlin*, **1**:16–25.

BENINDE, J. 1937. 'Zur Naturgeschichte des Rothirsches.' *Monogr. Wildsäugetiere*, Bd. **4**. Leipzig.

1940. 'Die Fremdblutkreuzung (sog. Blutauffrischung) beim deutschen Rotwild.' *Z. f. Jagdkde.*, *Sonderheft. Neudamm u. Berlin.*

BENSON, S. B. 1935. 'Geographic variation in *Neotoma lepida* in Arizona.' *Occas. Pap., Mus. Zool. Univ. Michigan*, No. 317.

BERG, E. S. 1926. *Nomogenesis or evolution determined by law.* London.

1933. 'Übersicht der Verbreitung der Süßwasserfische Europas.' *Zoogeogr.*, **1**:107–208.

1933. 'Die bipolare Verbreitung der Organismen und die Eiszeit.' *Zoogeogr.*, **1**:449–84.

1935. 'Über die amphiboreale (diskontinuierliche) Verbreitung der Meeresfauna in der nördlichen Hemisphäre.' *Zoogeogr.*, **2**:393–409.

BERGER, H. 1900. 'Experimentell-anatomische Studien über die durch den Mangel optischer Reize veranlaßten Entwicklungshemmungen im Occipitallappen des Hundes oder der Katze.' *Arch. f. Psychiatrie*, **33**:521–67.

1937. 'Physiologische Begleiterscheinungen psychischer Vorgänge.' In: Bumke and Foerster, *Handbuch der Neurol.*, **2**:492–526.

1938. 'Das Elektrencephalogramm des Menschen.' *Nova Acta Leopoldina*, **6**, B, No. 39. Halle (S).

BERGMANN, C. 1847. 'Über die Verhältnisse der Wärmeökonomie der Thiere zu ihrer Größe.' *Götting. Stud.*, **1**. Abt., 595–708.

BERGMANN, C., and R. LEUCKART. 1885. *Anatomisch-physiologische Übersicht des Thierreichs.* 2nd ed. Stuttgart.

BERINGER, C. CH. 1951. *Die Urwelt. Die Geschichte der Erde und des Lebens.* Stuttgart.

BERLIOZ, J. 1944. *La Vie des Colibris.* Paris.

BERNAL, J. D. 1951. *The physical basis of life.* London.

BERTALANFFY, L. VON. 1941. 'Stoffwechseltypen und Wachstumstypen.' *Biol. Zentralbl.*, **61**:510–32.

1951. *Theoretische Biologie.* 2 vols. 2nd ed. Berne. (1st ed., Berlin, 1932 and 1942.)

1944. 'Bemerkungen zum Modell der Elementareinheiten.' *Naturwiss.*, **32**:26–32.

1949. *Das biologische Weltbild.* Vol. 1. Berne.

1951. 'Metabolic types and growth types.' *Amer. Nat.*, **85**:111–17.

BEURLEN, K. 1932. 'Funktion und Form in der organischen Entwicklung.' *Natur-wiss.*, 20:73–80.

1937. *Die stammesgeschichtlichen Grundlagen der Abstammungslehre.* Jena.

BEUTNER, R. 1938. *Life's beginning on the earth.* Baltimore.

BIERENS DE HAAN, J. A. 1940. *Die tierischen Instinkte und ihr Umbau durch Erfahrung.* Leiden.

BIRCH, L. C. 1954. 'Experiments on the relative abundance of two sibling species.' *Austral. J. Zool.*, 2:66–74.

BIRCH, L. C., T. PARK, and M. B. FRANK. 1951. 'The effect of intraspecies and interspecies competition of the fecundity of two species of flour beetles.' *Evolution*, 5:116–32.

BIRUKOW, G. 1949. 'Entwicklung des Tages- und des Dämmerungssehens im Auge des Grasfrosches (*Rana temporaria* L.).' *Z. vergl. Physiol.*, 31:322–47.

BLAIR, F. 1948. 'Population density, life span and mortality rates of small mammals in the Blue Grass Meadow and Blue Grass Field Associations of Southern Michigan.' *Amer. Midland Naturalist*, 40: 395–419.

BLAIR, W. F., and W. E. HOWARD. 1944. 'Experimental evidence of sexual isolation between three forms of mice of the cenospecies *Peromyscus maniculatus*.' *Contrib. Lab. Vert. Biol. Univ. Mich.*, No. 26.

BLANK, H. 1934. 'Tiergröße und Stoffwechsel.' *Pflügers Arch.*, 234:310–17.

BODENHEIMER, F. S. 1931. 'Der Massenwechsel in der Tierwelt. Grundriß einer allgemeinen tierischen Bevölkerungslehre.' Atti 11. Congr. Intern. Zool. Padova (1930); *Arch. Zool. Ital.*, 26:98–111.

BOETTICHER, H. VON. 1940. 'Die Formenkreisbildung bei den Raken, Gattung *Coracias Linnaeus*.' *Senckenbergiana*, 22:362–9.

1941. 'Artenkreise oder Gattungen und Untergattungen.' *Z. f. Naturwiss.* (Halle), 94:52–60.

1944. 'Die Verwandtschaftsbeziehungen der afrikanischen Papageien (Poice-phalus und Agapornis).' *Zool. Anz.*, 145:10–27.

BOHN, G. 1903. 'Sur les mouvements oscillatoires des *Convoluta roscoffensis*.' *Compt. Rend. Acad. Sci., Paris*, 137:576–8.

BOK, S. T. 1936. 'The branching of the dendrites in the cerebral cortex.' *Proc. Kon. Ak. Wetensch. Amsterdam*, 39:1209–18.

BOK, S. T., and M. J. VAN ERP TAALMAN KIP. 1939. 'The size of the body and the size of the number of the nerve cells in the cerebral cortex.' *Acta Neerland. Morphol.*, 3:1–22.

BÖKER, H. 1927. 'Über die Ontogenese hochdifferenzierter anatomischer Konstruktionen' (*Verh. anatom. Ges.* 36). *Anat. Anz.* 63, Erg.-Heft, 96–108.

1935. 'Artumwandlung durch Umkonstruktion, Umkonstruktion durch aktives Reagieren der Organismen.' *Acta biotheoret.*, A, 1:17–34.

1936. 'Was ist Ganzheitsdenken in der Morphologie.' *Z. ges. Naturwiss.*, 253–76.

BOLK, L. 1926. *Das Problem der Menschwerdung.* Jena.

BOVERI, T. 1904. *Ergebnisse über die Konstitution der chromatischen Substanz des Zellkerns.* Jena.

BOVEY, P. 1941. 'Contribution à l'étude génétique et biogéographique de *Zygaena ephialtes* L.' *Rev. Suisse Zool.*, 48:1–90.

BOXBERGER, F. VON. 1953. 'Vergleichende Untersuchungen über das visuelle Lernvermögen bei weißen Ratten und weißen Mäusen.' *Z. f. Tierpsychol.*, 9:433–51.

BRAESTRUP, F. W. 1945. 'Progressive clines.' *Nature*, 156:337–9.

BRAMSTEDT, F. 1935. 'Dressurversuche mit *Paramaecium caudatum* und *Stylony-chia mytilus*.' *Z. f. vergl. Physiol.*, 22:490–516.

BRANDT, A. 1867. 'Sur le rapport du poids du cerveau à celui du corps chez différents animaux.' *Bull. Soc. Imp. Naturalistes, Moscow*, 40:525–43.

BRAUN, H. 1952. 'Über das Unterscheidungsvermögen unbenannter Anzahlen bei Papageien.' *Z. f. Tierpsychol.*, 9:40–91.

BREHM, A. 1912–16. 'Die Säugetiere.' In: *Brehms Tierleben*, ed. by L. Heck and M. Hilzheimer, 4. Leipzig and Vienna.

BREIDER, H. 1936. 'Eine Erbanalyse von Artmerkmalen geographisch vikariierender Arten der Gattung *Limia*. Z. *Vererbungsl.*, 71:441–99.

BRESSLAU, E. 1928–33. 'Turbellaria.' In: Kükenthal-Krumbach, *Handbuch der Zool.*, 2, I. Hälfte (1), 52–320, Berlin and Leipzig.

BRETSCHER, A., and P. TSCHUMI. 1951. 'Gestufte Reduktion von chemisch behandelten Xenopus-Beinen.' *Rev. Suisse Zool.*, 58:391–8.

BREUNING, S. 1932–6. 'Monographie der Gattung *Carabus* L.' *Bestimm. Tab. europ. Coleopt.* H. 104–10. Troppau.

1943. 'Beitrag zur Wertung der Geschlechtsorgane für die Systematik.' *Z. Morph. Ökol.*, 39: 523–6.

BRINKMANN, R. 1929. 'Statistisch-biostratigraphische Untersuchungen an mitteljurassischen Ammoniten über Artbegriff und Stammesentwicklung.' *Abh. Ges. Wiss. Göttingen, Math. Nat. Kl.*, N.F. 13.

BROCH, H. 1924. 'Hydroida.' In: Kükenthal-Krumbach, *Handbuch der Zool.*, 1. Berlin.

BRONN, H. G. 1853. *Allgemeine Einleitung in die Naturgeschichte*. Stuttgart.

BROOKS, J. L. 1950. 'Speciation in ancient lakes.' *Quart. Rev. Biol.*, 25:131–76.

BROOM, R., and G. W. H. SCHEPERS. 1946. 'The South-African fossil ape-men. The Australopithecinae.' *Transvaal Mus. Mem.* 2.

BROWN, GRAHAM. 1927. 'Die Großhirnhemisphären.' In: Bethe, Embden, *et al.*, *Handbuch der normal. u. pathol. Physiol.*, 10:418–522. Berlin.

BRÜCKE, E. T. VON. 1934. 'Gehirnphysiologie, allgemeine und vergleichende.' Handwörterb. *Naturwiss.*, 2nd ed., 4:810–50. Jena.

BRUMMELKAMP, R. 1938. 'Über das Verhältnis der Oberfläche des Frontalhirns zu derjenigen des ganzen Gehirns bei höheren Affen und Menschen.' *Proc. Kon. Ak. Wetensch. Amsterdam*, 41:1127–33.

1939. 'Das Wachstum der Gehirnmasse mit kleinen Cephalisierungssprüngen (sog. $\sqrt{2}$-Sprüngen) bei den Rodentiern.' *Acta Neerland. Morphol.*, 2:188–94; '. . . bei den Ungulaten', 260–7, '. . . bei Amphibien und Fischen', 268–71.

1940. 'Number of cortex cells and spinal cord length in mice, rats, guinea pigs and rabbits.' *Acta Neerland. Morphol.*, 3:278–81.

1946. 'On the dependence of the weight of the brain on the 2/9, resp. the 5/9 power of the weight of the body.' *Kon. Nederl. Ak. Wetensch. Proc.*, 49, No. 6.

BUCCIANTE, L., and E. DE LORENZI. 1930. 'Rapporti numerici fra cellule multipolare a cellule dei coni e bastoncelli in animali di differente mole somatica.' *Monit. Zool. Ital.*, 41:103–11.

BUCHNER, P. 1930. *Tier und Pflanze in Symbiose*. 2nd ed. Berlin.

BUCK, A. DE, E. SCHOUTE, and N. H. SWELLENGREBEL. 1934. 'Crossbreeding experiments with Dutch and foreign races of *Anopheles maculipennis*.' *Riv. Malariol.*, 13:1–29.

BUCKMANN, S. S. 1906. 'Brachiopod homoeomorphy.' *Quart. J. Geol. Soc.*, 62:433–55.

1909. *Yorkshire Type Ammonites*. 1. London.

BUDDENBROCK, W. VON. 1924–8. *Grundriß der vergleichenden Physiologie*. 2 vols. Berlin.

1934. 'Über die kinetische und statische Leistung großer und kleiner Tiere und ihre Bedeutung für den Gesamtstoffwechsel.' *Naturwiss.*, 22:675–80.

BÜDELER, W. 1951. 'Sterne, die wir nicht sehen können.' *Natur und Volk (Frankfurt a.M.)*, 81:36–42.

BUMKE, O. 1923. 'Über die materiellen Grundlagen der Bewußtseinserscheinungen.' *Psychol. Forsch.*, 3:272–81.

BÜNNING, E. 1935. 'Sind die Organismen mikrophysikalische Systeme?' *Erkenntnis*, 5:337–47.

1943. 'Quantenmechanik und Biologie.' *Naturwiss.*, 31:194–7.

BURGEFF, H. 1941. 'Konstruktive Mutationen bei Marchantia.' *Naturwiss.*, 29:289–99.

BURKHARDT, L. 1931. 'Über Bau und Leistung des Auges einiger amerikanischer Urodelen.' *Z. vergl. Physiol.*, 15:637–51.

BUTENANDT, A. 1953. 'Biochemie der Gene und Genwirkungen.' *Verh. Ges. Dtsch. Naturforsch. u. Ärzte*, 97:43–52.

BUXTON, P. A. 1938. 'The formation of species among insects in Samoa and other oceanic islands.' *Proc. Linn. Soc. London*, sess. 150, 264–7.

BUYTENDIJK, F. J. J. (After 1937). *Wege zum Verständnis der Tiere.* Zürich and Leipzig.

CAIN, A. J., and P. M. SHEPPARD. 1954. 'Natural selection in Cepaea.' *Genetics*, 39:89–116.

CALLAN, H. G., and H. SPURWAY. 1951. 'A study of meiosis in interracial hybrids of the newt *Triturus cristatus.*' *J. Genetics*, 50:235–49.

CARL, J. 1940. 'Un "cercle de races" en miniature chez les diplopodes de l'Inde méridionale.' *Arch. Sci. Phys.*, 22:227–33.

CARRUTHERS, R. G. 1910. 'On the evolution of *Zaphrentis delanouei* in Lower Carboniferous times.' *Quart. J. Geol. Soc.*, 66:523–38.

CASPERSSON, T. 1941. 'Studien über Eiweißumsatz der Zelle.' *Naturwiss.*, 29:33–43.

CASPERSSON, T., and K. G. THORSSON. 1953. 'Virus und Zellstoffwechsel.' *Verh. Ges. Dtsch. Naturforsch. u. Ärzte*, 97:68–75. Berlin, Göttingen, Heidelberg.

CASTLE, W. E. 1931. 'Size inheritance in rabbits.' *J. Exper. Zool.*, 60:325–38.

1932. 'Body size and body proportions in relation to growth and natural selection.' *Science*, 76:365–6.

1936. 'Size inheritance in mice.' *Amer. Nat.* 70:209–17.

1941. 'Size inheritance.' *Amer. Nat.*, 75:488–98.

CASTLE, W. E., and P. W. GREGORY. 1929. 'The embryological basis of size inheritance in the rabbit.' *J. Morph. and Physiol.*, 48:81–103.

CESNOLA, A. P. 1904. 'Preliminary note on the protective value of colour in *Mantis religiosa.*' *Biometrica*, 3:58–9.

CHILD, C. M. 1915. *Senescence and rejuvenescence.* Chicago.

CLARK, F. H. 1941. 'Correlation and body proportions in mature mice of the genus *Peromyscus.*' *Genetics*, 26:283–300.

CLAUS, C., K. GROBBEN, and A. KÜHN. 1932. *Lehrbuch der Zoologie.* 10th ed., Berlin and Vienna.

COLBERT, E. A. 1948. 'Evolution of the horned dinosaurs.' *Evolution*, 2:145–63.

1955. *Evolution of the Vertebrates. A history of the backboned animals through time.* New York and London.

COLOSI, G. 1923. 'Note sopra alcuni eufillopodi. IV. Thriops cancriformis e le sue forme.' *Atti Soc. Ital. Mus. Civ. Milano*, 62:75–87.

CONKLIN, E. G. 1912. 'Body size and cell size.' *J. Morph.*, 23:159–88.

CONSTANCE, L. 1937. 'A systematic study of the genus *Eriophyllum Lay.*' *Univ. Calif. Publ. Bot.*, 18:69–136.

COPE, E. D. 1884. *Progressive and regressive evolution among vertebrates.*

1896. *Primary factors of organic evolution.* Chicago.

CORDEIRO, A. R., and T. DOBZHANSKY. 1954. 'Combining ability of certain chromosomes in *D. willistoni* and invalidation of the "wild-type" concept.' *Amer. Nat.*, 88:75–86.

COTT, H. B. 1940. *Adaptive coloration in animals.* London.

CRONAU, B. C. 1902. *Der Jagdfasan, seine Anverwandten und Kreuzungen.* Berlin.

CROZIER, W. J. 1918. 'Assortative mating in a nudibranch mollusc *Chromodoris Zebra* H. *J. Exper. Zool.,* **27.**

1923. 'Selective coupling of gammarids.' *Biol. Bull.,* **45.**

CUÉNOT, L. 1941. *Invention et finalité en biologie.* Paris.

1951. *L'évolution biologique.* Paris.

CUVIER, G. 1801. *Leçons d'anatomie comparée.* 2 vols. Paris.

DACQUÉ, E. 1935. *Organische Morphologie und Paläontologie.* Berlin.

DAHR, P. 1939. 'Über Blutgruppen bei Anthropoiden.' *Z. Morph. Anthrop.,* **38:**38–45.

DALE, F. H. 1939. 'Variability and environmental responses of the kangaroo rat, *Dipodomys heermanni saxatilis.*' *Amer. Midland Naturalist,* **22:**703–31.

DARWIN, C. 1859. *On the origin of species by means of natural selection.* 2 vols. London.

DARWIN, E. 1794–98. *Zoonomia, or, the Laws of organic life.* London.

DAY, M. F. 1950. 'The histology of a very large insect, *Macropanesthia rhinocerus* Sauss (Blattidae). *Australian J. Scient. Research, B. Biol.,* 3:61–75.

DECUGIS, H. 1941. *Le vieillissement du monde vivant.* Paris.

DEHLINGER, U., and E. WERTZ. 1942. 'Biologische Grundfragen in physikalischer Betrachtung.' *Naturwiss.,* 30:250–53.

DELACOUR, J., and C. VAURIE. 1950. 'Les mésanges charbonnières (Revision de l'espèce *Parus major*).' *L'Oiseau et Rev. Franç. d'Ornithol.,* 20:91–121.

DELBRÜCK, M., and W. T. BALLEY. 1947. 'Induced mutations in bacterial viruses.' *Cold Spr. Harbor Sympos.,* 11:33–7.

DEMENTIEV, G. P. 1936. 'Zur Frage der Grenzen der systematischen Kategorien Art und Unterart.' *Zool. J. Moskau,* 15:82–95. (Russian, German summary.)

1938. 'Sur la distribution géographique de certains oiseaux paléartiques au point de vue de quelques questions générales de systematique.' *Proc. 8th Intern. Ornith. Congr. Oxford* (1934), 243–59.

DEMOLL, R. 1930. *Die Silberfuchszucht.* Munich.

DENZER, H. 1935. *Vergleichend-messende Untersuchungen an Säugetieren (Dissertation).* Würzburg.

DÉPÉRÉT, C. 1907. *Les transformations du monde animal.* Paris.

DETWILER, S. R. 1931. 'Heteroplastic transplantations of embryonic spinal cord segments in *Amblystoma.*' *J. Exper. Zool.,* 60:141–71.

1938. 'Heteroplastic transplantation of somites.' *J. Exper. Zool.,* 79:361–75.

DEVAUX, E. 1928. 'La déchéance des aptitudes reproductives chez les animaux au cours de l'évolution.' *Rev. Sci. Paris,* 66:173–77.

DICE, L. R. 1931. 'The occurrence of two subspecies of the same species in the same area.' *J. Mammal.,* 12:210–13.

1935. 'A study of racial hybrids in the deermouse *Peromyscus maniculatus.*' *Occas. Pap. Mus. Zool. Univ. Mich.,* No. 312.

1939. 'Variation in the deermouse (*Peromyscus maniculatus*).' *Contrib. Lab. Vert. Gen. Univ. Mich.,* No. 12.

1940. 'Relationships between the wood mouse and the cotton mouse in Eastern Virginia.' *J. Mammal.,* 21:14–23.

1947. 'Effectiveness of selection by owls of deermice (*Peromyscus maniculatus*) which contrast in color with their background.' *Contrib. Lab. Vert. Biol. Univ. Mich.,* No. 34.

DIEBSCHLAG, E. 1940. 'Psychologische Beobachtungen über die Rangordnung bei der Haustaube.' *Z. f. Tierpsychol.,* 4:173–88.

DIENER, K. 1917. Mitteil. *Geol. Ges. Vienna,* 9.

DOBZHANSKY, T. 1937. *Genetics and the origin of species.* New York. (3rd ed., 1951.)

1939. 'Genetics of natural populations. IV. Mexican and Guatemalan populations of *Drosophila pseudoobscura.*' *Genetics,* 24:391–412.

1948. 'Genetics of natural populations. XVI. Altitudinal and seasonal changes produced by natural selection in certain populations of *Drosophila pseudoobscura* and *Drosophila persimilis.*' *Genetics,* 33:158–76.

DOBZHANSKY, T., and F. N. DUNCAN. 1933. 'Genes that affect early developmental stages of *Drosophila melanogaster.*' *Arch. f. Entw. Mech.,* 130:109–30.

DOBZHANSKY, T., and C. EPLING. 1944. 'Contributions to the genetics, taxonomy, and ecology of *Drosophila pseudoobscura* and its relatives.' *Carnegie Inst. Wash. Publ.,* No. 554.

DOBZHANSKY, T., and P. C. KOLLER. 1938. 'An experimental study of sexual isolation in *Drosophila.*' *Biol. Zentralbl.,* 58:589–607.

DOBZHANSKY, T., and B. SPASSKY. 1944. 'Genetics of natural populations. XI. Manifestation of genetic variants in *Drosophila pseudoobscura* in different environments.' *Genetics,* 20: 270–90.

DOBZHANSKY, T., and C. C. TAN. 1936. 'Studies on hybrid sterility. III. A comparison of the gene arrangement in two species, *Drosophila pseudoobscura* and *Drosophila miranda.*' *Z. ind. Abstamm. u. Vererbungsl.,* 72:88–114.

DOEDERLEIN, L. 1887. 'Phylogenetische Betrachtungen.' *Biol. Zentralbl.,* 7:394–402.

1902. 'Über die Beziehungen naheverwandter "Thierformen" zueinander.' *Z. Morphol. Anthrop.,* 4:394–442.

DOHRN, A. 1875. *Der Ursprung der Wirbeltiere und das Prinzip des Funktionswechsels.* Leipzig.

DOLLO, L. 1893. 'Les lois de l'évolution.' *Bull. Belge. Géol.,* 7:164–7.

DONALD, H. P. 1936. 'On the genetical constitution of *Drosophila pseudoobscura,* race A.' *J. Genetics.,* 33:103–22.

DONALDSEN, H. H. 1924. 'The rat.' *Mem. Wistar Inst. Anat. and Biol.,* No. 6. Philadelphia.

DOSSE, G. 1937. 'Vergleichende Gewichtsuntersuchungen am Vogelskelett.' *Zool. Jb. Anat.,* 63:299–350.

DOUTT, J. K. 1942. 'A review of the genus *Phoca.*' *Ann. Carnegie Mus.,* 29:61–125.

DRIESCH, H. 1909. *Philosophie des Organischen.* 2 vols. Leipzig.

1932. 'Eugenio Rignanos Lehre vom Organischen in ihrer Entwicklung.' *Scientia,* 51:71–8.

DROSIHN, J. 1933. 'Über Art- und Rassenunterschiede der männlichen Kopulationsapparate von Pieriden (Lep.).' *Entom. Rundschau,* 50.

DUBININ, N. P., and G. G. TINIAKOV. 1947. 'Inversion gradients and selection in ecological races of *Drosophila funebris.*' *Amer. Nat.,* 81:148–53.

DUBOIS, E. 1898. 'Über die Abhängigkeit des Hirngewichtes von der Körpergröße.' *Arch. Anthrop.* (D), 25:1–28.

1930. 'Die phylogenetische Großhirnzunahme, autonome Vervollkommnung der animalen Funktionen.' *Biologia Genetica,* 6:247–92.

1940. 'The fossil remains discovered in Java by Dr. G. H. R. von Koenigswald.' *Proc. Acad. Wetensch. Amsterdam,* 43.

DU BOIS REYMOND, E. 1884. *Über die Grenzen des Naturerkennens.* Leipzig. (1st ed., 1872.)

DUERDEN, J. E. 1919. 'Crossing the North African and South African Ostrich. I.' *Genetics,* 8.

DUNCKER, H. 1929. 'Farbenvererbung bei Buntvögeln.' *Vögel ferner Länder,* 3.

DÜRER, A. 1622. *Van de menschlijcke Proportion.* Arnheim. (1st ed., 1527.) Quoted from E. von Eickstedt, 1942–3.

DUSSER DE BARENNE, J. G. 1937. 'Physiologie der Großhirnrinde.' In: Bumke und Foerster, *Handb. d. Neurol.,* 2:268–319. Berlin.

EAST, E. M. 1936. 'Genetic aspects of certain problems of evolution.' *Amer. Nat.*, **70**:143–58.

ECONOMO, C. VON, and G. N. KOSKINAS. 1925. *Die Cytoarchitektonik der Hirnrinde des erwachsenen Menschen.* Vienna and Berlin.

EDINGER, L. 1912. *Einführung in die Lehre vom Bau und den Verrichtungen des Nervensystems.* 2nd ed. Leipzig.

EDINGER, T. 1929. 'Die fossilen Gehirne.' Ergebn. Anat. und Ent. gesch. III. Abt., *Z. Ges. Anat.*, **28**.
1948. 'Evolution of the horse brain.' *Mem. Geol. Soc. Amer.*, No. 25.

EGGERS, F. 1939. 'Phyletische Korrelation bei der Flügelreduktion von Lepidopteren.' *7th Internat. Kongr. Entom.* Berlin (1938), 694–711.
1940. 'Unabhängige Korrelationen als Grundlage für Evolutionsforschung.' *Forsch. und Fortschr.*, **16**:255–58.

EGGERS, F., and J. GOHRBRANDT. 1938. '*Hypogymna morio* L. – Ein Sonderfall in der Gesetzmäßigkeit phyletischer Korrelationen?' *Zool. Jb.*, *Abt. f. Syst.*, **71**:265–76.

EHRENBERG, K. 1925. 'Die ontogenetische Entwicklung des Höhlenbärenskelettes.' *Palaeont. Z.*, **7**:48–52.
1928. 'Über Artwandlung und Artbenennung.' *Biol. gen.*, **4**:695–712.
1932. 'Das biogenetische Grundgesetz in seiner Beziehung zum biologischen Trägheitsgesetz.' *Biol. gen.*, **8**:547–66.
1939. 'Änderungen der Umwelt und Wandlungen der Tierwelt im Laufe der Erdgeschichte.' *Forsch. u. Fortschr.*, **15**:431–3.

EIBL-EIBESFELDT, I. 1950. 'Beiträge zur Biologie der Haus- und Ährenmaus nebst einigen Beobachtungen an anderen Nagern.' *Z. f. Tierpsychol.*, **7**:556–87.

EIMER, G. H. T. 1880–1901. *Die Entstehung der Arten.* 3 vols. Leipzig.

EIMER, T., and C. FICKERT. 1899. 'Die Artbildung und Verwandtschaft bei den Schwimmvögeln nach deren Zeichnung dargestellt.' *Nova acta Abh. Leop. Carol. dtsch. Ak. Naturf.*, 77, No. 1.

EISENTRAUT, M. 1926. 'Das geographische Prinzip in der Systematik der Ascidien.' *Zool. Anz.*, **66**:171–79.
1929. 'Die Variation der balearischen Inseleidechse *Lacerta lilfordi* Gthr.' *Sitz. ber. Ges. Naturforsch. Freunde Berlin*, 24–36.
1930. 'Beitrag zur Eidechsenfauna der Pityusen und Columbreten.' *Mitt. Zool. Mus. Berlin*, **16**:397–410.
1932. 'Biologische Studien im bolivianischen Chaco. II, Über die Wärmeregulation beim Dreizehenfaultier (*Bradypus tridactylus* L.); IV, Die Wärmeregulation beim Kugelgürteltier (*Tolypeutes conurus* Js. Geoffr.).' *Z. vergleich. Physiol.*, **16**:39–47; **18**:174–85.
1935. 'Die Entwicklung der Wärmeregulation beim jungen Igel.' *Biol. Zentralbl.*, **55**:45–53.
1950. *Die Eidechsen der spanischen Mittelmeerinseln und ihre Rassenaufspaltung im Lichte der Evolution.* Berlin.

EKMAN, S. 1917–20. 'Studien über die marinen Relikte der nordeuropäischen Binnengewäser. VI.' *Int. Rev. Hydrobiol.*, **8**:477–528.
1935. *Tiergeographie des Meeres.* Leipzig.
1940. 'Begründung einer statistischen Methode in der regionalen Tiergeographie.' *Nova Acta R. Soc. Scient. Upsaliensis*, ser. IV, **12**, No. 2, 117 pp.

ELLER, K. 1936. 'Die Rassen von *Papilio machaon* L.' *Abh. Bayr. Ak. Wiss.*, N.F., Heft 36.
1937. 'Zur Rassen- und Artfrage in dem Formenkreis von *Papilio machaon* L., *Z. angew. Entom.*, **24**:145–49.

ELTON, C. 1930. *Animal ecology and evolution.* London.

ENGEL, H. 1934. 'Über Echinodermen aus der Nordsee und dem Nordatlantik.' *Zool. Anz.*, **107**:23–30.

ENGELS, W. L. 1936. 'An insular population of *Peromyscus maniculatus* subsp. with mixed racial characters.' *Amer. Midland Naturalist*, **17**:776–80.

EPHRUSSI, B. 1953. *Nucleo-cytoplasmic relations in microorganisms.* Oxford.

ERHARDT, A. 1935. 'Systematische und geographische Verbreitung der Gattung *Opisthorchis*.' *Z. Parasitenkunde*, **8**:188–225.

ERLENBACH, F. 1938. 'Experimentelle Untersuchungen über den Blutzucker bei Vögeln.' *Z. vergl. Physiol.*, **26**:121–61.

ERRINGTON, P. L. 1943. 'An analysis of mink predation upon muskrats in North Central United States.' *Iowa State Coll. Agric. Exp. Station. Res. Bull.*, **320**:797–924.

FAHRENHOLZ, H. 1936. 'Die Nomenklatur der Läuse des Menschen.' *Arch. f. Naturgesch.*, N.F., **5**:663–7.

FEDERLEY, H. 1920. 'Die Bedeutung der polymeren Faktoren für die Zeichnung der Lepidopteren.' *Hereditas*, **1**.

— 1937. 'Fusion zweier Chromosomen als Folge einer Kreuzung.' *Acta Soc. Fauna Flora Fennica*, **60**:685–95.

FENTON, C. L. 1931. 'Studies of evolution in the genus *Spirifer*.' *Publ. Wagner Free Inst.*, **2**. Philadelphia.

FISCHEL, W. 1938. *Psyche und Leistung der Tiere.* Berlin.

FISCHER, E. 1933. 'Genetik und Stammesgeschichte der menschlichen Wirbelsäule.' *Biol. Zentralbl.*, **53**:203–20.

FISHER, R. A. 1930. *The genetical theory of natural selection.* Oxford.

— 1931. 'The evolution of dominance.' *Biol. Rev.*, **6**:345–68.

— 1936. 'The measurement of selective intensity.' *Proc. Roy. Soc. London*, B, **121**:58–62.

FISHER, R. A., and E. B. FORD. 1947. 'The spread of a gene in natural conditions in a colony of the moth *Panaxia dominula* L.' *Heredity*, **1**:143–74.

FITCH, H. S. 1940. 'A biogeographical study of the Ordinoides Artenkreis of garter snakes (Genus *Thamnophis*).' *Univ. Calif. Publ. Zool.*, **44** (No. 1):1–150.

FORBES, W. T. M. 1928. 'Variation in *Junonia lavinia* (Lep. Nymphalidae).' *J. New York Entom. Soc.*, **36**:305–20.

FORD, E. B. 1945. 'Polymorphism.' *Biol. Rev.*, **20**:73–88.

— 1949. 'Early stages in allopatric speciation.' In: Jepsen, Simpson, Mayr, *Genetics, Paleontology and Evolution*, 309–14. Princeton.

— 1955. *Moths.* London.

FOX, W. 1951. 'Relationships among the garter snakes of the *Thamnophis elegans* Rassenkreis.' *Univ. Calif. Publ. Zool.*, **50**:485–530.

FRANZ, H. 1929. 'Morphologische und phylogenetische Studien an *Carabus* L. und den nächstverwandten Gattungen.' *Z. wiss. Zool.*, **135**:163–213.

— 1941. 'Untersuchungen über die Bodenbiologie alpiner Grünland- und Ackerböden.' *Forschungsdienst*, **11**:355–68.

— 1949. 'Zur Kenntnis der Rassenbildung bei Käfern der ostalpinen Fauna.' *Zbl. Gesamtgeb. Entom.*, **3**:3–23.

FRANZ, S. S. 1926. *The evolution of an idea. How the brain works.* Los Angeles.

FRANZ, V. 1909. 'Das Vogelauge.' *Zool. Jb. Anat. u. Ontog.*, **28**:73–278.

— 1920. *Die Vervollkommnung in der lebenden Natur. Eine Studie über ein Naturgesetz.* Jena.

— 1924. *Geschichte der Organismen.* Jena.

— 1935. *Der biologische Fortschritt. Die Theorie der Organismen-geschichtlichen Vervollkommnung.* Jena.

FRENZEL, G. 1937. 'Die systematische Stellung des adriatischen Alcyonium.' *Notiz. Dtsch. Ital. Inst. Rovigno*, **2** (No. 6):1–15.

FREY, M. VON. 1926. 'Die Tangorezeptoren des Menschen.' In: A. Bethe, G. von Bergmann et al., Handb. der normal. u. pathol. Physiol., 11:94–130. Berlin.

FREY-WISSLING, A. 1938. Submikroskopische Morphologie des Protoplasmas und seiner Derivate. Berlin.

FRIEDRICH-FREKSA, H. 1940. 'Bei der Chromosomenkonjugation wirksame Kräfte und ihre Bedeutung für die identische Verdopplung von Nucleoproteinen.' Naturwiss., 28:376–9.

—— 1954. 'Die stammesgeschichtliche Stellung der Virusarten und das Problem der Urzeugung.' In: G. Heberer, Evolution der Organismen, 2nd ed., 278–301. Stuttgart.

FRIEDERICHS, K. 1930. Die Grundfragen und Gesetzmäßigkeiten der land- und forstwirtschaftlichen Zoologie, insbesondere der Entomologie. Berlin.

FRISCH, K. VON. 1938. 'Über die Bedeutung des Sacculus und der Lagena für den Gehörsinn der Fische.' Z. vergl. Physiol., 25:703–47.

FUHRMANN, O. 1928–33. 'Cestoidea.' In: Kükenthal-Krumbach, Handbuch der Zool., 2, I. Hälfte (2), 141–416. Berlin and Leipzig.

FÜRBRINGER, M. 1870. Die Knochen und Muskeln der Extremitäten bei den schlangenähnlichen Sauriern. Leipzig.

GÄBLER, H. 1939. 'Häufigkeit der Farbvarietäten der Nonne (Lymantria monacha L.).' Z. wiss. Zool., 152:1–11.

GABRITSCHEWSKY, E. 1924. 'Farbenpolymorphismus und Vererbung mimetischer Varietäten der Fliege Volucella bombylans und anderer "hummelähnlicher" Zweiflügler.' Z. ind. Abstamm. u. Vererbl., 32:321–53.

GALILEI, G. 1718. 'Discorsi e dimonstratione matematiche.' In: 2. Florence.

GARSTANG, W. 1922. 'The theory of recapitulation: a critical restatement of the biogenetic law.' J. Linn. Soc. London, 35:81–101.

GATES, R. R. 1940. 'The nature of genic differences.' Boll. Soc. Ital. Biol. Spec., 15: 175–83.

GAUDRY, A. 1896. Essai de paléontologie philosophique. Paris.

GAUSE, G. F. 1934. 'Experimental studies on the struggle for existence in Paramaecium caudatum, Paramaecium aurelia and Stylonychia mytilus.' Zool. J., 12. (Russian, English summary.)

—— 1935. 'Experimental demonstration of Volterra's periodic oscillations in the numbers of animals.' J. Exper. Biol., 12:44–8.

—— 1935. 'Experimentelle Untersuchungen über die Konkurrenz zwischen Paramaecium caudatum und Paramaecium aurelia.' Arch. Protistenkunde, 84:207–24.

GAUSE, G. F., N. P. MSARAGDOVA, and A. A. WITT. 1936. 'Further studies of interaction between predators and prey.' J. Anim. Ecol., 5:1–18.

GAWRILOW, W. 1941. 'Entdifferenzierung von Zellen in verschiedenen explantierten Geweben.' Protoplasma, 35:321–29.

GEITLER, L. 1939. 'Der Bau der polyploiden Somakerne der Heteropteren durch Chromosomenteilung ohne Kernteilung.' Chromosoma, 1:1–22.

GELEI, J. VON. 1936. 'Das erregungsleitende System der Ciliaten.' Compt. Rend. 12th Congr. Internat. Zool. Lisbonne (1935), 1:174–207.

—— 1939a. 'Die Gliederung des Neuronensystems der Ciliaten.' Math. Naturwiss. Anz., 58:950–75. (Hungarian, German summary.)

—— 1939b. 'Vollkommene Sinneselemente bei den höheren Ciliaten. II. Studie über die Sinnesorganellen von Aspidisca-Arten, Allgemeines. Math. Naturwiss. Anz., 58:474–518. (German, Hungarian summary.)

GEOFFROY ST. HILAIRE, E. 1822. Philosophie Anatomique. Paris.

GEORGE, T. N. 1933. Palingenesis and paleontology. Biol. Reviews, 8:107–35.

GERHARDT, U. 1944. 'Sexualbiologie und Morphologie.' In: Die Gestalt, 17: Halle (S.).

GEYR VON SCHWEPPENBURG, H. 1935. 'Zur Systematik der Gattung *Larix*.' *Mitt. d. Dendrol. Ges.*, No. 47.

GHIGI, A. 1931. 'Ibridismo e specie nuove.' *Arch. Zool. Ital.* (Rend. 11th Congr. Int. Zool.), **16**:114–27.

1938. 'Génétique et systématique des pintades huppés (Guttera).' *Proc. 8th Int. Ornith. Congr.* (1934), 392–8.

GIESELER, W. 1943. 'Die Fossilgeschichte des Menschen.' In: G. Heberer, *Evolution der Organismen*, 615–82. Jena.

GILDEMEISTER, E., E. HAAGEN, and O. WAIDMANN. 1939. *Handbuch der Viruskrankheiten*. 2 vols. Jena.

GISLÉN, T. 1938. 'Evolutional series towards death and renewal.' *Ark. f. Zool.*, **26**, series A, No. 16.

GLASS, B., and H. L. PLAINE. 1950. 'The immediate dependence of the action of a specific gene in *Drosophila melanogaster* upon fertilization.' *Proc. Nat. Ac. Sci.*, **36**:627–34.

GOETSCH, W. 1940. *Vergleichende Biologie der Insekten-Staaten*. Leipzig.

1948. 'Der Einfluß von Vitamin T auf Gestalt und auf Gewohnheiten der Insekten.' *Österr. Zool. Z.*, **1**:193–274.

GOLDSCHMIDT, E. 1952. 'Heterochromatic polysomy in *Gryllotalpa gryllotalpa* L.' *J. Genet.*, **50**:361–83.

1953. 'Multiple sex-chromosome mechanisms and polyploidy in animals.' *J. Genet.*, **51**:434–40.

GOLDSCHMIDT, R. 1920. *Mechanismus und Physiologie der Geschlechtsbestimmung*. Berlin.

1924. 'Erblichkeitsstudien an Schmetterlingen. IV.' *Z. ind. Abstamm. u. Vererbl.*, **34**:229–44.

1924–33. 'Untersuchungen zur Genetik der geographischen Variation, I–VII.' *Roux' Arch.*, **101** (1924); **116** (1929); **126** (1932); **130** (1933).

1927. *Physiologische Theorie der Vererbung*. Berlin.

1935. 'Geographische Variation und Artbildung.' *Naturwiss.*, **23**:169–76.

1940. *The Material Basis of Evolution*. New Haven.

1948. 'Ecotype, ecospecies and macroevolution.' *Experientia*, **4**.

1951. 'New heteromorphoses in *Drosophila melanogaster* Meig.' *Pan-Pacific Entomologist*, **27**:1–11.

1952. 'Evolution, as viewed by one geneticist.' *Amer. Scientist*, **40**:84–135.

1953. 'Experiments with a homoeotic mutant, bearing on evolution.' *J. exper. Zool.*, **123**:79–114.

GOLDSTEIN, K. 1927. 'Die Lokalisation in der Großhirnrinde. Nach den Erfahrungen am kranken Menschen.' In: A. Bethe, Embden, *et al.*, *Handb. der normal. u. pathol. Physiol.*, **10**. Berlin.

GOOSSEN, H. 1949. 'Untersuchungen an Gehirnen verschieden großer, jeweils verwandter Coleopteren- und Hymenopteren-Arten.' *Zool. Jb., Abt. allg. Zool.*, **62**:1–64.

GÖSSWALD, K. 1941. 'Rassenstudien an der Roten Waldameise *Formica rufa* L.' *Z. angew. Entomol.*, **28**:62–124.

1942. 'Art- und Rassenunterschiede bei der Roten Waldameise.' *Naturschutz*, **23**:109–15.

1944. 'Rassenstudien an der Roten Waldameise im Lichte der Ganzheitsforschung.' *Anz. f. Schädlingskde.*, **20**:1–8.

GOETHE, J. W. VON. 1817–22. *Zur Morphologie, Bildung und Umbildung organischer Naturen*. 2 vols. (Preface of 1807.)

GOTTSCHEWSKI, G., and C. C. TAN. 1938. 'The homology of the eye color genes in *Drosophila melanogaster* and *Drosophila pseudoobscura* as determined by transplantation. II.' *Genetics*, **23**:221–38.

GÖTZ, B. 1951. 'Die Sexualduftstoffe an Lepidopteren.' *Experientia*, **7**:406–18.

GÖTZ, P. 1931. 'Ozon der Atmosphäre – ein Grenzgebiet.' *Forsch. u. Fortschr.*, **7**:282–83.

GÖTZE, E. 1938. 'Bau und Leben von Caecum glabrum (Montagu).' *Zool. Jb., Abt. Syst.*, **71**:55–122.

GRABOWSKY, U. 1939. 'Experimentelle Untersuchungen über das angebliche Lernvermögen von *Paramaecium*.' *Z. f. Tierpsychol.*, **2**:265–82.

GRANT, V. 1949. 'Pollination systems as isolating mechanisms in Angiosperms.' *Evol.*, **3**:82–97.

GRANVIK, H. 1923. 'Contributions to the knowledge of the East African ornithology,' *J. f. Ornith.*, **71**:Sonderheft.

GREEN, C. V. 1930. 'Inheritance in a mouse species cross.' *Amer. Nat.*, **64**:540–44.

1931. 'Size inheritance and growth in a mouse-species cross (*Mus musculus* × *Mus bactrianus*). IV.' *J. Exper. Zool.*, **59**:247–63.

1934. 'An analysis of size genes.' *Amer. Nat.*, **68**:275–78.

GREENWOOD, P. H. 1951. 'Evolution of the African Cichlid fishes. The *Haplochromis* species-flock in Lake Victoria.' *Nature*, **167**:19–20.

GREGOR, J. W. 1946. 'Ecotypic differentiation.' *The New Phytologist*, **45**:254–70.

GREGORY, P. W., and H. GOSS. 1933. 'Glutathione concentration and hereditary body size. 2nd part.' *J. Exper. Zool.*, **66**:155–73.

GREGORY, W. K. 1934. *Man's place among the anthropoids*. Oxford.

1935. 'Reduplication in evolution.' *Quart. Rev. Biol.*, **10**:272–89.

1936. 'On the meaning and limits of irreversibility of evolution.' *Amer. Nat.*, **70**:517–28.

GROSS, F. 1932. 'Untersuchungen über die Polyploidie und die Variabilität bei *Artemia salina*.' *Naturwiss.*, **51**:962–67.

GROSS, W. 1939. 'Ein rezenter Crossopterygier entdeckt!' *Naturwiss.*, **27**:226–27.

GRÜNEBERG, K. 1943. *The genetics of the mouse*. Cambridge Univ. Press.

HAACKE, W. 1893. *Gestaltung und Vererbung*. Leipzig.

HAARDICK, H. 1941. 'Wachstumsstufen in der Embryonalentwicklung des Hühnchens.' *Biol. gen.*, **15**:30–74.

HAASEBESSELL, G. 1936. 'Polyploidie?' *Arch. Rass. u. Gesellsch. biol.*, **29**:377–84.

HADORN, E. 1949. 'Begriffe und Termini zur Systematik der Letalfaktoren.' *Arch. Jul.-Klaus-Stift. Vererb. Sozialanthrop.*, **24**:105–13.

1953. 'Genetik und Entwicklungsphysiologie.' *Verh. Ges. Dtsch. Naturforsch. u. Ärzte*, **97**:37–43.

HACKENBRUCH, P. 1888. *Experimentelle und histologische Untersuchungen über die Kompensations-Hypertrophie der Testikel*. (Dissertation.) Bonn.

HAECKEL, E. 1866. *Generelle Morphologie der Organismen*. 2 vols. Berlin.

1874. *Anthropogenie oder Entwicklungsgeschichte des Menschen*. Leipzig.

HAECKER, V. 1918. *Entwicklungsgeschichtliche Eigenschaftsanalyse* (Phänogenetik). Jena.

1925. 'Aufgaben und Ergebnisse der Phänogenetik.' *Bibliogr. Genetica*, **1** s'Gravenhage.

1925. *Pluripotenzerscheinungen. Synthetische Beiträge zur Vererbungs- und Abstammungslehre*. Jena.

HALDANE, J. B. S. 1924. 'A mathematical theory of natural and artificial selection.' *Trans. Cambridge Philos. Soc.*, **23**:19–41.

1932. *The causes of evolution*. New York and London.

1940. *New paths in genetics*. New York and London.

1948. 'Man's evolution: past and future.' In: G. L. Jepsen, *Genetics, Paleontology and Evolution*, 405–18. Princeton.

1955. 'The measurement of natural selection.' Atti 9. Congr. Intern. Genet. (1953). *Caryologia* (Florence) Suppl., 480–87.

HALL, E. R. 1940. 'Supernumerary and missing teeth in wild mammals of the orders Insectivora and Carnivora, with some notes on disease.' *J. Dental Research*, **19**:103–42.

HALLER, A. VON. 1762. *Elementa physiologiae corporis humani*. Vol. IV. Lausanne.

HANDLIRSCH, A. 1926–38. 'Insecta.' In: Kükenthal-Krumbach, *Handbuch der Zool.*, **4**. Berlin.

HANSON, F. B., and F. HEYS. 1930. 'A possible relation between natural (earth) radiation and gene mutations.' *Science*, **71**:43–44.

HANSTRÖM, B. 1928. *Vergleichende Anatomie des Nervensystems der wirbellosen Tiere*. Berlin.

HARDE, K. W. 1949. 'Das postnatale Wachstum cytoarchitektonischer Einheiten im Großhirn der Weißen Maus.' *Zool. Jb., Abt. Anat.*, **70**:225–68.

1955. 'Quantitative Differenzen cytoarchitektonischer Einheiten im Großhirn indischer Sciuriden-Arten von stark unterschiedener Körpergröße.' *Zool. Jb., Abt. Physiol.*, **66**:179–97.

HARMS, J. W. 1934. *Wandlungen des Artgefüges*. Tübingen.

HARTERT, E. 1910–38. *Die Vögel der paläarktischen Fauna*. With suppl. by E. Hartert and F. Steinbacher. 4 vols. Berlin.

HARTMANN, M. 1927. *Allgemeine Biologie*. Jena. (3rd ed., 1947.)

1937. *Philosophie der Naturwissenschaften*. Berlin.

HASSMANN, M. 1952. 'Vom Erlernen unbenannter Anzahlen bei Eichhörnchen (*Sciurus vulgaris*).' *Z. f. Tierpsychol.*, **9**:294–321.

HAYES, C. 1952. *The ape in our house*. London.

HEBERDEY, R. F. 1933. 'Die Bedeutung der Eiszeit für die Fauna der Alpen.' *Zoogeogr.*, **1**:353–412.

HEBERER, G. 1940. 'Jüngere Stammesgeschichte des Menschen.' In: Bauer-Just, *Handbuch der Erbbiol. d. Menschen*, **1**:584–644.

1943. 'Das Typenproblem in der Stammesgeschichte.' In: Heberer, *Evolution der Organismen*, 545–85. Jena.

1944. 'Das Neandertalerproblem und die Herkunft der heutigen Menschheit.' *Jenaische Z. f. Med. u. Naturwiss.*, **77**:263–89.

1951. *Neue Ergebnisse der menschlichen Abstammungslehre*. Göttingen.

1952. 'Die Fortschritte in der Erforschung der Phylogenie der Hominoidea.' *Ergebn. Anat. Entwickl.-gesch.*, 499–637.

HECKMANN, O. 1943. 'Milchstraße – Nebel – Weltall.' *Forsch. u. Fortschr.*, **19**:339–42.

HEDIGER, H. 1952. 'Seltene tropische Tiere und ihre Haltung in Zoologischen Gärten Nordamerikas.' *Acta tropica*, **9**:97–124.

HEDING, S. 1942. 'Holothurioidea. Pt. II.' *Danish Ingolf-Exp.*, **4**, pt. 3. Copenhagen.

HEINROTH, O. 1922. 'Die Beziehungen zwischen Vogelgewicht, Eigewicht, Gelegegewicht und Brutdauer.' *J. f. Ornith.*, **70**:172–285.

HELLER, M. 1928. 'Zur Kenntnis der Schutzfärbung bei Insekten.' *Zool. Anz.*, **78**:13–21.

HELLMICH, W. 1934a. 'Die Eidechsen Chiles, insbesondere die Gattung Liolaemus.' *Abh. Bayr. Ak. Wiss., Math. Nat. Abt.*, N.F., Heft 24.

1934b. 'Zur näheren Analyse der geographischen Variabilität.' *Forsch. u. Fortschr.*, **10**:358–59.

HELMCKE, J. G. 1940. 'Die Brachiopoden der Deutschen Tiefsee-Expedition.' *Wiss. Ergebn. Tiefsee-Exp. Valdivia*, **24**:215–316.

HEMMINGSEN, A. M. 1950. 'The relation of standard (basal) energy metabolism to total fresh weight of living organisms.' *Reports Steno Memorial Hospital*, **4**. Copenhagen.

HEMPELMANN, F. 1926. *Tierpsychologie vom Standpunkte des Biologen*. Leipzig.

HENKE, K. 1936. 'Versuche einer vergleichenden Morphologie des Flügelmusters der Saturniden auf entwicklungsphysiologischer Grundlage.' *Nova Acta Leopoldina* (Halle, S.), N.F. **4**: No. 18.

1938. Review of: W. F. Reinig, *Elimination und Selection. Biol. Zentralbl.*, **58**:553–5.

1950. 'Über Ordnungsvorgänge in der Spätentwicklung der Insekten.' *Rev. Suisse Zool.*, **55**:319–37.

1947–8. 'Einfache Grundvorgänge in der tierischen Entwicklung.' *Naturwiss.*, **34**:149–57, 180–6; **35**:176–81, 203–11, 239–46.

HENKE, K., and G. KRUSE. 1941. 'Über Feldgliederungsmuster bei Geometriden und Noctuiden und den Musterbauplan der Schmetterlinge im allgemeinen.' *Nachr. Ak. Wiss. Göttingen, Math. Phys. Kl.*, Heft 3, 138–96.

HENNIG, E. 1929. 'Von Zwangsablauf und Geschmeidigkeit in organischer Entfaltung.' *Rektoratsreden Tübingen*, No. 26, 3–48. Tübingen.

1932. *Wesen und Wege der Paläontologie.* Berlin.

1944. 'Organisches Werden, paläontologisch gesehen.' *Paläont. Z.*, **23**:281–316.

1947. 'Der Übergang vom Saurier zum Säuger.' *Naturwiss.*, **34**: 246–49.

HERING, E. 1870. *Über das Gedächtnis als eine allgemeine Funktion der organischen Materie.* Vienna.

L'HÉRITIER, P., Y. NEEFS, and G. TEISSIER. 1937. 'Apterisme des insectes et sélection naturelle.' *Compt. Rend. Acad. Sci. Paris*, **204**:907–9.

HERMANN, G. 1935. 'Die naturphilosophischen Grundlagen der Quantenmechanik.' *Naturwiss.*, **23**:718–21.

HERRE, W. 1948. 'Zur Abstammung und Entwicklung der Haustiere. II. Betrachtungen über vorgeschichtliche Wildschweine Mitteleuropas.' *Verh. Dtsch. Zool. Kiel*, 324–33.

HERRE, W., and F. RAWIEL. 1939. 'Vergleichende Untersuchungen an Unken.' *Zool. Anz.*, **125**:290–99.

HERRICK, C. J. 1934. 'Neurobiotaxis in the Corpus striatum.' *Psychiatr. Bl.*, **38**:419–25.

HERSKOWITZ, I. H. 1955. 'The production of mutations in *Drosophila melanogaster* with chemical substances administered in sperm baths and vaginal douches.' *Genetics*, **40**:76–89.

HERSHEY, A. D. 1947. 'Spontaneous mutations in bacterial viruses.' *Cold Spr. Harbor Symp.*, **11**:67–77.

HERTER, K. 1934. 'Studien zur Verbreitung der europäischen Igel.' *Arch. f. Naturgesch.* N.F., **3**:313–82.

1935. 'Igelbastarde (*Erinaceus roumanicus* ♂ × *E. europaeus* ♀).' *Sitz.-Ber. Ges. Naturforsch. Freunde Berlin*, 118–21.

HERTER, K., and W. R. HERTER, 1954. 'Die Verbreitung der Kreuzkröte (*Bufo calamita Laur.*) und der Wechselkröte (*Bufo viridis Laur.*) in Europa.' *Zool. Beiträge* (Berlin), N.F. **1**:203–18.

HERTWIG, P. 1936. 'Artbastarde bei Tieren.' In: Baur and Hartmann. *Handbuch der Vererbungswiss.*, **2** (B). Berlin.

HERZBERG, K. 1939. 'Die bis heute sichtbar zu machenden Virusarten als Einzelgebilde und in ihrer Beziehung zur Zelle.' *Arch. f. exp. Zellforsch.*, **22**:445–52.

HESS, W. R. 1948. 'Vegetative Funktionen und Zwischenhirn (Suppl. IV der *Helv. Physiol. Acta*). Basel.

HESSE, R. 1908. 'Das Sehen der niederen Tiere.' Jena.

1921. 'Das Herzgewicht der Wirbeltiere.' *Zool. Jb., Abt. Physiol.*, **38**:243–364.

1929. 'Die Stufenleiter der Organisationshöhe der Tiere.' *Sitzungsber. Preuß. Ak. Wiss., Phys. Math. Kl.*, 3.

1935. *Tierbau und Tierleben in ihrem Zusammenhang betrachtet.* 2nd ed. **1**, Jena.

1935. 'Der Haushalt der Insekten.' *Sitzungsber. Preuß. Ak. Wiss., Phys. Math. Kl.*, **12**:167–73.

HETSCH, H., and H. SCHLOSSBERGER. 1942. *Experimentelle Bakteriologie und Infektionskrankheiten.* 9th ed. Berlin and Vienna.

HEUTS, M. J. 1947. 'The phenotypical variability of *Gasterosteus aculeatus* L. populations in Belgium.' *Verh. Kon. Vlaam. Ac. Wetensch. Belgie,* 9, No. 25.
1947. 'Experimental studies on adaptive evolution in *Gasterosteus aculeatus* L.' *Evolution,* 1:89–102.

HEYDEMANN, F. 1943. 'Die Bedeutung der sogenannten Dualspecies (Zwillingsarten) für unsere Kenntnis der Art- und Rassenbildung bei Lepidopteren.' *Stettiner Entom. Ztg.,* 104:116–42.

HEYNE, A., and O. TASCHENBERG. 1908. *Die exotischen Käfer in Wort und Bild.* Eßlingen and Munich.

HILTERMANN, H. 'Populationen in ihrer Bedeutung für die Paläontologie und Stratigraphie.' *Erdöl u. Kohle,* 4:244–9.

HILTERMANN, H., and W. KOCH. 1950. 'Taxonomie und Vertikalverbreitung von Bolivinoides-Arten im Senon Nordwestdeutschlands.' *Geol. Jb.,* 64:595–632.

HINSCHE, G. 1932. 'Untersuchungen zur Entwicklungsgeschichte tierischer Bewegungen.' *Anat. Anz.,* 74:439–54.

HINTON, M. A. C. 1926. *Monograph of the Voles and Lemmings (Microtinae).* 1. London.

HISAW, F. L. 1925. 'The influence of the ovary on the resorption of the pubic bones of the pocket gopher, *Geomys bursarius* (Shaw).' *J. Exper. Zool.,* 42:411–42.

HOARE, C. A. 1943. 'Biological races in parasitic protozoa.' *Biol. Rev.,* 18:137–44.

HOBHOUSE, L. T. 1915. *Mind in evolution.* London.

HOFMANN, F. B. 1926. 'Der Geruchssinn beim Menschen.' In: Bethe, von Bergmann, *et al., Handbuch der norm. u. path. Physiol.,* 11:253–305. Berlin.

HOFSTEN, N. VON. 1951. 'The genetic effect of negative selection in man.' *Hereditas,* 37:157–265.

HOLDHAUS, K. 1933. 'Die europäische Höhlenfauna in ihren Beziehungen zur Eiszeit.' *Zoogeogr.,* 1:1–53.

HOLLY, M., H. MEINKEN, and A. RACHOW. *Die Aquarienfische in Wort und Bild.* Stuttgart.

HOLMES, F. O. 1948. 'Order Virales. The filtrable viruses.' In: Bergey's *Manual of determinative bacteriology.* Baltimore.

HOLTFRETER, J. 1934. 'Formative Reize in der Embryonalentwicklung der Amphibien, dargestellt an Explantationsversuchen.' *Arch. exper. Zellforsch.,* 15.

HOMEYER, B. 1951. 'Die Ontogenese cytoarchitektonischer Einheiten im Vorderhirn von *Triturus vulgaris* L.' *Zool. Jb., Abt. allg. Zool.,* 63:25–63.

HOOPER, E. T. 1936. 'Geographical variation in wood rats of the species *Neotoma fuscipes.*' *Univ. Calif. Publ. Zool.,* 42:213–46.

HORNBOSTEL, VON. 1931. 'Über Geruchshelligkeit.' *Pflügers Arch.,* 227:517–38.

HORTON, J. H. 1939. 'A comparison of the salivary gland chromosomes in *Drosophila melanogaster* and *Drosophila simulans.*' *Genetics,* 24:234–43.

HOVANITZ, W. 1941. 'Parallel ecogenotypical color variation in butterflies.' *Ecology,* 22:259–84.
1948. 'Ecological segregation of inter-fertile species of *Colias.*' *Ecology,* 29:461–69.
1950. 'The biology of Colias butterflies. II. Parallel geographical variation of dimorphic color phases in North American species.' *Wasmann J. Biol.,* 8:197–219.
1953. 'Polymorphism and evolution.' *Sympos. Soc. Exper. Biol.,* No. 7, 238–53.

HUBBS, C. L. 1943. 'Criteria for subspecies, species and genera, as determined by researches on fishes.' *Ann. New York Acad. Sci.,* 44:109–21.

HUBBS, C. L., L. C. HUBBS, and R. E. JOHNSON. 1943. 'Hybridization in nature between species of Catostomid fishes.' *Contrib. Lab. Vert. Biol. Univ. Mich.,* No. 22.

HUBBS, C. L., and R. R. MILLER. 1943. 'Mass hybridization between two genera of Cyprinid fishes in the Mohawe Desert, California.' *Pap. Michigan Acad. Sci.,* 28:343–78.

HUENE, F. FRH. VON. 1940. 'Die stammesgeschichtliche Gestalt der Wirbeltiere – ein Lebensablauf.' *Paläont. Z.*, **22**:55–62.

— 1943. 'Die Korrelation im phyletischen Auftreten der Pflanzen und der Tetrapoden.' *Neues Jahrbuch Min. Geol.* B, 28–31.

HUNDERTMARK, A. 1938. 'Über das Luftfeuchtigkeitsunterscheidungsvermögen und die Lebensdauer der drei in Deutschland vorkommenden Rassen von *Anopheles maculipennis.*' *Z. angew. Entom.*, **25**:125–41.

HUTCHINSON, G. E., and S. D. RIPLEY. 1954. 'Gene dispersal and the ethology of the Rhinocerotidae.' *Evolution*, **8**:178–9.

HUTT, F. B. 1949. *Genetics of the fowl.* New York, Toronto and London.

HUXLEY, J. S. 1932. *Problems of relative growth.* London.

— 1939. 'Clines: an auxiliary method in taxonomy.' *Bijdr. Dierk.*, editor, **27**:491–520.

— 1940. *The new systematics.* London.

— 1940. *The uniqueness of man.* London.

— 1942. *Evolution, the modern synthesis.* New York and London.

— 1955. 'Morphism and evolution.' *Heredity*, **9**:1–52.

HUXLEY, T. H. 1863. *Evidence as to man's place in nature.* New York

HYATT, T. 1894. 'Phylogeny of an acquired characteristic.' *Proc. Am. Phil. Soc.*, **32**:349–640.

IWANOFF, E., and J. PHILIPTSCHENKO. 1916. 'Beschreibung von Hybriden zwischen Bison, Wisent und Hausrind.' *Z. ind. Abstamm. u. Vererbl.*, **16**.

JACKSON, C. M. 1913. 'Postnatal growth and variability of the body and the various organs in the albino rat.' *Amer. J. Anat.*, **15**:1–68.

JACOBI, A. 1931. 'Das Rentier.' *Zool. Anz.*, **96**, Ergänzungs-Bd. Leipzig.

JAECKEL, O. 1902. *Über verschiedene Wege phylogenetischer Entwicklung.* Jena.

JANZER, W. 1950. 'Versuche zur Entstehung der Höhlentiermerkmale.' *Naturwiss.*, **37**: 286.

JARVIK, E. 1942. 'On the structure of the snout of Crossopterygians and lower Gnathostomes in general.' *Zool. Bidr. Uppsala*, **21**.

JAWORSKI, E. 1938. 'Untersuchungen über Rassenbildung bei Anthozoen.' *Thalassia*, **3**:1–57.

JENSEN, S. 1941. 'On subspecies and races of the lesser sand eel (*Ammodytes lancea* S. Lat.).' *Biol. Medd. Danske Vidensk. Selsk.*, **16**:1–33.

JERISON, H. J. 1955. 'Brain to body ratios and the evolution of intelligence.' *Science*, **121**:447–9.

JESCHIKOV, J. J. 1933. 'Zur Theorie der Rekapitulation.' *Zool. J.* (*Moscow*), **12**:57–76. (Russian, German summary.)

— 1937. 'Zur Rekapitulationslehre.' *Biol. gen.*, **13**:67–100.

JOHANSEN, H. 1944. 'Die Vogelfauna Westsibiriens. II.' *J. f. Ornith.*, **92**:1–105.

JOHNSEN, A. 1930. *Über den Unterschied von Mineralien und Lebewesen.* Berlin.

JORDAN, H. J. 1929. *Allgemeine vergleichende Physiologie der Tiere.* Berlin and Leipzig.

JORDAN, K. 1905. 'Der Gegensatz zwischen geographischer und nicht geographischer Variation.' *Z. wiss. Zool.*, **83**:151–210.

— 1929. 'On the geographical variation of the pine hawk-moth, *Hyloicus pinastri.*' *Novit. Zool.*, **36**:243–9.

JORDAN, K., H. FRUHSTORFER, *et al.* 1927. 'Die indoaustralischen Tagfalter.' In: Seitz, *Großschmetterlinge der Erde*, **9**. Stuttgart.

JORDAN, P. 1941. 'Über die Spezifität von Antikörpern, Fermenten, Viren, Genen.' *Naturwiss.*, **29**:89–100.

— 1944. 'Zum Problem der Eiweiß-Autokatalysen.' *Naturwiss.*, **32**:20–6.

— 1945. *Die Physik und das Geheimnis des organischen Lebens.* Braunschweig.

JULL, A. M., and J. P. QUINN. 1931. 'The inheritance of body weight in the domestic fowl.' *J. Heredity*, **22**:283–94.

JUST, G. 1934a. 'Das wandelnde Blatt und seine Anpassungserscheinungen.' *Forsch. u. Fortschr.*, **10**:157–8.

1934b. 'Zur Phylogenese von Anpassungscharakteren.' *Verh. Dtsch. Zool. Ges.*, **36**:126–33.

1936a. 'Über die Phylogenese spezialisierter Anpassungen.' *Compt. Rend. 12th Congr. Int. Zool.* Lisbon (1935), 35–50.

1936b. 'Weitere Untersuchungen zur Phylogenese spezialisierter Anpassungen.' *Verh. Dtsch. Zool. Ges.*, **38**:336–43.

KAFKA, G. 1914. *Einführung in die Tierpsychologie.* Leipzig.

KAGELMANN, G. 1951. 'Studien über Farbfelderung, Zeichnung und Färbung der Wild- und Hausenten.' *Zool. Jb., Abt. allg. Zool.*, **62**:513–630.

KALABUCHOV, N. J. 1938. 'On the influence of the temperature of the environment upon growth of mice (*Mus musculus* L.).' *Bull. Soc. Natural. Moscou*, Sect. Biol., N.S., 218–22.

KATZ, D., and A. TOLL. 1923. 'Die Messung von Charakter- und Begabungsunterschieden bei Tieren (Versuche mit Hühnern),' *Z. f. Psychol. u. Physiol. d. Sinnesorgane*, **93**.

KAUFMANN, L. 1927. 'Recherches sur la croissance du corps et des organes du pigeon.' *Biol. gen.*, **3**:105–18.

KAUFMANN, R. 1933. 'Variationsstatistische Untersuchungen über die "Artabwandlung" und "Artumbildung" an der oberkambrischen Trilobitengattung Olenus Dalm.' *Abh. Geol. Paläont. Inst. Greifswald*, **10**:1–54.

1934. 'Exakt nachgewiesene Stammesgeschichte.' *Naturwiss.*, **22**:803–7.

KAURI, H. 1954. 'Über die systematische Stellung der europäischen grünen Frösche *Rana esculenta* L. und *R. ridibunda* Pall.' *Lunds Univ. Arsskr.*, N.F., Avd. 2, **50**, No. 12.

KENDEIGH, S. C. 1934. 'The role of environment in the life of birds.' *Ecol. Monographs*, **4**:299–417.

KENDEIGH, S. C., and S. P. BALDWIN. 1937. 'Factors affecting yearly abundance of passerine birds.' *Ecol. Monographs*, **7**:91–124.

KESTNER, O. 1934. 'Über die Oberflächenregel des Stoffwechsels.' *Pflügers Arch.*, **234**:290–301.

KESTNER, O., and R. PLAUT. 1924. 'Physiologie des Stoffwechsels.' In: Wintersteins, *Handbuch der vergleich. Physiol.*, **2**: 2. Hälfte, 901–1102. Jena.

KIEFER, F. 1952. 'Copepoda Calanoida und Cyclopoida.' In: *Explor. Parc. Nat. Albert, Mission H. Damas*, Fasc. **21**. Brussels.

KING, J. C. 1947. 'Interspecific relationships within the guarani group of *Drosophila*.' *Evolution*, **1**:143–53.

KIPP, F. 1936. 'Studien über den Vogelzug in Zusammenhang mit Flügelbau und Mauserzyklus.' *Mitteil. über d. Vogelwelt*, **35**:49–80.

KIRIAKOFF, S. G. 1947. 'Le cline, une nouvelle catégorie systématique intraspécifique.' *Ann. Soc. Entom. Belg.*, **83**:130–9.

KLAATSCH, H. 1902. 'Entstehung und Entwicklung des Menschengeschlechts.' In Krämer, *Weltall und Menschheit*, **2**.

KLAATSCH, H., J. ANDRÉE, H. WEINERT, and J. LECHLER. 1936. *Das Werden der Menschheit und die Anfänge der Kultur.* Berlin.

KLATT, B. 1913. 'Über den Einfluß der Gesamtgröße auf das Schädelbild.' *Arch. f. Entw. Mech.*, **36**:387–471.

1919. 'Zur Methodik vergleichender metrischer Untersuchungen, besonders des Herzgewichtes.' *Biol. Zbl.*, **39**:406–21.

1941. 'Kreuzungen an extremen Rassetypen des Hundes (Bulldogge–Windhundkreuzungen). *Z. mensch. Vererb. u. Konstitl.*, **25**:28–93.

1948. 'Messend-anatomische Untersuchungen an gegensätzlichen Wuchsformtypen.' *Arch. f. Entw. Mech.*, **143**:573–92.

KLATT, B., and H. VORSTEHER. 1923. 'Studien zum Domestikationsproblem. Part II.' *Bibl. Genetica*, **6**. Leipzig.

KLATT, B., and H. OBOUSSIER. 1951. 'Weitere Untersuchungen zur Frage der quantitativen Verschiedenheiten gegensätzlicher Wuchsformtypen beim Hund.' *Zool. Anz.*, **146**:223–40.

KLATT, G. T. 1901. 'Über den Bastard von Stieglitz und Kanarienvogel.' *Arch. f. Entwicklungsmech.*, **12**:414–53; 471–528.

KLEINENBERG, N. 1886. 'Die Entstehung des Annelids aus der Larve von *Lopadorhynchus.' Zeitschr. wiss. Zool.*, **44**:1–227.

KLEIST, K. 1934. 'Gehirn-Pathologie.' In: Schjerning, *Handbuch der ärztlichen Erfahrungen im Weltkriege*, **4**:343–1408. Leipzig.

KLEMM, W. 1939. 'Zur rassenmässigen Gliederung des Genus *Pagodulina* clessin.' *Arch. f. Naturgesch.*, N.F., **8**:198–261.

KNETSCH, H. 1939. 'Die Korrelationen in der Ausbildung der Tympanalorgane, der Flügel, der Stridulationsapparate und anderer Organsysteme bei den Orthopteren.' *Arch. f. Naturgesch.*, N.F., **8**:1–69.

KNIPPER, K. 1939. 'Systematische, anatomische, ökologische und tiergeographische Studien an südosteuropäischen Heliciden (*Moll. Pulm.*).' *Arch. f. Naturgesch.*, N.F., **8**:327–517.

KNOERZER, A. 1940. 'Der saharo-sindische Verbreitungstypus bei der ungeflügelten Tenebrioniden-Gattung Mesostena.' *Riv. Biol. Colon.*, **3**:1–133.

KOBELT, W. 1881. 'Die sizilianische Iberus.' *Jb. Dtsch. Malakozool. Ges.*, **8**:50–67.

KOCHS, W. 1897. 'Versuche über die Regeneration von Organen bei Amphibien.' *Arch. mikrosk. Anat.*, **49**:441–61.

KOEHLER, O. 1939. 'Diskussion zu den Vorträgen Alverdes-Bramstedt. Nachschrift zur Diskussion zu Nr. 7, 8.-Abschlußwort zum Paramaecien-Vortrag (9).' *Verh. Dtsch. Zool. Ges.*, 127–32, 463–70, 473–75.

1941. 'Tauben, Wellensittiche und Dohlen erlernen unbenannte Anzahlen.' *Forsch. u. Fortschr.*, **17**:156–9.

1941. 'Von Erlernen unbenannter Anzahlen bei Vögeln.' *Naturwiss.*, **29**:201–18.

1943. ' "Zähl"-Versuche an einem Kolkraben und Vergleichsversuche an Menschen.' *Z. f. Tierpsychol.*, **5**:575–712.

1950. ' "Zählende" Vögel und vorsprachliches Denken.' *Verh. Dtsch. Zool. Ges.*, 1949, 219–38.

KOENIGSWALD, G. H. R. VON. 1939. 'Neue Menschenaffen- und Vormenschen funde.' *Naturwiss.*, **27**:617–22.

KÖHLER, E. 1935. 'Über die Variabilität des Ringmosaikvirus (X-Virus) der Kartoffel.' *Naturwiss.*, **23**:828–30.

KÖHLER, W. 1921. *Intelligenzprüfungen an Menschenaffen.* Berlin.

KOKEN, E. 1896. 'Die Gastropoden der Trias um Halstatt.' *Jb. Geol. Reichsanst.*, *Wien*, **46**.

KORNMÜLLER, A. E. 1935. 'Die bioelektrischen Erscheinungen architektonischer Felder der Großhirnrinde.' *Biol. Rev.*, **10**:383–426.

1937. *Die bioelektrischen Erscheinungen der Hirnrindenfelder.* Leipzig.

KORSCHELT, E. 1924. *Lebensdauer, Altern und Tod.* 3rd ed. Jena.

KORSCHINSKY, S. 1899. 'Heterogenesis und Evolution.' *Naturwiss. Wochenschr.*, **14**:273–8.

KOSSWIG, C. 1937. 'Betrachtungen und Experimente über die Entstehung von Höhlentiermerkmalen.' *Der Züchter*, **9**:91–101.

1948. 'Genetische Beiträge zur Präadaptationstheorie.' *Rev. Fac. Sci. Univ. Istanbul*, sér. B, **13**: fasc. 3, 176–209.

1948. 'Homologe und analoge Gene, parallele Evolution und Konvergenz.' *Commun. Fac. Sci. Univ. Ankara*, **1**:126–77.

1949. 'Phänomene der regressiven Evolution im Lichte der Genetik.' *Comm. Fac. Sci. Univ. Ankara*, **2**:110–50.

KOSSWIG, C., and L. KOSSWIG. 1936. 'Über Augenrück- und Mißbildung bei Asellus aquaticus cavernicolus.' *Verh. Dtsch. Zool. Ges.*, 274–81.

1940. 'Die Variabilität bei Asellus aquaticus, unter besonderer Berücksichtigung der Variabilität in isolierten unter- und oberirdischen Populationen.' *Rev. Fac. Sc. Univ. Istanbul*, B., **5**.

KRABBE, K. H. 1942. *Studies on the morphogenesis of the brain of lower mammals*. Copenhagen.

KRAMER, G. 1949. 'Macht die Natur Konstruktionsfehler?' *Wilhelmshavener Vorträge*, Heft 1.

1949. 'Über Inselmelanismus bei Eidechsen.' *Z. induct. Abstamm. u. Vererbl.*, **83**:157–64.

KRAMER, G., and R. MERTENS. 1938. 'Zur Rassenbildung bei West-istrianischen Inseleidechsen in Abhängigkeit von Isolierungsalter und Arealgröße.' *Arch. f. Naturgesch.*, N.F., **7**:189–234.

KREITMANN, L. 1927. 'L'acclimatisation du Lavaret de Bourget dans le Lac Léman et sa relation avec la systématique des Corégones.' *Verh. Int. Ver. Limnol.*, **4**.

KRIEG, H. 1937. 'Luxusbildungen bei Tieren unter besonderer Berücksichtigung der luftlebenden Wirbeltiere.' *Zool. Jb., Abt. Syst.*, **69**:303–18.

KRIES, J. VON. 1898. *Über die materiellen Grundlagen der Bewußtseinserscheinungen*. Tübingen and Leipzig.

KROGH, C. VON. 1943. 'Die Stellung des Menschen im Rahmen der Säugetiere.' In: G. Heberer, *Evolution der Organismen*, 589–614. Jena.

KRÖLLING, O. 1934. 'Anatomie und Entwicklungsgeschichte der Metapodien bei den Equiden im Lichte des biogenetischen Grundgesetzes.' *Verh. Zool.-Bot. Ges. Wien*, **84**:38–41.

KRUMBIEGEL, J. 1931. 'Das sogenannte Kompensationsgesetz Goethes betr. Korrelation von Kopfwaffen und Oberzähnen.' *Z. Säugetierkde*, **6**:186–202.

1938. 'Physiologisches Verhalten als Ausdruck der Phylogenese.' *Zool. Anz.*, **123**:225–40.

1941. 'Die Persistenz physiologischer Eigenschaften in der Stammesgeschichte.' *Z. f. Tierpsychol.*, **4**:249–58.

KRUMSCHMIDT, E. 1956. 'Postnatale Wachstumsgradienten von Hirnteilen bei Haushuhnrassen unterschiedlicher Körpergröße.' *Z. Morphol. Ökol. Tiere*, **45**:113–45.

KUHN, E. 1948. 'Der Artbegriff in der Paläontologie.' *Ber. schweiz. paläontol. Ges.*, 27. Jahresvers., 389–421; Ecl. geol. Helv., **41**, No. 2.

KUHN, O. 1933. 'Über morphogenetische Schilddrüsenhormonwirkungen in frühen Entwicklungsstadien.' *Nachr. Ges. Wiss. Göttingen, Math. phys. Kl.*, **6**, No. 7.

KUHN, O. 1937. *Die fossilen Reptilien*. Berlin.

1938. *Die Phylogenie der Wirbeltiere auf paläontologischer Grundlage*. Jena.

1939. *Die fossilen Amphibien*. Berlin.

1939. *Die Stammesgeschichte der wirbellosen Tiere im Lichte der Paläontologie*. Jena.

KÜHN, A. 1932. 'Entwicklungsphysiologische Wirkung einiger Gene von Ephestia kühniella.' *Naturwiss.*, **20**:947–77.

1935. 'Physiologie der Vererbung und Artumwandlung.' *Naturwiss.*, **23**:1–10.

1950. *Grundriß der Vererbungslehre*. Heidelberg.

KÜHNE, K. 1934. 'Die Symmetrieverhältnisse und die Ausbreitungszentren in der Variabilität der regionalen Grenzen der Wirbelsäule des Menschen.' *Z. Morph. u. Anthrop.*, **34**:191–206.

KÜHNE, W. G. 1943. 'The dentary of Tritylodon and the systematic position of the Tritylodontidae.' *Ann. Mag. Nat. Hist.* (ser. 11), **10**:589–601.

KULLENBERG, B. 1946. 'Om fågellätenas biologiska funktion.' *Vår Fågelvärld*, **5**:49–64. (English summary.)

1948. 'Der Kopulationsapparat der Insekten aus phylogenetischem Gesichtspunkt. Festschr. Nils von Hofsten.' *Zool. Bidr. Uppsala*, **25**:79–90.

KUPKA, E. 1950. 'Die Mitosen- und Chromosomenverhältnisse bei der großen Schwebrenke, *Coregonus wastmanni* (Bloch) des Attersees. *Österr. Zool. Z.*, **2**:605–23.

LA BARRE, W. 1954. *The human animal.* Chicago.

LACK, D. 1940. 'Variation in the introduced English sparrow.' *Condor*, **42**:239–41.

1947. *Darwin's finches.* Cambridge.

LAMARCK, J. DE. 1809. *Philosophie zoologique.* Paris.

LAMBRECHT, K. 1933. *Handbuch der Paläornithologie.* Berlin.

LAMPRECHT, H. 1944. 'Die genisch-plasmatische Grundlage der Artbarriere.' *Agri hortique genetica*, **2**:75–142.

1948. 'Zur Lösung des Artproblems. Neue und bisher bekannte Ergebnisse der Kreuzung *Phaseolus vulgaris* L. × *coccineus* L. und reziprok.' *Agri hortique genetica*, **6**:87–141.

1949. 'Systematik auf genischer und zytologischer Grundlage.' *Agri hortique genetica*, **7**:1–28.

LANCEFIELD, D. E. 1929. 'A genetic study of two races of physiological species in *Drosophila obscura.*' *Z. ind. Abst. u. Vererbl.*, **52**:287–317.

LANDAUER, W. 1931. 'Studies on the creeper fowl. III.' *J. Genet.*, **25**:367–94.

1934. 'Studies on the creeper fowl. VI.' *Bull. Storrs Agric. Exp. Stat.*, No. 193.

1945. 'Recessive rumplessness of fowl with kyphoscoliosis and supernumerary ribs.' *Genetics*, **30**:403–28.

1946. 'Genetic aspects of physiology.' In: Howell's *Textbook of Physiology.* 15th ed. (Fulton), Ch. 56. Philadelphia and London.

LANDAUER, W., *et al.* 1946–7. 'Insulin induced rumplessness of chickens.' *J. exper. Zool.*, **101**:41–50; **102**:1–22; **105**:279–316, 317–28.

LANGE, W. 1941. 'Die Ammonitenfauna der Psiloceras-Stufe Norddeutschlands.' *Palaeontographica*, **93**: Abt. A (Stuttgart).

LANKESTER, E. RAY. 1870. 'On the use of the term homology.' *Ann. Mag. Nat. Hist. (London)*, **6**.

LAPIQUE, L. 1898. 'Sur la relation du poids de l'encephale au poids du corps.' *Compt. Rend. Soc. Biol.*, ser. X, **5**.

1903. 'Sur la relation entre la longueur de l'intestin et la grandeur de l'animal.' *Compt. Rend. Soc. Biol. Paris*, **55**:29–30.

1907. 'Les poids encephaliques en fonction du poids corporal entre individus d'une même espèce.' *Bull. Soc. Anthrop., Paris*, ser. V, **8**:313–45.

1908. 'La grandeur relative de l'œil et l'appréciation du poids encephalique.' *Compt. Rend. Acad. Sci. Paris*, **147**:209–12.

LARTET, E. 1868. 'De quelques cas de progression organique vérifiables dans la succession des temps géologiques sur les mammifères de même famille et de même genre.' *Compt. Rend. Acad. Sci. Paris*, **66**:1119–22.

LASAREFF, N. J. 1938. 'On the question of the difference in the grade of inductive ability of the eye cup of different Amphibia species.' *Tr. Inst. exp. Morphogen.*, **6**:97–105. (Russian, English summary.)

LATIMER, H. B. 1925. 'The relative postnatal growth of the systems and organs of the chicken.' *Anat. Rec.*, **29**:367; **31**:233–53.

LATTIN, G. DE. 1939. 'Über die Evolution von Höhlentiercharakteren.' *Sitzungsber. Ges. Naturf. Freunde Berlin*, 11–41.

1949. 'Über die Artfrage in der *Hipparchia semele* L.-Gruppe.' *Entom. Z.*, **59**, No. 15–17.

LAUTH, H. 1935. 'Verbreitung von Körper- und Eigewicht bei Hausrassen.' *Z. Tierzucht*, **31**:271–310.

LAVINK, J. 1948. 'Sterne mit unsichtbaren Begleitern.' *Naturwiss.*, **35**:118–20.

LEBEDKIN, S. 1937. 'The recapitulation problem. I and II.' *Biol. gen.*, **13**:391–417; 516–94.

LE GROS CLARK, W. E. 1949. *History of the primates*. Brit. Mus. Nat. Hist., London.

LEHMANN, F. E. 1938. 'Die morphologische Rekapitulation des Grundplans bei Wirbeltierembryonen und ihre entwicklungsphysiologische Bedeutung.' *Vierteljahresschr. Naturforsch. Ges. Zürich*, **83**:187–92.

LEHMANN, O. 1906. 'Flüssige Kristalle und die Theorien des Lebens.' *Vers. Dtsch. Naturforsch. u. Ärzte*, Stuttgart.

1907. 'Flüssige Kristalle und ihre Analogien zu den niedrigsten Lebewesen.' *Kosmos*, Stuttgart, 5–13, 36–40.

LEICHTENTRITT, B. 1919. 'Die Wärmeregulation neugeborener Säugetiere und Vögel.' *Z. f. Biol.*, **69**: N.F. 51.

LEINEMANN, K. 1904. *Über die Zahl der Facetten in den zusammengesetzten Augen der Coleopteren*. Dissertation (Münster). Hildesheim.

LEMCHE, H. 1935. 'The primitive colour-pattern on the wings of insects and its relation to the venation.' *Vidensk. Meddel. Dansk Naturhist. Foren.*, **99**:45–64.

1937. 'Studien über die Flügelzeichnung der Insekten. I.' *Zool. Jb., Abt. Anat.*, **63**:183–288.

LERNER, I. M. 1954. *Genetic homeostasis*. Edinburgh and London.

LEVI, G. 1909. 'I gangli cerebrospinali.' *Arch. Ital. Anat. Embriol.*, **7**, Suppl. 1–392.

1925. Wachstum und Körpergrösse. *Ergebn. Anat. u. Entwicklungsgesch.*, **26**:87–342.

LEVI, R., and E. SACERDOTE. 1934. 'Ricerche quantitative sui sistema nervoso di *Mus musculus*.' *Monit. Zool. Ital.*, **45**:162–72.

L'HÉRITIER, P., and G. TEISSIER. 1937. 'Une expérience de sélection naturelle.' *Compt. Rend. Acad. Sci. Paris*, **204**:907–9.

LEWIS, E. B. 1955. 'Pseudoallelism and the gene concept.' Atti 9. Congr. Intern. Genet. (1953). Part 1. *Caryologia*, **6**, Suppl. 100–5.

LI, Y. C. 1927. 'The effect of chromosomal aberrations on development in *Drosophila melanogaster*.' *Genetics*, **12**:1–58.

LILLIE, R. S. 1945. *General biology and philosophy of organism*. Chicago.

LINDROTH, C. H. 1935. 'The boreo-britisch Coleoptera.' *Zoogeogr.*, **2**:579–634.

LINDSAY, R. B. 1948. 'Grenzprinzipien in der modernen Physik.' *Universitas*, **3**:1475–80.

LINKE, O. 1933. 'Morphologie und Physiologie des Genitalapparates der Nordseelittorinen.' *Wiss. Meeresuntersuch*, N.F., **19**, No. 5.

LINZBACH, A. J. 1950. 'Die Muskelfaserkonstante und das Wachstumsgesetz der menschlichen Herzkammern.' *Virchows Arch.*, **318**:575–618.

LIPPMANN, E. O. VON. 1933. *Urzeugung und Lebenskraft*. Berlin.

LLOYD MORGAN, C. 1896. *Animal behaviour*. London.

LOCKE, J. 1894. *An essay concerning human understanding*, ed. by A. C. Fraser. 2 vols., Oxford. (1st ed., 1690.)

LOEWENTHAL, H. 1923. 'Cytologische Untersuchungen an normalen und experimentell beeinflußten Dipteren (*Calliphora erythrocephala*).' *Arch. f. Zellforsch.*, **17**:86–101.

LÖNNBERG, E. 1929. 'The development and distribution of the African fauna in connection with and depending upon climatic changes.' *Arkiv f. Zool.*, **21**A (No. 4):1–33.

1939. 'Notes on some relict forms of *Cottus quadricornis* L., recently found in Swedish fresh-water lakes.' *Arkiv f. Zool.*, **31 B**, No. 2.

LORENZ, K. 1937. 'Über den Begriff der Instinkthandlung.' *Folia biotheor*, 2:18–50.

1940. 'Vergleichende Verhaltensforschung.' *Verh. Dtsch. Zool. Ges.*, **41**:69–102 (1939).

1943. 'Die angeborenen Formen möglicher Erfahrung.' *Z. f. Tierpsychol.*, 5:235–409.

LORKOVIČ, Z. 1928. 'Analyse der Speziesbegriffe und der Variabilität der Spezies auf Grund von Untersuchungen einiger Lepidopteren.' *Act. Soc. Sci. Nat. Croaticae*, **38**:1–64.

1941. 'Die Chromosomenzahlen in der Spermatogenese der Tagfalter.' *Chromosoma*, 2:111–91.

LUDWIG, W. 1940. 'Selektion und Stammesentwicklung.' *Naturwiss.*, 28:689–705.

1941. 'Zur evolutorischen Erklärung der Höhlentiermerkmale durch Allelelimination.' *Biol. Zentralbl.*, 62:447–55.

1942. 'Über die Rolle des Mutationsdrucks bei der Evolution.' *Biol. Zentralbl.*, 62:374–79.

1943. 'Die Selektionstheorie.' In: Heberer, *Evolution der Organismen* 479–520. Jena.

1950. 'Zur Theorie der Konkurrenz. Die Annidation (Einnischung) als fünfter Evolutionsfaktor.' In: *Neue Ergebnisse und Probleme der Zoologie* (Klatt-Festschr.), 516–37. Leipzig.

1954. 'Die Selektionstheorie.' In: Heberer, *Evolution der Organismen*. 2nd ed. Stuttgart.

LÜERS, H. 1937. 'Ein Beitrag zur vergleichenden Genetik an Hand des dominanten Bobbed im Y-Chromosom von *Drosophila funebris*.' *Z. ind. Abstamm. u. Vererbl.*, **74**:70–90.

LUKIN, E. I. 1937. 'On the substitution of non-hereditary variations by hereditary ones from the point of view of the selection theory.' *Uchen. Zap. Khark. Univ.*, 6–7. (Ukrainian, English summary.)

LULL, R. S. 1925. *Organic evolution*. New York.

LUMER, H. 1940. 'Evolutionary allometry in the skeleton of the domesticated dog.' *Amer. Nat.*, **74**:439–67.

LUNDMARK, K. 1930. *Das Leben auf anderen Sternen*. Leipzig.

LUNTZ, A. 1926. 'Untersuchungen über den Generationswechsel der Rotatorien I.' *Biol. Zentralbl.*, **46**.

LYNN, W. G. 1941. 'The embryology of *Eleutherodactylus nubicola*, an anurea which has no tadpole stage.' *Carnegie Inst. Washington Publ.*, **541**.

LYNN, W. G., and B. LUTZ. 1946. 'The development of *Eleutherodactylus nasutus* Lutz.' *Bol. Mus. nac. Rio de Janeiro*, N.S., No. 79.

MCCARTY, M. 1946. 'Chemical nature and biological specificity of the substance inducing transformation of pneumococcal types.' *Bacter. Reviews*, **10**:63–71.

MCDOUGALL, K. D., and W. MCDOUGALL. 1931. 'Insight and foresight in various animals.' *J. Comp. Psychol.*, **11**:237–73.

MCFARLANE-BURNET, F. 1946. *Virus as organism*. Cambridge, Mass.

MACH, E. 1865. 'Untersuchungen über den Zeitsinn des Ohres.' *Sitzungsber. Wiener Ak.*, **51**.

MCTAGGART COWAN, J. 1935. 'A distribution study of the *Peromyscus sitkensis* group of white-footed mice.' *Univ. Calif. Publ. Zool.*, **40** (No. 13): 429–538.

MAGNAN, A. 1911. 'Le foie et sa variation en poids chez les oiseaux.' *Bull Mus. Hist. Nat. Paris*, **17**:492–3.

MAMPELL, K. 1945. 'Analysis of a mutator.' *Genetics*, **30**:496–505.

MANEA, 1894. 'Rapporto fra il peso dei reni ed il peso e la superficie del corpo dei cani.' *Giorn. di Medic. Vet.*, **43**.

MANGOLD, O. 1935. 'Kombination verschiedener Geschmacksqualitäten zur Untersuchung des chemischen Sinnes des Regenwurms.' *Naturwiss.*, **23**:472–4.

1955. 'Die Entwicklung kleinster und größter Keime von *Triton alpestris* bei Fütterung und Hunger.' *Arch. Entwickl. Mech.*, **147**:373–404.

MANGOLD, O., and H. WAECHTER. 1953. 'Der Einfluß ungünstiger äußerer Bedingungen während der ersten Entwicklungsphasen auf die Ausgestaltung der Larven von *Triton alpestris*.' *Naturwiss.*, **40**:328–34; 595–9.

MAŘAN, J. 1927. 'The study of the rudiments of the wings by the genus *Pterostichus, Poecilus, Abax* and *Molops* (Col. Carabidae).' *Sborn. ent. Odd. nár. Mus. Praze.*, **54**:121–40.

MARGALEF, R. 1949. 'Importancia de la neotenia en la evolución de los crustáceos de aqua dulce.' *Publ. Inst. Biol. Aplicada (Barcelona)*, **6**:41–51.

1955. 'Temperatura, dimensiones y evolución.' *Publ. Inst. Biol. Aplicada (Barcelona)*, **19**.

MARK, R. 1930. 'Untersuchungen an Bastarden zwischen Kanarien und Wildfinken.' *Z. wiss. Zool.*, **137**:476–549.

MAROLD, E. 1939. 'Versuche an Wellensittichen zur Frage des "Zähl"-vermögens.' *Z. f. Tierpsychol.*, **3**:170–223.

MARSH, O. C. 1874. 'Small size of the brain in tertiary mammals.' *Amer. J. Sci. and Arts*, ser. 3, **8**:66–7.

MARSHALL, J. T. 1948. 'Ecologic races of song sparrows in the San Francisco Bay region.' *Condor*, **50**:193–250.

MARSHALL, W. H. 1940. 'A survey of the mammals of the islands in Great Salt Lake, Utah.' *J. Mammal.*, **21**:144–59.

MASCHKOWZEFF, A. A. 1936. 'Endogener und exogener Faktorenwechsel der embryonalen Entwicklung der Ontogenese und Phylogenese.' *Bull. Acad. Sci. URSS (Biol.)*, 945–97. (Russian, German summary.)

MATHEW, W. D. 1914. 'Time ratios in the evolution of mammalian phyla.' *Science*, N.S., **40**:232–5.

1915. 'Climate and evolution.' *Ann. New York Acad. Sci.*, **24**:171–318.

MATVEIEV, B. S. 1932. 'Zur Theorie der Rekapitultion. Über die Evolution der Schuppen, Federn und Haare auf dem Wege embryonaler Veränderungen.' *Zool. Jb., Abt. Anat.*, **55**:555–80.

MAUREL, E. 1902. 'Rapport du poids du foie au poids total de l'animal.' *Compt. Rend. Acad. Sci. Paris*, **135**:1002–5.

1903. 'Rapport du poids total et à la surface totale de l'animal.' *Compt. Rend. Soc. Biol. Paris*, **55**:196–8.

MAYR, E. 1932. 'Birds collected during the Whitney South Sea expedition.' *Amer. Mus. Novit.*, **20**:1–22; **21**:1–23.

1942. *Systematics and the origin of species.* New York.

1947. 'Ecological factors in speciation.' *Evolution*, **1**:263–88.

1950. 'Taxonomic categories in fossil hominids.' *Cold Spr. Harbor Symp. Quant. Biol.*, **15**:109–18.

1954. 'Geographic speciation in tropical echinoids.' *Evolution*, **8**:1–18.

MAYR, E., and C. VAURIE. 1948. 'Evolution in the family Dicruridae (Birds).' *Evolution*, **2**:238–65.

MEGUŠAR, F. 1907. 'Die Regeneration der Koleopteren.' *Arch. Entw. Mech.*, **25**:148–234.

MEISE, W. 1928. 'Rassenkreuzungen an den Arealgrenzen.' *Verh. Dtsch. Zool. Ges.*, 96–105.

1933. 'Scorpiones (Norweg. Zool. Exp. Galapagos Isls. 1925).' *Meddel. Zool. Mus. Oslo*, No. 39, 25–43.

1936. 'Zur Systematik und Verbreitungsgeschichte der Haus- und Weidensperlinge, *Passer domesticus* (L.) und *hispaniolensis* (T.).' *J. Ornith* **84**:631–72.

MEISSNER, M. 1924. 'Die Schilddrüse beim Zwerghund.' Z. Anat. Entwickl. Gesch., 70:598–600.

MEIXNER, J. 1933–6. 'Coleopteroidea.' In: Kükenthal-Krumbach, Handbuch der Zool., 4, Hälfte 2, Teil 1, 1037–1348. Berlin.

1939. 'Probleme der Rassendifferenzierung, aufgezeigt an Arten der Laufkäfergattung Trechus.' 7th Internat. Kongr. Entom. Berlin (1938), 303–18.

MELL, R. 1929. Grundzüge zu einer Ökologie der chinesischen Reptilien und einer herpetologischen Tiergeographie Chinas. Berlin and Leipzig.

1943. 'Inventur und ökologisches Material zu einer Biologie der südchinesischen Pieriden.' Zoologica, 36, Lief. 6, Heft 100. Stuttgart.

MENDHEIM, H. 1943. 'Beiträge zur Systematik und Biologie der Familie der Echinostomatidae.' Arch. f. Naturgesch., N.F., 12:175–349.

MERTENS, R. 1924. 'Ein Beitrag zur Kenntnis der melanotischen Inseleidechsen des Mittelmeeres.' Pallasia, 2:40–52.

1928. 'Zur Naturgeschichte der europäischen Unken (Bombina).' Z. Morph. u. Ökol., 11:613–23.

1931. 'Ablepharus boutonii (Desjardin) und seine geographische Variation.' Zool. Jb., Abt. Syst., 61:63–210.

1932. 'Über düster gefärbte Inseldeidechsen des Lago Maggiore.' Zool. Anz., 101:106–11.

1933. 'Zwerg- und Riesenformen unter Amphibien.' Senckenbergiana, 15:1–4.

MERTENS, R., and L. MÜLLER. 1940. 'Die Amphibien und Reptilien Europas.' Abh. senckenb. naturf. Ges., 451.

METZ, C. W., and E. G. LAWRENCE. 1938. 'Preliminary observations on Sciara hybrids between S. ocellaris and S. reynoldsi (Diptera).' J. Hered., 29:179–86.

MEUNIER, K. 1951. 'Korrelation und Umkonstruktion in den Größenbeziehungen zwischen Vogelflügel und Vogelkörper.' Biol. gen., 19:403–14.

MEYER, A. 1932–3. 'Acanthocephala.' In: Bronns, Klass u. Ordn. des Tierreichs, 4, Abt. 2, Buch 2. Leipzig.

MICHAELIS, P. 1949. 'Die Abänderungen des plasmatischen Erbgutes.' Z. indukt. Abstamm. u. Vererbl., 83:36–85.

1949. 'Über Plasmon-inducierte Genlabilität.' Naturwiss., 36:220.

1954. 'Cytoplasmic inheritance in Epilobium and its theoretical significance.' Advances in Genetics, 6:287–401.

MIČULICZ-RADECKI, M. VON, 1949. 'Betrachtungen zur Stammesgeschichte der Wildtauben.' Verh. Dtsch. Zool. (Mainz), 1949, 55–63.

MIJSBERG, M. A. 1931. 'Die Phylogenie der menschlichen Niere.' [Verh. Anat. Ges. (Amsterdam, 1930).] Anat. Anz., 71: Erg.-Heft, 248–51.

MILLER, A. H. 1939. 'An analysis of some hybrid populations of juncos.' Condor, 41:211–14.

1941. 'Speciation in the avian genus Junco.' Univ. Calif. Publ. Zool., 44:173–434.

1949. 'Some concepts of hybridization and intergradation in wild populations of birds.' Auk, 66:338–42.

MILLER, G. S. 1912. Catalogue of the mammals of Western Europe. London.

MILLER, S. L. 1953. 'A production of amino acids under possible primitive earth conditions.' Science, 117:528–9.

MILNE-EDWARDS, H. 1851. Introduction à la zoologie générale. Paris.

MÖLLER, A. 1950. 'Die Struktur des Auges bei Urodelen verschiedener Körpergröße.' Zool. Jb., Abt. allg. Zool., 62:138–82.

MOLLISON, T. 1936. 'Die serologischen Beweise für eine chemische Epigenese in der Stammesgeschichte des Menschen.' Arch. f. Rassenbiol., 30:457–68.

1938. 'Über den Begriff der Differenzierung im morphologischen und biochemischen Sinne.' Anthropol. Anz., 15, Sonderheft, 38–42.

MOORE, J. A. 1946. 'Incipient intraspecific isolating mechanism in *Rana pipiens*.' *Genetics*, 31:304–26.

1949. 'Geographic variation of adaptive characters in *Rana pipiens* Schreber.' *Evolution*, 3:1–24.

1954. 'Geographic and genetic isolation in Australian Amphibia.' *Amer. Nat.*, 88:65–74.

MORDVILKO, A. 1937. 'Artbildung und Evolution.' *Biol. gen.*, 12:245–98.

MOREAU, R. E. 1930. 'On the age of some races of birds.' *Ibis* (12), 6:229–39.

MULLER, H. J. 1939. 'Reversibility in evolution, considered from the standpoint of genetics.' *Biol. Rev.*, 14:261–80.

1946. 'Age in relation to the frequency of spontaneous mutations in *Drosophila*.' *Year Book Amer. Phil. Soc.*, 1945, 150–3.

MÜLLER, A. 1926. 'Über Konvergenzerscheinungen am Schädelskelett der Vögel.' *Verh. u. Mitt. Siebenbürg. Ver. Naturwiss.* Hermanstadt, 75–6: 1–16 (1925-6).

MÜLLER, H. 1952. 'Bau und Wachstum der Netzhaut des Guppy (*Lebistes reticulatus*).' *Zool. Jb., Abt. Physiol.*, 63:275–324.

MÜLLER, H. G. 1936. 'Untersuchungen über spezifische Organe niederer Sinne bei rhabdocoelen Turbellarien.' *Z. vergl. Physiol.*, 23:253–92.

MÜLLER-BÖHME, H. 1935. 'Beiträge zur Anatomie, Morphologie und Biologie der "Großen Wühlmaus".' *Arb. Biol. Anstalt Berlin*, 21:363–453.

NACHTSHEIM, H. 1936. 'Erbliche Zahnanomalien beim Kaninchen.' *Z. Züchtungskde.*, 11:273–87.

1941. 'Das Porto-Santo-Kaninchen.' *Die Umschau*, 45:151–4.

1943. 'Ergebnisse und Probleme der vergleichenden und experimentellen Erbpathologie. Jenaische Z.' *Naturwiss.*, 76:81–108.

NAEF, A. 1917. *Die individuelle Entwicklung organischer Formen als Urkunde ihrer Stammesgeschichte*. Jena.

1931. 'Phylogenie der Tiere.' In: Baur-Hartmann, *Handbuch der Vererbungswiss.*, 3. Lief. 13. Berlin.

NÄGELI, C. VON. 1884. *Mechanisch-physiologische Theorie der Abstammungslehre*. Munich and Leipzig.

NALEPA, A. 1917. 'Die Systematik der Eriophyiden, ihre Aufgabe und ihre Arbeitsmethode.' *Verh. Zool.-Bot. Ges. Wien*, 67:12–38.

NEWELL, N. D. 1949. 'Phyletic size increase, an important trend, illustrated by fossil invertebrates.' *Evolution*, 3:103–24.

NIETHAMMER, G. 1940. 'Die Schutzanpassung der Lerchen.' In: Hoesch and Niethammer; Vogelwelt Deutsch-Südwestafrikas. *J. f. Ornith.*, 88. Sonderheft.

NILSSON, N. H. 1930. 'Synthetische Bastardierungsversuche in der Gattung *Salix*. *Lunds Univ. Arsskr.*, N.F., 2:27, No. 4.

NOBIS, G. 1949. 'Vergleichende und experimentelle Untersuchungen an heimischen Schwanzlurchen.' *Zool. Jb., Abt. Anat.*, 70:333–96.

NOLTE, A. 1953. 'Die Abhängigkeit der Proportionierung und Cytoarchitektonik des Gehirns von der Körpergröße bei Urodelen.' *Zool. Jb., Abt. Physiol.*, 64:429–598.

NOVIKOFF, M. 1929. 'Das Prinzip der Analogie als Grundlage der vergleichenden Morphologie.' *10th Congr. Intern. Zool. Budapest* (1927), 301–21.

1930. *Das Prinzip der Analogie und die vergleichende Anatomie*. Jena.

1938. 'Das Studium der Homomorphien als wissenschaftliche Methode.' *Biol. gen.*, 14:85–110.

ÖKLAND, F. 1937. 'Die geographischen Rassen der extramarinen Wirbeltiere Europas.' *Zoogeogr.*, 3:389–484.

OMODEO, P. 1952. 'Cariologia dei Lumbricidae.' *Caryologia*, 4:173–275.

OPARIN, A. J. 1938. *The origin of life*. New York. (Originally published in Russian, Moscow and Leningrad.)

OSBORN, H. F. 1905. 'The ideas and terms of modern philosophical anatomy.' *Science*, N.S., **21**:959–61.

1906. 'The causes of extinction of mammalia.' *Amer. Nat.*, **40**:769–95; 829–59.

1922. 'Migrations and affinities of the fossil proboscideans of Eurasia, North and South America, and Africa.' *Amer. Nat.*, **56**:448–55.

1930. *Ursprung und Entwicklung des Lebens*. Stuttgart.

1934. 'Aristogenesis, the creative principle in the origin of species.' *Amer. Nat.*, **68**:193–235.

1935. 'The thirty-nine distinct lines of Proboscidean descent, and their migration into all parts of the world except Australia.' *Proc. Amer. Phil. Soc. Philadelphia*, **74**:273–85.

1942. *A monograph on the discovery, evolution, migration, and extinction of the mastodonts and elephants of the world*. **2**. New York.

OTTOW, B. 1950. 'Die erbbedingte Osteogenesis dysplastico-exostotica der ausgerotteten flugunfähigen Riesentaube Pezophaps solitaria der Mascareninsel Rodriguez.' *K. svenska Vetensk. Akad. Handl.*, 4th ser., **1**, No. 9.

PADOUR, L. 1950. 'Differenzen endokriner Drüsen bei Säugetieren verschiedener Körpergröße, mit besonderer Berücksichtigung der Langerhansschen Inseln.' *Zool. Jb., Abt. allg. Zool.*, **62**:102–37.

PAGENSTECHER, W. 1909. *Die geographische Verbreitung der Schmetterlinge*. Jena.

PALMER, J. G. 1898. 'The danger of introducing noxious animals and birds.' *Yearbook U.S. Dept. Agric.*

1937. 'Geographic variation in the mole *Scapanus latimanus*.' *J. Mammal.*, **18**:280–314.

PARK, T. 1948. 'Experimental studies of interspecies competition. I. Competition between populations of the flour beetles *Tribolium confusum* Duval and *Tribolium castaneum* Herbst.' *Ecol. Monogr.*, **18**:265–308.

PARR, A. E. 1926. *Adaptiogenese und Phylogenese. Zur Analyse der Anpassungserscheinungen und ihrer Entstehung*. Berlin.

PARROT, C. 1894. 'Über die Größenverhältnisse des Herzens bei Vögeln.' *Zool. Jb., Abt. Syst.*, **7**:496–522.

PARTMANN, W. 1948. 'Untersuchungen über die komplexe Auswirkung phylogenetischer Körpergrößenänderungen bei Dipteren.' *Zool. Jb., Abt. Anat.*, **69**:507–58.

PASEWALDT, G. 1888. *Experimentelle und histologische Untersuchungen über die kompensatorische Hypertrophie der Ovarien*. (Dissertation.) Bonn.

PASTEUR, L. 1862. 'Die in der Atmosphäre vorhandenen organisierten Körperchen, Prüfung der Lehre von der Urzeugung.' *Ostwalds Klassiker der exakten Wissensch.*, No. 39. Leipzig. (French original: Ann. de Chimie et Phys., 3, sér. 64.)

PAX, F. 1932. 'Beitrag zur Kenntnis der japanischen Dörnchenkorallen.' *Zool. Jb., Abt. Syst.* **63**:407–50.

1936. 'Anthozoa.' In: G. Grimpe and A. Remane, *Tierwelt der Nord- und Ostsee*, Lief. 30. Leipzig.

1942. 'Die Crustaceen der deutschen Mineralquellen.' *Abh. Naturforsch. Ges. Görlitz*, **33**:87–130.

PEARSON, K. 1903. 'Assortive mating in man.' *Biometrica*, **2**.

PEARSON, O. H. 1950. 'The metabolism of hummingbirds.' *Condor*, **52**:145–52.

PEDEN, K., and G. VON BONIN. 1947. 'The neocortex of Hapale.' *J. Compar. Neurol.*, **86**:37–64.

PEMBREY, M. S. 1895. 'The effect of variations in external temperature upon the output of carbonic acid and the temperature of young animals.' *J. Physiol.*, **18**.

PENNERS, A. 1922. 'Die Furchung von *Tubifex rivulorum* Lam.' *Zool. Jb., Abt. Anat.*, **43**:323–68.

1923. 'Die Entwicklung des Keimstreifs und die Organbildung bei *Tubifex rivulorum*.' *Zool. Jb., Abt. Anat.*, **45**:251–308.

1929–30. 'Entwicklungsgeschichtliche Untersuchungen an marinen Oligochäten. I. Furchung, Keimstreif, Vorderarm und Urkeimzellen von Peloscolex benedeni Udekem. II. Furchung, Keimstreif und Keimbahn von *Pachydrilus* (Lumbricillus) *lineatus* Müll.' *Z. wiss. Zool.*, **134**:307–44 (1929); **137**:55–119 (1930).

PENSO, G. 1955. 'Cycle of phage development within the bacterial cell.' *Protoplasma*, **45**:251–63.

PETERS, J. L. 1937. *Check-list of birds of the world*. Vol. III. Cambridge.

PETERSEN, B. 1947. 'Die geographische Variation einiger Fennoskandischer Lepidopteren.' *Zool. Bidr. Uppsala*, **26**:329–531.

1950. 'The relation between size of mother and number of eggs and young in some spiders and its significance for the evolution of size.' *Experientia*, **6**:96–8.

PEYER, B., and E. KUHN. 1928. 'Die Kopulation von *Limax cinereoniger* Wolf.' *Vierteljahresschr. Naturforsch. Ges. Zürich*, **73**:485–521.

PFLÜGER, E. 1875. 'Über die physiologische Verbrennung in den lebendigen Organismen.' *Pflügers Arch.*, **10**.

PHILIPTSCHENKO, J. 1927. *Variabilität und Variation*. Berlin.

PIÉRON, H. 1908. 'Les facteurs des mouvements périodiques des Convoluta dans leur habitat naturel.' *Bull. Mus. Hist. Nat.*, **14**.

PIROCCHI, B. 1951. 'Hochendemische Copepoden- und Cladoceren-Lokalformen im Karst.' *Arch. f. Hydrobiol.*, **45**:245–53.

PITTIONI, B. 1941. 'Die Variabilität des *Bombus agrorum* F. in Bulgarien.' *Mitt. Kgl. Naturwiss. Inst. Sofia*, **14**:238–311.

PLATE, L. 1904. 'Gibt es ein Gesetz der progressiven Reduktion der Variabilität?' *Arch. Rassenbiol.*, **1**:641–55.

1910. 'Vererbungslehre und Descendenztheorie.' *Festschr. 60. Geburtstag R. Hertwigs*, **2**:589–93. Jena.

1913. *Selektionsprinzip und Probleme der Artbildung*. 4th ed. Leipzig and Berlin.

1922–4. *Allgemeine Zoologie und Abstammungslehre*. **1**. Jena, 1922; **2**, 1924.

1925. *Die Abstammungslehre*. Jena.

1928. 'Über Vervollkommnung, Anpassung und die Unterscheidung von niederen und höheren Tieren.' *Zool. Jb., Abt. allg. Zool.*, **45**:745–98.

1932–3. *Vererbungslehre mit besonderer Berücksichtigung des Menschen*. 2nd ed. Jena.

1933. 'Sexualität und allgemeine Probleme.' *Vererbungslehre*. **2**. Jena.

POLL, H. 1911. 'Über Vogelmischlinge.' *Verh. 5th Int. Ornith. Kongr.* (1910), 399–468.

POMPECKY, J. F. 1925. *Umwelt, Anpassung und Beharrung im Lichte erdgeschichtlicher Überlieferung*. Berlin.

POPOFF, W. W. 1935. 'Gleichzeitige experimental-embryologische und genetische Untersuchungen einer Organogenese.' *Bull. Acad. Sci. URSS*, No. 8–9, 1237–44. (Russian, German summary.)

PORTER, K. R. 1941. 'Diploid and androgenetic haploid hybridization between two forms of *Rana pipiens* Schreber.' *Biol. Bull.*, **80**:238–64.

POULTON, E. B., and C. B. SANDERS. 1898. 'An experimental inquiry into the struggle for existence in certain common insects.' *Report 68th Meet. Brit. Assoc. Adv. Sci.*, 906–9.

PRZIBRAM, H. 1908. 'Die "Scherenumkehr" bei dekapoden Crustaceen.' *Arch. f. Entw. Mech.*, **25**:266–343.

1917 'Wachstumsmessungen an Sphodromantis bioculata. III. Länge regenerierender und normaler Schreitbeine.' *Arch. f. Entw. Mech.*, **43**:1–19.

1927. 'Entwicklungsmechanik der Tiere.' In: C. Oppenheimer and L. Pincussen, *Tabulae Biol.*, **4**:216–347.

QUENSTEDT, W. 1929. 'Die Entwicklungsgeschwindigkeit des Lebens in der geologischen Zeitfolge.' *Zbl. Min., Abt.* B, 513–32.

1930. 'Die Anpassung an die grabende Lebensweise in der Geschichte der Solenomyiden und Nuculaceen.' *Geol. und Paläont. Abh.*, N.F., **18**, Heft 1. Jena.

QUIRING, D. P. 1939. 'Notes on an African elephant (*Elephas loxodonta africana*).' *Growth*, **3**:9–13.

RAJEWSKI, B. N., and N. W. TIMOFÉEFF-RESSOVSKY. 1939. 'Höhenstrahlung und die Mutationsrate von *Drosophila melanogaster*.' *Z. ind. Abst. u. Vererbl.*, **77**:488–500.

RAMME, W. 1931. 'Verlust und Herabsetzung der Fruchtbarkeit bei macropteren Individuen sonst brachypterer Orthopterenarten.' *Biol. Zentralbl.*, **51**:533–40.

REEVE, E. C. R., and J. HUXLEY. 1945. 'Some problems in the study of allometric growth.' In: W. E. le Gros Clark and P. B. Medawar, *Essays on growth and form*. Oxford.

REINHARDT, F. 1944. 'Einseitige Ernährung und Entwicklungsstörungen beim Rippenmolch.' *Forsch. u. Fortschr.*, **20**:232–4.

REINIG, W. F. 1938. *Elimination und Selektion.* Jena.

1939. 'Besteht die Bergmannsche Regel zu Recht?' *Arch. f. Naturgesch.*, N.F., **8**:70–88.

REMANE, A. 1928. 'Exotypus-Studien an Säugetieren. I. Zur Definition der systematischen Kategorie Aberration oder Exotypus.' *Z. Säugetierkde.*, **3**:64–79.

1932. 'Archiannelida.' In: Grimpe-Wagler, *Tierwelt der Nord- und Ostsee*, 6, a, Lief. 22. Leipzig.

1939. 'Der Geltungsbereich der Mutationstheorie.' *Zool. Anz.*, *Suppl.*, **12**:206–20.

1951. 'Besteht ein Zusammenhang zwischen der Formverwilderung (Typolyse) und dem Aussterben der Tierstämme?' *Schr. Naturwiss. Ver. Schleswig-Holstein*, **25**:14–19.

1952. *Die Grundlagen des natürlichen Systems der vergleichenden Anatomie und der Phylogenetik.* Leipzig.

RENSCH, B. 1923. 'Über Konvergenzerscheinungen im Vogelreich und ihre stammesgeschichtliche Bedeutung.' *Verh. Dtsch. Zool. Ges.*, **28**:31–32.

1924. 'Das Dépérétsche Gesetz und die Regel von der Kleinheit der Inselformen als Spezialfall des Bergmannschen Gesetzes und ein Erklärungsversuch desselben.' *Z. ind. Abstamm. u. Vererbl.*, **35**:139–55.

1925. 'Untersuchungen zur Phylogenese der Schillerstruktur.' *J. f. Ornith.*, **73**:127–47.

1925. 'Das Problem des Brutparasitismus bei Vögeln. *Sitzungsber. Ges. Naturf. Freunde Berlin*, 55–69.

1927. 'Schwingenfärbung schillernder Vögel und geschlechtliche Zuchtwahl.' *Zool. Anz.*, **70**:93–9.

1928. 'Inselmelanismus bei Mollusken.' *Zool. Anz.*, **78**:1–4.

1928. 'Grenzfälle von Rasse und Art.' *J. f. Ornith.*, **76**:222–31.

1929. *Das Prinzip geographischer Rassenkreise und das Problem der Artbildung.* Berlin.

1931. 'Die Vogelwelt der Kleinen Sunda-Inseln Lombok, Sumbawa und Flores.' *Mitt. Zool. Mus. Berlin*, **17**:451–637.

1932. 'Über die Abhängigkeit der Größe, des relativen Gewichtes und der Oberflächenstruktur der Landschneckenschalen von den Umweltfaktoren.' *Z. Morph. u. Ökol.*, **25**:757–807.

1934. *Kurze Anweisung für zoologisch-systematische Studien.* Leipzig.

1934. 'Die Molluskenfauna der Kleinen Sunda-Inseln.' III. *Zool. Jb., Abt. Syst.*, **65**:389–422.

1935. 'Umwelt und Rassenbildung bei warmblütigen Wirbeltieren.' *Arch. f. Anthropol.*, N.F., **23**:326–33.

1936. 'Studien über klimatische Prallelität der Merkmalsausprägung bei Vögeln und Säugern.' *Arch. f. Naturgesch.*, N.F., **5**:317–63.

1937. 'Untersuchungen über Rassenbildung und Erblichkeit von Rassenmerkmalen bei sizilischen Landschnecken.' *Z. ind. Abstamm. u. Vererbl.*, **72**:564–88.

1938. 'Bestehen die Regeln klimatischer Parallelität bei der Merkmalsausprägung von homöothermen Tieren zu Recht?' *Arch. f. Naturgesch.*, N.F., **7**:364–89.

1939. 'Klimatische Auslese von Größenvarianten.' *Arch. f. Naturgesch.*, N.F., **8**:89–129.

1939. 'Typen der Artbildung.' *Biol. Rev.*, **14**:180–222.

1939. 'Über die Anwendungsmöglichkeit zoologisch-systematischer Prinzipien in der Botanik.' *Chronica Botanica*, **5**:46–9.

1941. ' "Elimination" oder Selektion bei der Girlitzausbreitung?' *Ornithol. Monatsber.*, **49**:94–104.

1943. 'Studien über Korrelation und klimatische Parallelität der Rassenmerkmale von Carabus-Formen.' *Zool. Jb., Abt. Syst.*, **76**:103–70.

1943. 'Die biologischen Beweismittel der Abstammungslehre.' In: Heberer, *Evolution der Organismen*, 57–85. Jena.

1944. 'Die paläontologischen Evolutionsregeln in zoologischer Betrachtung.' *Biol. gen.*, **17**:1–55.

1948. 'Organproportionen und Körpergröße bei Säugetieren und Vögeln.' *Zool. Jb., Abt. allg. Zool.*, **61**:337–412.

1948. 'Histological changes correlated with evolutionary changes of body-size.' *Evolution*, **2**:218–30.

1949. 'Biologische Gefügegesetzlichkeit.' In: *Das Problem der Gesetzlichkeit*, Joachim-Jungius-Ges., 117–37. Hamburg.

1949. 'Histologische Regeln bei phylogenetischen Körpergrößenänderungen.' *Symposium sui fattori ecologici e genetici della speciazione negli animali. La Ricerca Scientif.*, Suppl.

1950. 'Die Abhängigkeit der relativen Sexualdifferenz von der Körpergröße.' *Bonner Zool. Beitr.*, **1**:58–69.

1951. 'Complex effects of evolutionary changes in body-size.' Symposium. *Australian and New Zealand Association for the Advancement of Science*. Brisbane, 1951.

1952. *Psychische Komponenten der Sinnesorgane. Eine psychophysische Hypothese.* Stuttgart.

1952. 'Klima und Artbildung.' *Geol. Rundschau*, **40**:137–52.

1954. 'The relations between the evolution of central nervous functions and the body size of animals.' In: J. S. Huxley: *Evolution as a process*, 181–200. London.

1954. 'Neuere Untersuchungen über transspezifische Evolution.' *Verhandl. Dtsch. Zool. Ges.* (1952), 379–408. Leipzig.

1954. 'Die phylogenetische Abwandlung der Ontogenese.' In: Heberer, *Evolution der Organismen*, 2nd ed., 103–30. Stuttgart.

1955. 'Hirngröße und Lernfähigkeit.' *Arbeitsgemeinsch. f. Forschung Nordrhein-Westf., K. Arnold-Festschr.*, 597–601. Cologne and Opladen.

1956. 'Increase of learning capability by increase of brain size.' *Amer. Nat.*, **90**:81–95.

1956. 'Relative Organmaße bei tropischen Warmblütern.' *Zool. Anz.*, **156**:106–24.

1958. 'Die Abhängigkeit der Struktur und der Leistungen tierischer Gehirne von ihrer Größe.' *Naturwiss.*, **45**:145–54; 175–80.

RENSCH, B., and R. ALTEVOGT. 1953. 'Visuelles Lernvermögen eines Indischen Elefanten.' *Z. f. Tierpsychol.*, **10**:119–34.

RENSCH, I. 1934. 'Systematische und tiergeographische Untersuchungen über die Landschneckenfauna des Bismarck-Archipels.' *Arch. f. Naturgesch.*, N.F., **3**:445–88.

REY, E. 1892. 'Altes und Neues aus dem Haushalte des Kukkucks.' *Zool. Vortr.*, Pt. 11. Leipzig.

RIBBERT, H. 1894. 'Beiträge zur kompensatorischen Hypertrophie und zur Regeneration.' *Arch. f. Entw. Mech.*, 1:69–90.

RICHARDS, O. W., and A. J. KAVANAGH. 1945. 'The analysis of growing forms.' In: W. E. le Gros Clark and P. B. Medawar, *Essays on growth and form.* Oxford.

RICHARDSON, D. 1931. 'Effects of heteroplastic transplantations of the ear in Amblystoma.' *Proc. Soc. Exper. Biol. Med.*, 28:998–9.

1932. 'Some effects of heteroplastic transplantation of the ear vesicle in Amblystoma.' *J. Exper. Zool.*, 63:413–45.

RICHET, C. 1891. 'Poids du cerveau, de la rate et du foie chez les chiens de différentes tailles.' *Comp. Rend. Soc. Biol.*, sér. 9, 3:405–15.

RIDDLE, O. 1927. 'Studies on thyroids.' *Endocrinology*, 11:161–72.

1938. 'Prolactin, a product of the anterior pituitary, and the part it plays in vital processes.' *Scientific Monthly*, 47:97–113.

RIDDLE, O., D. E. CHARLES, and G. E. CAUTHEN. 1932. 'Relative growth rates in large and small races of pigeons.' *Proc. Soc. Exper. Biol. and Med.*, 29:1216–20.

RIETZ, E. DU. 1930. 'The fundamental units of biological taxonomy.' *Svensk Botan. Tidskr.*, 24:333–428.

RIPLEY, S. D., and H. BIRCKHEAD. 'On the fruit pigeons of the *Ptilinopus purpuratus* group.' *Amer. Mus. Novitates*, No. 1192.

RIZKI, M. S. M. 1951. 'Morphological differences between two sibling species, *Drosophila pseudoobscura* and *Drosophila persimilis*.' *Proc. Nat. Acad. Sci.*, 37:156–9.

ROBB, R. C. 1935. 'A study of mutations in evolution.' *Genetics*, 31:39–46; 47–52.

ROBINSON, B. 1748. *A dissertation on the food and discharge of human bodies.* London.

ROMER, A. S. 1933. *Man and the vertebrates.* Chicago.

RÖNSCH, G. 1954. 'Entwicklungsgeschichtliche Untersuchungen zur Zelldifferenzierung am Flügel der Trichoptere *Limnophilus flavicornis* Fabr.' *Z. Morph. u. Ökol.*, 43:1–63.

ROSA, D. 1899. *La riduzione progressiva della variabilita a i suoi rapporti col' estinzione e coll' origine delle specie.* Turin. (German ed., Jena, 1903.)

1931. *L'Ologénèse. Nouvelle théorie de l'évolution.* Paris.

ROTMANN, E. 1933. 'Die Rolle des Ektoderms und Mesoderms bei der Formbildung der Extremitäten von Triton. 2. Operationen im Gastrula- und Schwanzknospenstadium.' *Arch. f. Entw. Mech.*, 129:85–119.

1935. 'Der Anteil von Induktor und reagierendem Gewebe an der Entwicklung des Haftfadens.' *Arch. f. Entw. Mech.*, 133:193–224.

1939. 'Der Anteil von Induktor und reagierendem Gewebe an der Entwicklung der Amphibienkeime.' *Arch. f. Entw. Mech.*, 139:1–49.

ROTMANN, E., and T. J. MACDOUGALD. 1936. 'Die Struktur normaler und heteroplastisch transplantierter Epidermis von *Triton taeniatus* (und *palmatus*) und *cristatus* nach der Metamorphose.' *Verh. Dtsch. Zool. Ges.*, 38:88–96.

RUBNER, M. 1883. 'Über den Einfluß der Körpergröße auf Stoff und Kraftwechsel.' *Z. f. Biol.*, 19:312–96.

RUEDEMANN, R. 1918. 'Paleontology of arrested evolution.' *New York State Mus. Bull.*, No. 196 (XIII. Rep. Direct. 1916.)

RÜGER, L. 1943. 'Die absolute Chronologie der geologischen Geschichte als zeitlicher Rahmen der Phylogenie.' In: Heberer, *Evolution der Organismen*, 183–218. Jena.

RUMJANCEV, B. F. 1928. 'Die Variabilität von *Limnaea stagnalis*.' *Trav. Soc. Nat. Leningrad*, 63. (Russian, German summary.)

RUNNSTRÖM, S. 1936. 'Die Anpassung der Fortpflanzung und Entwicklung mariner Tiere an die Temperaturverhältnisse verschiedener Verbreitungsgebiete.' *Bergens Mus. Årb.* 1936, *Nat. Rekke*, No. 3.

RUSKA, H. 1950. *Virus. Eine kurze Zusammenfassung der Kenntnisse über das Virusproblem.* Darmstadt.

RUSKA, H., and K. POPPE. 1947. 'Morphologische Beziehungen zwischen filtrierbaren Mikroorganismen und großen Virusarten.' *Z. f. Naturforsch.*, **2b**:35–6.

RUTZ, H. 1948. 'Wachstum, Entwicklung und Unverträglichkeitsreaktionen bei Artchimären von Triton.' *Rev. Suisse Zool.*, **55**:623–74.

SALFELD, H. 1921. 'Kiel- und Furchenbildung auf der Schalenaußenseite der Ammonoideen in ihrer Bedeutung für die Systematik und Festlegung von Biozonen.' *Zbl. f. Min. Geol.*, 343–7.

SALOMONSEN, F. 1931. 'Diluviale Isolation und Artenbildung.' *Proc. 7th Int. Ornith. Kongr. Amsterdam*, 413–38.

SALT, G. W. 1952. 'The relation of metabolism to climate and distribution in three finches of the genus *Carpodacus.*' *Ecol. Monogr.*, **22**:121–52.

SARS, G. O. 1899. *An account of the Crustacea of Norway.* 2. Bergen.

SCHÄFER, E., and R. MEYER DE SCHAUENSEE. 1939. 'Zoological results of the 2nd Dolan expedition to western China and eastern Tibet 1934–1936. 2. Birds.' *Proc. Acad. Nat. Hist. Philadelphia* (1938), **90**:185–260.

SCHÄFER, H. 1940–2. *Elektrophysiologie.* 2 vols. Vienna.

SCHIEMANN, K. 1939. 'Vom Erlernen unbenannter Anzahlen bei Dohlen.' *Z. f. Tierpsychol.*, **3**:292–347.

SCHIERMANN, G. 1930–4. 'Studien über Siedelungsdichte im Brutgebiet. I and II.' *J. f. Ornith.*, **78**: 138–80 (1930); **82**:455–86 (1934).

SCHILDER, F. A. 1950. 'Körpergröße und Organzahl der Organismen.' *Hallische Monogr.*, No. 18. Halle (S.).

SCHILDER, M., and F. A. SCHILDER. 1939. 'Prodrome of a monography of living Cypraeidae.' *Proc. Malacol. Soc. London*, **23**:119–231.

SCHILLER, P. VON. 1933. 'Intersensorielle Transposition bei Fischen.' *Z. vergl. Physiol.*, **19**:304–9.

—— 1943. 'Umwegversuche an Elritzen.' *Z. f. Tierpsychol.*, **5**:101–31.

SCHILLING, E. 1951. 'Metrische Untersuchungen an den Nieren von Wild- und Haustieren.' *Z. Anat. Entw.-gesch.*, **116**:67–95.

SCHINDEWOLF, O. H. 1925. 'Entwurf einer Systematik der Perisphincten.' *Neues Jb. Mineral., Beilage*, **52 B**:309–43.

—— 1936. *Paläontologie, Entwicklungslehre und Genetik. Kritik und Synthese.* Berlin.

—— 1937. 'Geologisches Geschehen und organische Entwicklung.' *Bull. Geol. Inst. Uppsala*, **27**:321–40.

—— 1940. 'Zur Theorie der Artbildung.' *Sitzungsber. Ges. Naturforsch. Freunde Berlin*, 1940, 368–85.

—— 1942. 'Evolution im Lichte der Paläontologie. Bilder aus der Stammesentwicklung der Cephalopoden.' *Jenaische Z. f. Med. u. Naturwiss.*, **75**:324–86.

—— 1946. 'Zur Kritik des "Biogenetischen Grundgesetzes".' *Naturwiss.*, **33**:244–9.

—— 1950. *Grundfragen der Paläontologie.* Stuttgart.

—— 1950. *Der Zeitfaktor in Geologie und Paläontologie.* Stuttgart.

SCHLABRITZKY, E. 1953. 'Die Bedeutung der Wachstumsgradienten für die Proportionierung der Organe verschieden großer Haushuhnrassen.' *Z. Morph. u. Ökol. Tiere*, **14**:278–310.

SCHMALHAUSEN, J. J. 1925. 'Über die Beeinflussung der Morphogenese der Extremitäten vom Axolotl durch verschiedene Faktoren.' *Arch. f. Entw. Mech.*, **105**:483–500.

—— 1949. *Factors of evolution. The theory of stabilizing selection.* Philadelphia and Toronto.

SCHMIDT, H. 1918. *Geschichte der Entwicklungslehre.* Leipzig.

1925. 'Neotenie und beschleunigte Entwicklung der Ammoneen.' *Paläont. Z.,* 7:198–205.

1935. *Einführung in die Paläontologie.* Stuttgart.

SCHMIDT, W. J. 1952. 'Wie entstehen die Schillerfarben der Federn?' *Naturwiss.,* 39:313–18.

SCHNECKE, C. 1941. 'Zwergwuchs beim Kaninchen und seine Vererbung.' *Z. menschl. Vererb. u. Konstitl.,* 25:425–57.

SCHNEIDER, C. 1935. 'Lepidopterologische Beobachtungen.' *Jahreshefte Ver. Vaterl. Naturk. Württ.,* 91:125–7.

SCHNEIDER, C., and A. WÖRZ. 1939. 'Die Lepidopterenfauna von Württemberg.' *Jahreshefte Ver. Vaterl. Naturk. Württ.,* 95:231–87.

SCHOLANDER, P. F. 1955. 'Evolution of climatic adaptation in homeotherms.' *Evolution,* 9:15–26.

SCHRAMM, G. 1943. 'Über die Spaltung des Tabakmosaikvirus in niedermolekulare Proteine und die Rückbildung hochmolekularen Proteines aus den Spaltstücken.' *Naturwiss.,* 31:94–6.

1943. 'Über die Struktur des Tabakmosaikvirus.' *Forsch. u. Fortschr.,* 19:225–6.

1948. 'Zur Chemie des Mutationsvorganges beim Tabakmosaik-Virus.' *Z. f. Naturforsch.,* 3b:320–8.

1953. 'Chemie der Viren.' *Verh. Ges. Dtsch. Naturforsch. u. Ärzte,* 97 (1952), 61–8.

SCHRAMM, G., G. BERGOLD, and H. FLAMMERSFELD. 1946. 'Zur Konstitution der Hefenukleinsäure.' *Z. f. Naturforsch.,* 1:328–36.

SCHRAMM, G., and M. WIEDEMANN. 1951. 'Größenverteilung des Tabakmosaik-virus in der Ultrazentrifuge und im Elektronenmikroskop.' *Z. f. Naturforsch.,* 6b:379–83.

SCHUH, F. 1937. 'Gedanken über die bei der tierischen Entwicklung hervortretenden Entwicklungsrichtungen.' *Paläont. Z.,* 19:117–26.

SCHULZ, A. H. 1931. 'Man as a primate.' *Scientific Monthly,* 33.

SCHULZ, C. 1951. 'Die relative Größe cytoarchitektonischer Einheiten im Groß-hirn der Weißen Ratte, Weißen Maus und Zwergmaus.' *Zool. Jb., Abt. allg. Zool.,* 63:64–106.

SCHUSTER, O. 1950. 'Die klimaparallele Ausbildung der Körperproportionen bei Poikilothermen.' *Abh. Senckenberg. Naturforsch. Ges.,* Abh. 482. Frankfurt a. M.

SCHWANITZ, F., and H. SCHWANITZ. 1955. 'Eine Großmutation bei *Linaria maroccana* L. mut. *gratioloides.' Beitr. Biol. Pflanz.,* 31:473–97.

SCHWARZ, O. 1936. 'Über die Systematik und Nomenklatur der europäischen Schwarzkiefern.' *Notizbl. Botan. Garten u. Mus. Berlin-Dahlem,* 13:216–43.

SCHWARZBACH, M. 1950. *Das Klima der Vorzeit.* Stuttgart.

SCHWEGLER, E. 1941. 'Ein Beitrag aus der Stammesgeschichte der Jura-Belem-niten zur Frage "Zufall oder Richtung in der organischen Entwicklung?"' *Paläont. Z.,* 22:169–80.

SCHWERDTFEGER, F. 1944. *Die Waldkrankheiten. Ein Lehrbuch der Forstpatho-logie und des Forstschutzes.* Berlin.

SCHWIND, J. L. 1932. 'Further experiments on the limb and shoulder girdle of Amblystoma.' *J. Exper. Zool.,* 63:345–63.

SCOTT, W. B. 1936. 'The laws of mammalian evolution.' *Scientific Monthly,* 43:421–9.

SEIDENBERG, R. 1950. *Posthistoric man.* Univ. North Carolina Press, Chapel Hill.

SEILER, J. 1925. 'Zytologische Vererbungsstudien an Schmetterlingen.' *Arch. Vererb. Sozialanthrop.,* 1:63–117.

1936. 'Neue Ergebnisse aus der Kreuzung parthenogenetischer Schmetterlinge mit Männchen zweigeschlechtlicher Rassen.' *Verh. Dtsch. Zool. Ges.*, **38**:147–50. (Zool. Anz., Suppl. 9.)

1938. 'Ergebnisse aus einer Kreuzung einer diploidparthenogenetischen Solenobia triquetrella mit Männchen einer bisexuellen Rasse.' *Rev. Suisse Zool.*, **45**:405–12.

SEMON, R. 1905. *Die Mneme als erhaltendes Prinzip im Wechsel des organischen Geschehens.* 1st ed. Leipzig.

ŠENGÜN, A. 1945. 'Experimente zur sexuell-mechanischen Isolation.' *Rev. Fac. Sci. Univ. Istanbul*, B., **9**: Pt. 4, 239–53.

SEWERTZOFF, A. N. 1931. 'Studien über die Reduktion der Organe der Wirbeltiere.' *Zool. Jb., Abt. Anat.*, **53**:611–700.

1931. *Morphologische Gesetzmäßigkeiten der Evolution.* Jena.

SHEPHERD, R. H., D. A. SHOLL, and A. VIZOSO. 1949. 'The size relationships subsisting between body length, limbs and jaws in man.' *J. Anat.*, **83**:296–302.

SHOLL, D. 1948. 'The quantitative investigation of the vertebrate brain and the applicability of allometric formulae to its study.' *Proc. Royal. Soc.*, B, **135**:243–58.

SIBLEY, C. G. 1954. 'Hybridization in the red-eyed towhees of Mexico.' *Evolution*, **8**:252–90.

SICKENBERG, O. 1931. 'Morphologie und Stammesgeschichte der Sirenen.' *Palaeobiologica*, **4**:405–44.

SIEBER, R. 1936. 'Die Cancellariidae des niederösterreichischen Miozäns.' *Arch. f. Molluskenkde.*, **68**:65–115.

1937. 'Die Fasciolariidae des niederösterreichischen Miozäns.' *Arch. f. Molluskenkde.*, **69**:138–60.

SIERP, H. 1939. 'Physiologie.' In: Fitting *et al. Lehrbuch der Botanik für Hochschulen.* 20th ed. Jena.

SIMPSON, G. G. 1944. *Tempo and mode in evolution.* New York.

1949. *The meaning of evolution.* New Haven.

1953. *The major features of evolution.* New York.

SIMROTH, H., and H. HOFFMANN. 1928. 'Pulmonata.' In: Bronns, *Klass. u. Ordn. des Tierreichs*, 3, Abt. II, Buch 2. Leipzig.

SINNOT, E. W., and L. C. DUNN. 1935. 'The effect of genes and the development of size and form.' *Biol. Rev.*, **10**:123–51.

SLEPTSOV, M. M. 1939. 'On the finding of rudiments of the hind flippers in the Black Sea Dolphin (*Delphinus delphis* La.).' *Zool. J. (Moskow)*, **18**:351–66.

SMITH, P. E., and E. C. MCDOWELL. 1931. 'The differential effect of hereditary mouse dwarfism on the anterior-pituitary hormones.' *Anat. Rec.*, **50**:85–93.

1931. 'An hereditary anterior-pituitary deficiency in the mouse.' *Anat. Rec.*, **46** (1930):249–57.

SNELL, G. D. 1929. 'Dwarf, a new Mendelian recessive character in the housemouse.' *Proc. Nat. Acad. Sci.*, **15**:733–4.

SNELL, O. 1891. 'Die Abhängigkeit des Hirngewichtes von dem Körpergewicht und den geistigen Fähigkeiten.' *Arch. Psychiatrie*, **23**:436–46.

SNETHLAGE, H. 1928. 'Meine Reise durch Nordostbrasilien. II.' *J. f. Ornith.*, **76**:503–87.

SNOO, K. DE. 1942. *Das Problem der Menschwerdung im Lichte der vergleichenden Geburtshilfe.* Jena.

SONNEBORN, T. M. 1950. 'The cytoplasm in heredity.' *Heredity*, **3**:11–36.

SOVERI, J. 1940. 'Die Vogelfauna von Lammi, ihre regionale Verbreitung und Abhängigkeit von den ökologischen Faktoren.' *Acta Zool. Fennica*, **27**:1–176.

SPASSKY, B., S. ZIMMERING, and T. DOBZHANSKY. 1950. 'Comparative genetics of *Drosophila prosaltans*.' *Heredity*, 4:189–200.

SPEMANN, H. 1943. *Forschung und Leben*. Stuttgart.

SPENCER, H. 1855. *Principles of psychology*. London.

SPENCER, W. 1949. 'Gene homologies and the mutants of *Drosophila hydei*.' In: Jepsen *et al.*, *Genetics, Paleontology and Evolution*. Princeton.

SPERBER, J. 1944. 'Studies on the mammalian kidney.' *Zool. Bidr. Uppsala*, 22:249–432.

SPETT, G. 1929. 'Zur Frage der Homogamie und Pangamie bei Tieren. Untersuchungen an einigen Coleopteren.' *Biol. Zentralbl.*, 49:385–92.
— 1931. 'Entwicklung der Artunterschiede in der postembryonalen Ontogenese zweier Arten der Gattung Chorthippus (Orth.).' *Arch. f. Entw. Mech.*, 124:241–72.
— 1932. 'Gibt es eine partielle sexuelle Isolation unter den Mutationen und der Grundform von *Drosophila melanogaster* Meig.?' *Z. ind. Vererb. u. Abstamm.* L., 60:63–83.

SPIEGELMANN, S. 1945. 'Physiological competition as a regulatory mechanism in morphogenesis.' *Quart. Rev. Biol.*, 20:121–46.

SPIESS, E. B. 1950. 'Experimental populations of *Drosophila persimilis* from an altitudinal transect of the Sierra Nevada.' *Evolution*, 4:14–33.

SPIETH, H. 1947–9. 'Sexual behavior and isolation in *Drosophila*. I and II.' *Evolution*, 1:17–31 (1947); 3:67–81 (1949).

SPINA FRANCA NETTO, A. 1951. 'Dimensioni e numero dei neuroni in relazione alla mole somatica.' *Z. f. Zellforsch.*, 36:222–34.

SPINOZA, B. DE. 1677. 'Ethica.' In: *Opera quotquae reperta sunt*. Ed. by J. van Vloten and J. P. R. Land. 3rd ed. Den Haag, 1914.

SPURWAY, H. 1953. 'Genetics of specific and subspecific differences in European newts.' *Sympos. Soc. Exper. Biol.*, No. VII, 200–37.

STALKER, H., and H. L. CARSON. 1948. 'An altitudinal transect of *Drosophila robusta* Sturtevant.' *Evolution*, 2:295–305.

STANDFUSS, M. 1909. 'Hybridationsexperimente.' *Proc. 7th Int. Congr. Cambridge, Mass.*, 57–73.

STEBBINS, G. L., jr. 1950. *Variation and evolution in plants*. New York.

STEGMANN, B. 1934. 'Über die Formen der großen Möven ("subgenus *Larus*") und ihre gegenseitigen Beziehungen.' *J. f. Ornith.*, 82:340–80.

STEIN, G. 1938. 'Biologische Studien an deutschen Kleinsäugern.' *Arch. f. Naturgesch.*, N.F., 7:477–513.

STEINBACHER, G., and J. STEINBACHER. 1943. 'Über die Entstehung und das Alter von Vogelrassen.' *Zool. Anz.*, 141:141–7.

STEINBÖCK, O., and B. AUSSERHOFER. 1950. 'Zwei grundverschiedene Entwicklungsabläufe bei einer Art (*Prorhynchus stagnatilis* M. Sch., Turbellaria).' *Arch. f. Entw. Mech.*, 144:155–77.

STEINER, H. 1942. 'Bastardstudien bei Pleurodeles-Molchen. Letale Fehlentwicklung in der F_2-Generation bei artspezifischer Kreuzung.' *Arch. Jul.-Klaus-Stift. f. Vererbl. Sozialanthrop.*, 17:428–32.
— 1945. 'Über letale Fehlentwicklung der zweiten Nachkommenschaftsgeneration bei tierischen Artbastarden.' *Arch. Jul.-Klaus-Stift. f. Vererb. Sozialanthrop.*, Erg. Bd., 20:236–51.
— 1948. 'Einige tiergeographische Aspekte zur Frage der modifikatorischen oder genotypischen Differenzierung der Coregonen in den Gewässern des Alpennordrandes.' *Rev. Suisse Zool.*, 55:338–46.

STEINER, H., and G. ANDERS. 1946. 'Zur Frage der Entstehung von Rudimenten. Die Reduktion der Gliedmaßen von *Chalcides tridactylus* Laur.' *Rev. Suisse Zool.*, 53:537–46.

STEINIGER, F. 1937. ' "Ekelgeschmack" und visuelle Anpassung einiger Insekten.' *Z. wiss. Zool.*, A, **149**:221–57.

1937. 'Beobachtungen und Bemerkungen zur Mimikryfrage.' *Biol. Zentralbl.*, **57**:47–58.

1938. 'Die Genetik und Phylogenese der Wirbelsäulenvarietäten und der Schwanzreduktion.' *Z. menschl. Vererb. u. Konstitl.*, **22**:583–668.

STERN, C. 1954. 'Genes and developmental patterns.' *Caryologia*, 6, Suppl., 355–69.

1955. 'Gene action.' In: Willier, Weiss, and Hamburger, *Analysis of development.* Philadelphia and London.

STERNFELD, R. 1913. *Die Reptilien und Amphibien Mitteleuropas.* Leipzig.

STIRTON, R. A. 1947. 'Observations on evolutionary rates in hypsodonty.' *Evolution*, **1**:32–41.

STOCKARD, C. R. 1921. 'Developmental rate and structural expression.' *Amer. J. Anat.*, **28**:115–277.

1934. 'Internal constitution and genic factors in growth determination.' *Cold Spr. Harbor Symp. on Quant. Biol.*, **2**:118–27.

STOLPE, M., and K. ZIMMER. 1939. 'Der Schwirrflug des Kolibri im Zeitlupenfilm.' *J. f. Ornith.*, **87**:136–55.

STOLTE, H. A. 1937. 'Cellularbiologie und Ganzheitsbiologie, Biostatik und Biodynamik.' *Z. ges. Naturwiss.*, **3**:113–28.

STORER, T. I., and P. W. GREGORY. 1934. 'Color aberrations in the pocket gopher and their probable genetic explanation.' *J. Mammal.*, **15**:300–12.

STRESEMANN, E. 1919. 'Über die europäischen Gimpel.' *Beitr. Zoogeogr. palaearkt. Reg.*, Heft 1, 25–56. Munich.

1919. 'Über die Formen der Gruppe Aegithalos caudatus und ihre Kreuzungen.' *Beitr. Zoogeogr. palaearkt. Reg.*, Heft 1, 1–24. Munich.

1926. 'Übersicht über die Mutationsstudien I–XXIV und ihre wichtigsten Ergebnisse.' *J. f. Ornith.*, **74**:377–85.

1927–34. 'Aves.' In: Kükenthal-Krumbach, *Handbuch der Zool.*, 7th vol., 2nd Hälfte. Berlin and Leipzig.

1931. 'Die Zosteropiden der indoaustralischen Subregion.' *Mitt. Zool. Mus. Berlin*, **17**:201–38.

1943. 'Ökologische Sippen-, Rassen- und Artunterschiede bei Vögeln.' *J. f. Ornith.*, **91**:305–24.

STRESEMANN, E., and N. W. TIMOFÉEFF-RESSOVSKY. 1947. Artentstehung in geographischen Formenkreisen. I. Der Formenkreis *Larus argentatus–cachinnans–fuscus.' Biol. Zentralbl.*, **66**:57–76.

STROMER, E. 1944. 'Gesicherte Ergebnisse der Paläozoologie.' *Abhandl. Bayr. Akad. Wissensch.*, N.F., Pt. 54. Munich.

STROUHAL, H. 1939. 'Landasseln aus Balkanhöhlen.' *Zool. Anz.*, **125**:181–90.

STUBBE, H. 1938. 'Genmutation. I. Allgem. Teil.' *Handbuch der Vererbungswissenschaft*, **2**. Berlin.

1940. 'Neue Forschungen zur experimentellen Erzeugung von Mutationen.' *Biol. Zentralbl.*, **60**:113–29.

STUBBE, H., and F. VON WETTSTEIN. 1941. 'Über die Bedeutung von Klein- und Großmutation in der Evolution.' *Biol. Zentralbl.*, **61**:265–97.

STURTEVANT, A. H., and T. DOBZHANSKY. 1936. 'Observations on the species related to new forms of *Drosophila affinis* with descriptions of seven.' *Amer. Nat.*, **70**:574–84.

1936. 'Geographical distribution and cytology of "sex-ratio" in *Drosophila pseudoobscura* and related species.' *Genetics*, **21**:473–90.

STURTEVANT, A. H., and C. C. TAN. 1937. 'The comparative genetics of *Drosophila pseudoobscura* and *Drosophila melanogaster.' J. Genetics*, **34**:415–32.

SUMNER, F. B. 1930. 'Genetic and distributional studies of three subspecies of *Peromyscus.' J. Genetics*, **23**:277–376.

1932. 'Genetic, distributional, and evolutionary studies of the subspecies of deer mice (*Peromyscus*).' *Bibliogr. Genet.*, **9**:1–106.

1935. 'Evidence for the protective value of changeable coloration in fishes.' *Amer. Nat.*, **69**:245–66.

SUOMALAINEN, E. 1941. 'Vererbungsstudien an der Schmetterlingsart *Leucodonta bicoloria.' Hereditas*, **27**:313–18.

1947. 'Parthenogenese und Polyploidie bei Rüsselkäfern (Curculionidae).' *Hereditas*, **33**:452–6.

SVÄRDSON, G. 1945. 'Chromosome studies on Salmonidae.' *Reports Swedish Inst. Fresh Water Fishery Research*, **23**:1–151.

SVILHA, A. 1935. 'Development and growth of the prairie deer mouse *Peromyscus maniculatus bairdii.' J. Mammal.*, **16**:109–16.

SURBECK, G. 1920. 'Beitrag zur Kenntnis der Schweizerischen Coregonen.' *Festschrift Zschokke*, No. 15. Basel.

SYLVESTER-BRADLEY, C. S. 1951. 'The subspecies in palaeontology.' *Geol. Magazine*, **88**:88–102.

SZALAI, T. 1936. 'Der Einfluß der Gebirgsbildung auf die Evolution des Lebens.' *Paläont. Z.*, **18**:113–22.

TATE, G. H. H. 1941. 'A review of the genus *Hipposideros* with special reference to Indo-Australian species.' *Bull. Amer. Mus. Nat. Hist.*, **78**:353–93.

1941. 'Review of *Myotis* of Eurasia.' *Bull. Amer. Mus. Nat. Hist.*, **78**:537–65.

TAVANI, G. 1951. 'Tierkrankheiten der Vorzeit.' *Umschau*, **51**:304–5.

TEISSIER, G. 1931. 'Recherches morphologiques et physiologiques sur la croissance des insectes.' *Trav. Stat. Biol. Roscoff*, **9**:29–238.

1934. 'Dysharmonies et discontinuités dans la croissance.' *Actual. Sci. Industr. Paris*, **95**:39.

1939. 'Biométrie de la cellule animale et végétale.' *Tab. Biologicae*, **19**(I):1–64.

TEST, A. R. 1946. 'Speciation in limpets of the genus *Acmaea.' Contrib. Lab. Vert. Zool. Univ. Mich.*, No. 31.

THIEL, M. E. 1930. 'Untersuchungen über den Einfluß der Abwässer von Hamburg-Altona auf die Verbreitung der Arten der Gattung Sphaerium in der Elbe bei Hamburg.' *Internat. Revue Hydrobiol.*, **24**:467–84.

1936-8. 'Scyphomedusae.' In: Bronns, *Klass. u. Ordn. des Tierreichs*, **2**, Abt. II, Buch. 2. Leipzig.

THIENEMANN, A. 1928. 'Die Felchen des Laacher Sees.' *Zool. Anz.*, **75**:226–34.

1939. 'Grundzüge einer allgemeinen Ökologie.' *Arch. für Hydrobiol.*, **35**:267–85.

1940. 'Unser Bild der lebenden Natur.' 90–1. *Jahresber. Naturhist. Ges. Hannover*, 27–51.

THOMAS, R. H., and J. S. HUXLEY, 1927. 'Sex ratio in pheasant species-crosses.' *J. Genetics*, **18**:233–46.

THOMPSON, W. D'ARCY. 1917. *Growth and Form*, Cambridge.

THORPE, W. H. 1930. 'Biological races in insects and allied groups.' *Biol. Rev.*, **5**:177–212.

1940. 'Ecology and the future of systematics.' In: Huxley, *The new systematics*, 341–64. Oxford.

TICEHURST, C. B. 1938. *A systematic review of the genus Phylloscopus*. London, British Museum.

TIELECKE, H. 1940. 'Anatomie, Phylogenie und Tiergeographie der Cyclophoriden.' *Arch. f. Naturgesch.*, N.F., **9**:317–71.

TILNEY, F. 1927. *The brain from ape to man. A contribution to the study of human evolution.* 2 vols. New York.

TIMOFÉEFF-RESSOVSKY, H. 1931. 'Über phänotypische Manifestierung der polytopen (pleiotropen) Genvariation Polyphaen von *Drosophila funebris.*' *Naturwiss.*, **19**:765–68.

TIMOFÉEFF-RESSOVSKY, N. W. 1937. *Experimentelle Mutationsforschung in der Vererbungslehre.* Dresden and Leipzig.

—— 1939. 'Genetik und Evolution.' *Z. ind. Abst.- u. Vererbl.*, **76**:158–218.

—— 1940. 'Zur Frage über die "Eliminationsregel": Die geographische Größenvariabilität von *Emberiza aureola* Pall.' *J. f. Ornith.*, **88**:334–40.

—— 1940. 'Eine biophysikalische Analyse des Mutationsvorganges.' *Nova Acta Leopoldina*, N.F., **9**:209–40.

TIMOFÉEFF-RESSOVSKY, N. W., K. G. ZIMMER, and L. DELBRÜCK. 1935. 'Über die Natur der Genmutation und der Genstruktur.' *Nachr. Ges. Wiss. Göttingen. Math. phys.* Kl. VI, Biol., N.F., **1**:189–245.

TINBERGEN, N. 1942. 'An objective study of the innate behaviour of animals.' *Bibl. biotheoret.*, **1**:39–98.

—— 1951. *The study of instinct.* London.

TINBERGEN, N., and A. C. PERDECK. 1950. 'On the stimulus situation releasing the begging response in the newly hatched Herring Gull chick (*Larus argentatus argentatus* Pont.).' *Behaviour*, **3**:1–39.

TONOLLI, V. 1949. 'Isolation and stability in populations of high altitude diaptomids.' *La Ricerca Scientif.*, **19**, Suppl.

TSCHUMI, P. 1954. 'Konkurrenzbedingte Rückbildungen der Hinterextremität von *Xenopus* nach Behandlung mit einem Chloraethylamin.' *Rev. Suisse Zool.*, **61**:177–270.

TURESSON, G. 1926. 'Die Bedeutung der Rassenökologie für die Systematik und Geographie der Pflanzen.' *Repert. Spec. nov. Regn. veg.*, Suppl. 41.

—— 1929. 'Zur Natur und Begrenzung der Arteinheiten. *Hereditas*, **12**:323–34.

UBISCH, L. VON. 1923. 'Das Differenzierungsgefälle des Amphibienkörpers und seine Auswirkungen.' *Arch. f. Entw. Mech.*, **52**:641–70.

—— 1933. 'Keimblattchimären.' *Naturwiss.*, **21**:325–9.

—— 1950. 'Undersökelser over Pleuronectider.' *Rep. Norweg. Fish. a. Marine Invest.*, **9**, No. 10. Bergen.

UEXKÜLL, J. VON. 1909. *Umwelt und Innenwelt der Tiere.* Berlin.

ULE, C. H. 1925. 'Über Auftreten und Verbreitung von *Amphidasis betularia* L. f. *carbonaria* Jord. auf dem Kontinent.' *Intern. Entom. Z.*, **18**:258–63; **19**:74–76; 137.

URBAHN, E. 1913. *Abdominale Duftorgane bei weiblichen Schmetterlingen.* (Dissertation.) Jena.

USCHMANN, G. 1939. *Der morphobiologische Vervollkommnungsbegriff bei Goethe und seine problemgeschichtlichen Zusammenhänge.* Jena.

VANDEL, A. 1934. 'La parthénogenèse géographique.' *Bull. Biol. France Belg.*, **68**:419–63.

—— 1949. *L'homme et l'évolution.* Paris.

VASSEUR, E. 1952. 'Geographic variation in the Norwegian sea-urchin, *Strongylocentrotus.*' *Evolution*, **6**:87–100.

VAUGIEN, L. 1949. 'Relation biométrique interspécifique dans la croissance des oiseaux.' *Compt. Rend. Acad. Sci. Paris*, **228**:206–7.

VAURIE, C. 1949. 'A revision of the Dicruridae.' *Bull. Amer. Mus. Nat. Hist.*, **93**:201–342.

VAVILOV, N. J. 1922. 'The law of homologous series in variation.' *J. Genet.*, **12**.

VERHOEFF, K. W. 1938. 'Chilopoden-Studien zur Kenntnis der Epimorphen.' *Zool. Jb., Abt. Syst.*, **71**:339–88.

VERRIER, M. L. 1950. 'Poecilogenie, ecologie et répartition géographique chez les Ephémères.' *Compt. Rend. Acad. Sci. Paris*, **230**:1794–6.

VERSLUYS, O., O. POETZL, and K. LORENZ. 1939. *Hirngröße und hormonales Geschehen bei der Menschwerdung.* Vienna.

VERWORN, M. 1910. *Die Entwicklung des menschlichen Geistes.* 1st ed., Jena, 1910; 4th ed., 1920.

1912. 'Die zellular-physiologischen Grundlagen des Abstraktionsprozesses,' *Z. allg. Physiol.,* **14.**

1919. *Die Mechanik des Geisteslebens.* 4th ed. Leipzig.

VIETS, K. 1926. 'Indische Wassermilben.' *Zool. Jb., Abt. Syst.,* **52:**369–94.

VILLEE, C. A. 1942. 'The phenomenon of homoeosis.' *Amer. Nat.,* **76:**494–506.

VOGT, C. 1863. *Vorlesungen über den Menschen.*

VOGT, C., and B. HOFER. 1909. *Die Süßwasserfische von Mitteleuropa.* Halle (S.).

VOIPIO, P. 1950. 'Evolution at the population level with special reference to game animals and practical game management.' *Papers on Game Research,* **5.** Helsinki.

1954. 'Über die gelbfüssigen Silbermöven Nordwesteuropas.' *Acta Soc. Fauna Flora Fennica,* **71.** (No. 1):1–56.

VOLKELT, H. 1914. 'Über die Vorstellungen der Tiere.' *Arb. z. Entw. psychol.,* **1,** Pt. 2.

WACHSMUTH, C., and F. SPRINGER. 1897. 'The North American Crinoidea Camerata.' *Mem. Mus. Comp. Zool.,* Harvard, 20, 21.

WADDINGTON, C. H. 1948. 'The concept of equilibrium in embryology.' *Folia Biotheoret.,* 3:127–38.

WALLACE, A. R. 1890. Darwinism. 2nd ed. New York.

WALTER, E. 1913. *Unsere Süßwasserfische.* Leipzig.

WALTER, W. G. 1953. *The living brain.* London.

WALTHER, J. 1908. *Geschichte der Erde und des Lebens.* Leipzig.

WARNECKE, G. 1938. 'Über die taxonomische Bedeutung der Genitalarmatur der Lepidopteren.' *7th Int. Kongr. Entom. Berlin,* 461–81.

WATSON, D. M. S. 1952. 'Is "macroevolution" reality?' *Trans. N. York Acad. Sci.,* ser. 2, **14:**302–3.

WEBER, H. 1933. *Lehrbuch der Entomologie.* Jena.

1939. 'Zur Fassung und Gliederung eines allgemeinen biologischen Umweltbegriffes.' *Naturwiss.,* 27:633–44.

WEBER, H. H. 1941. 'Über Empfindungspotentiale.' *Forsch. u. Fortschr.,* 17:305.

WEBER, M. 1927–8. *Die Säugetiere.* 2nd ed. 2 vols. Jena.

WEDEKIND, R. 1914. 'Beiträge zur Kenntnis der oberkarbonischen Goniatiten.' *Mitt. Mus. Essen,* Pt. 1.

1920. 'Über Virenzperioden (Blüteperioden).' *Sitzungsber. Ges. Beförd. Naturw. Marburg.*

1927. 'Umwelt, Anpassung und Beeinflussung, Systematik und Entwicklung im Lichte erdgeschichtlicher Überlieferung.' *Sitzungsber. Ges. Beförd. Naturw. Marburg,* 62:237–45.

WEHRLI, H. 1938. '*Anchitherium aurelianense* Cuv. von Steinheim am Albuch.' *Paläeontographica,* **8,** Teil 7. Stuttgart.

1940. 'Gerichtete oder ungerichtete Entwicklung?' *Forsch. u. Fortschr.* 16:322–4.

WEIDEL, W. 1953. 'Entwicklung und Problematik der Virusforschung.' *Verh. Ges. Dtsch. Naturforsch. u. Ärzte,* 97:56–61. Berlin, Göttingen, Heidelberg.

WEIDENREICH, F. 1931. 'Über Umkehrbarkeit der Entwicklung.' *Paläont. Z.,* **13:**177–86.

1940. 'Man or ape?' *Natur. Hist.,* **45.**

1947. 'The trend of human evolution.' *Evolution,* 1:121–136.

WEIGELT, J. 1943. 'Paläontologie als stammesgeschichtliche Urkundenforschung.' In: G. Heberer, *Evolution der Organismen,* 131–82. Jena.

WEINERT, H. 1932. *Ursprung der Menschheit.* Stuttgart.
    1947. *Menschen der Vorzeit.* 2nd ed. Stuttgart.

WEISMANN, A. 1913. *Vorträge über Deszendenztheorie.* 3rd ed. 2 vols. Jena.

WELCKER, H., and A. BRANDT. 1902. 'Gewichtswerthe der Körperorgane bei dem Menschen und den Thieren.' *Arch. Anthrop.* (D), **28**:1–89.

WELLENSIEK, U. 1953. 'Die Allometrieverhältnisse und Konstruktionsänderungen bei dem kleinsten Fisch im Vergleich mit etwas größeren verwandten Formen.' *Zool. Jb., Abt. Anat.,* **73**:187–228.

WENZ, W. 1938. 'Gastropoda.' In: Schindewolf, *Hdb. Paläozool.,* **6**, Lief. 1. Berlin.

WESENBERG-LUND, C. 1904–8. *Plankton investigations of the Danish lakes.* I and II. Copenhagen.

WESTENHÖFER, M. 1942. *Der Eigenweg des Menschen.* Berlin.

WESTOLL, T. S. 1938. 'Ancestry of the tetrapods.' *Nature* (London), **141**:127–8.

WETTSTEIN, R. VON. 1898. *Grundzüge der geographisch-morphologischen Pflanzensystematik.* Jena.

WEYER, F. 1937. 'Rassenforschung bei Stechmücken.' *Naturwiss.,* **25**:529–35.

WHITE, M. J. D. 1954. *Animal ecology and evolution.* 2nd ed. Cambridge.

WHITMAN, C. O. 1919. *Inheritance, fertility and the dominance of sex and color in hybrids of wild species of pigeons.,* ed. by O. Riddle. Washington.

WIENER, N. 1948. *Cybernetics.* New York.

WILBERT, H. 1953. 'Experimentelle Untersuchungen über Materialkompensation bei der Entwicklung von Insekten.' *Z. f. Morph. u. Ökol. d. Tiere,* **41**:350–71.

WILLOUGHBY, R. R., and C. M. POMERAT. 1932. 'Homogamy in the toad.' *Amer. Nat.,* **66**:223–34.

WILSON, E. B. 1903. 'Notes on the reversal of asymmetry in the regeneration of the chelae in *Alpheus heterochelis.' Biol. Bull.,* **4**:197–201.

WILSON, E. O. 1955. 'A monograph revision of the ant genus *Lasius.' Bull. Mus. Comp. Zool. Harvard Coll.,* **113**:1–205.

WINKEL, K. 1951. 'Vergleichende Untersuchungen einiger physiologischer Konstanten bei Vögeln aus verschiedenen Klimazonen.' *Zool. Jb., Abt. Syst.,* **80**:256–76.

WITELEY, M. A., and K. PEARSON. 1900. 'Data for the problem of evolution in man. A first study of the variability and correlation of the hand.' *Proc. Roy. Soc.,* **65**.

WOHLFAHRT, T. A. 1939. 'Untersuchungen über das Tonunterscheidungsvermögen der Elritze (*Phoxinus laevis* Agas.). *Z. f. vergl. Physiol.,* **26**:570–604.

WOLTERECK, R. 1931. 'Beobachtungen und Versuche zum Fragenkomplex der Artbildung.' I. *Biol. Zentralbl.,* **51**:231–53.
    1940. *Ontologie des Lebendigen.* Stuttgart.

WOLTERSDORFF, W., and M. RADOVANOVIC. 1938. *Triturus alpestris reiseri* Wern. und *T. alpestris* Laur., vergesellschaftet im Proskoško-See.' *Zool. Anz.,* **122**:23–30.

WOOD, A. E. 1937. 'Parallel radiation among the geomyoid rodents.' *J. Mammal.,* **18**:171–6.

WRIGHT, S. 1932. 'The roles of mutation, inbreeding, cross-breeding, and selection in evolution.' *Proc. 6th Int. Congr. Genet.,* **1**.
    1940. 'The statistical consequences of Mendelian heredity in relation to speciation.' In: Huxley, J.; *The New systematics,* 161–83. Oxford.

WRINCH, D. M. 1936. 'On the molecular structure of chromosomes.' *Protoplasma,* **25**:550–69.

WURMBACH, H. 1954. 'Steuerung von Wachstum und Formbildung durch Wirkstoffe. IV. Die Wirkung von Thymusfraktionen auf das Wachstum von Kaulquappen.' *Arch. f. Entw. Mech.,* **147**:79–118.

YAEGER, C. H., T. S. PAINTER, and R. M. YERKES. 1940. 'The chromosomes of the chimpanzee.' *Science*, 74–5.

YERKES, R. M. 1945. *Chimpanzees, a laboratory colony*. New Haven.

YERKES, R. M., and A. W. YERKES. 1929. *The great apes. A Study of anthropoid life*. New Haven.

ZELENY, C. 1905. 'Compensatory regulations.' *J. Exper. Zool.*, 2:1–102.

ZEUNER, F. E. 1946. *Dating the past. An Introduction to geochronology*. London.

ZIEHEN, T. 1913. *Erkenntnistheorie auf psycho-physiologischer und physikalischer Grundlage*. Jena.

1915. *Die Grundlagen der Psychologie*. 2 vols. Leipzig and Berlin.

1921. 'Die Beziehungen der Lebenserscheinungen zum Bewußtsein.' *Abhandl. zur theoret. Biol.*, Pt. 13. Berlin.

1924. *Leitfaden der physiologischen Psychologie*. 12th ed. Jena.

1934–9. *Erkenntnistheorie*. Vol. 1 (1934); vol. 2 (1939). Jena.

ZILSEL, E. 1935. 'P. Jordans Versuch, den Vitalismus quantenmechanisch zu retten.' *Erkenntnis*, 5:56–64.

ZIMMER, C. 1930. 'Untersuchungen an Diastyliden (Ordnung Cumacea).' *Mitt. Zool. Mus. Berlin*, 16:583–658.

ZIMMER, C., and B. RENSCH. 1928. 'Vögel, Aves.' In: Brohmer, Ehrmann, and Ulmer, *Tierwelt Mitteleuropas*, 7, Lief. 2. Leipzig.

ZIMMER, K. G., and N. W. TIMOFÉEFF-RESSOVSKY. 1942. 'Über einige physikalische Vorgänge bei der Auslösung von Genmutationen durch Strahlung.' *Z. ind. Abst.- und Vererbl.*, 80:353–72.

ZIMMERMANN, K. 1934. 'Zur Genetik der geographischen Variabilität von *Epilachna chrysomelina* F.' *Entom. Beih. Berlin-Dahlem*, 1:86–90.

1936. 'Die geographischen Rassen von *Epilachna chrysomelina* F. und ihre Beziehungen zu *Epilachna capensis* Thunbg.' *Z. ind. Abst.- u. Vererbl.*, 71:527–37.

ZIMMERMANN, S. 1932. 'Über die Verbreitung und die Formen des Genus *Orcula* Held in den Ostalpen.' *Arch. f. Naturgesch.*, N.F., 1:1–56.

ZIMMERMANN, W. 1938. *Vererbung "erworbener Eigenschaften" und Auslese*. Jena.

1935. 'Rassen- und Artbildung bei Wildpflanzen.' *Forsch. u. Fortschr.*, 11:272–4

ZUCKERMANN, S., *et al.* 1950. 'Growth and form (Royal Society discussion).' *Nature* (London), 165:952–5.

ZUMPT, F. 1941. 'Die Rassenfrage bei *Anopheles maculipennis* Meigen.' *Z. f. Parasitenkde.*, 12:372–87.

Author Index

Abel 19, 20, 83, 88, 91, 95, 112, 115, 123, 124, 178, 186, 202, 203, 221, 226, 234, 235, 259, 273, 301
Adam 25
Aeppli 7
Ågrell 139
Alcmaeon 328
Alexander 11
Allee 43
Allers 346
Alpatov 52, 179
Altevogt 164, 165
Alverdes 334, 340
Amadon 106
Anaximander 282
d'Ancona 7, 11, 12, 25
Anders 223
Andrews 125
d'Arcy Thompson 135, 169, 170
Aristotle 334
Armbruster 26
Arndt 25, 109, 123, 298, 341
Arrhenius 310
Audova 239
Auerbach 6
Augustine, St. 328
Ausserhofer 250
Autrum 337

Backman 162
von Baer 57, 98, 203, 249, 250, 256, 265, 282, 291
Bailey 313
Baker 25, 50
Baldi 25
Baldwin 106, 188
Balinsky 246
Baltzer 200, 246, 248
Bargmann 173
Batak 214
Bates 49
Bauer 3, 7
Beadle 4
Becker 264
Beecher 232
de Beer 253, 254, 260, 262
Beheim-Schwarzbach 52

Beklemishev 49
Benazzi 7, 25
Benick 91
Beninde 87
Benson 138
Berg 24, 58, 88, 90
Berger 240, 347
Bergmann, A. 21
Bergmann, C. 134, 145, 160
Bergson 356
Berkeley 323, 353
Bernal 316
von Bertalanffy 129, 133, 160, 311, 315, 318
Beurlen 1, 57, 92, 98, 99, 103, 113, 123, 126, 191, 205, 232, 249, 272, 284, 298, 300, 316, 356
Bierens de Haan 347
Birch 11
Birckhead 32
Birukow 248
Blair 11, 51
Blank 160
Blasche 21
Blum 316
Bodenheimer 122
Böker 129, 131, 188, 189, 190
von Boetticher 32
Bohn 334
Bohr 319
du Bois Reymond 358
Bok 154, 294
Bolk 260, 304, 305
von Bonin 303
Boschko-Stepanenko 179
von Boxberger 165
Bovey 26
Bozler 346
Braestrup 43
Bramstedt 334
Brandt 135, 142, 145
Braun 165, 341
Breider 32
Bresslau 271
Bretscher 182
Breuning 24, 52
Brinkmann 89, 262
Broch 252

Bronn 282
Brooks 56
Broom 276, 301
Brown 325, 332
von Brücke 303, 325
Brummelkamp 142, 144, 153, 294, 303
Buchner 70
de Buck 49
Buckman 259
von Buddenbrock 160, 161, 211, 215, 284, 290, 343
Bünning 320
Bütschli 72
Bumke 325
Burgeff 104
Burkhardt 151
Butenandt 3
Buxton 108
Buytendijk 331, 340, 345

Cain 11
Callan 26, 32, 53
Carruthers 19
Carson 11, 49
Castle 167, 168, 211
Cauthen 168, 244
Cesnola 11
Charles 168, 244
Child 234
Clark 167, 191
Colbert 83, 209
Colosi 25
Condillac 323
Conklin 154, 213
Constance 50
Cope 57, 94, 133, 144, 166, 204, 206, 237, 272, 301
Cordeiro 8
Cott 11
Cuénot 10, 318
Cuvier 98, 142

Dacqué 57, 98, 110, 123, 124, 191, 198, 206, 268, 284, 300
Dahl 28
Darwin, C. 12, 57, 82, 90, 93, 95, 98, 120, 180, 203, 282, 285, 289, 291, 292, 300, 301, 328
Darwin, E. 282
Day 156
Decugis 236
Dehlinger 315
Delacour 28, 29
Delbrück 313
Dementiew 27
Democritus 336
Demoll 26
Denzer 145
Dépérét 93, 204, 206, 207
Descartes 323

Detwiler 246
Devaux 237
Dice 11, 26, 30, 31, 138
Diebschlag 343
Diener 206
Dobzhansky, vii, 2, 3, 4, 6, 8, 9, 10, 11, 12, 14, 26, 49, 51, 53, 192, 193, 217, 252, 253, 306
Döderlein 25, 204
Doflein 201
Dohrn 275
Dollo 123
Donaldson, 173
Dosse 148
Doutt 91
Driesch 310, 318, 356
Drosihn 52
Dubinin 49
Dubois 135, 136, 142, 144, 293, 294, 295, 304
Duncan 252, 253
Duncker 26
Dunn 169
Dürer 169
Dusser de Barenne 325

East 3, 53
von Economo 325, 332
Edinger, L. 352
Edinger, T. vii, 260, 288, 333
Eggers 183
Ehrenberg 20, 110, 226, 254, 255, 256
Eibl-Eibesfeldt 164
Eimer 37, 169, 180, 204, 260, 336
Eisentraut 25, 36, 37
Ekman 25, 85, 239
Eller 41, 52
Elton 9, 12, 122
Empedokles 318
Engel 25
Ephrussi 188
Epling 53
Erhardt 25
Erlenbach 161
van Erp Taalman Kipp 154, 294
Errington 11

Fahrenholtz 49, 91
Fechner 81, 352, 356
Federley 26
Fenton 234
Fickert 169
Fischer, E. 131
Fisher 5, 12, 22
Fitch 32
Ford 9, 12, 20, 22, 43
Fox 49
Frank 11
Franz, H. 52, 122
Franz, S. 325

Franz, V. 246, 282, 283, 285, 291
Frenzel 25
von Frey 337
Friedrich 252
Friedrich-Freksa 312, 315
Friedrichs 121
Fuhrmann 105, 108, 271
Fürbringer 223

Gäbler 20
Gabritschewsky 26
Galileo 134
Garstang 241
Gaudry 57, 93, 109, 204, 283, 292, 296
Gause 11, 121
Gawrilow 127, 128
von Gelei 334
Geitler 7
Geoffroy St Hilaire 180
Gerhardt 231
Geyr von Schweppenburg 25
Ghigi 54
Gieseler 301
Gildemeister 312
Gislén 232
Glass 4
Goethe 180, 282, 291, 301, 359
Goetsch 157, 176
Goldschmidt, E. 7
Goldschmidt, R. 26, 28, 32, 55, 56, 106, 167,
    244, 245, 266, 271, 272
Goldstein 325, 332
Goossen 156, 157, 172
Goss 168
Gösswald 48
Gottschewsky 193
Götz 51, 238
Götze 71, 175
Graber 182
Grabowsky 334
Granvik 117
Green 167
Greenwood 56
Gregor 50
Gregory 126, 168, 194, 291, 302
Grenacher 72
Groebbels 213, 214
Gross, F. 7
Gross, W. 267, 272
Grünberg 65
Grüneberg 131, 168

Haacke 204
Haager 312
Haardick 169
Hackenbruch 181
Hadorn 4, 14
Haeckel 57, 81, 98, 112, 113, 237, 259, 282,
    291, 301, 356

Haecker, 187, 193, 241, 256
Haempel 276
Haldane vii, 3, 4, 12, 13, 316
Hale 186
von Haller 134, 142, 145, 149
Hammond 331
Handlirsch 84
Hanson 6
Harde 151, 153, 294
Harms 132
Hartert 23
Hartmann 263, 310, 318, 320
von Hartmann 356
Hassmann 341
Hayes 304
Heberdey 88
Heberer 144, 301
Heckmann 80
Hediger 304
Heding 25
Heinrich 161, 162
Heinroth 145
Heisenberg 319
Heller 11
Hellmich 36
Helmcke 25
Helmholtz 310
Hemmingsen 160
Hempelmann 252, 350, 351
Henke 40, 196, 197, 249
Hennig 18, 58, 83, 86, 99, 110, 116, 204, 232,
    269, 283
Heraclitus 81
Hering 329
l'Héritier 10, 11
Herre 32, 91
Herrmann 321
Hershey 313
Herskowitz 6
Herter 26, 33, 333
Hertwig, O. 336
Hertwig, P. 53
Hertwig, R. 336
Hess 332
Hesse 72, 135, 145, 179, 201, 277, 278, 286
Heuts 49
Heyne 226
Hilgendorf 19
Hiltermann 19, 217
Hilzheimer 161, 276
Hinsche 259
Hinton 138
Hobhouse 334
Hoernes 19
Hofer 157, 158
Hofmann 337
von Hofsten 305
Holdhaus 88
Holmes 312

Holtfreter 127, 128
Homeyer 134, 153
Hooper 138
Hornbostel 324
Horton 53
Hovanitz 32, 46
Howard 51
Hubbs 32, 54
von Huene 58, 98, 103, 111, 205
Hume 323
Hundertmark 49
Hutchinson 202, 237
Hutt 168
Huxley, J. vii, 2, 3, 26, 32, 43, 55, 133, 135, 169, 170, 188, 284, 285, 297, 298, 302, 307
Huxley, T. H. 301
Hyatt 232

Jackson 145, 148
Jaeckel 116
Janzer 225
Jarvik 269
Jaworski 25
Jennings 340
Jensen 50
Jerison 294
Jeshikov 257
Johannsen 27, 30, 32, 54
Johnsen 310
Jordan H. J. 309, 311
Jordan, K. 25, 35, 52
Jordan, P. 315, 316, 319
Just 200

Kafka 334
Kagelmann 60
Kalabuchov 168
Kant 323, 345, 352
Katz 343
Kaufmann, L. 148
Kaufmann, R. 17, 18, 198, 199
Kauri 33
Kelvin 310
Kendeigh 122
Kestner 160
Keys 6
Kiefer 25
Kikuth 312
Kipp 201
Kiriakoff 43
Klatt 125, 135, 136, 145, 147, 169, 170
Kleinenberg 272
Kleist 294, 325, 332
Klemm 90
Knoerzer 32
Kobelt 194
Kober 110
Koch 19, 217

Kochs 181
Koehler, O. 165, 334, 340, 341, 342, 345, 347
Koehler, W. 304
Koken 203
Koller 51
Kopsch 303
Kornmüller 325, 332
Korschelt 163
Korschinsky 283, 300
Koskinas 325, 332
Kosswig 10, 193, 225
Krabbe 248, 249
Kramer 37, 68, 91
Kreitmann 27
Krieg 229
von Kries 325
von Krogh 301
Krölling 186, 225
Krumbiegel 46, 161, 180, 216, 225, 259
Krumschmidt 143, 153
Kruse 197
Kühn, A. 4, 5, 177
Kuhn, E. 17, 231
Kuhn, O. 83, 264
Kühne, K. 131
Kühne, W. G. 269
Kullenberg 4, 52
Kupka 7

La Barre 302
Lack vii, 12, 91, 108
Lamarck 187, 282, 291, 300, 301, 328, 334
Lambrecht 99
Lamprecht 8, 14
Lancefield 53
Landauer 131, 133, 224
Lange 19, 209
Lankaster 202
Lapique 135, 136, 142, 145
Lartet 292
Lasareff 246
Latimer 148
de Lattin 33, 35, 225
Laurence 53
Lauth 168
Lebedkin 257
Lechter 301
Le Gros Clark 301
Lehmann 258, 310
Leichtentritt 211
Leinemann 163, 215
Lemche 147
Lerner 297
Levi 154, 155, 179, 213
Lewis 5
Leydig 336
Li 242
Lillie 1, 318
Lindner 65

Lindroth 91
Lindsay 321
Linné 301
Linzbach 154
Lippmann 309
Lloyd Morgan 188, 347
Locke 323, 328
Lönnberg 90, 236
Lorenz 305, 344
von Lorenz 186
Lorkovič 7, 26
Lotka 11
Löwenthal 154, 213
Ludwig 9, 12, 58, 182, 225
Lüers 250
Lukin 188
Lull 232
Lundmarck 310
Luntz 263
Lutz 251
Lynn 251

Mach 81, 163, 323, 337, 356
Magnan 135, 145
Mampell 4
Manea 145
Mangold 134, 211, 337
Mařan 224
Margalef 46, 261
Mark 125
Markham 303
Marold 341
Marsh 288, 292
Marshall 47, 91
Maschkowzeff 240
Matthew 110, 273
Matveiev 258
Maurel 135, 145
Mayr vii, 2, 3, 21, 25, 27, 28, 30, 32, 34, 47,
  49, 144, 304
McCarty 8
McDougald 245
McDowell 168
McTaggart Cowan 138
Megušar 181
Meinhert 65
Meise 25, 27, 32, 54
Meissner 145
Meixner 52, 62
Mell 46
Mendheim 108
Mertens 24, 34, 35, 37, 90, 91
Metz 53
Meunier 144, 178
Meyer 109
Meyer de Schauensee 40
Michaelis 4, 8
Miculicz-Radecki 62
Mijsberg 260

Miller, A. H. vii, 27, 54
Miller, G. S. 138, 139
Miller, R. R. 32, 54
Miller, S. L. 316
Milne-Edwards 282, 291
Möller 150, 151, 163
Mollison 290
Moore 26, 32, 46
Mordvilko 106, 107
Moreau 91
Müller, A. 202
Müller, F. 246
Müller, H. 151
Muller, L. 4, 24, 124

Nachtsheim 26, 90, 168
Naef 246, 266
von Naegeli 57, 204, 283, 300
Nagel 236
Neder 156
Neefs 10
Neumayr 19, 260
Nevill, George 259
Newell 206, 217
Niethammer 47
Nobis 32
Nolte 153, 154
Novikoff 71, 200

Oboussier 145, 170
Oekland 23
Omodeo 7
Oparin 316
Osborn 1, 57, 103, 115, 120, 204, 219, 234,
  236, 301
Ottow 234, 238

Padour 145, 154
Pagenstecher 59
Palmer 138, 235
Paracelsus 81
Park 11
Parr 266
Partmann 156, 163, 172, 215
Pasewaldt 181
Pasteur 309
Pauly 356
Pavlov 260
Pax 6, 25
Pearson, K. 51, 179
Pearson, O. H. 172, 173
Peden 303
Penners 242, 256
Penso 313
Perdeck 231
Peters 33
Petersen 26, 40, 46, 157
Peyer 231

Pfeffer 278
Pflüger 316
Philiptschenko 26, 58
Pieron 334
Pirocchi 25
Pittioni 46
Plaine 4
Planck 319
Plate 72, 106, 112, 125, 204, 230, 266, 277, 278, 283, 285, 291, 298, 317, 336, 352
Poll 26
Pomerat 51
Pompecki 95
Popoff 245, 246
Poppe 312
Porsch 107
Porter 26
Poulton 11
Przibram 163, 181, 268

Quenstedt 93, 104, 110, 115
Quinn 168
Quiring 137, 145, 147

Rachow 157, 158
Radovanovic 26
Rajewsky 6
Ramme 182
Rauber 303
Rawiel 32
Reetz 165
Reinig 10, 39
Remane 22, 58, 127, 167, 177, 200, 217, 234, 266
Remington-Kellogg 124
Rensch, J. 39
Ribbert 181
Richardson 246
Richet 135, 145, 160, 352
Richter 310
Riddle 132, 145, 168, 244
du Rietz 25, 50
Riggs 202
Ripley 32, 237
Rizki 53
Robb 137
Robinson 134, 145
Roemer 26
Romer 220
Rönsch 248
Rosa 54, 58, 98, 112, 113
Rose 153
Rotmann 245
Roux 180, 310
Rowlands 168
Rubner 160
Ruedemann 95
Rüger 82
Rumjancew 19

Runnström 50
Ruska 302, 312
Rutz 244, 248

Salfeld 196
Salomonsen 45, 90
Salt 46
Sars 64
Saxena 46
Schäfer, E. 40
Schäfer, H. 332
Schäffer 117
Scheminsky 347
Schepers 301
Schewiakoff 72, 278
Schiermann 121, 165
Schilder 25, 104, 148, 215
von Schiller 324
Schilling 137, 155
Schindewolf 58, 86, 89, 98, 99, 104, 110, 198, 199, 205, 254, 260, 261, 262, 266, 271, 272
Schlabritzky 142, 145, 147, 148, 154, 213
Schmalhausen 2, 133, 181, 188, 297
Schmidt, H. 19, 240, 301, 309, 310
Schmidt, W. J. 130
Schnecke 168
Schneider 20, 142
Scholander 46
Schoute 49
Schramm 302, 312, 314
Schulz 151, 244
Schuster 46, 110, 144, 222
Schütte 157
Schwanitz 104
Schwarz 25
Schwarzbach 95, 104, 238
Schwegler 205
Schwerdtfeger 157
Schwind 146
Scott 102, 110
Seidenberg, 307
Seiler 26, 53
Seitz 25
Semon 329
Šengün 52
Sewertzoff 222, 224, 241, 244, 246, 247, 253, 256, 266, 283, 298, 299
Shankara 81
Sheppard 11
Sholl 142, 294
Sibley 54
Sieber 89
Sierp 310
Simpson vii, 2, 10, 17, 87, 88, 95, 99, 102, 137, 237, 268, 284, 297, 298
Sinnot 169
Sleptsow 125
Smith 43, 168, 303
Snell 131, 135, 142, 145, 168

Snethlage 117
Sonneborn 8, 188
Spassky 11, 193
Spencer 193, 298, 323, 328, 336
Sperber 155
Spett 52, 244
Spiegelman 181
Spiess 11
Spieth 51
Spina Franca Netto 154, 155
Spinoza 81, 323, 328, 355
Springer 232
Spurway 26, 32, 53
Stalker 11, 49
Standfuss 32
Stanley 313
Stebbins 2, 3, 25, 50, 55, 104
Stegmann 30
Stehlin 186
Stein 163
Steinbacher, F. 23
Steinbacher, G. 90
Steinbacher, J. 90
Steinböck 250
Steiner 14, 32, 49, 223
Steiniger 11, 131
Stern 106
Sternfeld 157, 158
Stirton 221
Stockard 145, 181
Stolte 129
Storer 194
Stresemann 35, 47, 90, 161
Stromer 102, 110, 296
Stubbe 3, 104
Sumner 11, 26
Suomalainen 7, 26
Surbeck 27
Svärdson 7
Svilha 26
Swellengrebel 49
Sylvester-Bradley 19, 281
Szalay 110

Tan 192, 193
Taschenberg 226
Tavani 238
Teissier 10, 11, 133, 154, 181, 213
Test 49
Thiel 123, 251
Thienemann 27, 121
Thomas 26, 32
Thompson 310
Thorpe 50
Ticehurst 30
Tielecke 172
Timoféeff-Ressovsky 3, 5, 6, 9, 22, 30, 40,
    129, 130
Tinbergen 11, 231, 344

Tiney 334
Tiniakov 49
Toll 343
Tonolli 25
Trendelenburg 335
Tschetverikov 9
Tschumi 181, 182
Turesson 50

von Ubisch 243, 244, 248
von Uexküll 334
Ule 20, 21
Urbahn 51
Uschmann 282

Vandel 7, 318
Vasseur 25
Vaugien 144
Vaurie 28, 29, 144
Vavilov 193
Venzlaff 213
Verrier 250
Versluys 295, 304, 365
Verworn 81, 303, 346, 348, 354
Viets 25
Villee 4
Vogt 157, 158, 301
Voipio 30, 41
Volkelt 345
Volterra 11
Vorsteher 145

Waagen 18, 106
Wachsmuth 232
Wächter 134
Waddington 297
Waldmann 312
Wallace 237
Walter E. 157, 158
Walter, W. G. 331
Walther, J. 99, 110
Warnecke 52
Watson 276, 292
Weber, H. 84, 129, 139, 324
Weber, M. 99, 186, 230
Wedekind 57, 99, 103, 259, 260
Wehrli 18, 205
Weidenreich 125
Weigelt 267
Weinert 301
Weismann 197
Weiss 181
Welcker 135, 145
Wellensiek 153, 172, 175, 244
Wenz 85, 86, 99
Wertz 315
Wesenberg-Lund 25
Westenhöfer 301
Westoll 269, 270

von Wettstein 104
Weyer 48, 49
White 3, 7
Whiteley 179
Whitman 26
Wiedemann 302
Wiedersheim 292
Wilbert 181, 182
Williston 291
Willoughby 51
Wilson, E. B. 181
Wilson, E. O. 33
Winkel 46
Woltereck 56, 58, 300, 356
Woltersdorff 26
Wood 201
Wörz 20
Wright 12, 27, 37
Wundt 81, 323, 336, 352, 356

Wurmbach 132
Wyckoff 303

Yerkes 304, 347

Zeleny 184
Zeuner 82, 91
Zhelochovtsev 49
Ziehen 81, 300, 306, 320, 323, 325, 326, 329,
    334, 337, 340, 342, 343, 345, 347, 348, 349,
    352, 353, 354, 356
Zimmer 25, 117
Zimmering 193
Zimmermann, K, 26
Zimmermann, St. 90
Zimmermann, W. 25, 284
Zittel 91, 232, 233
Zumpt 49

General Index

abbreviation of development 241, 259, 260
Ablepharus 34, 35, 36
abstractions 165, 325, 339 ff.
Acanthocephala 109
Acarapis 48
acceleration of development, 240, 255, 259, 271
Accipiter 22
Acipenser 257
Acmaea 49
acme 113
Acrocephalus 52, 185
Actinia 25
Adapis 186, 208
adaptation 236
adaptiogenesis 266
addition to final stages 66, 256 ff., 260, 264
adult preponderance 260
Aegithalos 27
Aepyornis 179, 217
Africanthropus 20
age of categories 83 ff.
age of individuals 162, 163, 218
Alaptus 175
albinism 193
Allen's rule 39, 43, 46, 144
allometric exponent 136, 147
allometric growth 133 ff., 166, 167, 169, 190, 198, 218 ff., 226, 228 ff., 253, 294, 362
allopolyploidy 14, 54
alternation of generations 262
Amara 88
Amblypoda 120
Amblyrhiza 207
Ambystoma 246, 260
Ammodytes 50
Ammonites 18
Amphidasis 20, 21, 89
Amphidromus 30
Amphilina 270
anaboly 66, 241, 247, 253, 257, 265
anagenesis 97, 281 ff., 305
Anaplasmidae 312, 313
anastrophes 99
Anchitherium 18, 205
Andrena 164
Anguilla 89

Anguis 222, 223, 224, 299
anisomerism 291
Anopheles 49
Antedon 94
antennae, correlations 179
anterio-posterior development 169
Anthracomyia 19
Antirrhinum 104
antlers 217, 254, 258
aphanisia 225
Aphodius 164
Apis 164
Apodemus 149
apogenesis 268
Archaeoceti 124, 236, 269
Archaeopteryx 68, 210, 271, 288
archallaxis 66, 241, 242, 244, 265
Archiannelida 175, 177, 198, 210, 217
Arietites 195, 209
aristogenesis 57, 103, 204
Armeria 25
aromorphosis 266, 283
Artemia 7
Artenkreis 30, 32, 35, 87
Arvicola 140
asexual reproduction 66
associations 326
assortative mating 51
Astrapotherium 228, 230
atavism 125, 186, 240
Atherina 253
Atta 176
attention 326
auditory ossicles 275
Aurelia 251
Australian fauna 111
Australopithecus 301, 332, 362
autogenesis 317 ff.
autonomous forces 57, 98, 99, 113, 191, 205, 236, 282, 284, 309, 317
autonomy 298 ff.

Babirussa 67, 228, 230
back mutations 6, 125
bacteria 312, 313
bacteriophages 313
Balaena 68

412

balance in biocoenosis 121, 122, 123
Baluchitherium 178
barbules of feathers 130
Bartonella 312
Basilosaurus 207
Belone 247, 253
Bergmann's rule 39, 41, 43, 45, 46, 144, 188, 304
biocoenosis 121
biogenetic rule 226, 241, 266
bionomogenesis 317 ff.
bipinnaria 179
birds of paradise 60, 61
birds of prey 120
Bison 26, 208
Blatta 157
blood cells 214
Bolivinoides 19
Bombina 90, 246
borderline cases of races, 30, 50
Brachiopoda 113, 114
Brachiosaurus 73, 120, 178, 179
bradymorphism 240
brain 142, 143, 151, 152, 157, 163, 167, 172, 240, 248, 249, 256, 260, 288, 293 ff., 296, 302, 303, 306, 332
brain development 248, 249
brain of insects 157, 172
brain-size 142, 167, 172, 260, 288, 293 ff., 302
Broca's region 303, 306, 348
Brontosaurus 73, 179
Brontotherium 221, 228, 229, 232
Bufo 51, 246
bulldogs 170
Bunolophodon 20, 88, 208

Caecum 171, 172, 175, 212
Calliphora 163, 215
Cancellaria 19, 89
Capra 91
capybara 154, 155
Carabus 24, 31, 35, 36, 41, 42, 46, 163, 225
Carausius 181, 182
Carcinus 181
Carduelis 54, 125
Cartesian transformations 169
Caspian fauna 110
Cassis 209
Castor 186
Castoroides 207, 216
Catops 164
Catostomus 54
causality 319, 320, 356, 357
cave bear 254, 255
cave-dwelling animals 69, 88, 225, 254, 255
Cavia 154, 174
cells, of cortex 153, 154, 155
cell-number 179
cell-size 213

centralization 282, 291
Cepaea 193, 194, 247, 248
cephalization 144, 293
Cerithium 209
Cervus 20, 87, 91
cestods 108
Chalcides 222, 238, 299
chelae 181
Chen 22
chimeras 243
Chiton 300
Chlamydotherium 216
Choleva 91
Choleopus 142
Chortippus 244
Chromodoris 51
Chrysochloris 203
Clymenia 198, 199
ciliate bands 179
Ciona 25
Clathrina 251
cleavage 66, 250, 251
clines 43
Clupea 90
co-adaptation 127
Cochlostyla 119
coelacanths 117
Coeloplana 270
coenogenesis 250
Coereba 20
Coliuspasser 22
columniform legs 73, 178, 179
compensation 180 ff., 191
complexity 289 ff., 299
concepts 306, 325
conceptions 326, 328, 329, 339, 340, 341, 342, 343, 349, 352, 358
conclusions 326, 344 ff., 352
cones, numbers 151, 152, 175
consciousness 319, 320, 322 ff., 345, 350, 351, 352, 359
conservative prestages 266
constructive genes 129, 166, 191, 267
convergence 68, 191, 200, 205
Contarinia 156
Cope's rule 133, 144, 166, 190, 206–12, 215, 217, 218, 219, 221, 239, 293, 302
Copris 164, 183
Coregonus 27, 90
Corethra 253
cornea, size 149
corpora pedunculata 157
cortex 143, 151, 152, 153, 155, 167, 215, 332, 333
cortical regions 143
Corvus 27
Cottus 90
counting ability 341, 342
creeper fowl 224

Creodonta 115
Crepidula 244
Cricetus 22
crinoids 113, 114
Crioceras 233
Crossopterygia 270
Ctenoplana 270
cuckoo's eggs 48
Culex 48, 253
culture, human 306, 307, 362
Cyclophoridae 172
Cypraeidae 25, 62, 104
Cysticetes 312, 313
Cystidium 302

Darwin's finches 11, 108
dating the past 82
decrease of body-size 170 ff., 209
decrease of variability 112, 113
de-differentiation 127, 128
degeneration 230, 231, 232 ff.
dendrites 164, 346
density of breeding birds 121, 122
Deroceras 284
detour experiments 347
detours of development 257, 258, 275
deviation 66, 246 ff., 260, 264, 265
diametagensis 260
diaptomids 25
Diastylis 25
Diatryma 179
Dicrurus 144, 228
Didunculus 179
Dikerogammarus 51
Dilepididae 108
Dinocyon 207
Dinophilus 175, 217
Dinornis 179, 217
Dinotherium 208, 220
Diodon 169
Diphyllobothrium 105
Diplodocus 120, 179
Diplogonoporus 105
Diplommatina 172
Dipodomys 28
Diprotodon 206
Dipterus 115
directed evolution 57, 97, 205, 279
Diurodrilus 175
division of labor 282, 283, 291
Dixippus 200
Doedicurus 206, 216
doubling of neurons 304
Drepanididae 106, 107
Drosophila 4, 5, 6, 10, 26, 49, 51, 54, 105, 106, 129, 157, 163, 192, 215, 242, 249, 250, 252, 264
Dryopithecus 208
duplications of organs 105

Dutch spotting 194
dwarf animals 168, 171, 193
Dytiscus 156

Echinococcus 262
Echinocyamus 243
ecological races 16, 47, 50
ectogenesis 317 ff.
egg-number 157, 158, 159, 172, 216
Elephas 87, 113, 179, 208, 216, 219, 228, 273
Eleutherodactylus 66, 251, 265
elimination 10, 39, 40, 41, 43, 45
Elpistostege 269, 270
Emberiza 40
emotions 327, 328, 333
endomitosis 7
engrams 324
Engraulis 88
Entamoeba 50
Enteroconcha 70
Enteroxenos 70
Eohippus 88, 260, 273, 288, 333
Eotherium 113, 269
epacme 98
Ephestia 264
Epilachna 26
Epimys 154, 174
Epinephelus 89
epistasis 260
Equus 137, 186, 260, 288
Eremina 30, 195
Erinaceus 26, 90, 174
Eriocheir 123
Eumeces 223, 299
Eumorphoceras 240
eunomia 226
Eunotosaurus 269
Eupomatus 251
Eusmilus 202
Evotomys 140, 149
excessive growth 226
excessive structures 202, 229, 230, 231
excretory organs 201
exotypes 22
expressivity 5
explosive evolution 89, 92, 95, 97, 98 ff., 106, 107, 109, 110, 205, 235, 273, 361
extinction 205, 232, 234 ff., 296
extracerebral chains of associations 307
eyes, evolution 71, 72, 277, 278
eyes, reduction 225
eye-size 148, 149, 150, 215

facial bones 137, 141
feather length 167
feelings 323, 327, 333, 339, 352, 358
fertility 31
fetalization 260, 304
Fissurellidae 114

flourishing radiation 99, 103, 104, 112
flower-pickers 210
fluctuations of population 3, 9
foot of man 302
Foraminifera 217
forced development 68 ff.
forest-birds 117, 118, 119
Formica 48
frizzle-fowl 133
Funambulus 153

Gadus 90
Galapagos Islands 108
Galerida 47, 91
Gammarus 51
gaps of fossil record 18, 267, 268, 271, 280
Gasterosteus 49
Gastrotricha 177, 210
gastrulation 66, 251
gene flow 14
genepistasis 260
genetic races 16
Gennaeus 54
Gentiana 25
geographic races 16, 18, 22 ff., 43, 52, 55
geographic subgenus 30
Geospiza 108
Geotrupes 164
gerontism 232
Gerris 7
gestation 161
giant animals 119, 120, 177, 178, 179, 206, 207, 208, 209, 216, 217, 226, 227, 235, 237, 238
Gigantostraca 113, 114
Gila 54
giraffes 144, 207, 229
Gloger's rule 39, 40, 43
Glyptodon 206, 216
globuli-cells 156
Goniatites 260
gorilla, foot 302
Guttera 54
Gyraulus 19

hair length 167
Halicore 113
hallucinations 324
Haltica 210
Hamites 232, 233
Hapale 303
Hawaian islands 106, 107, 108
head-bones, number 292
head-shape 141
head-size 139, 167, 173
heart-size 136, 145, 146, 147, 172, 212
Helicella 195
Helix 37, 38
Hemignathus 106, 107

hemichromes 65, 66
herbivorous animals 177, 178
Heterandria 153, 172, 175, 244
heterochronism 66, 135, 246
heterogony 262, 263
heteromorphism 240
heteroploidy 7
heterosis 8, 12, 217
hibernation of butterflies 59
Hipparchia 35
Hippolais 90
Hippopotamus 207
histological structures 149, 154
historical races 16, 17
hologenesis 58
homeostasis 297
Homo 20, 87, 208, 301, 362
homogamy 14
homoiothermy 75, 95, 238
homologous genes 192, 193, 197, 203
homology 192, 200, 202
homozygosity 37
honey creepers 106, 107
hopeful monsters 106, 271
Hoplophoneus 219, 221
hormones 132, 165
hornbills 219
horns of antelopes 60
hovering mechanisms 175
hummingbirds 172
Hyaenodon 115
hybrid races 16, 31, 32, 54, 55
hybrids of species 53, 245
hybrid sterility 53
Hydrochoerus 154, 155
Hyla 32
Hyloicus 35
hylopsychism 352, 355
Hymenolepidae 50, 108
hypermorphosis 253
Hyracotherium 137, 260, 273

Iberus 195
Ichthyostegopsis 270
ideas 323, 340, 341
idioadaptation 266
Iguanodon 73, 120
imaginations 327, 349, 352, 358
imitation 347
improvement 282, 283, 287, 288, 297, 299
incubation of birds 59, 161, 162
inertia 204, 226
Inoceramus 209
Inostranzewia 202
insight 344 ff.
instincts 164, 176, 306
intensity of evolution 112 ff.
intestine length 148
iridescent colors 130

irreversibility 97, 123
island races 10, 36, 37
isocortex 153
isolation 3, 13, 14, 15, 55, 305
Isopoda 63, 64
iteration 97, 191, 198, 199, 203

Jaculus 74
Jaymotius 83
judgements 326, 344 ff., 352
jumping animals 74
Junco 27, 54
Junonia 30
juvenile preponderance 259

Kallima 200
kidney size 145, 146, 212
kidney structure 215
kladogenesis 97, 281, 317
Koalemus 206
Kosmoceras 17, 89, 196, 233, 262

Lacerta 36, 37, 91
Lamellibranchia 114
Lamblia 105
Langerhans' cells 154
Larus 30
larvae of water-insects 253
Latimeria 267
law of the unspecialized 94, 204, 237, 272,
    285, 301, 302
leaf-like butterflies 200
learning 164, 165, 330, 331, 334
leaves of plants 63, 69
Lebistes 151, 165, 175, 244
legs, size 222, 223
lemurs 111
lens, size 149, 150, 215
lens, development 245
Leptoteredra 270
Leucochroa 195
Levantina 195
life, characteristics 310 ff.
life expectancy 162, 163, 218
life, origin 309 ff., 362
Liolaemus 36
Limax 231
Limia 32
Limosa 145
limits of size 170, 171, 172, 191
Limulus 86
Linaria 104
Lingula 82, 94
links between groups 269 ff.
lipopalingenesis 259
liquid crystals 310
Littorina 171, 253
Lituites 261

liver size 145, 146, 167, 212
lizards 145
localization of engrams 324
Lophius 253
Lophura 54
Loris 174
lower limit of body size 171, 172, 191
Lucanus 228
lungs 240, 276
Luscinia 31, 90, 183, 185
Lutianus 89
Lymantria 20, 26, 32, 55, 244, 245
Lyonetta 175
Lyrurus 164
Lytoceras 232

Machairodus 185, 219
Macrocephalites 262
macro-evolution 1, 276
macro-mutation 99, 103, 104, 105, 106
Macropanesthia 156
Macropus 276
Macroscaphites 232, 233
Manticoceras 89
Marchantia 104
marmot 141
marsupials 111
marten 141
Meckel's cartilage 276
Megaceros 207, 212, 216, 219, 228
Megalonyx 206, 216
Megaladapis 186, 208, 217, 219
Megamys 207
Megatherium 206, 217
Melania 19, 183
melanism 22, 37, 38
Melolontha 156, 157, 160
Melospiza 27
memory 329, 330, 331, 339
mental images 322 ff., 339 ff.
Merychippus 186, 259
Merycodus 207
Mesaxonia 116
Mesozoa 177, 270
metabolism 160, 213, 216, 218
metagensis 251, 262
Metridium 25
Metrioptera 182, 183
microphysical processes 319, 320, 321
Micromys 153
Microscelis 34
modification 26, 187, 240
Moeritherium 186, 207, 219, 220, 269, 273
molarization 221
Monarcha 34
Monomorium 164
moods 327
Moulinsia 172
mouse 141, 149, 154, 164, 167, 174

Murella 26, 38, 194, 195
Murex 184
Mus 141, 149, 154, 164, 167, 174
Mustelus 279
mutation 3, 4, 15, 18, 33, 106, 187, 240, 279, 304, 317, 361
mycetomes 69
Mycetozoa 270
Mylodon 206, 217

nanism 131, 168, 171, 193
Nannipus 208
Nautilus 94
Neanderthal man 301, 303, 329
Necrophorus 164
Neides 200
neomorphosis 259, 272, 280
Neopilina 271, 300
neostriatum 153
neoteny 66, 258, 261
Nereis 251, 252
nerve-cells 144, 293, 346
new organs 266 ff., 274 ff.
nomogenesis 58
Notoryctes 203
Nototherium 217
nucleoproteins 315 ff.
Nucula 93, 104, 115
Nummulites 93

old lakes 56
Olenus 17, 18, 198, 199
omission of final stages 259
ommatidia, number 163, 176, 215
ontogeny 66, 133 ff., 161, 166, 167, 239 ff., 250 ff., 263 ff.
Ophiodes 222, 223
Ophiomorus 299
Ophisaurus 222, 224, 225, 299
Opisthocomus 188, 189, 190
Orbitoides 93
Orcula 90
organic evolution 188
organic time 163
Orohippus 186
Orthagoriscus 155, 169, 170
Orthoceras 209
orthogenesis 57, 97, 133, 137, 198, 203, 204, 205, 218, 219, 221, 226, 229
orthoselection 190, 204, 225, 262
Oryctolagus 90
Osteolepis 292
Ostrea 209
Otala 37, 38
Otiorhynchus 7
Otis 164, 231
overspecialization 26, 204, 205, 221
Oxyaena 115
oxygen consumption 63, 148, 160, 172, 173

Pachyaena 207
Pachydiscus 209
Pachycephala 34
Pachydrilus 242, 243
Pachys 20, 21, 89
paedomorphosis 260
Pagodulina 90
Palaeomastodon 219
Palaeopropithecus 208, 217
Palaeotragus 207
palingenesis 66, 258 ff.
Panaxia 22
panmixia 14
panspermia hypothesis 310
Pantolambda 185
Pantosteus 54
Papilio 41
Papuina 38, 39, 119
Paracentrotus 50
parallel evolution 191 ff.
parallel mutation 193
parallel selection 198, 201
parallelism, psychophysiological 354 ff.
Paramaecium 284
parasites 70, 71, 109, 298
Parechinus 243
Pareiasaurus 120
paripotency 193
Parnassius 52
Parus 28, 29, 30, 44, 154, 213
Passer 27, 30, 50, 54, 91
Passerella 47
Patella 251
pectoral fins 276
Pediculus 49
perfection 283, 284, 285, 300
Peripatus 270
Periophthalmus 132
Peromyscus 26, 30, 31, 34, 51, 138, 167, 191
persistent types 92, 93, 95
Petasiger 109
Pezophaps 238
Phacochoerus 185
Phasianus 26, 32
phenocopy 26, 188
phenogenetics 241, 256
Phiomia 219, 220
Phyllobates 172
Phylloscopus 30
physiological races 16
Picus 31
Pinus 25
Pipilo 30, 54
Pithecanthropus 20, 208, 301, 302, 303, 304, 329
pituitary dwarfism 131, 193
placental mammals 112
plasmagenes 188
Plasmodium 50

plasmatic mutations 8
plasticity 27, 289, 295 ff., 300, 304
plastology 197
pleiotropism 12, 130, 133, 179, 190, 267
Plethodontidae 175
Pleurodeles 132
Pleuronectes 89
Pleurotomariidae 94, 114, 209
pluteus 243
Podiceps 164
poecilogony 250
Poecilus 224
Polygordius 176
polyisomerism 291
polymorphism 12, 13, 42
polyploidy 7
population size 3, 9, 12, 15, 95
Portunion 70
Primula 54
progressive adaptation 113
progressive evolution 163, 281 ff., 358
prolactin 131
prolongation of development 259
Prorhynchus 250
Prosobranchia 114
protective coloration 200
proterogenesis 66, 241, 258 ff., 260, 261, 262
Protocetus 95, 124, 207, 269
Protodrilus 175
Protopterus 276
Psiloceras 195
Psylla 50
Pteranodon 74, 221, 225, 232
Pterodina 263
Ptiliidae 171
Ptychopoda 5
Pulmonata 114
Pulsatilla 25
pulse rate 160
Punctum 172
Pyrotherium 207
Pyrrhocoris 179
Pyrrhula 28, 54, 90
Python 68, 223

quantitative mutation 291

Rana 246, 267
Rangifer 87
Raphanobrassica 55
Rassenkreis 23, 24, 25, 28, 29, 32, 33, 34, 36, 87, 90, 304
rat 141, 164
rates of evolution 268, 361
rationalization 289, 291 ff., 299, 300
Ratufa 153
real mutations 58, 267
reasoning 348, 349
recapitulation 241, 257, 266

recombination 3, 8, 14
rectigradation 204
reduction of organs 186, 210, 222, 299
regressions 298, 299
Rensch's rule 43
respiration of insects 65
respiration rate 160
retina 150, 151, 152
retardation of evolution 240, 259
Rhinoceros 237
Rhipidura 20, 22
robot 330, 331
rods, number 151, 152
Rubus 54

Salamandra 151, 152, 153, 211, 246, 258
Salix 54
Sarmatian Sea 110
Saxoceras 19, 209
Scarabaeus 164
schizotypic splitting 56
Sciurus 138
selection 3, 12, 14, 15, 43, 45, 190, 203, 225, 279, 304, 305, 361
self 326, 341, 343
senility 234
sensations 323, 324, 333, 335 ff., 343, 349 ff. 358
sense organs 163
Sepia 25
Seps 68, 222, 223
serial organs 148
Serinus 40, 41, 125
sessile animals 73
sexual armatures 52
sexual dimorphism 157, 159, 160
sexual maturity 304
sexual reproduction 66
sexual selection 11
sexuality 14
shells of snails 62
Simbirskites 260
Sinanthropus 20, 303, 329
Sinopa 115
Siphateles 54
Sirenia 113
skeleton weight 148
smallest animals 119, 173, 174
Smerinthus 32
Smilodon 185, 219, 221
Solenobia 7, 53
Solenomya 93, 115
somatogenic induction 187
Sorex 52, 174
spatiality 338
Sphaeriidae 171
Sphenodon 94
Spartina 55
specialization 112 ff., 236

speech center 303, 306, 348
speed of evolution 82, 112 ff., 268
spinescence 232
Spirifer 234
Spiroceras 233
spontaneous generation 309
Stegocephalia 270
Stegosaurus 229, 232
Stentor 340
Stephanoceras 19
Sthenurus 206, 217
streamlined bodies 73
stridulating organs 62
Strombus 209
Struthio 231
subordination of parts 291
superspecies 30
Sycon 251
Sylvia 22, 201
symphysis of pubic bones 173, 174
synorganization 127, 167
systemic laws, 128, 129, 137, 165, 167, 191, 319, 359

tachygenesis 259
tachymorphism 240
tachytely 92
Taenia 105, 262
Talpa 138, 174, 203, 216, 249
tarsiers 111
tarsus 171
taxonomy 176
tectum opticum 153
teeth 69, 201, 202, 218
temporality 339
Tendipes 253
Tenebrio 181
terrestrial life 109
territoriality 13
Tetrao 164
Thamnophis 49
Theromorpha 198, 269
Thomomys 194
Thylacinus 235
thyroxin 132, 265
Thysanoptera 171
Tiaropsis 342, 343
time signatures 198
tissue-cultures 127, 128
Titanotheriidae 115, 120, 207, 219, 228, 229, 235
tobacco mosaic virus 302
tobacco necrosis virus 303

Tortrix 157
Toxodon 217
Troglodytes 45
transplantations 245, 246
trematodes 108
trends 57 ff.
Tridacna 209
Trigla 155, 253
Triceratops 73, 120, 221, 229, 232
Trichoglossus 31, 33, 34
Trichoniscus 7
Triturus 32, 134, 151, 152, 175, 211, 245, 246, 267
Tritylodon 269
Trypanosoma 50
Tubifex 242, 243, 256
Turrilites 232, 233
typogenesis 99
typostrophism 266
Tyrannosaurus 120, 178

Uintatherium 185, 202
undirected evolution 57 ff.
unoccupied niches 108, 111
upper size limit 177, 191
Uronemus 115
Ursus 87, 216, 254, 255, 256

Vanellus 145
vertebral column 124, 131, 299
vesicular eyes, 71, 72, 265, 277, 278
Vespa 156, 157, 164
Vespertilio 249
vestigial organs 68, 222, 223, 224, 225
vestigial physiological characters 259
virus 302, 303, 312 ff.
vitalism 318, 356
viviparity 175, 176
Viviparus 19, 251
Volborthella 208
volition 323, 327, 333, 343, 352, 358

weasel 141
whales 112, 126, 236
Williston's rule 291
wing-pattern of butterflies 196, 197
wing-shape of birds, 201

Xiphophorus 165, 175, 244

Zaphrentis 19
Zeuglodon 124